Springer Geophysics

The Springer Geophysics series seeks to publish a broad portfolio of scientific books, aiming at researchers, students, and everyone interested in geophysics. The series includes peer-reviewed monographs, edited volumes, textbooks, and conference proceedings. It covers the entire research area including, but not limited to, geodesy, planetology, geodynamics, geomagnetism, paleomagnetism, seismology, and tectonophysics.

More information about this series at http://www.springer.com/series/10173

V. I. Ferronsky

Nuclear Geophysics

Applications in Hydrology, Hydrogeology, Engineering Geology, Agriculture and Environmental Science

Springer

V. I. Ferronsky
Water Problems Institute of the Russian
Academy of Sciences
Moscow
Russia

Every effort has been made to contact the copyright holders of the figures and tables which have been reproduced from other sources. Anyone who has not been properly credited is requested to contact the publishers, so that due acknowledgment may be made in subsequent editions.

Springer Geophysics
ISBN 978-3-319-12450-6 ISBN 978-3-319-12451-3 (eBook)
DOI 10.1007/978-3-319-12451-3

Library of Congress Control Number: 2014959274

Springer Cham Heidelberg New York Dordrecht London

Printed on acid-free paper

Springer is part of Springer Science+Business Media (www.springer.com)

Preface

Progress in geoengineering, hydrology and environmental sciences as a whole closely depends on the development of new methods of investigation, which are based on the latest achievements in adjacent branches of science such as physics, chemistry, electronics, geophysics and so on. This book provides the fundamentals and field applications of nuclear techniques in engineering geology, hydrogeology, civil engineering, agriculture and environmental study. The book is intended for scientists and researchers, teachers, students and postgraduates engaged in field studies for solving scientific and practical problems in applied geology and hydrology for civil, road, power, airfield, harbour engineering, drainage and irrigation.

Nuclear geophysics is a branch of applied geophysics, where the nuclear methods for study of physical-chemical and geoengineering properties of rocks and soils at their geological exploration are considered. Despite their relatively short history, nuclear methods have achieved worldwide popularity in science and technology. Because of their exceptionally advantageous features, they have found many practical applications whose range is continuously expanding.

The development of nuclear methods for solving geoengineering and hydrological problems depends first of all on the development of reliable and objective methods for estimating the physical parameters of grounds and rocks and of studying the dynamics and geochemistry of underground waters under natural conditions. It is not long since methods based only on the study of samples taken from excavations were used both in estimating physical properties (density, moisture content and porosity) of rocks in regional studies and in surveys for suitable building areas.

The development of field methods for determining the physical properties of grounds began with the use of gamma-rays and neutrons, which are now regarded as uniquely reliable tools for estimating the density and the moisture content of grounds under natural conditions. Italian physicist Bruno Pontecorvo was the founder of nuclear geophysics. In 1941 he was the first to propose neutron well logging for geological exploration of oil and gas fields. Since that time nuclear techniques have started to develop for the geological exploration of mineral resources, engineering geology and hydrogeology.

In 1950, Bernhard and Berdan of Rutgers University (USA) were the first to use gamma-ray absorption as a method of determining the density of grounds. The back-scattering of gamma-rays and of neutrons in 1950 was used by Belcher at Cornell University in the USA to investigate the density and the moisture content of grounds respectively. These methods were subsequently developed by many researchers in Germany, UK, France, former USSR, Poland and other countries.

The first gamma-ray density and neutron moisture gauges were used to estimate the suitability of various areas for civil engineering projects in the former USSR, e.g., at construction of the Volga and Nurek hydroelectric power stations for monitoring of density at the ground dam pouring out. They were also applied in geoengineering surveys in Western Siberia, in hydromelioration studies of irrigation schemes in Central Asia, in Southern Ukraine and many other areas in different countries.

The manufacture of neutron moisture and gamma-ray density gauges for the investigation of grounds and rocks was started in the USA, UK, France and the former USSR by firms such as Nuclear Chicago, Nuclear Enterprises, Russian Enterprise 'Isotope' and others. In 1966 the working group of the Coordinating Council of the International Hydrological Decade decided to recommend gamma-ray and neutron methods for determining the density and moisture content of soils and grounds as satisfactory and reliable for extensive practical use.

The successes in the development and application of neutron and gamma-ray methods in engineering geology and hydrogeology have not, however, led to the solution of a very important problem of cardinal improvement of the field methods of investigation in these applied sciences. When engineering geology surveys for suitable constructional areas are being carried out, it is essential to have data on the properties of the ground at depths down to 15–20 m.

So far, this can only be done by the nuclear well logging technique. This method is being widely used in industrial and prospecting geophysics. Many publications have appeared on this subject, including monographs by E.M. Filippov (1962), V.I. Ferronsky et al. (1968; 1969), C.G. Clayton (1983), IAEA (1968, 1971; 1981, 1983, 1999) and K. Froehlich (2010).

However, nuclear well logging, which was being used in prospecting geology and especially in prospecting for oil and gas fields, has not turned out to be suitable for engineering geological studies. The accuracy of the method was insufficient to determine the physical characteristics of grounds, owing to the interference by the very existence of the well casing.

The optimal solution of the problem of the nuclear logging method in engineering geology for studies of loose deposits down to a depth of 20–25 m was put forward by V.I. Ferronsky (1969). The idea of his approach was as follows. To avoid the influence of interference due to construction of the well, the penetration logging method (including gamma, gamma-gamma and neutron-neutron logging) is used. In this case the logging probe, which is mounted on the tip of the drill rod, is forced into unconsolidated deposits by an axial load, which is applied by a vehicle-mounted hydraulic-mechanical system. The method is most readily applicable in solving engineering geological problems and for hydrogeological investigations of aquifers in both saturated and unsaturated zones.

Practical applications of advanced original nuclear logging techniques and methods are presented in the book. These techniques are the penetration logging methodology and facilities, where the logging sonde is sunk into the friable deposits under investigation by a vehicle-mounted hydraulic device. The logging sonde is penetrated to depths about 40 m in loose formations at rates of up to 2 m/min on land and near shore marine areas. Besides having obvious operational advantages such as fast penetration, this method also avoids some of the well construction problems. The sonde is in direct contact with the medium and thus gives increased accuracy and reliability for interpretation.

Radioactive and stable isotopes, which are the constituents of water molecules, or migrate with them, have been used widely in recent years in a number of countries to investigate the motion of water on a regional scale. These isotopes include above all tritium (half-life of decay $T_{1/2}$ = 12.32 year), carbon-14 ($T_{1/2}$ = 5730 year), deuterium and oxygen-18. Natural tritium as a constituent of water molecules, chemical properties of which are practically indistinguishable from hydrogen, is particularly widely employed in investigations. Tritium and carbon-14 are used to solve many hydrogeological problems related to water movement in saturated and unsaturated zones.

Hydrogeological processes occurring over longer periods of time are being investigated with the aid of natural radioactive carbon-14. Because its half-life of decay is equal to 5730 year, it can be used to cover very long periods of time during the motion of water in nature, including the glacial period.

Some information related to other natural radioactive isotopes in groundwaters like ^{7}Be, ^{10}Be, ^{22}Na, ^{26}Al, ^{32}Si, ^{32}P, ^{33}P, ^{36}Cl, ^{37}Ar, ^{39}Ar and radioactive isotopes of the uranium-thorium series is presented in the book. But investigation of the regularities on the distribution of these isotopes in natural waters is limited by technical difficulties of sampling, concentration and measurement of the corresponding samples.

The use of stable isotopes of deuterium and oxygen-18 enables us to establish the interrelations between water-bearing horizons, the supply sources for groundwaters and the connections between open reservoirs and other sources of effluents and the origin of individual water-bearing horizons and also to investigate the conditions under which glaciers are formed and moved.

Systematic measurements of precipitation and of the abundance of natural isotopes in precipitation are essential for the success of regional hydrogeological and hydrological studies using naturally occurring isotopes. The first condition is satisfied by using the data supplied by the hydrometeorologic service supported by an appropriate national network of stations. To satisfy the second condition, the International Atomic Energy Agency (IAEA), since 1953, is collecting and publishing the data obtained from the WMO/IAEA Isotopes-in-Precipitation Network.

It must be stressed the role of the International Atomic Energy Agency in providing the Secretariat for the Working Group on Nuclear Techniques in Hydrology of the UNESCO International Hydrological Decade (1965–1974), regular scientific symposia and systematic research coordinated programs on this subject. All these

IAEA actions create an efficient basis for fruitful scientific cooperation of the specialists in nuclear techniques from different countries.

The use of artificial and natural radioactivity as a tracer in groundwater as well as mathematical modelling for interpretation of experimental data are also discussed.

In this book we systematically review and generalise the results obtained by the author and his colleagues from many countries during the field and laboratory studies concerned with the development and application of radioactive sources and tracer techniques in geoengineering and hydrology.

Particular attention is paid to the range of validity of these methods and to the solution of practical problems. The basic physics of radioisotopes and emitted radiation, as well as the interaction of radiation with matter, methods of radiation detection, radiation hazard problems and many others are briefly discussed in the book.

The author is grateful to his colleagues V.A. Polyakov, V.S. Goncharov, V.T. Dubinchuk, T.A. Gryaznov, L.V. Selivanov, B.P. Krovopuskov, D.M. Lantsman, A.I. Avsyuk, Yu.B. Seletsky, V.M. Maslennikov, A.K. Priymachuk and V.I. Demchenko for many years of fruitful co-operation in the field, with laboratory investigations and providing design and preparation of the penetration logging equipment. The author especially wishes to thank B. Malashenkov for his assistance in preparation of the manuscript for the book.

References

Clayton CG (1983) Nuclear geophysics. Pergamon Press, Oxford

Ferronsky VI, Danilin AI, Dubinchuk VT et al (1968) Radioactive investigative methods in engineering geology and hydrogeology. Atomizdat, Moscow

Ferronsky VI (1969) Penetration logging methods for engineering geological investigation. Nedra, Moscow

Filippov EM (1962) Applied nuclear geophysics. USSR Academy of Sciences Publ House, Moscow

Froehlich K (ed) (2010) Environmental radionuclides: tracers and timers of terrestrial processes. Elsevier, Amsterdam

International Atomic Energy Agency (1968) Guidebook on nuclear techniques in hydrology. IAEA, Vienna

International Atomic Energy Agency (1971) Nuclear well logging in hydrology. Technical report series No 126, IAEA, Vienna

International Atomic Energy Agency (1981) Stable isotope hydrology. In: Gat JR, Gonfiantini R (eds). Technical reports series No 210, IAEA, Vienna

International Atomic Energy Agency (1983) Guidebook on nuclear techniques in hydrology. IAEA, Vienna

International Atomic Energy Agency (1999) Nuclear geophysics and its applications. IAEA, Vienna

Acknowledgements

Chapters 1, 2, 3, 4, 5, 6, 7, 8, 12, and 13 are based on the publications below and figures and tables in these chapters were reproduced with the kind permission of the copyright holder from these publications.

V.I. Ferronsky, A.I. Danilin, V.T. Dubinchuk et al (1968) Radioisotope Investigative Methods in Engineering Geology and Hydrogeology. Atomizdat, Moscow (in Russian)

V.I. Ferronsky, A.I. Danilin, V.T. Dubinchuk et al (1969) Radioisotope Investigative Methods in Engineering Geology and Hydrogeology. US AEC, Springfield (Translation of Russian edition)

V.I. Ferronsky (1969) Penetration logging methods for engineering geological investigation. Nedra, Moscow (in Russian)

V.I. Ferronsky, A.I. Danilin, V.T. Dubinchuk et al (1977) Radioisotope Investigative Methods in Engineering Geology and Hydrogeology (Second Edition). Atomizdat, Moscow (in Russian)

V.I. Ferronsky (1969) Penetration Logging Methods in Engineering Geology. Nedra, Moscow (in Russian)

V.I. Ferronsky and Gryaznov TA (1979) Penetration Logging. Nedra, Moscow (in Russian)

International Atomic Energy Agency (1971) Nuclear Well Logging in Hydrology. Technical report series No 126, IAEA, Vienna

V.I. Ferronsky, V.A. Polyakov (1983) Isotopy of the Hydrosphere. Nauka, Moskva (in Russian)

The figures and tables of Chap. 9, 10, and 11 were reproduced from the book: V.I. Ferronsky and V.A. Polyakov, Isotopes of the Earth's Hydrosphere, 2012, Springer, Dordrecht

Contents

Part IV Other Applications

Chapter 1
Introduction: Fundamentals of Nuclear Physics

Abstract A short overview of the discovery of natural, artificial and cosmogenic radioactivity and nuclear particles (proton, neutron and electron) is presented. The discovery of stable isotopes, their abundance in the Earth's crust, parameters of sub-atomic particles and the binding energy of nuclei are discussed. Physics of nuclear reactions, sources of radioactivity, the law of radioactive decay and attenuation of radiation by matter are considered. The principal ideas on measurement techniques and health hazards are stated. Non-solved problems of the nuclear structure and the nature of strong nuclear interactions are discussed.

1.1 Natural Stable and Radioactive Isotopes

It was assumed during ancient times and up to the end of the nineteenth century that the smallest particle of an element is an atom, which was the main subject of study in atomic physics. But as time passed, a 'magic 1895–1905 decade' appeared in physics.

In 1895 German physicist W. Roentgen discovered X-rays produced by the Crookes' cathode tube. In 1896 French physicist A. Becquerel discovered the natural radioactivity of uranium decayed products. In 1897 English physicist J. Thompson in the Cavendish laboratory identified the electron, the fundamental particle of matter. In 1898 Pierre and Maria Curie discovered radioactivity of polonium and radium. Later on, in addition to the elements of the thorium, uranium and actinium series, natural radioactivity was discovered for kalium-40, rubidium-87, samarium-82 and others, which did not belong to the trans-uranium series.

In 1934 F. and I. Joliot-Curie first obtained artificial radioactive elements. In 1911, in experiments on scattering of α-particles by a matter, E. Rutherford discovered the atomic nucleus and in 1918 he identified the proton. He observed scattering of α-particles by two large angles and assumed that the charge of an atom is located in its centre. The fact of existence of the atomic nucleus was not immediately accepted. Only after N. Bohr introduced his quantum theory of the atom and G. Moseley experimentally showed the shift of the dark-line roentgen spectrum of different atoms was the atomic nucleus considered in physics. Later on in 1934 the neutron was identified by G. Chadwick. However, even today, an acceptable theory of the atomic nucleus structure does not exist.

V. I. Ferronsky, *Nuclear Geophysics*, Springer Geophysics,
DOI 10.1007/978-3-319-12451-3_1

In 1913, English radio-chemist F. Soddy proposed the term 'isotopes' for identifying atoms of the same element but having different masses. It was done in order to determine places of the decay products of the radioactive elements in Mendeleev's table of chemical elements. The famous English physicians J. Thomson and F. Aston played an important role in the discovery and study of stable isotopes. In 1911, Thompson developed the method of parabolas for determination of the ratio of the particle charge to its mass by recording mass spectra. Based on this, using the apparatus designed in the Cavendish Laboratory, Thomson discovered the neon isotopes of mass number 22, which was reported in January 1913. In 1919, Aston proved the existence of the neon isotopes and soon found the isotopes of chlorine and mercury. Several years later, more than 200 isotopes of various elements were discovered, with the exception of those of hydrogen and oxygen that were then considered as 'simple' elements.

In 1929, W. Giauque and H. Johnston, applying the new techniques of research by absorption air spectra, discovered the isotopes of oxygen of mass numbers 17 and 18. In 1931, Johnston, on the basis of the free electrons and protons rule developed from analysis of established isotopes, concluded that hydrogen isotopes with masses 2 and 3 should exist. Soon after this in 1932, Urey, Brickwedde and Murthy, using the mechanism of isotope fractionation, detected the Balmer's lines of deuterium in the enriched spectrum of gaseous hydrogen using the spectral method. A detailed history of isotope discovery and isotope studies has been reported by Aston in his monograph (Aston 1942).

Atoms are the smallest amount of matter that retain the properties of an element. Atoms are composed of smaller particles that are protons, neutrons and electrons. At present, there are 105 different elements on the Earth, 90 of them exist in nature and 15 are man-made. The total number of protons of an atom is equal to the number of electrons and is given the symbol Z. The number of non-charged neutrons in a nucleus is given the symbol N. The mass number of the nucleus is the total number of protons and neutrons, which is given the symbol $A=Z+N$. Each of the chemical elements has a unique atomic number because the atoms of different elements contain a different number of protons and a different combination of protons and neutrons. Not all combinations of numbers of protons and neutrons are possible. Up to now, about 2500 specific nuclides with unique combinations of neutrons and protons have been identified in the solar-system matter. Figure 1.1 shows the general picture of the world of nuclei.

The long-lived nuclei present on the Earth are shown by black squares Unbounded combinations of N and Z lie outside the lines marked "last proton/neutron unbound".

The majority of combinations of the protons and neutrons appear to form unstable nuclei, which are therefore short-lived or not formed at all. It is assumed in astrophysics that the pre-solar cloud, from which ~4.5 billion years ago the Sun, the planets, satellites and meteorites were formed, initially consisted of ^{1}H (~75 %) and ^{4}He (~25 %). These two elements were formed in the primordial Universe during the "big bang" at the cosmologic temperature $kT=60$ keV. All other elements, which are ~2 %, were formed in the stars by nucleosynthesis. Namely,

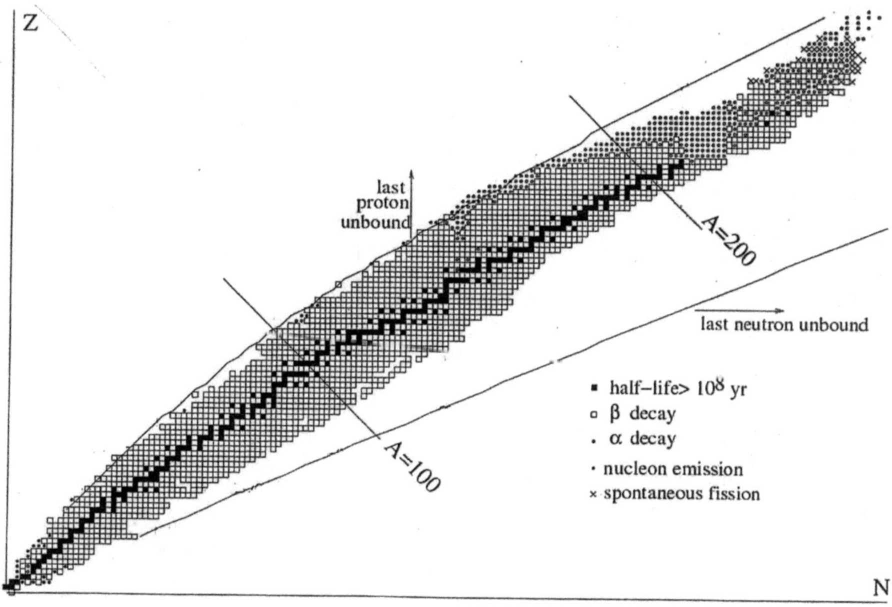

Fig. 1.1 A map of nuclei (Basdevant et al. 2005)

elements with A<60 were formed by ^{4}He burning at exothermic reactions. As a result, peaks in solar-system abundances are observed for the elements ^{4}He, ^{12}C, ^{16}O, ^{20}Ne, ^{28}Si, ^{32}S, ^{36}Ar, ^{40}Ca and ^{56}Fe. And creation of elements with A >60 is assumed to have occurred at the star nucleosynthesis by slow and fast neutron capture. The number of neutrons in a nucleus with Z protons can be in some limit different because the nucleus stability is determined by the proton-neutron ratio. The elements with odd Z do not have more than two stable isotopes. The elements with even Z do have more than two stable isotopes. Tin has 10 (maximum) stable isotopes, xenon has 9 and cadmium and tellurium have 8 stable isotopes. The content of the isotopes in the Earth's crust is practically limited by 10 elements from 8 to 26 Z (Table 1.1).

The other elements of the periodic system are represented by a negligible value of the total mass.

The size and mass of atoms are very small and use of normal measuring units for them is inconvenient. Dalton in 1803 introduced the mass of hydrogen as the atomic mass unit (*amu*). In 1906 it was decided to use the oxygen scale of *amu*. But after the three oxygen isotopes (^{16}O, ^{17}O and ^{18}O) were discovered, the carbon scale (1/12 of ^{12}C) was accepted as *amu*, which is equal to $1.66 \cdot 10^{-24}$ g. Thus, properties of the three subatomic particles are as follows (Table 1.2).

The unit for energy is the electron-volt (eV), which is the amount of energy acquired by a single electron when it overcomes a potential difference of 1 V: $1\ \text{eV} = 1.602 \cdot 10^{-19}$ J.

Table 1.1 Stable isotopes composing the Earth's crust

Element	Atomic number, Z	Mass content, %
O	8	47.0
Si	14	29.5
Al	13	8.05
Fe	26	4.65
Ca	20	3.30
Na	11	2.50
K	19	2.50
Mg	12	1.87
Ti	22	0.45
Mn	25	0.10

Table 1.2 Properties of the three subatomic particles

Particle	Location	Charge	Mass (amu)
Neutron	Nucleus	none	1.008665
Proton	Nucleus	+1	1.007277
Electron	Shells around nucleus	−1	0.000548

The atomic radii are approximately the same for all atoms and equal to $r_a = \sim 2 \cdot 10^{-10}$ m. The nucleus does not have a sharp outer boundary. The relationship between the radius of a nucleus and the atomic mass number A is found in the form:

$$r_n = (1.25 \cdot 10^{-15}\,\mathrm{m})A^{1/3}, \tag{1.1}$$

where r_n is the radius of the nucleus (m); A is the atomic mass number (dimensionless).

In Table 1.3 the calculated radii values for some nuclides are shown.

Rutherford's planetary model of an atom was accounting for an attractive force to overcome the repulsive force between protons. At the same time, two forces are present in the nuclides. They are electrostatic forces between charged particles and gravitational forces between any two objects having masses. But the calculation for protons in the framework of classical physics shows that gravitational force is equal to $\sim 10^{-24}$ N and electrostatic force is $\sim 10^{-12}$ N.

Table 1.3 The radii values of some light, intermediate and heavy nuclides

Nuclide	Radius of nuclide (cm)	Nuclide	Radius of nuclide (cm)
${}^{1}_{1}H$	$1.25 \cdot 10^{-15}$	${}^{178}_{72}Hf$	$7.01 \cdot 10^{-15}$
${}^{10}_{5}B$	$2.69 \cdot 10^{-15}$	${}^{238}_{92}U$	$7.74 \cdot 10^{-15}$
${}^{56}_{26}Fe$	$4.78 \cdot 10^{-15}$	${}^{252}_{98}Cf$	$7.89 \cdot 10^{-15}$

Energy is a fundamental parameter of nuclei, which is responsible for their drives and equilibrium including nuclear reactions and radioactive decays. The equivalency between the mass and energy is a main principle in nuclear physics following from Einstein's theory of relativity ($E=mc^2$). The binding energy of a nuclide is the conception that considers a nuclide formed from protons (hydrogen atoms) and neutrons. It follows from this that the sum of the nuclide masses differs from the nuclear mass by the binding energy (Basdevant et al. 2005):

$$B(A,Z) = Nm_n c^2 + Zm_p c^2 - m(A,Z)c^2, \tag{1.2}$$

or

$$B(A,Z) = -E_B(A,Z), \tag{1.3}$$

where B(A, Z) is the positive nucleus energy and $E_B(A, Z)$ is the negative binding energy.

Table 1.4 presents calculated data of the binding energy for some nuclides and the average binding energy per nucleon (B/A). The table shows that, in general, nucleus binding energy increases with increased atomic number Z. But the criterion for nuclide stability appears to be the average binding energy per nucleon BE_{av}.

Figure 1.2 shows the binding energy per nucleon B/A as a function of A. It is seen here that, in light, nuclei binding energy is increased with A and reaches its maximum around the iron-nickel elements ($A \approx 50-60$). This energy can be released for practical purpose by fusion reactions on light elements. The binding energy for $12<A<225$ is $7.7<B/A<8.8$ (MeV), i.e., proportional to the number of nucleons.

Heavy nuclides have lower values of average binding energy but when some of them are unstable, like ^{235}U and ^{239}Pu, which split into two pieces, then fission energy releases. It is worth noting that we are far from a complete understanding of the nucleus structure and strong interactions in order to prevent nuclear "pile-up" (Basdevant et al. 2005).

Table 1.4 Binding energy and its average value for some nuclides

Nuclide	BE, MeV	BE_{av}, MeV
^{2}D	2.226	1.113
^{4}He	28.296	7.074
^{14}N	104.659	7.476
^{16}O	127.619	7.976
^{19}F	147.801	7.779
^{40}Ca	342.052	8.551
^{55}Mn	482.070	8.765
^{58}Fe	509.945	8.792
^{62}Ni	545.259	8.795
^{206}Pb	1622.340	7.875
^{238}U	1822.693	7.658

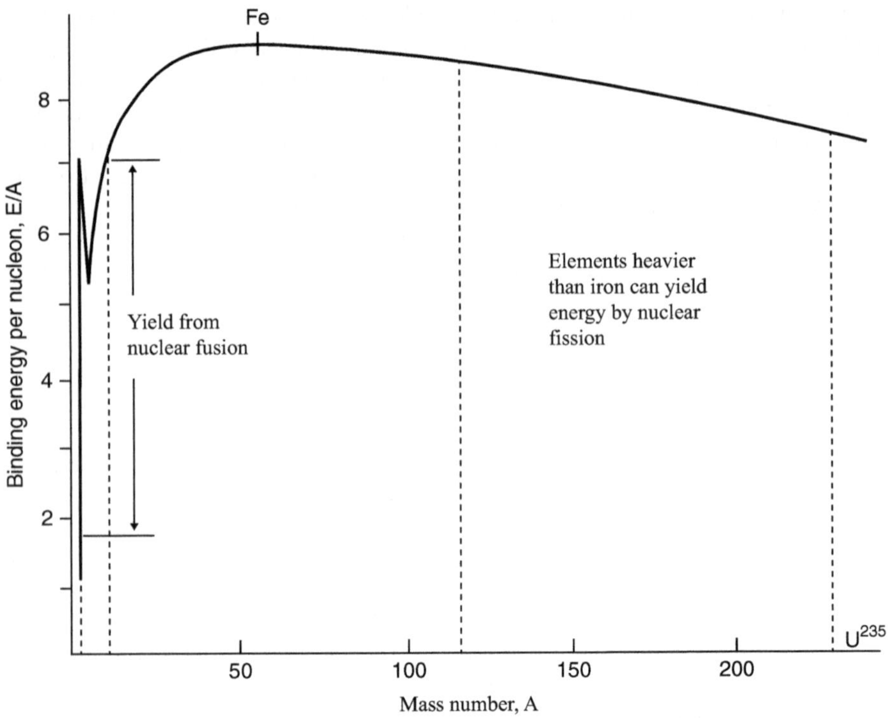

Fig. 1.2 Binding energy as a function of the number of elements (Basdevant et al. 2005)

Soon after the discovery of hydrogen and oxygen isotopes, studies were undertaken for the distribution of these isotopes in natural waters. The first works by Bleckney, Gould, Manian, Urey, Brodsky, Vernadsky, Vinogradov, Tays and Florensky were carried out in 1932–1942. They investigated river, lake, sea, rain and groundwaters. Thus, the first comparative data concerning hydrogen and oxygen isotope variations in natural waters were obtained. But many of these measurements were based on density techniques, leading to an inadequate accuracy.

Only in the 1940s, when more perfect mass spectrometers were designed by Dempster and Nir and especially after improvements by McKinny in 1950, did studies of the isotopic composition of natural waters become more common. Numerous data on the first isotope studies of natural waters and study results of the physical and chemical properties of heavy water and methods of its investigation were presented in the work of Kirshenbaum (1951). The early mass spectrometric studies, carried out by Kirshenbaum, Graff and Forstat in 1945, Silverman in 1951, Friedman in 1953 and Epstein and Mayeda in 1953, dealt with the distribution of deuterium and oxygen-18 in natural waters and other objects. These studies initiated systematic research of the natural regularities governing the distribution of the stable isotopes of hydrogen and oxygen in the hydrosphere and their application to the solution of scientific and practical problems. Great credit for the organisation and supervision of numerous studies on the hydrochemistry of hydrogen and oxygen in different countries must be given to researchers of the old school such as Rankama, Urey and Vinogradov, who made important contributions to the development of the geochemistry of isotopes as a whole.

Among the general natural principles it has been found that the water of the ocean, being the main reservoir of the hydrosphere, has a fairly homogeneous hydrogen and oxygen isotope composition. Thus Craig (1961) proposed it as a standard, relative to which all other measurements of deuterium and oxygen-18 content are carried out. It was found that the average isotopic composition of ocean water remains constant over geologic time. Slight regional deviations in the average isotope composition of oceanic masses of water are the result of the continuous evaporation of water from the oceanic surface, which continually enriches the upper layers with heavy isotopes. The isotopic composition of the ocean as a whole is maintained constant by the balance between evaporated water and precipitation (including continental runoff) and the persistent mixing of all oceanic waters. Experimental data on the distribution of hydrogen and oxygen isotope ratios in the ocean reinforce those of salinity distribution and provide a more reliable base for the study of the dynamics of water in the ocean on a global scale.

One problem related to the ocean and being solved with the help of isotope data, is that of paleotemperature reconstruction. Considerable progress has been made both on the basis of the analysis of oxygen isotopes in carbonates of fossil marine fauna (deposited during the past 730 million years) and chert paleothermometry, developed in recent years by Epstein and his co-workers. Progress has also been made on the longer time perspective with work on ancient geological epochs, including the Precambrian.

It should be pointed out that the isotope composition of the worlds' ocean water and that of the pore water of oceanic sediments and basement rocks in particular, has not been studied sufficiently and there is plenty of scope for further research.

The isotopic composition of atmospheric moisture and surface continental waters indicates a great variation resulting from considerable latitudinal and altitudinal temperature variations in time. The principles of the temporal and spatial distribution of the isotopic composition in atmospheric moisture and surface waters, related to these factors, provide a basis for the interpretation of isotopic data and the elucidation of the condition of global and regional water circulation in nature. At present, comprehensive factual material concerning the isotope composition of atmospheric precipitation has been obtained for the whole world. These data were obtained in the course of systematic measurements of deuterium and oxygen-18 concentrations in precipitation, sampled at the global IAEA/WMO network of stations. They have been published by IAEA and are available for use by specialists.

The picture of the distribution of the isotopic composition of groundwater is most indicative due to the complicated processes involved in the formation of groundwater in the past. The interpretation of the data on isotopic composition from the point of view of isotope fractionation provides the key to explaining the processes involved in the formation of groundwater over geologic time. The most important fact obtained from an examination of the isotopic composition of groundwater is that its formation in artesian basins has involved admixing, in different proportions, of the more recent meteoric waters with the ancient waters of marine origin. There is no conclusive evidence indicating the presence of juvenile water in groundwater. This is clearly demonstrated by investigations carried out in the regions of recent volcanism where, in all areas of the world, hydrothermal waters have been found to be similar to the waters of local atmospheric precipitation. However, it has been shown no less convincingly that in most of the investigated ore deposits of hydrothermal origin the majority of the original ore-bearing fluids are of meteoric origin.

The important role in unification of a procedure for presentation of the isotope content of hydrogen and oxygen and some other elements of natural objects involves the work that is undertaken by the Isotope Hydrology Section of the IAEA in intercomparison of the main isotope standards. In this work the leading mass spectrometry laboratories from various countries are involved. The numerical characteristics of the main standards, expressed in terms of per mille (‰), in present use are as follows:

		δD, ‰	$\delta^{18}O$, ‰
V-SMOW	Vienna standard of mean ocean water	0	0
SLAP	Standard of light Antarctic precipitation	−428	−55.5
GISP	Greenland precipitation ice standard	−189.8	−24.85
NBS–1	Distilled water of the Potomac River	−47.6	−7.94
NBS–1 A	Snow water from Yellowstone Park	−183.3	−24.33
Standard of groundwater from upper carboniferous sediments in Moscow		−94.5	−13.08

In this book all the data of isotopic composition of hydrogen and oxygen presented are relative to the SMOW standard. The isotope results of earlier works of other authors sited in the book were recalculated with respect to the SMOW standard.

The story of cosmogenic isotopes dates back to 1912, when the Austrian physician Victor Hess discovered cosmic-rays by effect of ionisation of air molecules. Later on it was found that the initial cosmic radiation of solar, galactic and metagalactic origin of higher energy particles (mainly protons), as a result of interaction with atomic nuclei of the air, produce secondary less energetic radiation. This secondary radiation in nuclear reactions with nuclei of air atoms produces a great number of cosmogenic radioactive isotopes. Tritium and radiocarbon are cosmogenic radioisotopes that have the most wide practical application in the study of natural water.

The search for tritium started in 1931, when Johnston predicted its existence among the hydrogen isotopes. But the attempts of a number of scientists (Aston among them), undertaken over a number of years and aimed at its discovery in an enriched hydrogen and natural water spectrum, failed. Rutherford, analysing these studies, came to the conclusion that either the amount of this isotope was too small or it does not exist at all.

While the search for natural tritium was unsuccessful, Olefant, Harteck and Rutherford using irradiation of deuterium-bearing compounds with deuterons, detected the tracks of particles with mass number 3, which could have been related to both tritium and helium. In the same year, this discovery was confirmed by Dees. At first, it was assumed that both the isotopes discovered were stable. But Bonner found in 1938 that the mass of ^{3}He is less than that of ^{3}H. The assumption of radioactive transformation of the tritium to helium was confirmed by Alvarez and Cornog in 1939. They made the first estimates of the half-life of tritium to be 150 ± 40 days. Later, they found it to be several years. In 1940, O'Neil and Goldhaber, studying the production of tritium using the reaction ^{6}Li(n,α)^{3}H, determined the half-life of tritium to be 12.1 year.

The search for radioisotope ^{14}C was carried out in parallel with the search for tritium. In 1934, Kurie (and later, in 1936, Bonner and Brubaker and Burcham and Goldhaber) irradiated air with neutrons in a chamber and discovered that radiocarbon is formed from nitrogen by the reaction ^{14}N(n, p)^{14}C. The search for the latter radioisotope was undertaken only after the discovery of the secondary neutron flux produced in the atmosphere by cosmic-rays, which was made by Korff and Libby in 1937–1939. For this purpose Libby designed a special neutron counter, being elevated by a balloon up to a height of 50,000 ft. After these investigations, Korff, in 1940, reported that the production of ^{14}C in the atmosphere is likely to proceed according to the (n, p) reaction on nitrogen under irradiation by secondary neutrons. In 1941, Cornog and Libby showed that in the (n, p) reactions on nitrogen the fast neutrons produce tritium and slow ones ^{14}C. In 1946, Libby detected radiocarbon in the atmosphere and came to the conclusion that the biosphere could be traced with it. Thus, radiocarbon dating was discovered. The half-life period of ^{14}C was not determined at once. At first it was assumed, by analogy with ^{35}S, to be equal to 4 months. Then it was considered to be 25,000 year and only later was found to be equal to 5730 year.

The search for natural tritium was continued simultaneously. The increasing sensitivity of the law background radiometric apparatus seemed to be advantageous. Tritium was detected almost simultaneously in 1950 by Faltings and Harteck in atmospheric hydrogen and in 1951 by Grosse, Johnston, Wolfgang and Libby in lake water. They found concentrations of tritium in atmospheric hydrogen $^3\mathrm{H/H} \approx 3.5 \cdot 10^{-15}$ and in lake water $^3\mathrm{H/H} \approx 3.5 \cdot 10^{-18}$ which appeared to be in agreement, to order of magnitude, with those established later. A detailed history of the tritium and radiocarbon techniques may be found in works by Libby (1957, 1973).

Tritium, being a perfect indicator of water motion, was immediately used by a number of researchers in the investigation of natural waters. But in the pre-nuclear test era only a few individual measurements of its content in natural waters were carried out. A large amount of tritium released into the atmosphere in the course of thermonuclear weapons tests increased its natural content by two orders of magnitude. Therefore, work aimed at re-estimating the equilibrium content and natural concentrations in natural waters is presently being carried out.

As to radiocarbon, its content has been increased by bombs, although by lesser amounts. The work by Münnich (1957) reported the first application of ^{14}C to the dating of natural waters.

From the investigations carried out on the basis of data on tritium concentrations in atmospheric precipitation, gathered over a number of years at the IAEA/WMO network of stations, it was concluded that atmospheric air masses mix rather thoroughly. Valuable evidence highlighting the distribution of tritium in atmospheric moisture has been provided by observing the decline of bomb-tritium ejected before 1963 in atmospheric precipitation. These observations have thrown light upon the general principles of the temporal and spatial distribution on a global scale of tritium in atmospheric moisture, oceans and surface waters. Similar studies for obtaining the radiocarbon distribution in natural waters have been undertaken. Thus, a good experimental and theoretical base involving the tritium and ^{14}C isotope tools has been developed for the solution of a wide range of scientific and practical problems related to the investigation of dynamics of water in the hydrosphere.

Radiometric methods are used for measurement of cosmogenic isotope radioactivity by means of low-level beta-spectrometers with scintillation detectors or by proportional counters with internal gas filling. Some other methods and techniques have been developed for measurement of natural concentrations of the cosmogenic isotopes.

Tritium concentration in natural objects is accepted to be expressed in tritium units (TU). 1 TU corresponds to the content of 1 atom of tritium per 10^{18} atoms of protium. Up to recently, it was assumed that the half-life period of the tritium decay is equal to 12.26 year. In September 1979, a group of experts gathered in the IAEA, Vienna discussed new experimental data of the National Bureau of Standards, USA and accepted a new value of the half-life period of the tritium decay equal to 12.430 year. The new decay constant $\lambda = 0.55764$ year^{-1} corresponds to the above value of the half-life period of tritium decay (standard NBS SRM–4526C). And also 1 TU = 7.088 counts/min · 1 kg^{-1} water or 1 TU = 3.193 pCi · кг$^{-1}$ of water. In 2000, as a result of the work of the IAEA program TRIC-2000, the half-life period of tritium decay was corrected to 12.32 year and the decay constant $\lambda = 0.05625$ year^{-1} (1 TU = 0.119 Bq per 1 kg of water).

Radioactivity of the other cosmogenic radionuclides is measured in the units of decay per minute on 1 kg or 1 t of water. The total activity of radionuclides is measured in curie (Ci) or in becquerel (Bq). 1 Ci is equal to $3.7 \cdot 10^{10}$ counts/s or $3.7 \cdot 10^{10}$ Bq.

Specific activity of radiocarbon in natural objects is measured in counts/min on 1 g of carbon. But this is not convenient because it needs to know the precise efficiency of the counting techniques. This form of expression of the ^{14}C content gives a rise to difficulty in obtaining consistent data for different laboratories having non-identical measurements. In order to unify measurements, it was proposed to express radiocarbon contents in per cent (or per mille) with respect to a standard of the modern carbon (Olsson 1979; Stuiver and Suess 1966). Because modern natural objects have variable ^{14}C specific activity, a standard of the radiocarbon is used for unification of the measurements. For this purpose the standard of NBS USA is used, which is the oxalic acid having a radioactivity of 14.3 count/min on 1 g of carbon ($1.176 \cdot 10^{-10}$% ^{14}C). The standard of the modern carbon is defined as a specific activity of ^{14}C in wood in 1950, after correction on isotope fractionation and presence in the atmosphere of the 'industrial' carbon dioxide (Stuiver and Suess 1966). Such a definition of a standard for the modern carbon corresponds to 0.950 (95%) of the oxalic acid activity of the NBS standard, i.e., 13.6 count/min on 1 g of carbon.

Some information related to other cosmogenic isotopes in the atmosphere and hydrosphere such as ^{7}Be, ^{10}Be, ^{22}Na, ^{26}Al, ^{32}Si, ^{32}P, ^{33}P, ^{36}Cl, ^{37}Ar, ^{39}Ar is presented in the book. Investigation of the regularities on distribution of these isotopes in natural waters is mainly limited by technical difficulties of sampling, concentration and measurement of the corresponding samples.

The radiogenic isotopes, the lifetime of which is comparable to age of the hydrosphere and the Earth as a whole, have, up to recent times, been used mainly for the purposes of geochronology. But the studies of their distribution in natural waters have shown them to be valuable indicators of the processes involved in the motion of natural waters, both on global and regional scales, for a large period of geologic time. The most fruitful in this respect was the trend, related to application of the effect of the disequilibrium ratio of $^{234}U/^{238}U$ in natural waters, which was discovered in 1953 by Cherdyntsev and Chalov (Cherdyntsev 1969; Chalov 1975).

The application of the disequilibrium of uranium and other heavy radioactive elements being rapidly delivered seems to be advantageous for discovering the main periods of evolution of the hydrosphere and for the solution of many practical problems.

Both the natural and artificial radioactive isotopes have found practical applications in nuclear geophysics. They are divided into four groups, namely, alpha, beta, gamma and neutron emitting radiation.

1.2 Nuclear Reactions and Sources of Radioactivity

Nuclear reaction is the process of collision of two nuclides or any nuclide and particle (proton, neutron, electron, α, β, γ and X-rays) with the transformation of at least one or more nuclides. Without the transformation the process is called scattering. The spontaneous disintegration of an unstable natural or artificial nuclide with emission of α and β particles or γ-rays is called radioactive decay. Experimental data indicate that nuclei, like atoms, possess a shell-structure with respect to energy distribution, which separates together with absorption or extraction of a neutron, proton or any other particle. The discontinuity in separation of the energy is due to the excess binding energy for nuclei (Basdevant et al. 2005). New information related to consideration of the models in the nuclear structure can be found in our works (Ferronsky et al. 2011; Ferronsky and Ferronsky 2013).

Nuclear reaction probability for a nuclide with a particle is considered in terms of the statistical concept and called the microscopic section. Its unit is the barn ($\sigma = 10^{-28}$ m^2). This is the area of a nuclide exposed to a given reaction. The nuclide cross-sections are calculated assuming a classical trajectory for the particle and using quantum perturbation theory on plane waves to treat particle collision and scattering due to electromagnetic and weak interactions.

The nuclear reactions, similar to chemical reactions, are written as follows:

$$^{A}_{Z}X + x \rightarrow {}^{A}_{Z}Y + y, \tag{1.4}$$

where X is the target's nucleus; x is the bombarding particle; Y is the final nucleus-product; y is the emitted particle; A and Z are the mass number and charge of the nucleus.

If the nucleus-product appears to be an exciting nucleus then the symbol A is written as A_m.

The following reactions can be real depending on the type of bombarding and emitting particles (Filippov 1962):

1. The reaction of α-capture by nucleus and neutron emission (α, *n* reaction):

$$ {}^{A}_{Z}X + \alpha \rightarrow {}^{A+3}_{Z+2}X + n, \tag{1.5}$$

   i.e., the mass is increased by 3 units and the charge by 2 units.
2. The reaction of a neutron capture by nucleus and gamma-quantum emission (*n*, γ radiation capture):

$$ {}^{A}_{Z}X + n \rightarrow {}^{A+1}_{Z}X + \gamma, \tag{1.6}$$

   as a result, the isotop of the target nucleus appears and gamma-quantum emission.
3. The reaction of a neutron capture by nucleus and proton emission (n, p reaction):

$$ {}^{A}_{Z}X + n \rightarrow {}^{A}_{Z-1}X + p, \tag{1.7}$$

   as a result, the isobar of the target nucleus appears with negative β^- particle emission. The reaction goes, as a rule, by fast neutrons.
4. The reaction of a neutron capture by nucleus and α-particle emission:

$$ {}^{A}_{Z}X + n \rightarrow {}^{A-3}_{Z-2}X + \alpha, \tag{1.8}$$

   i.e., the mass is decreased by 3 units and the charge by 2 units. The reaction goes, as a rule, by slow neutrons.
5. The reaction of a neutron capture by a nucleus that is split:

$$ {}^{A}_{Z}X + n \rightarrow {}^{A_1}_{Z_1}X + {}^{A_2}_{Z_2}X + in, \tag{1.9}$$

   two splinters (often radioactive) are created and two or three neutrons (i). The reaction goes both by slow and fast neutrons.
6. The reaction of neutron elastic or non-elastic scattering, which is accompanied by the neutron energy change:

$$ {}^{A}_{Z}X + n \rightarrow {}^{A_m}_{Z}X + n'. \tag{1.10}$$

   The exciting nucleus by X^{Am} by gamma-quantum emission comes back to initial state.

7. The reaction of a neutron capture by a nucleus accompanied by two neutrons emission

$$ {}^{A}_{Z}X + n \rightarrow {}^{A-1}_{Z}X + 2n, \quad (1.11) $$

as a result, the mass number of the target isotope increases by unit.
8. The nuclear photo effect reaction (the nucleus split and neutron emission):

$$ {}^{A}_{Z}X + \gamma \rightarrow {}^{A-1}_{Z}X + n, \quad (1.12) $$

at which the isotope of the initial element with a mass number of unit less is created.

There are many other nuclear reactions that will be discussed in the chapters below. During the nuclear reactions the binding energy Q releases. In this case the reaction is written as:

$$ {}^{A}_{Z}X + x \rightarrow {}^{A}_{Z}Y + y + Q. \quad (1.13) $$

In experiments with nuclear reactions we try to obtain information about the nuclear structure and to understand the physics of the energy release at the nuclear interaction.

Since the time of Earth's creation the short-lived nuclei have decayed. Only nuclei with a lifetime more of 4.5×10^9 year are living at present. These nuclei emitting α, n, β and γ radiation are called radiogenic or fossil. The unstable nuclei entering the Earth's atmosphere in the form of cosmic rays are called cosmogenic. Those unstable nuclei produced in nuclear reactors and scientific laboratories are called artificial.

The original heavy fossil radioactive nuclides are ^{238}U, ^{235}U and ^{232}Th and form three natural radioactive chains that proceed through a series of α and β decays three stable isotopes of lead ^{206}Pb, ^{207}Pb, and ^{208}Pb (see Figs. 11.1, 11.2, and 11.3).

The emitting α, n, β and γ radiation used in nuclear physics and geophysics can be found in the reference books.

1.3 Laws of Radioactive Decay and Attenuation of Radiation

Each radioactive nucleus decays independently of other nuclei. Hence, the mean value of the decay acts ΔN during the period of time Δt is proportional to the number of non-decayed nuclides N and time Δt:

$$ \Delta N = -\lambda N \, \Delta t, \quad (1.14) $$

where λ is the decay constant. The minus sign shows that the nuclei number decreased in time.

After differentiating and integration with respect to t from 0 to t and with respect to N from N_0 to N we have:

$$N = -N_0 e^{-\lambda t}. \tag{1.15}$$

After transformations one can obtain:

$$\lambda = \frac{\ln 2}{T} = \frac{0.693}{T}. \tag{1.16}$$

With radioactive decay the number of initial isotope N_1 decreases and the isotopes of product N_2 grows. The product N_2 is often also the radioactive daughter's isotope. The amount of isotopes N_1 and N_2 depends on their decay constants. If $\lambda_1 \ll \lambda_2$ then the radioactivity reaches the steady state value at $t > 5T_2$.

Duration of the radioactive isotope life τ is defined from (1.14) as:

$$\tau = \frac{1}{N_0}\int_0^\infty \lambda t N dt = \lambda \int_0^\infty t e^{-\lambda t} dt = \frac{1}{\lambda}. \tag{1.17}$$

The exponential law of radioactive decay gives only the mean value of the decayed nuclei. Deviation of the decay number Δn from the mean value $\bar{n}$ during a period of time is determined by Poisson's law:

$$P(n) = \frac{\bar{n}^n}{n!} e^{-\bar{n}}, \tag{1.18}$$

where P(n) is the probability of the registered acts of decays within the interval time of registration.

The deviation values n from their mean value $\bar{n}$ are characterised by the square deviation value σ that is defined by the expression:

$$\sigma = \pm\sqrt{\bar{n}}. \tag{1.19}$$

Particles and gamma radiation produced in nuclear reactions of decays interact with matter depending on their nature. At radiation of high energy passage through matter it interacts with electrons and nuclei of atoms as independent particles. If the energy of radiation is similar to the binding energy of the atoms and molecules then the scattering of the radiation on the atoms and molecules of matter should be considered.

The number of nuclei of rigid or liquid matter in the unit of volume is defined by the Avogadro number:

$$N_A = \frac{A_0 \rho}{A}, \tag{1.20}$$

where $A = 6.02 \times 10^{23}$; A is the atomic mass; and ρ is its density.

For gaseous matter the number of nuclei is determined by the Loschmidt number $L_0 = 2.683 \times 10^{19}$.

The number of electrons n_e in the unit of volume is determined by the product of the nuclei number and electrons in the atom Z. For the isotopes of elements from helium to iron Z/A = ½, the value n_e will be:

$$n_e = \frac{A_0 \rho}{2}. \tag{1.21}$$

The scattering by a particle scattering centre is characterised by the probability dP = p(E,Q)dQ (where p(E,Q) is the probability density as a function of the energy and solid angle in an infinitely small interval (Q, Q+dQ)).

The possible results of the scattering process are total absorption of the particle, elastic scattering and inelastic scattering. The two last effects are differed by safety of the scattering particles and their energy.

1.4 Measurement Techniques and Health Hazards

Nuclear physics is the science, techniques and technology of the last century. Its modern measurement instrumentation is based mainly on semiconductor electronics and digital computation.

Nuclear geophysics uses fundamentals of nuclear physics but applies specific measurement instruments, electronics and digital computation. Specific solving problems need new laboratory and field apparatus and instruments. Researches and engineers use both commercial and specially prepared instrumentation. Nuclear sources of α, β, n and γ radiation are prepared on reactors using appropriate nuclear reactions. Gas filled, proportional, scintillation and semiconductor detectors for registration of used radiation are commercially produced products. Neutron laboratory and well generators are now also manufactured. Penetration logging rigs and instrumentation both for land marine bottom sediment study are prepared by individual orders.

The use of radionuclide both for laboratory experiments and for field work must take into account consideration of health and safety. The International Atomic Energy Agency has prepared special guidelines to the safe handling of radioisotopes in different fields of application. Countries have established their own national regulations that must be taken into account in planning any work with radioisotopes.

References

Aston FW (1942) Mass-spectra and isotopes. Edward Arnold, London
Basdevant J-L, Rich J, Spiro M (2005) Fundamentals in nuclear physics. Springer, Heidelberg
Chalov PI (1975) Isotope fractionation of natural uranium. Frunze, Ilim

Cherdyntsev VV (1969) Uanium–234. Atomizdat, Moskva
Craig (1961) Standard for reporting concentrations of deuterium and oxygen-18 in natural waters. Science 133:1833–1834
Filippov EM (1962) Applied nuclear geophysics. USSR Academy of Sciences Publishing House, Moscow
Ferronsky VI, Denisik SA, Ferronsky SV (2011) Jacobi dynamics. Springer, Dordrecht
Ferronsky VI, Ferronsky SV (2013) Formation of the solar system. Springer, Dordrecht
Kirshenbaum I (1951) Physical properties and analyses of heavy water. McGraw-Hill, New York.
Libby WF (1967) History of radiocarbon dating. In: Radiocarbon dating and methods of low-level counting: proceedings of a symposium, IAEA, Vienna, pp 3–25
Libby WF (1973) History of tritium. In: Moghissi A, Carter M (eds) Tritium. Messenger graphics, Phoenix, pp 3–11
Münnich KO (1957) Messungen des ^{14}C-Gehaltes von hartem Grundwasser. Naturwissenschaften 44:32–34
Olsson IU (1979) The radiocarbon contents of various reservoirs. In: Radiocarbon dating. University of California Press, Berkely, pp 613–618
Stuiver M, Suess HE (1966) On the relations between radiocarbon dates and true sample ages. Radiocarbon 8:534–540

Part I
Use of Nuclear Techniques for Determination of Soil Properties

Chapter 2
Methods Based on the Absorption of Gamma-Ray Beams by Matter

Abstract Physical effects of a gamma-ray beam passing through matter as a basis for soil density determination is discussed. These effects for moderate-energy gamma rays are the photoelectric and Compton effects and production of electron-positron pairs in the electric field of the nuclei. In order to justify this general statement the narrow and broad gamma-ray beams were experimentally analysed. It is shown that for light elements and intermediate-energy of gamma rays the mass absorption coefficient μ is independent from the atomic number and consequently on the chemical composition of the material. This effect creates the physical basis for use of gamma-ray absorption for measuring the density of multi-component media like rock and soil material. The mass absorption coefficient for gamma-rays is an important parameter for absorbing media, governing the sensitivity of the gamma-ray absorption method, the degree to which such measurements represent the medium as a whole and the reliability of density measurements. Studies of moisture dynamics in soils and other ground materials, evaporation processes and also determination of the amount of water stored in the snow cover by the gamma-ray absorption method are discussed in this chapter.

2.1 Main Principles

Methods based on gamma-ray absorption are used most widely for measuring the in situ bulk density of soil and rock materials in geoengineering and hydrogeological studies. It is also used for the study of moisture dynamics in an unsaturated zone and for direct measurement of snow water content at the assessment of a water balance. The basic principle of these methods is that the absorption of a given gamma-ray beam in a particular material depends on the total mass between the source and detector and under certain conditions the transmitted gamma-ray intensity depends only on the bulk density of the absorbing medium (Babinets and Zvolsky 1961; Danilin 1957; Artsybashev 1965; Ferronsky et al. 1968, 1977; International Atomic Energy Agency 1968, 1983, 1999; Clayton 1983).

Since the absorption method can be used to determine the mass of the material along the path traversed by the gamma rays, it follows that in this sense the method is equivalent to weighing as a means of determining mass. Moreover, the ground or

V. I. Ferronsky, *Nuclear Geophysics,* Springer Geophysics,
DOI 10.1007/978-3-319-12451-3_2

rock material can be “weighed” with the aid of the gamma rays under natural conditions, i.e., in situ. Ground layers 40–50 cm thick can be explored in this way and the mean density for the entire layer can thus be determined. Hence, in contrast to the direct sampling method, the gamma-ray absorption method provides bulk density data that are more representative for the soil material. The advantages of the method are as follows:

a. The measurements can be carried out directly in the field with low expenditure of labour and time;
b. The density of any ground can be measured independently of its granulometric, mineralogical and chemical composition and also independently from its aggregate state, structural properties and texture;
c. The method can be used to determine the density of ground and rock material of any consistency (quicksand, water-saturated sand, clay, etc.).

It is well known that moderate-energy gamma rays undergo the following basic interactions with matter: (a) total absorption (photoelectric effect), (b) Compton scattering by electrons and (c) production of electron-positron pairs in the electric field of the nucleus:

$$\sigma = \sigma_{ph} + \sigma_c + \sigma_{pp}, \tag{2.1}$$

where σ_{ph}, σ_c and σ_{pp} are the partial cross-sections describing the above three basic interactions. The probability of interaction of a gamma ray with matter per unit path length is called the linear absorption coefficient, which is defined by:

$$\mu_0 = \sigma n_a, \tag{2.2}$$

where n_a is the number of atoms per unit volume of the material. The reciprocal of the linear absorption coefficient is defined as the mean free path:

$$\lambda = \frac{1}{\mu_0}. \tag{2.3}$$

The ratio of the linear absorption coefficient to the density of the medium is the mass absorption coefficient:

$$\mu = \frac{\mu_0}{\rho}. \tag{2.4}$$

The possible energy change when gamma rays interact with matter can be divided into three ranges. In each range one of the above three basic interactions is the most probable (Table 2.1). At low energies, gamma rays interact mainly through the photoelectric effect. Compton-effect predominates at intermediate energies and, finally, pair production is the leading process at high energies.

In the energy range in which Compton scattering predominates, the mass absorption coefficient μ is a very slowly varying function of the atomic number Z. In fact,

Table 2.1 Gamma-ray energy ranges in which one of the three basic interactions predominates. (Leypunsky et al. 1960)

Medium	Photoelectron effect (keV)	Compton effect	Pair production (MeV)
Air	$E \leq 20$	20 keV < E < 23 MeV	$E > 23$
Aluminium	$E \leq 50$	50 keV < E < 15 MeV	$E > 15$
Iron	$E \leq 120$	120 keV < E < 9.5 MeV	$E > 9.5$
Lead	$E \leq 500$	500 keV < E < 4.7 MeV	$E > 4.7$

$$\mu = \frac{\sigma_c n_a Z}{\rho} = \sigma_c Z \frac{A_0}{A}, \tag{2.5}$$

where Z is the charge of the nucleus; A is the atomic weight; A_0 is Avogadro's number; and ρ is the density of the material. In light nuclei, in which the number of protons and neutrons is the same, the Z/A ratio is constant and equal to 1/2. Therefore:

$$\mu \approx \frac{1}{2} \sigma_c A_0 Z = \text{const.} \tag{2.6}$$

The exception is hydrogen, for which Z/A = 1.

It follows from Fig. 2.1 that for light elements and intermediate-energy gamma rays the mass absorption coefficient μ is independent from the atomic number and

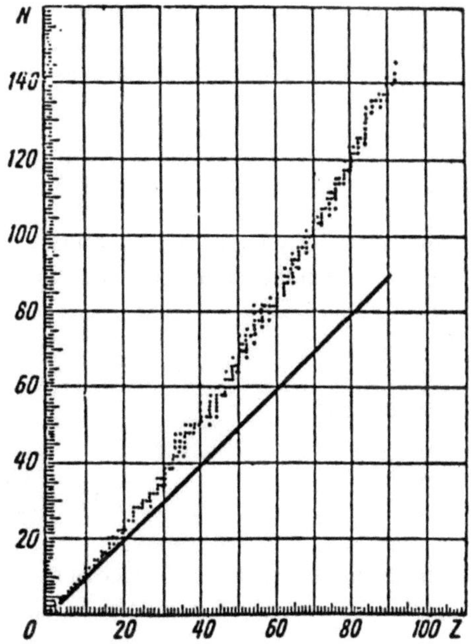

Fig. 2.1 Dependence on number of neutrons (*N*) per atom of atomic number Z for stable isotopes: ------—Z/A = 1/2; ····—Z/A < 1/2

consequently of the chemical composition of the material. This is the physical basis for use of gamma-ray absorption at intermediate energy in measurements of density of multi-component media such as rock and soil material.

2.2 Transmission of Narrow and Broad Gamma-Ray Beams Through Matter

The absorption of a narrow beam of gamma rays in a given material is described by exponential law:

$$I = I_0 e^{-\mu\rho x}, \tag{2.7}$$

where I_0 is the beam intensity at the point of observation in the absence of the absorbing material; I is the intensity in the presence of the absorbing material; ρ is the density of the medium in g/cm^3; and x is the thickness of the medium traversed by the beam in cm.

For a narrow gamma-ray beam originating in an isotropic and mono-energetic point source, the above absorption formula can be rewritten in the form:

$$I = \frac{Q_S}{4\pi R^2} \cdot e^{-\mu\rho x}, \tag{2.8}$$

where Q_S is the activity of the source in counting rate per second; R is the distance between the source and the point of observation.

In the absence of the absorbing medium the gamma-ray intensity at a point at a distance x from the source is given by:

$$I_0 = \frac{Q_S}{4\pi x^2}, \tag{2.9}$$

If the gamma-ray intensity is measured at a distance x by a suitable detector of efficiency ε, then the gamma-ray counting rate is given by:

$$n_\gamma = I_\varepsilon, \tag{2.10}$$

and the quantity:

$$n_0 = \frac{Q\varepsilon}{4\pi x^2} = I_0\varepsilon \tag{2.11}$$

is the counting rate in the absence of the absorbing medium. From Eq. (2.7) we find that the counting rate recorded by the detector can be written in the form:

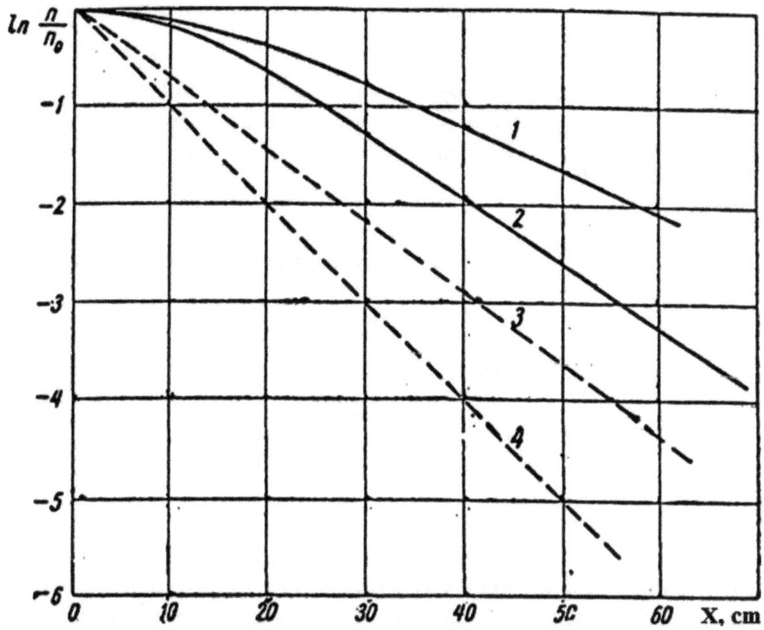

Fig. 2.2 Curves of gamma-ray beam attenuation from the point source of ^{60}Co (*1*, *3*) and ^{137}Cs (*2*, *4*) in a sand medium of 1.3 g/cm^3: *1* and *2* are for a broad beam of gamma rays; *3* and *4* are for a narrow beam of gamma rays. (Ferronsky et al. 1968)

$$n = n_0 e^{-\mu\rho x}, \tag{2.12}$$

and hence

$$\rho = \frac{\ln(n_0/n)}{\mu x}. \tag{2.13}$$

Therefore, by measuring the ratio n_0/n, which gives the degree of attenuation of the gamma-ray beam in the medium under investigation, we can determine the density of the medium if we know the mass absorption coefficient μ. It also follows from Eq. (2.12) that, if the absorption curve is plotted on a semi-logarithmic scale, the result is a straight line (Fig. 2.2).

Equations (2.7) and (2.12) are valid only in the case of narrow-beam geometry. This condition is difficult to achieve in practice because it requires the use of a large collimating system (Fig. 2.3a). A more common procedure in practice is to use a broad (Fig. 2.3b) or partially collimated gamma-ray beam that has passed through the medium but also multiply scattered secondary gamma rays. The semi-logarithmic graph of the gamma-ray intensity as a function of distance from the source will then lie above the straight line corresponding to the narrow-beam geometry and the slope of this curve will be increasing over the initial section of the curve (see Fig. 2.2).

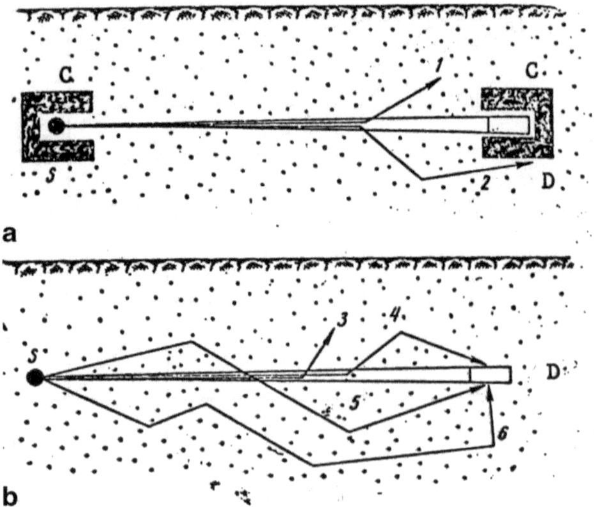

Fig. 2.3 Narrow (**a**) and broad (**b**) beam geometry: *S* is the gamma-ray source, *D* is the detector, *C* is the collimators, *1* and *3* are singly scattered gamma rays, *2*, *4*, *5*, *6* are the multiply scattered gamma rays. (Ferronsky et al. 1968)

Compton scattering is the main effect responsible for the appearance of secondary gamma rays. For example, in the case of 1 MeV gamma rays in water, the mean number of Compton scatterings is greater by a factor of about 14 than the mean number of photoelectric events.

In each interaction the gamma rays lose energy and these losses are on average much smaller for low-energy gamma rays. Secondary gamma rays must therefore accumulate in the low-energy part of the spectrum. This build up is restricted by the fact that as the energy of the scattered gamma rays decreases, there is a rapid increase in the photoelectric contribution to the overall absorption.

Since the secondary gamma rays have a broad energy distribution and can propagate in different directions, in effect "forgetting" their previous history, the broad-beam geometry problem cannot be regarded as solved if only the intensity at a given point in the medium is known. To obtain a complete description of the gamma-ray field we must know the spectral and angular distributions at a given point in the medium. It is frequently sufficient to have the integral characteristics of the gamma-ray field. As a rule, these are more readily determined experimentally. The attenuation of a broad gamma-ray beam is thus described by the expression:

$$I = BI_0 e^{-\mu\rho x}, \tag{2.14}$$

where B is the so-called build-up factor, which is usually determined experimentally. It depends on the energy of the primary gamma rays, the type of detector and geometry of the experiment.

A very fruitful practical device has been the use of the effective mass absorption coefficient beam μ_{eff} for a broad beam, in which case the absorption law is described by an expression similar to (2.12) except that μ is replaced by μ_{eff}. The effective mass absorption coefficient depends on the same parameters as the built-up factor B and is also determined experimentally (see below).

Measurements on a broad monochromatic gamma-ray beam of moderate energy and originating from a point source in an infinite homogeneous medium without heavy impurities show that:

1. As the distance from the source increases, there is an increase in the contribution of multiply scattered gamma rays to the total intensity.
2. Equilibrium in the energy and angular distributions is established at a certain distance from the source.
3. The maximum of the energy distribution of multiply scattered gamma rays occurs at about 0.05–0.08 MeV.

In the region where the scattering radiation reaches equilibrium energy distribution the absorption of a broad gamma-ray beam again follows the exponential law. This is indicated by the fact that absorption curves plotted on the semi-logarithmic scale again become a straight line (Fig. 2.2).

Finally, in the region of the equilibrium, i.e., where there is no substantial redistribution of the spectral components of the transmitted gamma rays, the value of μ_{eff} is practically constant and the spectral sensitivity of the detectors has little effect upon it. Thus, for a ^{60}Co point source the effective mass absorption coefficient is $\mu_{eff} \approx 0.037$–$0.040\ cm^2/g$ for different types of gas-filled and scintillation counters.

In the case of broad–beam geometry, even a small amount of absorbing material ensures that most of the recorded intensity consists of a softer secondary component for which photoelectric absorption is important. It follows that, in the case of rocks containing heavy-atom impurities (Fe, Cu, Zn, Pb, Sn etc.), the effective mass absorption coefficient μ_{eff} will depend on the concentrations of these elements and unambiguous density determinations become very difficult. These impurities absorb strongly the soft-scattered component as a result of the photoelectric effect, so that the spectral distribution of the radiation is distorted and the density measurements are subject to errors.

2.3 Mass Absorption Coefficients of Rocks

The mass absorption coefficient for gamma rays is an important parameter of absorbing media, governing the sensitivity of the gamma-ray absorption method, the degree to which such measurements represent the medium as a whole and the reliability of density measurements of soil and rock material by the method. The magnitude of this coefficient depends largely on the gamma-ray energy. Under certain conditions μ_{eff} is practically independent from the chemical composition of the rock and soil material. This is exceedingly important, because it ensures that the method is very general.

For gamma-ray energies of approximately 0–20 MeV and the most important elements in soil and rock materials, the mass absorption coefficients decrease with increasing gamma-ray energy (Table 2.2). A particularly rapid reduction in the value of this coefficient is observed between 0 and 0.5 MeV. This is due to the rapid reduction in the probability of photoelectric absorption with increasing gamma-ray energy.

Table 2.2 Mean mass absorption coefficient μ (cm^2/g) for ^{60}Co and ^{137}Cs sources in various rocks

Rock	^{60}Co	^{137}Cs
Soils	0.0565	0.0767
Igneous rocks	0.0559	0.0760
Sedimentary rocks	0.0563	0.0767
Sandstone	0.0560	0.0761
Clayey rocks	0.0563	0.0767
Limestone rocks	0.0563	0.0769
Pure sand	0.0563	–
Water	0.0627	–
Mean value of μ for the above rocks	0.0562	0.0765

Table 2.3 Values of μ of ^{60}Co and ^{137}Cs gamma rays for the most important rock-forming elements

Element	Si	Al	O	Na	Mg	K	Ca	Fe	C	H
^{60}Co	0.0560	0.0546	0.0566	0.0539	0.0559	0.0547	0.0558	0.0534	0.0566	0.112
^{137}Cs	0.0774	0.0764	0.0771	0.0734	0.0760	0.0752	0.0768	0.0727	0.0771	0.0144

Photoelectric absorption, which is very dependent on the atomic number, increases rapidly at low energies (0–100 keV) so that it is difficult to establish unambiguous values for the density of multi-component media.

We have used data given by Rukhin (1961) on the chemical composition of some rocks to calculate the value of μ for these materials. The values of μ were calculated on the basis of data published by Leypunsky et al. (1960). The absorption coefficient μ was found to be practically the same for these very different rocks (Artsybashev 1965).

In fact, the deviation of μ for ^{60}Co and ^{137}Cs gamma rays is only slightly dependent on the composition including water. Density measurements on soil and rock materials by the gamma-ray absorption method are therefore usually carried out with the natural moisture content.

While the Z/M ratio for elements making up the soil and rock material is 0.5, in the case of hydrogen this ratio is equal to unity. Hence the value of μ for hydrogen is higher by a factor of 2 than for the other elements in the soil material (see Table 2.3). It follows that the mass absorption coefficient of water is higher by 11 % as compared with the rock skeleton.

Tables 2.3 and 2.4 give the values of μ for the main chemical element and compound (sand-clay and carbonate) rocks. These values were calculated using the experimental data published by Leypunsky et al. (1960) interpolated to the energy of ^{60}Co and ^{137}Cs and applying a second- order polynomial and the Lagrange method.

The moisture content is the most important factor affecting the mass absorption coefficient and hence the accuracy of bulk density determination.

Table 2.4 Values of μ of ^{60}Co gamma rays and the most important compounds in sedimentary rocks

Compound	SiO_2	Al_2O_3	Fe_2O_3	FeO	CaO	MgO	K_2O	Na_{20}	CO_2	H_2O
μ	0.0563	0.0555	0.0544	0.0542	0.0561	0.0561	0.0554	0.0553	0.0586	0.0627

We can estimate the error that is introduced if one neglects the difference between the mass absorption coefficients of water and the soil skeleton when the bulk density ρ_w of moist soil material of thickness h_s is determined. In this case the expression (2.13) acquires the form:

$$\rho_w = \frac{\ln(n_0/n)}{\mu_s h_s}. \tag{2.15}$$

On the other hand, taking into account the difference between μ_w and μ_s, one has:

$$\rho_w = \frac{\ln(n_0/n)}{\mu_s h_s} - \frac{\rho_w h_w}{\mu_s h_s}. \tag{2.16}$$

Expressing the value of h_w ρ_s $h_s/100$, we obtain:

$$\rho_w = \frac{\ln(n_0/n)}{\mu_s h_s} - \frac{\mu_w - \mu_s}{\mu_s}\frac{w\rho_s}{100}, \tag{2.17}$$

where w is the natural moisture content (%).

The second term in this equation is a measure of the absolute error in the density measurement by the gamma-ray absorption method if we use the μ_s for moist soil. Consequently,

$$\Delta\rho = \frac{\mu_w - \mu_s}{\mu_s} \cdot \frac{w\rho_s}{100} = -0.11\frac{w\rho_s}{100}. \tag{2.18}$$

The relative systematic error is therefore given by:

$$\delta = \frac{\Delta\rho}{\rho} = -0.11w. \tag{2.19}$$

As the moisture content increases, this error will also increase. Thus, for a moisture content of w = 10, 20 and 30 % the systematic errors in the bulk density measurements will be 1.1, 2.2 and 3.3 %, respectively.

A narrow gamma-ray beam depends on the size and shape of the collimating apertures and also on the collimator materials. Any collimating system can readily be developed and used under stationary laboratory equipment for the determination of the density of rocks and other materials (Artsybashev 1965; Filippov 1962).

The use of collimation under field conditions, on the other hand, is practically impossible or involves major technological difficulties. In most cases, therefore, field measurements of the density of rocks by the gamma-ray absorption method are carried out with broad-beam geometry and either slight collimation or no collimation at all (Fig. 2.3).

This can be done in two ways. In the first method, use is made of (2.13) with μ_{eff} determined in a preliminary experiment for the given type of the density-measuring equipment. The second method involves the use of a calibration graph giving the intensity of the recorded radiation (or some function of it) versus the density.

The theoretical (calculated) values of μ_{eff} are the basic control data that one should try to obtain in the development of the measuring apparatus and method. The closer the value of μ_{eff} is to the theoretical, the more accurate will be the final density (other things being equal). It is therefore important to select the conditions of the measurement so that they ensure the maximum possible value for μ_{eff}.

Dubinchuk (1966) carried out an experimental study of the dependence of μ_{eff} on the mass ρx of material. Such experiments must be carried out on media that are homogeneous in density for a broad range of thickness, which is difficult to achieve in natural soil material. Homogeneous concrete plates were therefore made for these experiments. Their size was 50 × 50 × 5 cm and their density lay between 0.56 and 2.1 g/cm^3. The values of μ were calculated from the known chemical composition for ^{60}Co (0.056 cm^2/g) and ^{137}Cs (0.076 cm^2/g). Stacks of plates of the required thickness and density were placed between the radiation source and the detector and the transmitted gamma-ray intensity was recorded.

The gamma-ray counting rate was first determined without the absorbers (in air) and then with the absorber in position. The gamma-ray detector was a scintillation-counter spectrometer [NaI(Tl) phosphor of 30 × 40 mm, working in conjunction with an FEU-29 photomultiplier] incorporating a single-channel analyser, so that it was possible to record the pulse-height spectra transmitted by the absorbers (Fig. 2.4a, b).

The pulse height corresponding to the gamma rays is plotted along the horizontal axis and is normalised to the height of the photo-peak (total absorption peak) corresponding to direct radiation, i.e., it is given in units of $V/V_{0.66}$ for the ^{137}Cs and $V/V_{1.25}$ for ^{60}Co. For the sake of convenience, the gamma-ray spectra are normalised in the area. The primary gamma-ray photo-peaks are clearly seen on the right of the spectra recorded by the scintillation counter. These are due to the photoelectric effect in the scintillator. In the soft part of the spectrum there are peaks scattered component predominates in the recorded gamma radiation. For an absorber thickness of 3–4 mean free paths there is no substantial redistribution of the spectrum components, i.e., equilibrium is set up in the spectral composition.

Figure 2.5a and b show on a semi-logarithmic scale and relative units the gamma-ray absorption curves for ^{60}Co and ^{137}Cs obtained for different levels of amplitude discrimination. The absorber layer thickness in units of the mean free path, $Z = x/\lambda = \mu\rho x$, is plotted along the horizontal axis. It follows from Fig. 2.5 that the absorption curve obtained for a narrow beam of ^{60}Co gamma rays was the same to within the experimental error as the theoretical curve calculated from Eq. (2.12)

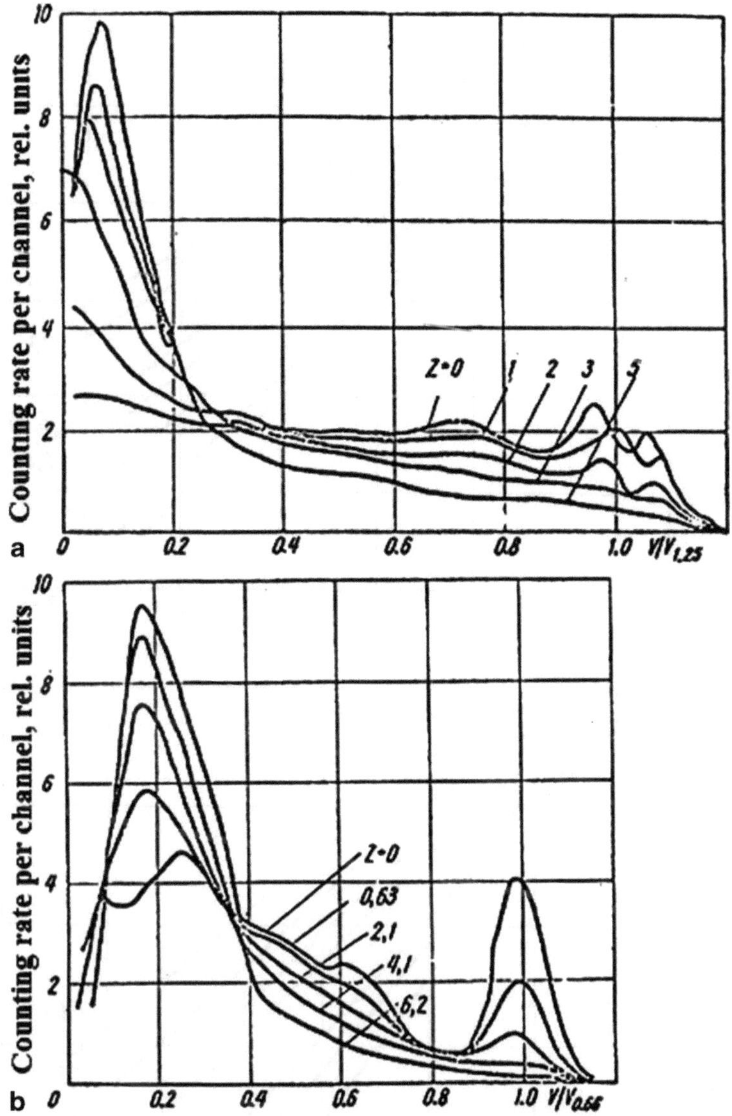

Fig. 2.4 Pulse-height spectra for ^{60}Co (**a**) and ^{137}Cs (**b**) gamma rays after passed through different thicknesses of absorber expressed in mean free paths ($Z=x/\lambda$). (Ferronsky et al. 1968)

(broken line). As the discrimination bias is increased there is an increase in the slope of the curves and, when the bias reaches the photo-peaks corresponding to the primary radiation, the absorption curves approach the theoretical relationship. This is explained by the fact that an increasing proportion of the multiply scattered radiation is removed from the recorded counting rate as the discriminator level is increased.

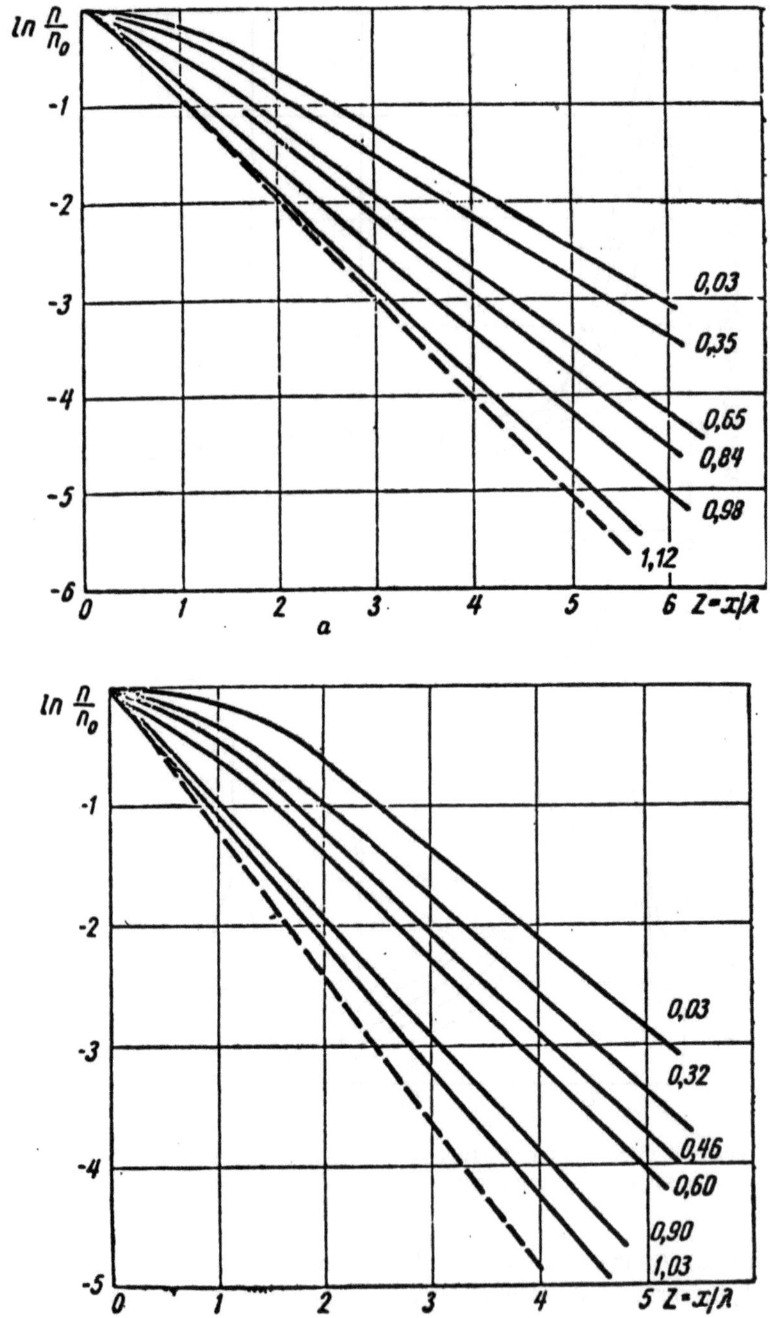

Fig. 2.5 Absorption of ^{60}Co (**a**) and ^{137}Cs (**b**) gamma rays for different discrimination levels given as $V/V_{1.25}$ and $V/V_{0.66}$ in relative units. (Ferronsky et al. 1968)

The values of μ_{eff} were determined experimentally under different conditions of absorption and detection. As expected, the minimum values of μ_{eff} are observed with zero amplitude discrimination, i.e., for the maximum contribution of multiply scattered radiation to the recorded intensity. As the discrimination level increases, there is an increase in μ_{eff}. The effective absorption coefficient μ_{eff} is found to be equal to μ when the layer thickness is up to 2 mean free paths (in the experiments approximately 20 cm) and the discrimination threshold reaches the photo-peak (curve 1.03 for ^{60}Co and curve 0.98 and 1.12 for ^{137}Cs). For thick layers and the same discrimination level, a 10% reduction in μ_{eff} was observed for ^{60}Co gamma rays and a 4% reduction for ^{137}Cs gamma rays. This is explained by the fact that the scintillation counter has a finite energy resolution. In our case the resolution amounts to about 12% (for $E = 0.66$ MeV). For thin absorbing layers, when most of the contribution to the scattered radiation is due to gamma rays scattered only a few times, this energy is substantially different from the primary gamma-ray energy and pulse-height discrimination at the photo-peak level completely cuts off the scattered component. For thicker absorbers there is a certain build-up of gamma rays scattered through small angles and remaining within the solid angle subtended by the detector at the source. The energies of these gamma rays are close to the primary and since the detector has a finite resolution, these gamma rays are recorded even with very high discrimination levels, leading to a reduction in μ_{eff}. The fact that this does occur is also indicated by the increasing asymmetry of the photo-peaks (Fig. 2.4a, b) as the thickness of the absorbing material increases. This is a basic limitation and unless suitable corrections are introduced, it is impossible to obtain the theoretical value of μ_{eff} for thick absorber and discrimination levels up to the left edge of the photo-peak.

We are now in position to draw the following conclusions with regard to the dependence of μ_{eff} on the mass ρx of the absorbing material. The most highly defined dependence is observed when there is either a low degree of discrimination or no discrimination at all. As the mass increases, the value of μ_{eff} increases asymptotically to a constant. The higher the discrimination levels the smaller is the slope and the weaker is the dependence of μ_{eff} on the mass of the absorber and consequently, on the thickness of the absorber layer for a given density (or on the density for a given thickness).

The weak dependence of μ_{eff} on the mass of the absorbing material between the source and the detector (when the absorber thickness is greater than about 3 mean free paths) is due to the onset of spectral equilibrium in the energy distribution of the transmitted radiation. The obtained data can be used to select the conditions under which μ_{eff} is constant within given limits.

Let us suppose now that the centre point of the bulk-density range of the soil material is ρ_0 (g/cm^3). If the ^{60}Co gamma rays are recorded without discrimination, the thickness of the absorbing material must be chosen from the condition $x_0 \approx S\lambda = 5\mu\rho_0$, i.e., at the centre of the most slowly-varying section of the dependence of μ_{eff} on the mass of the absorbing layer. If the range of variation of the measured density of the soil material lies between ρ_1 and ρ_2, where $x_0\mu\rho_1 = 4.$, $x_0\mu\rho_2 = 6$, then μ_{eff} can be assumed to be constant to within ±2.5%. Let us suppose that $\rho_0 = 1.5$ g/cm^3, in which case:

$$x_0 = \frac{5}{0.056 \cdot 1.5} \approx 60\text{cm};$$

$$\rho_1 = \frac{4}{x_0\mu} = \frac{4\mu\rho_0}{5\nu} - \frac{4}{5}\rho_0 = 1.2\text{g/cm}^3,$$

and

$$\rho_2 = \frac{6}{x_0\mu} = \frac{6}{5}\rho_0 = 1.8\text{ g/cm}^3$$

For the discrimination level of 0.9 the measurements can be carried out in a broader range of variation of the absorbing mass, for example, $Z = 4.5 \pm 1.5$. The coefficient μ_{eff} will then be constant to within $\pm 0.6\,\%$. For this case:

$$x_0 = \frac{\text{4ю5}}{0.056 \cdot 1.5} = 54\text{cm}, \quad \rho_1 = \frac{3}{4.5} \cdot 1.5 = 1\text{ g/cm}^3$$

$$\rho_2 = \frac{6}{4.5} \cdot 1.5 = 2\text{ g/cm}^3.$$

With a higher discrimination level, for example 1.03, it is possible to isolate two ranges for selecting the optimal absorption conditions: one for $x\mu\rho < 2$ and another for $x\mu\rho > 2$. In the first region μ_{eff} will be equal to the theoretical value to within $0.5\,\%$. In the second it will be constant to the same degree of accuracy (and will be equal to 0.955μ).

The densities of the majority of ground and sedimentary rock materials lie in the range 1.0–2.3 g/cm^3. For many types of ground material this range is narrower. For sand and sandy loams they may lie, for example, between 1.3 and 1.9 g/cm^3. It follows that for relatively short ranges of variation of density and a distance of at least 3 mean free paths, the value of μ_{eff} remains practically constant.

2.4 Sensitivity of the Method

Sensitivity is the most complete measure of how effective a particular physical principle is as the basis of a given method. In general, the sensitivity of the gamma-ray absorption method is the ratio of the "reaction" of the gamma-ray beam to a change in the density of the absorbing medium. It can be defined as the ratio of the increase in the gamma-ray counting rate recorded by the detector to the increase in the density. It follows that, in the limit, sensitivity is defined as the derivative of the counting rate with respect to density. Differentiating (2.12) with respect to ρ, we find that the sensitivity is given by:

$$\eta = \frac{dn}{d\rho} = (-\mu x)e^{-\mu\rho x}. \tag{2.20}$$

It follows from this expression that the sensitivity is a function of the thickness of the absorbing material, the density and the mass absorption coefficient for the particular gamma rays used. The absolute sensitivity $|\eta|$ decreases exponentially with increasing density. Consequently if a rate meter is used to determine the density, its scale will be necessarily non-linear because the sensitivity is a function of density.

The dependence of $|\eta|$ on the thickness x of the absorbing material is somewhat more complicated: for small values of x, when the exponential factor has little effect, the absolute sensitivity increases, it then reaches a maximum and finally falls almost exponentially. By differentiating (2.20) we can readily show the maximum of the function $|\eta| = |f(x)|$ occurs for:

$$x_{max} = \frac{1}{\mu\rho} = \lambda. \quad (2.21)$$

It follows that x_{max} decreases with increasing density and μ.

Since, in general, μ_{eff} is a function of the density and thickness of the absorbing material, we find from Eq. (2.13) that for broad-beam geometry:

$$\eta_{bb} = \frac{dn}{d\rho} = -\left[\frac{d\mu_{eff}}{d\rho}\rho x + \mu_{eff} x\right] n. \quad (2.22)$$

It was shown above that, by suitably choosing the thickness of the absorbing material and the method of recording, it is possible to ensure that μ_{eff} remains practically constant. In this case:

$$\frac{d\mu_{eff}}{d\rho} \approx 0.$$

The result is an expression analogous to equation (2.20):

$$\eta_{bb} = -\mu_{eff} x n_0 e^{\mu_{eff}\rho x} = -\mu_{eff} x n. \quad (2.23)$$

Sensitivity of the gamma-ray absorption method can also be measured in another way. Let us define the sensitivity as the ratio of the increase in $\ln(n/n_0)$ with respect to density, i.e., as the derivative of $\ln(n/n_0)$ with respect to density. For narrow-beam geometry we then have:

$$G = \frac{d}{d\rho}\ln(n/n_0) = -\mu x. \quad (2.24)$$

For broad-beam geometry, for which μ_{eff} is a function of ρ and x, one has correspondingly:

$$G = \frac{d}{d\rho}\ln(n/n_0) = -x\left[\rho\frac{d\mu_{eff}}{d\rho} + \mu_{eff}\right]. \quad (2.25)$$

Let G be defined as the logarithmic sensitivity of the method. This quantity is convenient for a number of reasons. Firstly, the calibration curves are frequently constructed in the form $\ln(n/n_0)=f(\rho)$. Secondly, G is independent of the activity of the source and some of the design features of the probes (thickness and material of the probe envelope).Thirdly, the logarithmic sensitivity is equal to the slope of the calibration curve $\ln(n/n_0)=f(\rho)$ (see Fig. 2.5a, b). For narrow-beam geometry and given absorber thickness x, the logarithmic sensitivity G is in general a function of the density and thickness of the absorber. It was shown above that, with a suitable chosen absorber thickness and discrimination level, the sensitivity G of a scintillation counter is practically independent of x and ρ. This is frequently the major advantage of the semi-logarithmic curve.

In practice, it is essential to exploit the following methods for increasing the sensitivity:

a. The use of low-energy gamma rays ensures high μ_{eff}. In this case, however, there are two important restrictions. As the gamma-ray energy decreases, the chemical composition of the rock has an increasing effect and the transmitted intensity is decreased.
b. In the case of broad-beam geometry it is convenient to use a scintillation spectrometer in combination with amplitude discrimination. The coefficient μ_{eff} then approaches the theoretical value.

A more precise parameter describing a given instrument is the sensitivity threshold. This can be defined as the absolute or relative increase in density, which is greater or equal to the absolute (σρ) or relative (δρ) error. In particular:

$$|\Delta\rho| \geq \sigma_\rho \quad \text{or} \quad \frac{\Delta\rho}{\rho} \geq \delta_\rho. \tag{2.26}$$

If we use the formula $n = n_0 e^{-\mu\rho x}$ to determine the density of the ground material, then the condition given by (2.26) with (2.20) substituted into it can be written in the following way:

$$|\Delta\rho| = \left|\frac{\Delta n}{n}\right| \geq \sigma_\rho, \quad \frac{\Delta\rho}{\rho} = \frac{1}{\mu\rho x}\left|\frac{\Delta n}{n}\right| \geq \delta_\rho.$$

If we use the functional relation $\ln(n/n_0)=f(\rho)$ we have:

$$|\Delta\rho| = \left|\frac{\Delta n \ln(n/n_0)}{G}\right| \geq \sigma_\rho, \qquad \frac{\Delta\rho}{\rho} = \left|\frac{\Delta \ln(n.n_0)}{1(n/n_0)}\right| \geq \delta_\rho.$$

Thus, if we take Eq. (2.8) into account, we obtain the following expression for the sensitivity threshold in the case of narrow-beam geometry:

$$\alpha = \left|\frac{d\upsilon}{\rho}\right|_{min} = \frac{1}{\mu\rho x}\sqrt{\frac{4\pi R^2 e^{\mu\rho x}}{Q_s \varepsilon T}} = \left(\frac{4\pi R^2}{Q_s \varepsilon T}\right)^{1/2} \frac{e^{1/2\mu\rho x}}{\mu\rho x} \tag{2.27}$$

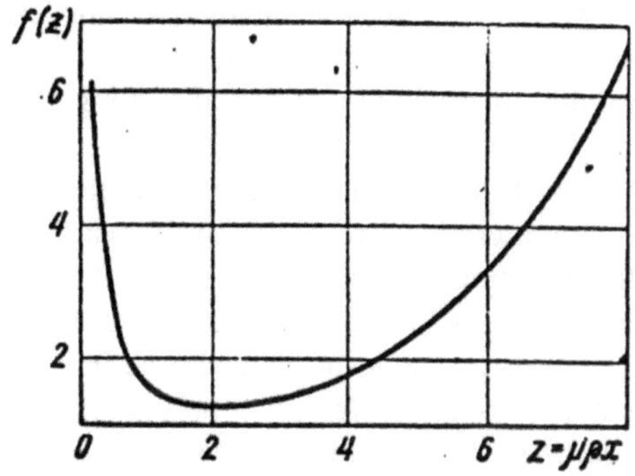

Fig. 2.6 Graph of the function $f(z)=e^{z/2}/z$

where R is the distance between the source and the detector and T is the time of the measurement in the case of discrete counting, or twice the time constant in the case of a rate meter.

In the case of density meters in which the source and detector are completely within the ground, or the detector is tightly pressed against the ground as the probe containing the source is inserted into it, it can be assumed that $R \approx x$, in which case:

$$\alpha' = \left[\frac{4\pi}{Q_S \varepsilon T}\right]^{1/2} \frac{e^{1/2\mu\rho x}}{\mu\rho}. \tag{2.28}$$

The above expressions indicate the conditions under which a given relative change in the bulk density $\Delta\rho/\rho$ will be greater than the relative error in rate-meter measurements in density determinations.

It is clear from the foregoing discussion that the sensitivity threshold decreases with increasing source activity, detector efficiency and time of measurement T. The quantity α plotted as a function of the thickness of the absorbing material has the deep minimum shown in Fig. 2.6. This figure shows the function $f(z)=e^{z/2}/z$, where $z=\mu\rho x$. It is clear that the minimum sensitivity threshold occurs at $z=2$, i.e., when $x=2/\mu\rho=z\lambda$. The optimum absorber thickness (with the statistical measurement errors taken into account) thus corresponds to 2 mean free paths.

If the distance between the source and the detector is equal to the thickness of the absorbing material, i.e., $R \approx x$, the sensitivity threshold given by (2.28) has extreme properties when it is plotted as a function of μ and ρ but has no extreme when it is plotted against the thickness of the absorbing material. For a given source activity, recording efficiency and time of measurement, the sensitivity threshold for density determinations increases with increasing absorber thickness.

The expressions given by (2.27) and (2.28) and the data of Fig. 2.6 enable us to determine the optimum working conditions for a given sensitivity threshold α. For broad-beam geometry these formulae remain the same except that μ is replaced by μ_{eff}, which is a function of x and ρ and use is made of function f(z) shown in Fig. 2.6.

The absorbing layer is usually not less than 20 cm thick. The minimum density of the ground and sedimentary rocks can be assumed to be 1.0 g/cm^3. The minimum thickness of the absorbing layer in these units (for ^{60}Co) is therefore $z_{min} = 1.0 \times 20 \times 0.056 = 1.12$. This means that it can be assumed that the working range of the gamma-ray density meter designed for measurements on ground and rock materials corresponds to the readily branch of the function f(z). The graph of this function can be readily used to obtain the value of $e^{1/2\mu\rho x}$ in (2.28). All that needs to be done is to multiply by x the values given by Fig. 2.6.

Let us consider the special case in which it is desired to determine the optimum parameters for a "feeler" type gamma-ray density meter in which the gamma-ray source is ^{60}Co and the detector is a scintillation counter incorporating a 30 × 40 mm NaI(Tl) crystal. For narrow-beam geometry, for narrow-beam geometry and zero discrimination, the efficiency is $\varepsilon = 0.4$, while for broad-beam geometry with a discrimination level at the left-hand edge of the photo-peak we have $\varepsilon - 0.08$ (Vartanov and Samoylov 1964). For narrow- or broad-beam geometry but with a high level of discrimination, it can be assumed that $\mu_{eff} = 0.056$ cm^2/g, whereas for broad-beam geometry without discrimination we shall use the value $\mu_{eff} = 0.030$ cm^2/g. When the density must be measured down to a depth of 0.5 m with a probe in intimate contact with the surface of ground, we can use equation (2.28). We shall assume that the density meter has a sensitivity threshold of at least $\alpha = 0$ (1 %) in the density range of 1.2–2.3 g/cm^3. In this case the function $x = f(z)$ increases with ρ and, therefore, if we ensure the necessary sensitivity α for a medium with a density corresponding to the upper limit of the above range (ρ_{max}), the sensitivity threshold will be at least as indicated (other things being equal).

The condition for the density-meter parameters can therefore be written in the form:

$$\left[\frac{4\pi}{Q_S \varepsilon T}\right]^{1/2} x f(z_{max}) = 0.01.$$

The most important parameter is the activity of the gamma-ray source:

$$Q_S = \frac{4\pi}{\varepsilon T} x^2 f^2 (z_{max}) \cdot 10^4.$$

In our case, $z_{max} = \mu \rho_{max} x_{max} = 0.056 \times 2.3 \times 50 = 6.4$. This corresponds to $f(z_{max}) = 4$.
Consequently,

$$Q_S = \frac{1}{T} \frac{4 \cdot 3.14 \cdot 50^2 \cdot 4^{2\cdot} \cdot 10^4}{0.4 \cdot 7} = \frac{1}{T} 1.8 \cdot 10^9 \text{ count/s.}$$

If we assume that T is 10, 30 and 60 s and use the factor 3.7×10^7 to convert the activity into millicuries (mCi), then since each ^{60}Co decay is accomplished by the emission of two gamma rays, we obtain:

$$Q_{10} = \frac{1.8 \cdot 10^9}{2 \cdot 3.7 \cdot 10^7 \cdot 10} = 2.45 \text{ mCi},$$

$$Q_{30} = 0.8 \text{ mCi and } Q_{60} = 0.4 \text{ mCi}$$

The sensitivity threshold is thus affected largely by two factors, namely, (1) the sensitivity of the method, i.e., the slope of the calibration curve and (2) the statistical errors of the rate-meter measurements.

When the mass thickness of the absorbing layer is small ($z = \mu\rho x < 1$), the sensitivity increases with increasing x while the sensitivity threshold decreases. For large thickness ($z > 2$), and the same source activity, the statistical errors increase together with the sensitivity threshold (see Fig. 2.6). The sensitivity threshold reaches a minimum in the intermediate range.

2.5 Deviations from the Mean Density

The gamma-ray absorption method yields the average density for a given volume of absorbing material. The volume over which the average is taken depends on the origin of the gamma rays recorded by the detector. When narrow or broad gamma-ray beams are used together with a high level of pulse-height discrimination, this volume is determined by the solid angle subtended by the detector at the source.

In the case of "feeler" type gamma-ray density gauges the volume over which the average is taken can be increased several-fold by using a group of counters radial arranged relative to the "feeler". This is used in the GP-0-50 density meter designed by the Russian Scientific Research Institute for Hydrogeology and Engineering Geology that incorporates four STS-6 counters. In this design the volume over which the density average is taken is $V = 2400 \text{ cm}^3$. Moreover, in this case of "feeler-type" density gauges the average density can be evaluated over the maximum possible volume by rotating the "feeler" about the axis. With this method of averaging the maximum volume for the GP-0-50 it turns out to be $V_{max} = (15\text{–}16) \times 10^3 \text{ cm}^3$. In this respect, the classical method of taking direct samples cannot compete with the gamma-ray density gauge.

When the densities are determined under broad-beam conditions, the volume over which the average density is evaluated becomes somewhat greater, since there is now an additional volume in which the scattered radiation received by the detector is produced.

Various researchers have shown that approximately 90 % of the recorded gamma-ray intensity arrives from the volume lying within the solid angle subtended by the detector at the source. The remaining fraction arrives from the volume lying within an axial distance from the beam axis of not more than 1 mean free path of the primary radiation. For example, for ^{60}Co and $\rho = 1.8 \text{ g/cm}^3$ this distance amounts to 10 cm.

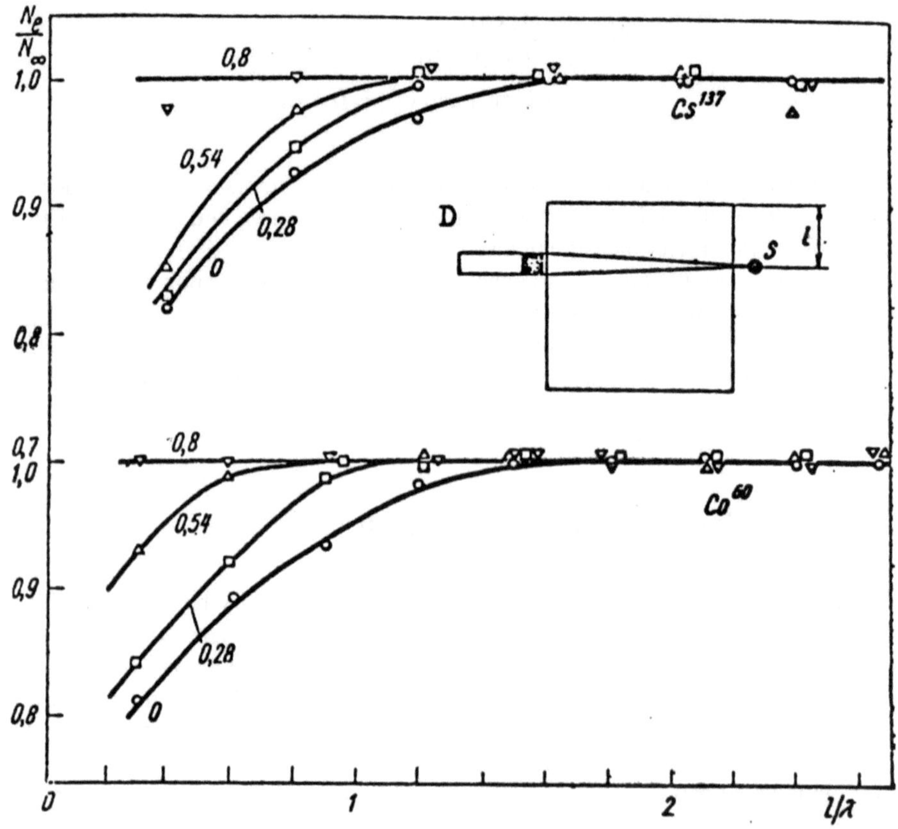

Fig. 2.7 The scheme and results on the effect of the separation boundary between two media in a broad-beam absorption experiment (^{60}Co gamma rays). The discrimination levels are shown in units of $V/V_{0.66}$ for ^{137}Cs and $V/V_{1.25}$ for ^{60}Co

The above conclusions are confirmed by an experiment on the effect of a separation boundary between two media (aid and absorber) on absorption measurements with the radiation directed along the boundary. This experiment and the final results are shown in Fig. 2.7. The measurements were carried out with a ^{60}Co source and the NaI(Tl) scintillator (30 × 40 mm, FEU-29). The equipment incorporated a pulse-height analyser. The absorbing medium was glass. As the distance between the beam and the separation boundary increased the recording density showed a slight increase but remained practically constant after $\ell \geq \lambda$ (7 cm). When the discriminator level was placed at the left-hand edge of the photo-peak, the scattered boundary had no effect on the recorded intensity because the scattered gamma rays were thus eliminated. The boundary had maximum effect in the absence of discrimination and thus, of course, corresponded to the maximum volume over which the average was taken. Since the recorded radiation arrived mainly from the region defined by the solid angle of the source–detector system, it is readily understood why the method has good spatial resolution and can be used to identify intercalations whose thickness is not less than the size of the detector.

Analysis of experimental and theoretical data on the density-averaging effect for narrow-beam geometry shows that this procedure results in the true mean density for the volume under examination.

In the case of broad-beam geometry and when μ_{eff} is independent of absorber thickness and density, the final results are also the true average density values. If, on the other hand, the absorber thickness is less than three mean free paths and the ground is highly layered in density, then the value of μ_{eff} will depend on the density and the results of measurements may differ from the true mean density. When the density variation in the medium under investigation is small and lies within the limits of the usual random variation, the effect of this factor on the final results will be small.

2.6 Determination of Soil Density by Gamma-Ray Absorption

The most widely used methods are the following:

a. Horizontal absorption of gamma rays between the source and detector at different depths in the ground (Fig. 2.8a, c);
b. Vertical absorption of gamma rays using a rod "feeler' inserted into the ground with the gamma-ray source placed at the lower end and the detector at the upper end (Fig. 2.8b).

In both methods the thickness of the ground layer examined is preferably 30–50 cm, since greater path lengths lead to a large reduction in the transmitted intensity, i.e., a much lower counting rate. This means that unless the source activity is increased, the statistical counting errors become greater. On the other hand, higher source activities are undesirable from the point of view of radiation hazard. Density determinations from gamma-ray absorption have been popular since 1955.

Most of the early work on bulk-density determination involved measurements of horizontal absorption of a gamma-ray beam between two parallel tubes. The gamma-ray detector was usually a gas-filled counter and the source was either ^{60}Co or ^{137}Cs. Ferronsky (1956) and Churayev (1965) replaced the two parallel tubes by U-shaped arrangements ("radioactive fork") consisting of two parallel pointed rods fixed on a frame and pressed into the ground to the required depth. The gamma-ray source was a ^{60}Co or ^{137}Cs preparation located at the lower pointed end of one of the rods and the detector (gas-filled counter) was placed at the end of the other rod (Fig. 2.8c). The counting equipment was first calibrated under laboratory conditions, using materials of known density, so that a graph was obtained of the counting rate as a function of density, or a preliminary determination was made of μ_{eff}.

It was found (Ferronsky 1956; Volarovich and Churayev 1960; Emelyanov 1962; and others) that the difference between the density of the ground obtained by the old

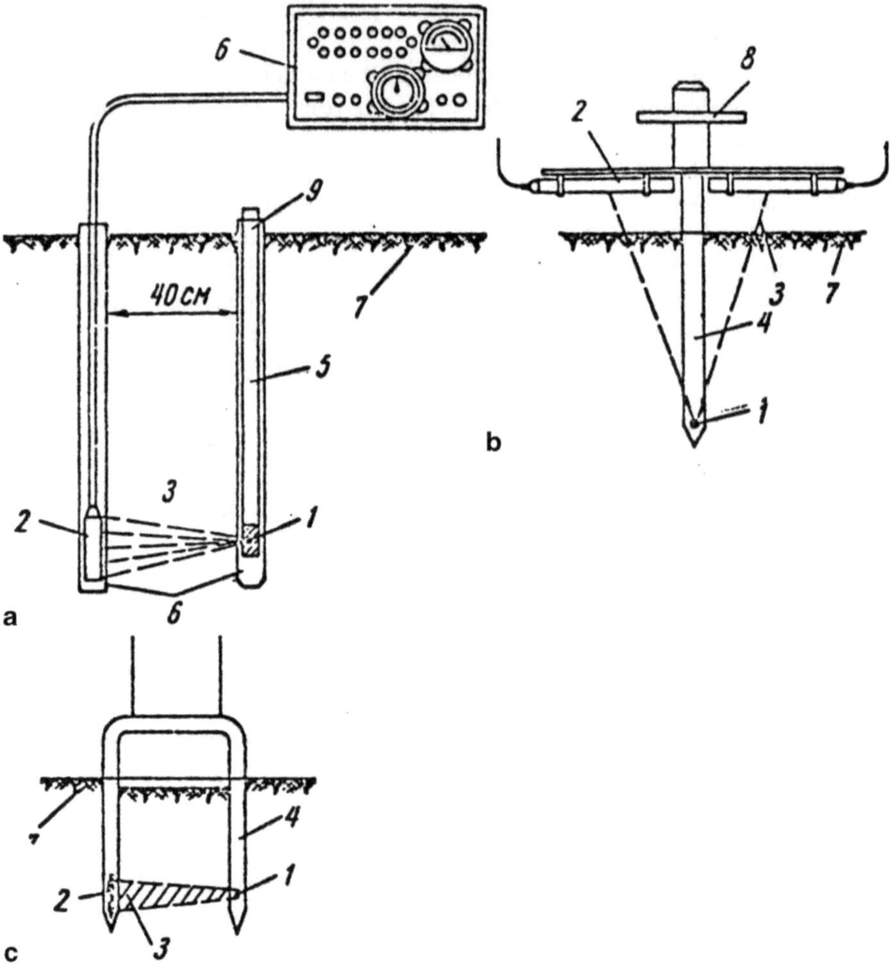

Fig. 2.8 a Horizontal beam; b vertical beam; c "radioactive fork": 1–gamma-ray source' *2*–detector; *3*–gamma-ray beam; *4*–fork; *5*–bag carrying the gamma-ray source; *6*–radiometer; *7*–ground; *8*–handle; *9*–tubes. (Ferronsky et al. 1968)

classic method on the one hand and the gamma-ray absorption method on the other is never more than 1.5–2.0 %.

We note that μ_{eff} obtained as a result of calibration of density meters increases when collimation is employed. Thus, Yemelyanov (1962) used an STS-5 counter in a lead envelope (screen) having a wall thickness of 1 cm and a window cut along the length of the cathode and found that the effective absorption coefficients for ^{60}Co gamma rays were as shown in Table 2.5 for various layer thicknesses of loam.

When the thickness of the ground material placed in the path of non-collimated gamma rays reaches a figure of the order of 30 cm (for densities in the range 1.5–1.6 g/cm^3),

Table 2.5 Values of (cm^2/g, according to Emelyanov)

Thickness of loam cm	Source without collimator	Source with collimator of diameter, cm	
		4.5	7.5
10	0.0303	0.0360	0.0380
20	0.0340	0.0390	0.0420
30	0.0360	0.0410	0.0470

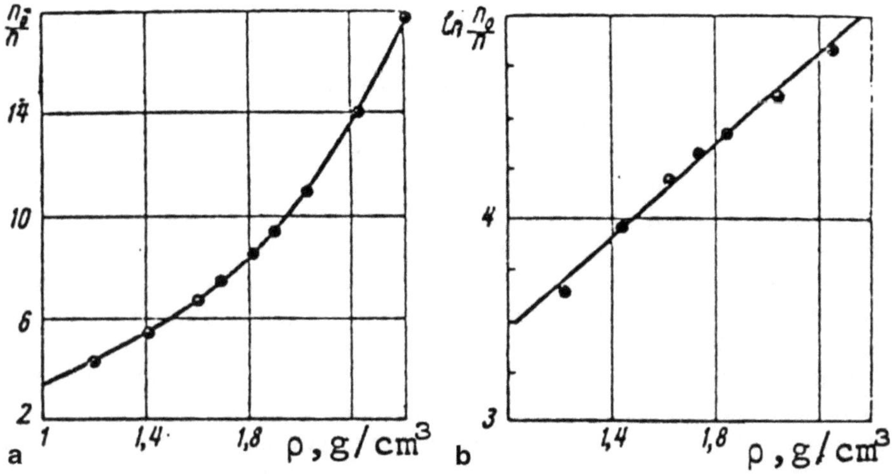

Fig. 2.9 The ratio n_0/n as a function of the density of the ground (**a**); the ratio n_0/n as a function of the density of the ground on a semi-logarithmic scale (**b**); $\mu = 0.04$, $h = 30$ cm. (Ferronsky et al. 1968)

the effective mass absorption coefficient increases only slightly after the thickness is increased further. For example, it was found that the coefficient changed by only 2.35 when the thickness of a layer of sand was increased from 28 to 48 cm.

When the mass of the absorber is in excess of 60–70 g/cm^2 the mass absorption coefficient remains practically constant and then ^{60}Co gamma rays are employed. It lies in the range 0.037–0.040 cm^2/g, depending on the geometry of the experiment and the resolution of the measuring apparatus.

One way to calibrate the gamma-ray density meters is to plot the intensity ratio n_0/n as a function of ρ for given absorber thickness (Fig. 2.9a). When plotted on a semi-logarithmic scale, this relation gives a straight line (Fig. 2.9b). This method eliminates errors due to source decay and drift in the measuring apparatus.

Less accurate results are obtained under field conditions when the calibration graph is given in the form $n = f(\rho)$. This method of calibration is of course used only for pioneer-type rate meters.

Our studies have shown (see also Emelyanov 1962) that μ_{eff} varies appreciably near the source of the ground when the latter is examined with a horizontal broad or partially collimated gamma-ray beam. The height of the ground layer above the horizontal gamma-ray beam for which μ_{eff} ceases amounts to 1–1.5 mean free paths

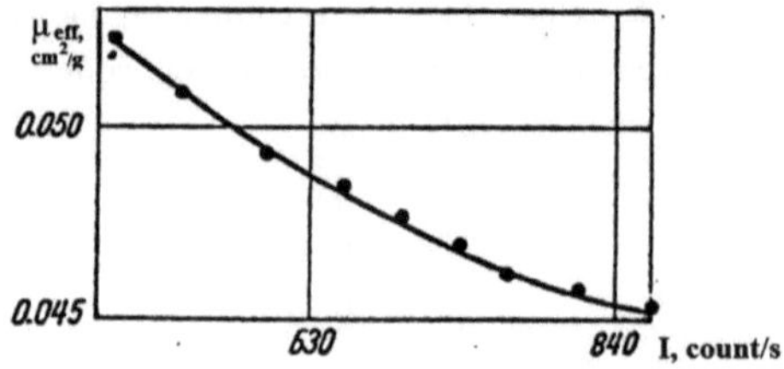

Fig. 2.10 Effective absorption coefficient plotted as a function of the counting rate obtained with an NaI(Tl). (Ferronsky et al. 1968)

(when $\rho = 1.5$ g/cm^3 and $\lambda = 14$–16 cm for ^{60}Co). When the uppermost layer of the ground (15 cm) is examined, the values of μ_{eff} must therefore be those obtained for this particular layer, or use must be made of gamma-ray density meters with "feelers" introduced vertically into ground.

Correct measurements of the gamma-ray counting rate n_0 in air are of great importance for the accurate determination of the density. Gamma rays emitted by a source are at the proximity of the Earth's surface and are partly scattered by it and some of them are intercepted by the detector. The instrument will therefore count not only the primary but also some of the surface-scattered gamma rays. Emelyanov has shown that, if the gamma-ray counting rate at a height of 90 cm above the surface is 100 %, then at a height of 5 cm it amounts to 107.8 %.

Since the calibration of gamma-ray density meters for ground materials of different density and moisture content is a very laborious operation, some workers have used values of μ_{eff} for different moisture-content ranges, e.g., 5–15 % and 15–30 %. According to Emelyanov's data (1962), the associated error in the measured density for a partially collimated ^{60}Co gamma-ray beam was not more than ±0.03 g/cm^3.

An increase in the sensitivity and accuracy of the method under field conditions was achieved by Nesterov (1959), who was the first to succeed in using crystal scintillator NaI(Tl), a measuring system in which provision was made for amplitude discrimination. The accuracy was about 1 % and the discrepancy with the data obtained by the cutting ring method (direct samples) reached 2 % in individual cases.

The value of μ_{eff} for ^{60}Co gamma rays obtained that time by Nesterov was close to the theoretical values and amounted to ~0.051 cm^2/g. The usual method was employed to determine this coefficient. The counting rate was first measured in the air gap between the tubes and then the measurements were repeated in the ground of known density. By measuring different counting rates in air as a function of the discrimination level and then repeating this in the ground, Nesterov obtained a series of values for μ_{eff} that he then used in practice.

Figure 2.10 shows the dependence between the counting rate in air and μ_{eff} obtained by Nesterov. It should be noted that Nesterov's result $\mu_{eff} = 0.051$ cm^2/g is in agreement with our own value.

Methods based on the use of parallel holes (Fig. 2.8a) are very restricted, because it is often difficult under field conditions to ensure that the holes are strictly parallel, especially at depths in excess of 1 m. In the case of compact ground materials the radioactive fork is difficult to insert into the ground and therefore preliminary apertures have to be made. For example, if the distance between tubes is changed from

Table 2.6 Comparative density data for a 30-cm ground layer, obtained with a GP-0-50 gamma-ray density meter and by the cutting ring method (direct sampling)

Density (g/cm³)		Discrepancy (%)	Density (g/cm³)		Discrepancy (%)
γ-method	Direct method		γ-method	Direct method	
1.40	1.41	0.0	1.65	1.64	0.6
1.74	1.75	0.6	1.52	1.50	1.3
1.68	1.67	0.6	1.47	1.50	2.0
1.60	1.66	3.6	1.48	1.51	2.0
1.83	1.82	0.5	1.68	1.69	0.6
1.56	1.57	0.6	1.90	1.98	4.0
1.56	1.58	1.2	1.81	1.83	1.08
1.54	1.55	0.6	1.83	1.85	1.08
1.71	1.64	4.3	1.82	1.80	1.1
1.72	1.66	3.6	2.05	2.07	0.9

40 to 39 cm, the counting rate changes by 5% for the same density. This change in the gamma-ray counting for $\mu_{eff}=0.050$ cm²/g causes an additional error in the density that amounts to ~1.6% for $\rho=1.5$ g/cm³. This error increases with increasing density of the ground and vice versa. To obtain accurate results it is necessary to ensure that the holes are strictly parallel and minimum possible gaps (less than 3 mm) exist between the hole walls on the one hand and the source and detector on the other.

As noted above, radioactive gauges are used in measurements of the density of the upper layers of friable deposits and exposed rocks. One of those was suggested by Danilin (1959) as a means of investigating the dynamics of the moisture content in soils. The types of gauges shown in Fig. 2.8c were used in engineering geological practice by Ferronsky (1956) and in the peat industry by Volorovich and Churaev (1960). The particular advantage to this method is that the gamma-ray source is introduced into the ground to a strictly-defined depth and the distance between the source and the gamma-ray counters is constant.

Table 2.6 shows comparative experimental data for loam under field conditions. These measurements were carried on a test area during the construction of the hydropower station on the Nureck River, using the gamma-ray density meter GP-0-50 and the direct sampling method. The results obtained by the two methods agree in most cases to within 0.6–2% and only in two cases out of 20 does the discrepancy rise to about 4%. Owing to the spatial variability of the ground density it is hardly surprising that these discrepancies do in fact occur.

The data given in Table 2.7 were obtained by measuring the absorption of a horizontal uncollimated gamma-ray beam in a 30-cm layer of loam compacted by a vibration.

It has already been noted that the use of scintillation counters with amplitude discrimination ensures that the measured values of μ_{eff} lie close to the theoretical

Table 2.7 Comparative density data for compacted Langarskii loam, obtained by horizontal gamma-ray absorption and the direct cutting ring

Depth (cm)	Density (g/cm³)		Depth (cm)	Density (g/cm³)	
	γ-method	Direct method		γ-method	Direct method
10	2.16	2.15	40	2.13	2.13
20	2.08	2.13	50	2.15	2.12
30	2.13	2.13	60	2.20	2.18

result and this improves the sensitivity of the method and the accuracy of the measured density. We investigated the performance of a scintillation gamma-ray density meter incorporating an NaI(Tl) phosphor (30 × 30 mm) mounted on an FEU-15 photomultiplier. The phosphor, the photomultiplier, an emitter follower and a linear amplifier are assembled together in a duralumin container. The pulses are recorded by an M-30M scaler, which incorporates a transistorised pulse-height discriminator, amplitude and time normalisers based on a trigger circuit. In this way the scaling factor was increased from 64 to 128.

The effective mass absorption coefficient μ_{eff} for ^{60}Co gamma rays was measured for ground and water using a source with an activity of 1.4 mg-eqt. of radium. The mass absorption coefficients for the ground material and for water were found to be $\mu_s = 0.0568$ and $\mu_w = 0.0625$ cm²/g respectively. The two values were thus close to the theoretical predictions.

Summing up, we may draw the following conclusions:

1. The density of the uppermost layer of friable deposits or exposed rocks down to a depth of 50 cm is best determined with the aid of gamma-ray density meters either of the radioactive fork or rod probe type, using pointer-indicator rate meters or portable scaling units. The latter can be used to carry out these measurements layer by layer, e.g., in layers of 0–30, 0–40 and 0–50 cm.
2. At depths between 30 and 150 cm the density of the ground material can be determined by absorption measurements using a horizontal gamma-ray beam but when this is done it is important to ensure that the holes or pipes are strictly parallel.
3. When the density of the uppermost ground layer is determined with gamma-ray density meters, the productivity (deduced from experimental data) increases by a factor of 20–30 as compared with the cutting-ring method. At the same time, the method is much less laborious.
4. The density values obtained for rock by the gamma-ray absorption method are much more representative of the medium as a whole than the values obtained with cutting rings.
5. The density of rocks at depths in excess of 150 cm is best determined by recording back-scattered gamma rays, since at such depths it is difficult to ensure that the holes are strictly parallel.

The gamma-ray absorption methods can be used: (a) to determine the density (bulk density) of soil down to 150–200 cm in the case of geoengineering investigations and also for the purpose of hydrogeology, hydrology and agrometeorology; (b) to investigate the compaction of freshly deposited ground material under dams, highways, airfields and so on and (c) to determine the density of ice and snow and also of various building materials.

2.7 Studies of Moisture Content Dynamics in Soil

It was shown above that the gamma-ray absorption method can be used to determine the mass of material (in g/cm^3) between the source and the detector. Consequently, the gamma-ray method enables us to effectively weigh the ground in situ.

By systematically examining a given ground layer in which the mass of the solid phase is kept constant, it is possible to investigate the moisture-content dynamics in this layer.

This method has the following undoubted advantages:

1. It can be used to investigate the moisture-content dynamics in undisturbed layers of ground material without taking direct samples for drying and without much effort.
2. The moisture content obtained in this way is very representative of the medium as a whole, because it automatically takes the necessary average, in contrast to the direct-sample method that yields the moisture content at a particular point.

The gamma-ray absorption method of measuring the moisture content of the ground was first put forward and investigated in the former USSR by Danilin (1955). The moisture content can be expressed in millimetres of water layer, or in weight–%. To calculate the amount of water h_w in a ground layer of thickness h_s with density ρ_s, we must rewrite (2.13) in the form:

$$h_w = \frac{\ln(n_0/n)}{\mu_w \rho_w} - \frac{\mu_s \rho_s h_s}{\mu_w \rho_w}. \tag{2.29}$$

Substituting $\rho_w = 1$ and $\mu_s/\mu_w \approx 1$, we have:

$$h_w = \frac{\ln(n_0/n)}{\mu_w} - \rho_s h_s. \tag{2.30}$$

The second term, $\rho_s h_s$, on the right-hand side of this equation is the mass of the ground skeleton. It can be assumed to be constant. The accuracy of h_w will then depend only on the accuracy of the counting rates n_0 and n and the accuracy of the measured value of μ_w for water.

To calculate the moisture content w of the ground in a given layer, we can rewrite (1.30) in the form:

$$h_w = \frac{w \rho_s h_s}{100}, \tag{2.31}$$

so that:

$$w = 100\left(\frac{\ln(n_0/n)}{\mu_w \rho_s h_s} - 1\right). \tag{2.32}$$

In practice, however, there is no need to determine n_0 for air. It is more convenient to take n_0 as the gamma-ray counting rate after the gamma ray has passed through a layer of the ground material having a known moisture content h_0 (in cm). The quantity d at any other time is then given by:

$$h_w = h_0 \pm \Delta h_w, \tag{2.33}$$

and:

$$\Delta h_w = \frac{\ln(n_0/n)}{\mu_w} \tag{2.34}$$

and hence, finally:

$$h_w = \Delta h_w \pm \frac{\ln(n_0/n)}{\mu_w}. \tag{2.35}$$

The moisture content (in %) can then be calculated from:

$$w = w_0 \pm \Delta w \tag{2.36}$$

where w_0 is the initial moisture content corresponding to n_0 and is determined only once by some other independent method, e.g., by the thermostatic-weighing method. The quantity Δw can be found from the formula:

$$\Delta w = \frac{100 \ln(n_0/n)}{\mu_w \rho_s h_s}, \tag{2.37}$$

in which case:

$$w = w_0 \pm \frac{100 \ln(n_0/n)}{\mu_w \rho_s h_s}. \tag{2.38}$$

Consequently, in order to be able to investigate the moisture-content dynamics in a ground layer of thickness h_s, we must know: (a) the initial moisture content w_0 corresponding to the initial gamma-ray counting rate n_0; (b) the density of the ground skeleton (solid phase) ρ_s in the absorbing layer; (c) the value of μ_w for the gamma rays of the given energy in water (μ_w).

The initial moisture content of the absorbing layer must be determined with the maximum possible accuracy by taking samples at different points throughout the layer in which systematic observations of the moisture content by gamma-ray absorption are being carried out.

The necessity of an accurate determination of the initial moisture content w_0 by the thermostatic-weighing method is dictated by the spatial variability of the moisture content of the ground.

The bulk density ρ_s of the ground skeleton in which the moisture content is to be determined by the gamma absorption method can be obtained as described above.

The gamma-ray density meter GP-0-50 was used to investigate moisture dynamics in the uppermost layer of the soil down to a depth of 50 cm. All that is necessary is to insert a tube having a diameter of 15–16 mm vertically into the soil. The rod probe is then inserted into the tube to strictly defined depths, for example 0–30, 0–40 and 0–50 cm. The number of such tubes in a given area depends on the particular problem and the required degree of reliability of the final moisture content.

Moisture dynamics in soils and other ground materials has been investigated by the gamma-ray absorption method by Danilin (1957), Zavelsky and Dobrovol'skaya (1962) and later on by many other researchers. Their works have shown that the moisture content of such materials can be determined to within 1–1.5 %, i.e., roughly to the same accuracy as by the thermostatic-weighing method. In the case of the aeration zone, the gamma-ray absorption method can be used to within at least the same accuracy as the thermostatic-weighing method.

Figure 2.11 gives a comparison of moisture content measurements by these two methods in the region of the Murgab River (Turkmenistan).

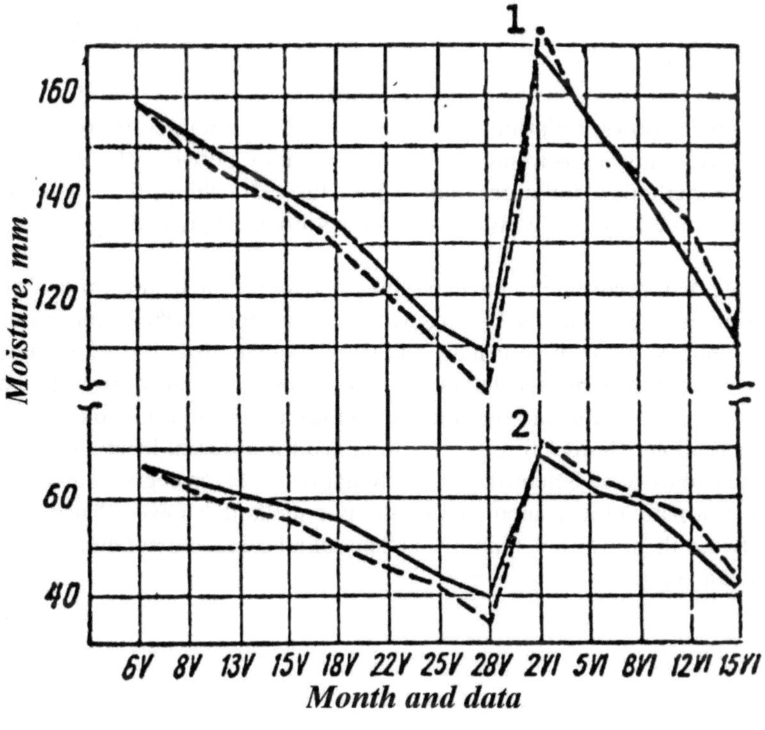

Fig. 2.11 Dynamics of moisture content in soils obtained by gamma-ray absorption (•) and thermostatic drying (○) methods

Table 2.8 Mean moisture content obtained by gamma-ray absorption and thermostatic drying

Soil layer (cm)	Gamma-ray absorption (mm H_2O)	Thermostatic drying (mm H_2O)
May 19, 1962		
0–25	96	90
0–50	170	166
0–75	240	237
0–100	315	306
May 28, 1962		
0–25	78	84
0–50	154	152
0–75	220	220
0–100	294	289
June 8, 1962		
0–25	78	73
0–50	148	148
0–75	209	209
0–100	281	277
June 15, 1962		
0–25	73	73
0–50	140	138
0–75	198	197
0–100	251	262

The figure shows variations in moisture content obtained by repeated thermostatic-weighing measurements. The discrepancy between the data obtained by the two methods is not more than 1–2 %. Table 2.8 shows the mean moisture content obtained by the gamma-ray absorption method and by thermostatic drying.

We note that, according to the experimental data for the 0–100 cm layer of soil, the maximum discrepancy is 11 mm, which is less than 1 wt% of moisture. If we compare the deviations of percentage moisture contents from the mean values, obtained as a result of four repeated measurements by gamma-ray absorption and by thermostatic drying, we find that the gamma-ray absorption method gives rise to smaller variations. Similar results were obtained for different locations. For example, at the Dubovskaya Scientific Research Station for Hydrology (Southern region, Russia), 158 parallel measurements by the two methods showed that the discrepancy between them was less than 2 % in 84 % of all cases and not more than 1 wt% in 52 % of all cases.

2.8 Determination of the Amount of Water Stored in Snow Cover

Snow cover and the determination of the amount of water stored in the cover are of major practical and scientific importance, especially in countries like Russia, Canada and USA where enormous territories are covered by snow over a considerable time of the year. Data on the snow cover and its water equivalent are necessary for calculations and estimates of the water-carrying capacity of rivers and of the accumulation of water in reservoirs, irrigation systems and moisture reservoirs of the soil. This information pertains to catastrophic phenomena such as floods and landslides. Observations of the state and thickness of the snow cover in such regions is essential for predicting snow slides producing damage to roads, bridges and other facilities.

Physically, the water equivalent of snow cover can be determined by gamma-ray absorption in the same way as in the case of ground material, except that when a vertical gamma-ray beam is used the final result gives directly the water equivalent in cm H_2O or in g/cm^2, since the mass of the snow for $\rho_w = 1$ is then equivalent to the thickness of a water layer in centimetres.

Special automatic telemetric devices are used to determine the water equivalent of the snow cover in inaccessible mountainous regions. These devices transmit the data by UNF of short-wave radios. They incorporate a collimator, the gamma-ray source, the detector (one or a number of Geiger counters), a rate meter or scaling unit, a coding mechanism, a programming device (usually a clock) and a radio transmitter with the appropriate supplies. An installation of this kind was developed by Doremus (1951).

Telemetric devices for the automatic determination of the water equivalent of snow cover are used in many countries. To obtain more accurate data on the mean water equivalent of snow, in USSR a number of STS-6 counter arrays are employed. They are held in metal counters and placed at a height of 4 m above the collimators that contain the ^{60}Co sources. The distance between the scaling-coding devices and the gamma-ray detector (STS-6 counters) is about 30 m in the radial direction, so that the system covers an area of about 3000 m^2. During the period of radio transmission of data on the water equivalent of the snow cover, the counters are successively connected for 2 min at a time to the scaling unit by the programming device and after each pulse-counting cycle (from each detector consisting of three STS-6 counters) the data are transmitted to the receiver in telegraphic code. A 15-watt shortwave transmitter is employed for the measurements of the water equivalent of snow cover obtained with three groups of counters set up at three different points and connected to a single scaling-coding device. Control measurements of the water equivalent of snow cover near the radiation source are carried out during the operation of the device. It is clear from the graphs that the two sets of results are in good agreement.

Table 2.9 Measurements of the water reserves of the snow cover in the Kzyl-Cha basin

No. of measurements	Depth of snow (cm)	Mean water reserves by γ-ray gauge (mm)	Mean water reserves byweighing snow gauge (mm)
40	136–414	229.3	249
11	121–130	369	374
6	77–138	408	407
12	66–138	165	165
9	52–138	363	339

The accuracy of each of the ^{60}Co sources was about 12 mg-eqt. of radium. The collimator sizes were as follows: d = 1000 mm, l = 120 mm, diameter of the collimating channel d = 8 mm. The initial counting rate n_0 had been determined for each detector (three STS-6 counters) before snowfall and the coefficient μ_{eff} for water had been determined in preliminary experiments. The values of this coefficient were found to be practically constant for water layers up to 400 mm and the given collimators. The scaling device incorporated 15 trigger units based on the TKh-48 thyratrons with a total capacity of 32,768 pulses. In practice, the gamma-ray count (n_0) was about 29,000 pulses per cycle.

Comparison of the results obtained by the automatic gamma-ray snow gauge and the snow weighing gauge did not exceed 5–7 %. The discrepancy between the gamma-ray results and the direct measurements of precipitation was not more then 2–3 %. Such discrepancies are fully acceptable if it is remembered that the snow is not deposited uniformly because of drifting and snow-storms.

For field measurements of the water equivalent of snow, Danilin (1955) developed a snow gauge consisting of a suitably graduated duralumin rod with a gamma-ray source (^{60}Co) at the lower pointed end and a gamma-ray counter enclosed in a metal envelope at the top.

Table 2.9 shows comparative data on the water reserves in snow cover, measured by the portable gamma-ray snow cover gauge in the region of the Kzyl-Cha River (Central–Asian region)

Even for the highly non-uniform snow cover in the mountainous regions of this district, the mean results obtained by the two methods were found to be quite close.

The density of snow at different depths can be determined by analogy with the density determination of ground materials in situ using a horizontal gamma-ray beam. The thickness of the absorbing snow layer that can be examined with a broad beam is the range of 100–200 cm.

It is best to employ scintillation counters with pulse-height discrimination in the portable gamma-ray snow gauges, so that a constant μ_{eff} close to the calculated (theoretical) value can be used. The effectiveness of snow surveys can be increased by factor of 25–30 with these portable gamma-ray snow gauges when the depth of the snow cover is in excess of 120 cm.

The gamma-ray absorption method can be used at representative points to investigate the reduction in the water reserves of the snow cover during the spring thaw and so to investigate its dynamics.

2.9 Studies of Evaporation Processes

Studies of the water balance in soil are of importance in connection with the processes responsible for the formation of groundwaters and their dynamics. One of the elements that affects the water balance in the ground is evaporation from the surface.

This type of problem can be investigated by gamma-ray absorption without weighting samples. Field experiments of this kind were carried out in 1957–1958 by Danilin (1959) at the Valdayskaya and Dubovskaya Scientific Research Hydrologic Laboratories and also at the 'Gigant State Farm' near Rostov in Russia.

The method employed was completely analogous to that used in moisture-content studies, with vertical, partially collimated, gamma-ray beams. A led collimator 8 cm in diameter and having an aperture of 8 mm was placed under a GGI–500–59 evaporator. The gamma-ray source was ^{60}Co with an activity of 6–7 mCi. An STS-6 counter working in conjunction with an M-30M scaler was placed above the evaporator containing the soil sample and at a strictly defined distance from it. The number of pulses counted in each cycle of measurement was about 49,000. The effective mass absorption coefficient for water μ_w was determined experimentally and was found to be 0.036 cm^2/g. The mean total amounts of evaporated water between July 9 and August 21, 1959 obtained by the gamma-ray absorption method and by weighing the evaporator were found to be the same (the discrepancy was 1.2 mm).

References

Artsybashev VA (1965) Gamma-ray method of density measurement. Atomizdat, Moscow

Babinets AE, Zvolsky ST (1961) Measurement of density and moisture content of grounds by radioisotope method. Ukraine Publ. House, Kiev

Churayev NV (1965) Application of isotopes and radiation sources in hydrology and hydrogeology. Atomic Energy 18:14–18

Clayton CG (1983) Nuclear geophysics. Pergamon Press, Oxford

Danilin AI (1955) Method of measuring the moisture content of soil using gamma-rays. Pochvovedenie 7:74–83

Danilin AI (1957) Application of nuclear radiation in hydrometeorology. Gidrometeoizdat, Leningrad

Danilin AI (1959) The use of gamma-rays in studies of the water regime in soils and snow cover. Izv, AN USSR, Ser. Geograph. 3:14–18

Doremus IA (1951) Telemetring system for radioactive snow gauge. Electronics 24:88–91

Dubinchuk VT (1966) Maximum working depth in studies of ground density by the gamma-gamma method. In: Proc.10th Conference of Junior Scientists and Engineers at VSEGINGEO, Moscow, pp 24–29
Emelyanov VA (1962) Gamma rays and neutrons in field soil-melioration studies. Atomizdat, Moscow
Ferronsky VI (1956) Checking the ground density at dam construction using the "Radioactive Fork" RV-2. AN USSR Center of Technical Information, Moscow
Ferronsky VI, Danilin AI, Dubinchuk VT et al (1968) Radioactive investigative methods in engineering geology and hydrogeology. Atomizdat, Moscow
Ferronsky VI, Danilin AI, Dubinchuk VT et al (1977) Radioactive investigative methods in engineering geology and Hydrogeology, 2nd edn. Atomizdat, Moscow
Filippov EM (1962) Applied nuclear geophysics. USSR Academy of Sciences Publ House, Moscow
International Atomic Energy Agency (1968) Guidebook on nuclear techniques in hydrology. IAEA, Vienna
International Atomic Energy Agency (1983) Guidebook on nuclear techniques in hydrology (1983 Edition). IAEA, Vienna
International Atomic Energy Agency (1999) Nuclear geophysics and its applications. IAEA, Vienna
Leypunsky OI, Novozilov BV, Sacharov VN (1960) Propagation of gamma-rays through matter. Fizmatgiz, Moscow
Nesterov VY (1959) Portable scintillation counter for gamma-ray measurement. VINITI AN SSSR, Moscow
Rukhin LB (1961) Fundamentals of lithology. Gostoptekhizdat, Moscow
Vartanov NA, Samoylov PS (1964) Practical methods in scintillation spectroscopy. Atomizdat, Moscow
Volarovich MP, Churayev NV (1960) Investigation of peat using radioactive isotopes. USSR Acd. Sci Publ. House, Moscow
Zavelsky FS, Dobrovol'skaya ND (1962) Methodologic recommendations for measurement of moisture-content dynamics in the aeration zone using gamma-ray absorption method. VSEGINGEO, Moscow.

Chapter 3
The Gamma-Ray Back-Scattering Method

Abstract The functional relationship between scattering gamma-radiation and the density of a medium is the basis of this discussed method. The chief parameters that govern the practical usefulness of a given method, namely, its sensitivity to density changes and its resolution in depth are considered. These parameters depend on the probe length (axial distance between the source and the detector) and the gamma-ray energy. There are three different types of gamma-ray density gauges: surface-type probe, well-logging probe and a probe on the rod inserting directly into the soil. The principal difference between the various types is in the design of the measuring probe. A number of designs for surface-type probes are available. Mass-produced logging equipment has been widely used by geophysicists for density measurements in wells. At a relatively small depth (up to 25 m) the density can be conveniently determined by inserting a rod probe directly into the ground. Technological aspects of measurement and calibration are discussed.

3.1 Principles and Range of Application

The use of the back-scattering method for the determination of soil density under field conditions is based on the functional relationship between the scattered gamma-ray intensity and the density of the back-scattered medium. In the most general case the scattered gamma-ray intensity will depend on the density of the medium, the energy of the primary gamma rays and the distance between the source and the detector. This dependence has an extreme character. In the initial part of the curve (ab) of $I_{\gamma\gamma}$ as a function of ρ (Fig. 3.1) the scattered intensity rises with increasing ρ. Over the region (bc) the intensity is a two-valued function of the density. A further increase of density is accomplished by a reduction in the scattered intensity (region cd) and beyond this the curve becomes very shallow.

The most convenient working regime is the initial part ab of the curve in which the method has maximum sensitivity to the density variation under otherwise equal conditions. By sensitivity we mean the ratio of the increase in the scattered intensity to the corresponding increase in the density of the investigated medium. However, the initial part of this curve cannot be used in practice, because it corresponds to a density change between 0 and 1 g/cm^3, whereas the density of ground and rock

V. I. Ferronsky, *Nuclear Geophysics*, Springer Geophysics,
DOI 10.1007/978-3-319-12451-3_3

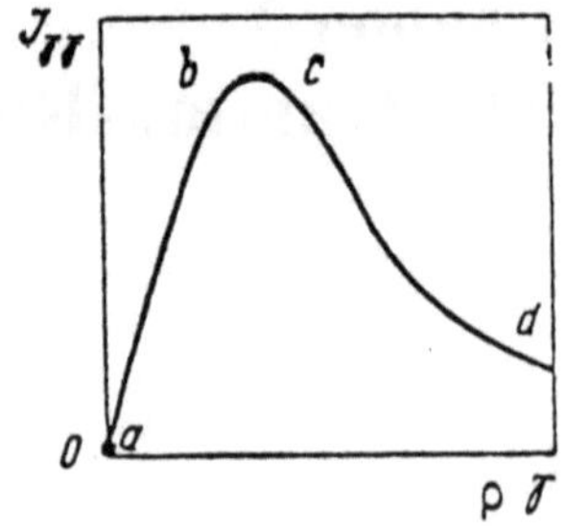

Fig. 3.1 Schematic dependence of the ground gamma-ray intensity $I_{\gamma\gamma}$ on the ground density at a certain distance from the gamma source

materials encountered in practice is at least 1.3–1.4 g/cm^3. The only exceptions are deposits of peat bogs, sapropels and certain other materials that have densities of less than 1 g/cm^3. Consequently, the falling branch of the curve, cd, must be used for the majority of ground and rock materials encountered in nature.

To determine the scattered gamma-ray intensity as a function of the density of the medium, we must consider the propagation of gamma rays in an infinite or semi-infinite medium and deduce the distribution function I(r,Q,E) for the scattered radiation at a distance r from the source. If we know the energy E and angular parameters of the detector sensitivity ε(E,Q), we can then obtain the required recorded gamma-ray intensity in the form:

$$I_{\gamma\gamma} = \int_V \int_0^E I(r,Q,E)\varepsilon(Q,E)dEdQ. \tag{3.1}$$

Depending on whether the medium is infinite or semi-infinite, the integrals are evaluated either over the whole space (4π) or half of it (2π). An exact analytic expression for the distribution function of Eq. (3.1) has not so far been obtained because of major mathematical difficulties. Certain simplifications have therefore had to be introduced and these have resulted in an approximate solution. Dyad'kin (1955) used the diffusion approximation to show that the scattered gamma-ray intensity is given by:

$$I_{\gamma\gamma} = B\rho^2 e^{-\rho L/L_0} \frac{I}{L}, \tag{3.2}$$

where B is a constant that depends on the source power; ρ is the density of the medium; L_0 is the diffusion length; and L is the probe length.

A more rigorous formula was obtained by Voskoboynikov (1957), again in the diffusion approximation:

$$I_{\gamma\gamma} = A\frac{Q\rho}{L}e^{-\mu\rho L}, \tag{3.3}$$

where A is a constant and Q is the source power.

Another solution, obtained by Filippov (1962), is:

$$I_{\gamma\gamma} = \frac{CQ}{4\pi L^b} e^{-\mu_m L}, \tag{3.4}$$

where C is a constant that takes into account gamma-ray absorption in the walls of the instrument, b is a coefficient that takes into account the length of the logging probe and μ_m is the mean absorption coefficient for the entire spectrum of the scattered gamma rays.

The above formulae are not, however, suitable for practical application although they do reflect correctly the general dependence of thc scattered intensity of the probe and medium parameters. The usual procedure in practice is to interpret measurements on the basis of data obtained with a particular probe and particular working conditions.

Experimental verification of the various approximate formulae using the data of Verbovenko and Fakidov (1958) has shown that the best results are obtained with the following empirical formula:

$$I_{\gamma\gamma} = k\rho e^{\lambda\rho}, \tag{3.5}$$

where k and λ are the constants that depend on the design of the probe.

Gamma-ray density meters based on the use of back-scattering consist of two main elements, namely, the probe and the recording unit. The probe is usually inserted directly into the ground. It incorporates the radiation source, the gamma-ray detector and the screen between the detector and the source (Fig. 3.2) to protect the detector from direct radiation. The whole assembly is enclosed in a case. The recorders of the gamma radiation are discrete-count radiation meters and rate meters.

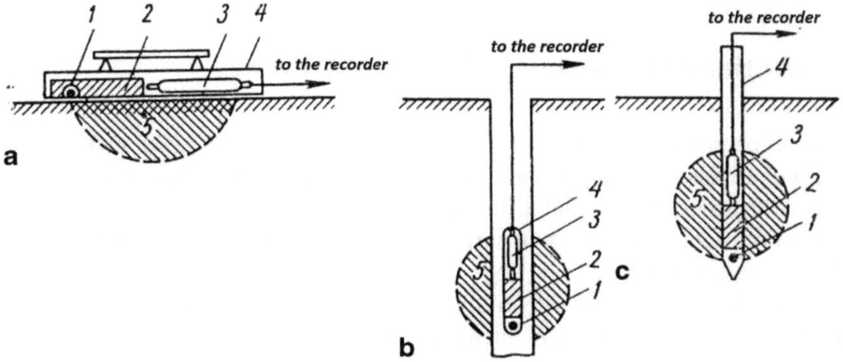

Fig. 3.2 Measurements of density of the ground by the back-scattering method: **a** Surface-type gamma-ray density meter; **b** density measurements in wells; **c** bulk-density measurements by direct insertion of the measuring probe into the ground; *1* gamma source; *2* screen; *3* gamma-ray detector; *4* body of the probe; *s* zone of measurement. (Ferronsky et al. 1968)

Depending on the type of nature of the problem, there are three basic arrangements for bulk-density measurements based on the back-scattering of gamma rays (Fig. 3.2).

In the first arrangements (Fig. 3.2a), used in the so-called surface density meters, the density can be determined for the surface layer of soil or exposed excavation area. The density can be found both at individual points and continuously over a given surface profile using gamma-gamma profiling. In the latter case the probe and the recorder equipment can be transported along the profile.

The scheme in Fig. 3.2b is used to determine the density of the ground in wells. The measurements can be taken either at individual points or continuously by gamma-gamma logging.

The scheme in Fig. 3.2c is essentially the same as the last arrangement, except that here the instrument is pressed into the ground rather than inserted into an existing or bored hole. This method has a number of important methodological, technical and economic advantages.

The choice of a particular method depends on the solving problem and the field conditions. In the case of surface deposits and exposed materials, it is best to use the surface-type density gauges, which are portable and do not require insertion of the probe into the soil. Surface deposits are of course extremely inhomogeneous in their properties, including the density and it is therefore necessary to estimate the suitability of a particular measuring method for taking the average of the density over a given volume of the soil.

Experiments have shown that the scheme in Fig. 3.2c gives the best result for the density at depth. A good result as far as accuracy and resolution are concerned can be obtained by careful scanning of a well if its diameter is not much greater than the diameter of the probe and the casing tube is thin-walled.

Measurements of density both by direct insertion of the probe into the ground and in boreholes have been successfully used, e.g., in geoengineering searches for construction sites, in hydrologic work and in civil engineering.

3.2 Optimal Parameters of Measuring Probe

The chief parameters that govern the practical usefulness of a given method are its sensitivity to density changes and its resolution in depth. Both these parameters depend on the probe length axial distance between the source and the detector and the gamma-ray energy. In the case of surface-density meters the probe length is usually called the measuring base.

As already indicated, the sensitivity of the method is characterised by the derivative of the scattered gamma-ray intensity with respect to the density, i.e.,:

$$\eta = \frac{dI_{\gamma\gamma}}{d\rho}. \tag{3.6}$$

It is clear that the sensitivity as defined here is zero at the peak of the $I_{\gamma\gamma} = f(\rho)$ curve (Fig. 3.1) and we can use Eq. (3.6) to determine the critical density for a given probe length for which the sensitivity is in fact zero:

$$\frac{dI_{\gamma\gamma}}{d\rho} = 0. \tag{3.7}$$

According to Volarovich and Churaev (1960), the critical density ρ_{cr} is given by:

$$\rho_{CK} = \frac{1}{\mu L}. \tag{3.8}$$

This expression is in good agreement with experimental data. It shows that, as the probe length increases, the peak of the $I_{\gamma\gamma} = f(\rho)$ curve shifts toward lower densities and vice versa. It follows that when low densities are measured it is best to use short probes and vice versa.

It is also clear from Eq. (3.8) that as the gamma-ray energy increases, i.e., μ decreases, the maximum of the $I_{\gamma\gamma} = f(\rho)$ curve shifts toward higher densities.

Dyad'kin (1955), Voskoboynikov (1957), Ochkur (1957), Filippov (1962) and others have shown that the sensitivity increases with increasing probe length. Extensive experimental work in connection with this problem has been carried out by many researchers.

The experiments in 2π geometry by Dubinchuk (1966) were carried out with a scintillation counter incorporating an NaI(Tl) phosphor of 30×40 mm. The scattering medium was in the form of specially prepared concrete plates ($50 \times 50 \times 5$ cm) of constant and known density, whose chemical composition was close to that of sandy loam. The density of the plates was varied in steps of 0.2 g/cm^3 in the range 0.6–2.2 g/cm^3. In addition, glass with a density of 2.6 g/cm^3 was employed as the scattering medium. Steps were taken to minimise the effect of scattering and to prevent direct gamma radiation from reaching the detector. The thickness of a given stack of plates of given density was not less than 3 mean free paths of the gamma quanta and the distance of the probe from the plate edges was at least two mean free paths.

Experimental data obtained in this way for ^{60}Co and ^{137}Cs sources (which are the sources used most widely in practice) and for probe lengths of 15–90 cm are shown in Fig. 3.3.

The sensitivity of the method was determined by measuring the slope of each curve. It is clear from the experimental data that the sensitivity was very low for small probe dimensions. However, the parameter increases with increasing probe length. To facilitate comparisons of sensitivity data for different probe designs and different source activities, we shall use the logarithmic sensitivity, which is defined as the ratio of the corresponding increment in the density.

Figure 3.4 shows the experimental results obtained by Dubinchuk for the logarithmic sensitivity as a function of probe length for ^{60}Co and ^{137}Cs sources. It is clear

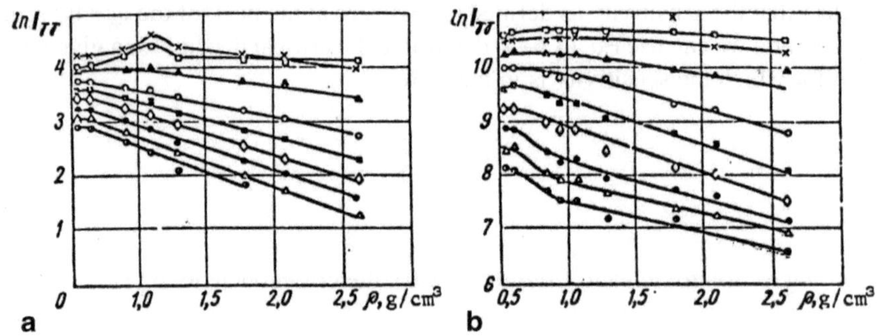

Fig. 3.3 Intensity of scattered gamma-rays as a function of density under 2π geometry conditions for ^{60}Co (**a**) and ^{137}Cs (**b**). The probe lengths are as follows (cm): □ – 15; X – 20; △ – 30; ○ – 40; □ – 50; ◇ – 60; ○ –70; ⊾ – 80; ⊕ – 90. (Ferronsky et al. 1968)

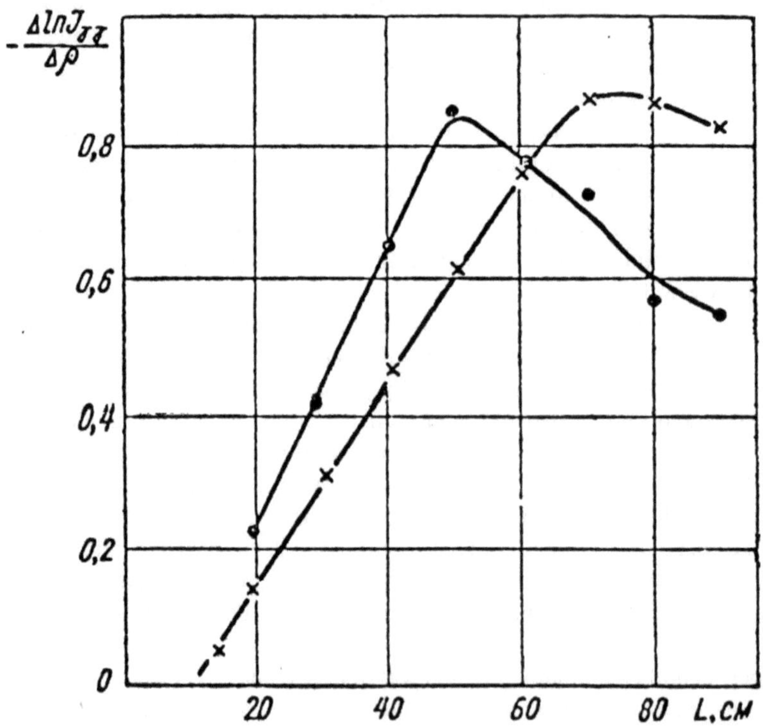

Fig. 3.4 Logarithmic sensitivity as a function of probe length under conditions of 2π geometry for ^{60}Co (— x — x —) and ^{137}Cs (— · — · —). (Ferronsky et al. 1968)

that the maximum sensitivity for ^{137}Cs is reached for probe lengths of 50 cm, whereas for ^{60}Co this length is 70–80 cm. This is confirmed by existing published data. Moreover, up to these probe lengths the sensitivity is a linear function of the length. It is also clear that the sensitivity for ^{137}Cs is considerably higher than for ^{60}Co. This conclusion is in complete agreement with the physical basis of this process, since μ is higher for softer gamma rays and hence the sensitivity must also be higher.

The second important parameter of the back-scattering method is the sensitivity with respect to the maximum working depth. When the density of the ground is measured, sensitivity to the maximum working depth is important for several reasons. Firstly, since the ground material is statistically inhomogeneous as far as density is concerned, it is important to obtain as accurate density data as possible. For this purpose the measured densities must cover the maximum possible volume of the ground material, so that an adequate average can be taken. On the other hand, depending on the particular problem in hand, it may be undesirable for the individual density inhomogeneities to be completely lost in the overall average. Finally, when the bulk density is measured inside casing tubes, the necessary maximum working depth can be chosen so as to minimise interference associated with the sinking and casing of wells.

Existing theoretical and calculated data on the maximum working depth (Filippov 1962; Bulatov and Leypunsky 1959) are very approximate, since they do not take fully into account scattered radiation and the assumed boundary conditions are not completely realistic. The problem can only be investigated reliably and completely by carrying out the necessary experiments.

Experimental work by Dubinchuk under 2π conditions has resulted in data on the maximum working depth as a function of the probe size, the density of the medium and the energy of primary gamma rays. The experiments were carried out under laboratory conditions using $50 \times 50 \times 5$ cm concrete plates of known density in the range 0.6–2.2 g/cm^3. The scattered gamma rays were recorded by a scintillation counter (30×40 mm NaI(Tl) phosphor and FEU-29 photomultiplier), incorporating a differential amplitude analyser (AADO) and the "Volna" scaler. The amplitude-energy resolution deduced from the width of the ^{137}Cs total absorption peak was about 10 %. The measurements were carried out by analysing the energy spectrum of the recorded radiation. An estimate was made of the spectral composition of the radiation, which is very important for the physical interpretation of experimental data and the choice of the optimal probe parameters. The sources were ^{60}Co and ^{137}Cs. The method was as follows. With a given probe length and given density of the concrete plates, the scattered gamma radiation was measured both without any discrimination (integral recording) and with the spectrometer window defining a particular pulse-height range. The thickness of the scattering medium was gradually increased beginning with 5 cm (single plate), until complete saturation was recorded, i.e., until a further increase in the thickness of the scattering medium had no effect on the recorded radiation.

The maximum working depth for given probe parameters, type of source and density value was determined from the curve representing the integral intensity of the recorded scattered radiation as a function of the thickness $h_{0.9}$ of the scattering medium in which 90 % of the maximum working depth ensures the necessary accuracy and represents with sufficient completeness the contributions of gamma-ray scattering from different layers of the medium.

It is clear from the experimental data given in Fig. 3.5 that the curves obtained for ^{60}Co and ^{137}Cs are generally similar. For probe lengths of less than 2–3 mean free paths of the gamma rays the maximum working depth increases with increasing probe length up to a certain limit. Further increase in the size of the probe results in a practically constant maximum working depth. This depth depends on the density of the medium and varies from 26 cm (density 0.6 g.cm³) to 11 cm (density 2.6 g/cm³) for ^{60}Co, whereas for ^{137}Cs it varies from 13 cm (density 0.6 g/cm³) to 5 cm (density 2.6 g/cm³).

These experimental data are in agreement with the theoretical studies of Bulatov and Leypunsky (1959). The latter authors have shown that the build-up factor for scattered gamma rays, which governs the depth of penetration of the scattered

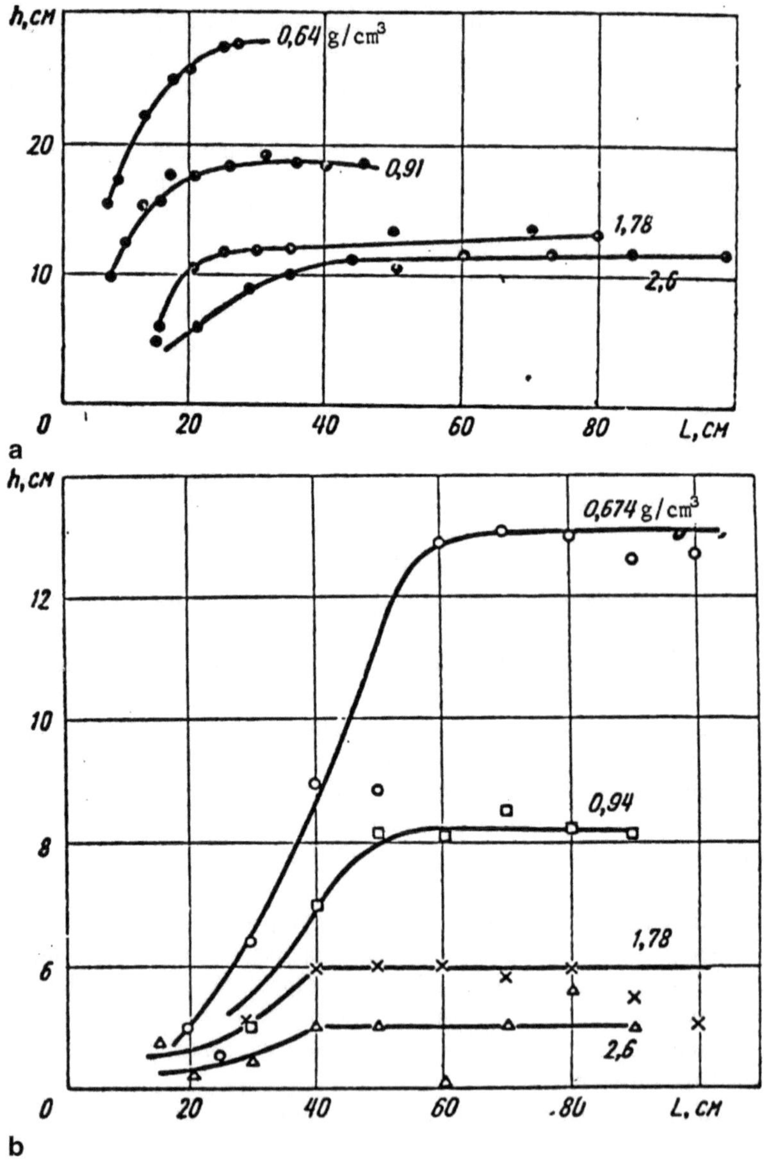

Fig. 3.5 Maximum working depth in the gamma-ray back-scattering method as function of the probe size and the density of the medium for 2π geometry and ^{60}Co (**a**) and ^{137}Cs (**b**) sources. (Ferronsky et al. 1968)

radiation, reaches its maximum value for light elements when the thickness of the reflector is equal to one or two mean free paths at the primary gamma-ray energy.

The dependence of the maximum working depth on the probe size can be explained as follows. For short source-detector distance, most of the received radiation has been scattered only two or three times. When the probe length is in excess of three mean free paths, multiply scattered gamma rays predominate and the spectral composition of the radiation reaches a relative equilibrium. This corresponds to the maximum working depth. Pulse-height spectra obtained for the scattered radiation have confirmed this conclusion. We note that an increase in the maximum working depth by about 20 % was achieved for ^{60}Co by recording the soft component of the scattered gamma rays ($E<200$ keV). Consequently, the multiply scattered component of the recorded radiation arrives from larger depths. This explains the certain increase in the maximum working depth with increasing probe size when multiply scattered gamma rays are recorded. Hence we have the important practical result that the maximum working depth can be varied by suitably choosing the part of the spectrum to be recorded.

The existing belief that the maximum working depth increases with increasing probe size has not been confirmed experimentally due to the fact that some workers extrapolate the results of experiments in the region where equilibrium has not yet been established in the spectrum of the recorded radiation, where the maximum working depth does in fact increase with increasing size.

The total $v_{0.9}$ of the scattering medium has been measured. Assuming cylindrical symmetry for the recorded scattered gamma radiation, we can use the formula:

$$v_{0.9} = \pi h_{0.9}^2 (L + h_{0.9}) \tag{3.9}$$

Least-squares analysis of the data has resulted in the following empirical formulae for the maximum working depth when the probe length is greater than three mean free paths:

for a ^{60}Co source

$$h_{0.9} = \frac{6.7}{\rho^2} + 10.5; \tag{3.10}$$

and for a ^{137}Cs source

$$h_{0.9} = \frac{3.8}{\rho^2} + 4.4. \tag{3.11}$$

These formulae are valid to within 10–15 % for densities in the range of 0.6–2.6 g/cm^3.

Interesting experiments on the maximum working depth under 4π geometry have been carried out at the Moscow Geological Prospecting Institute. The experiments were carried out under field conditions using a ^{60}Co sourse. The specimen block was cut from loam of density 1.88 g/cm^3 and a natural moisture content of 21.5 %. The height of the block was 1.5 m and the diameter was gradually

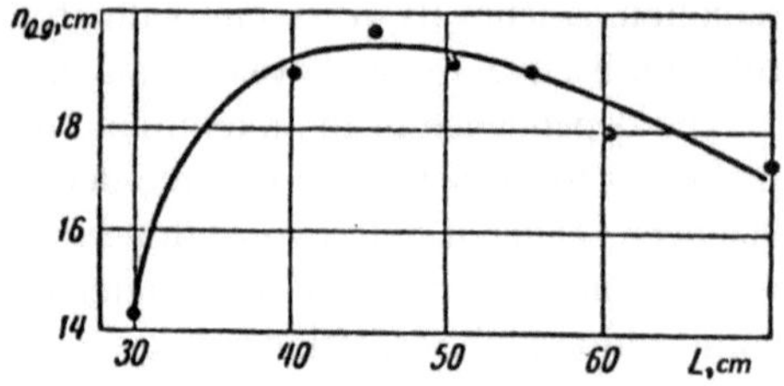

Fig. 3.6 Maximum working depth in the gamma-ray back-scattering method as a function of the probe length for 4π geometry and a ^{60}Co source (density of the ground 1.88 g/cm^3). (Ferronsky et al. 1968)

reduced from 70 cm. A 3-m and 89 mm hole in diameter was drilled through the centre of the block and the measurements were performed in the hole. When the maximum working depth was investigated, the length of the 50 mm diameter probe was varied from 30 to 70 cm. The maximum depth was defined as the effective radius within which 90% of the recorded radiation was scattered. It is clear from the data in Fig. 3.6 that as the probe length increases the maximum depth at first increases, then reaches a maximum and finally tends to decrease for probe lengths in excess of 55 cm. The last effect can be explained by the presence of a gap between the body of the probe and the whole wall but it may also be due to experimental errors.

Comparison of the results of these experiments with the above estimates of the maximum working depth for 2π geometry and the same values of the density and source energy shows that, in the second case, the maximum depth increases to some extent, reaching about 19 cm as compared with 12–13 cm. However, the general variation and the optical values are quite close in the two cases.

We can now generalise our discussion of the main parameters involved in density measurements by the back-scattering method, which are connected with the construction of the measuring probe, the source energy and the properties of the ground itself. Thus, depending on the nature and the requirements of the particular engineering problem, the sensitivity of the method and the maximum working depth can be varied within certain limits.

This means that the choice of the probe size and the type of source can be approached in a creative fashion and many errors associated with experimental conditions can be avoided. The detection of the soft component of the scattered gamma radiation is, however, always restricted by the limit at which the chemical composition of the ground material begins to have an appreciable effect through the predominance of photoelectric absorption.

3.3 Design of Gamma-Ray Density Gauges and the Range of Their Application

The principal difference between the various types of gamma-ray density meters is in the design of the measuring probe. The recording equipment is chosen not for the particular conditions of the problem but as the basis of the requirements of radiometric measurements, availability of supplies, convenience of taking the readings and so on.

3.3.1 Surface-Type Gamma-Ray Density Gauges

A number of designs for surface-type probes are available. The probe developed by Volarovich and Churaev (1960) was designed for density measurements of peat. The probe consists of a brass body and a supporting plate (Fig. 3.2a). It incorporates the source of radiation and a gas-filled counter (STS-6) that is connected by cable via an insulating collar to the radiometer. In practice, the plate is placed on the surface to be investigated. To reduce the effect of the plate material on the final data, a rectangular window is cut under the counter and is covered by thin aluminium foil in order to exclude any moisture penetration. The gamma-ray source can be moved inside the instrument, so that the distance between the source and detector can be varied within the range of 16.5–31 cm. A similar design was proposed by Filippov and Kuznetsov (1959).

The design of a surface-type gamma-ray gauge developed by Makarov and Basin (1961) differs from the probe described above by the fact that it incorporates five counters in a Perspex box that is readily demountable so that the counters can be easily replaced. Provision is also made for the insertion of a lead filter that cuts off the soft gamma-ray component. A preamplifier and a cathode follower are built into the handle of the instrument, so that standard scalers such as PS-100, B-2 and so on can be employed.

Surface-type gauges are also widely used in different countries such as the one designed and manufactured by Nuclear-Chicago in the USA (model P-22). The ^{137}Cs source is located in a cylindrical lead container that can be rotated so that the source is screened off while the device is transported. The detector consists of six halogen G.M. counters. A decatron field radiometer (type 2800) is recommended for measuring the scattered gamma-ray intensity. This unit consists of five decades and can be used to record up to 12,000 pulses/s.

3.3.2 Gamma-Ray Density Gauges Used in Wells

Mass-produced logging equipment has been widely used by geophysicists for density measurements in wells. This equipment consists of a probe that is inserted into well, a ground-surface control panel and a communication channel. It is, however, very inconvenient to use in practice, because when wells are examined down to a depth of some hundreds and occasionally thousands of metres, it is necessary to incorporate not only the radiation sources and detectors but also amplifying devices, emitter followers, converters and so on, so that the probe length may reach up to 1.5–2 m. The surface control panel is designed so that it produces directly the logging diagrams. Geoengineering problems are as a rule concerned with depths of 15–25 m, whilst many problems in hydromelioration and agriculture involve only the first few metres from the surface. Portable equipment is therefore essential for a broad range of problems in geoengineering, civil engineering and hydromelioration. In this section we shall consider the design of gamma-ray density gauges for well exploration, which can be used for density measurements at individual points within the well. Problems connected with the use of gamma scattering to obtain the depth distribution of the density are discussed in Chap. 5 in connection with radioactivity logging.

Figure 3.7 shows the well probe described by Ferronsky et al. (1968). The probe is designed for density measurements in small boreholes not more than 50 mm in diameter and 20 m deep. It does not incorporate any amplifying circuits and therefore its length is less than 0.5 m for a diameter of 30 m. The probe consists of a ^{60}Co source (3 mg-eqt. of radium) and a detector. The two are separated by a conical lead screen. The body of the probe consists of a duralumin tube. A rubber plug prevents the penetration of moisture into the probe. To prevent damage to the probe due to wall collapse or well deviation, the lower part of the probe is surrounded by a push-fit steel envelope 36 mm in diameter. When the probe gets stuck in the well the steel envelope remains but the probe itself can be withdrawn.

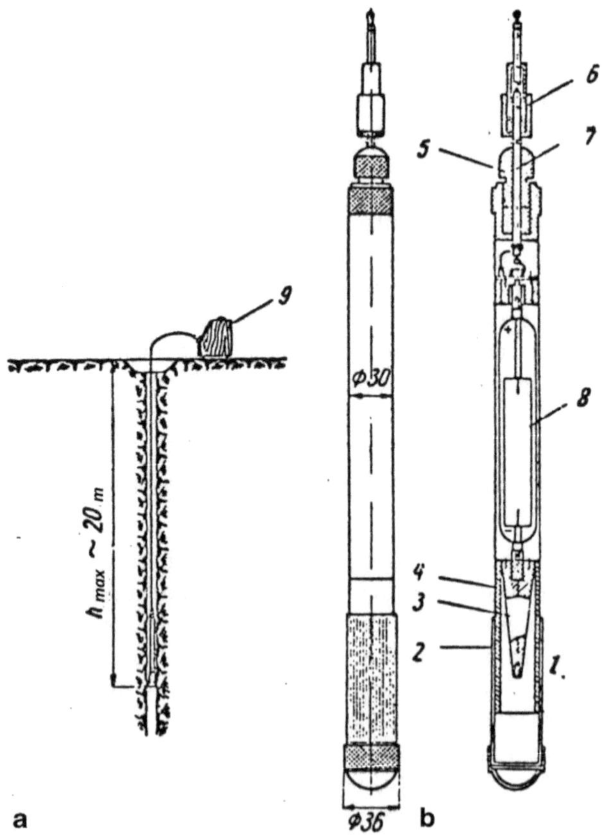

Fig. 3.7 Scheme of measurement (**a**) and design of the gamma-ray probe for use in wells (**b**): *1* source; *2* protective end-piece; *3* lead screen; *4* body of probe; *5* rubber plug; *6* plug; *7* cable; *8* halogen counter; *9* rate meter. (Ferronsky et al. 1968)

Another design of the GGP-2 gamma-ray density gauge, developed by Emelyanov (1962), includes a probe for density measurements at large depths, a portable rate meter and a calibration device. The probe consists of a ^{137}Cs source (1.5 mg-eqt. of radium) and four STS-5 counters that are separated from the source by a lead screen. The counters are also surrounded by a thin-walled lead filter that absorbs soft multiply-scattered gamma rays and thus eliminates the effect of the chemical composition of the medium on the final data. The counting rate is measured with a battery-operated M-30M scaler.

The gamma-ray probe (model P-20), manufactured by Nuclear-Chicago in the USA and designed for well measurements, has a diameter of 41 mm and incorporates a ^{137}Cs source, a lead screen, a set of halogen counters, an amplifier and a standard casing tube.

Before the probe is transported, it is inserted into a protective container in which it is held in position by screw-in stoppers. The gap in the casing tube does not exceed 2 mm.

3.3.3 Gamma-Ray Density Gauges for Direct Insertion into the Ground

At relatively small depths (up to 25 m) the density can be conveniently determined by inserting a rod probe directly into the ground. The obvious advantage of this is that no special holes have to be prepared in advance and this reduces the cost and increases the efficiency. When the probe is pushed in, a certain amount of compaction takes place and this introduces errors into the final results but this can be allowed for with sufficient accuracy by suitable calibration.

The principle of this device has been described by Ferronsky (1958). The prototype instrument was designed for density measurements at individual points by manually inserting the probe into the ground to a depth of 1–1.5 m. The probe (Fig. 3.2c) consists of a ^{60}Co source (up to 3 mg.-eqt. of radium) and a gas-filled STS-1 counter, separated by a distance of 40 cm in a steel tube with an internal diameter of 15 mm. A lead screen is placed between the detector and the source. The handle of the probe contains the pulse amplifier. This amplifier is unnecessary when the cable employed is short (a few metres). The scattered gamma rays are recorded by a battery-operated pulse counter based on cold cathode thyratrons as described by Korablev (1956). A general view of the probe and the pulse counter is shown in Fig. 3.8. The density is measured by pressing the probe into the ground to the required depth and measuring the gamma rays scattered by the surrounding medium at this point. The ground density is then found from a calibration curve ($I_{\gamma\gamma}=f(\rho)$) for the particular instrument, obtained by inserting it into media of known density.

Possible applications of the above instruments are governed by the depth of insertion of the probe (1–1.5 m) and by the possibility of manually insertion. The instrument is used in geoengineering studies in road dam and airfield construction and, for instance, for geotechnical control in civil engineering.

The same principle was used in the radioactivity probe designed for measuring the density of peat and developed by Volorovich and Churaev (1960). Since the resistance of peat to the insertion of the probe is quite low, the probe in fact can be pressed into a depth of a few metres.

3.4 Technological Aspects of Measurement and Calibration

The main aim of density measurements under field conditions is to obtain reliable, quantitative, density data representative for the given volume of the ground. Since the sensitivity and the maximum working depth of the gamma-ray back-scattering

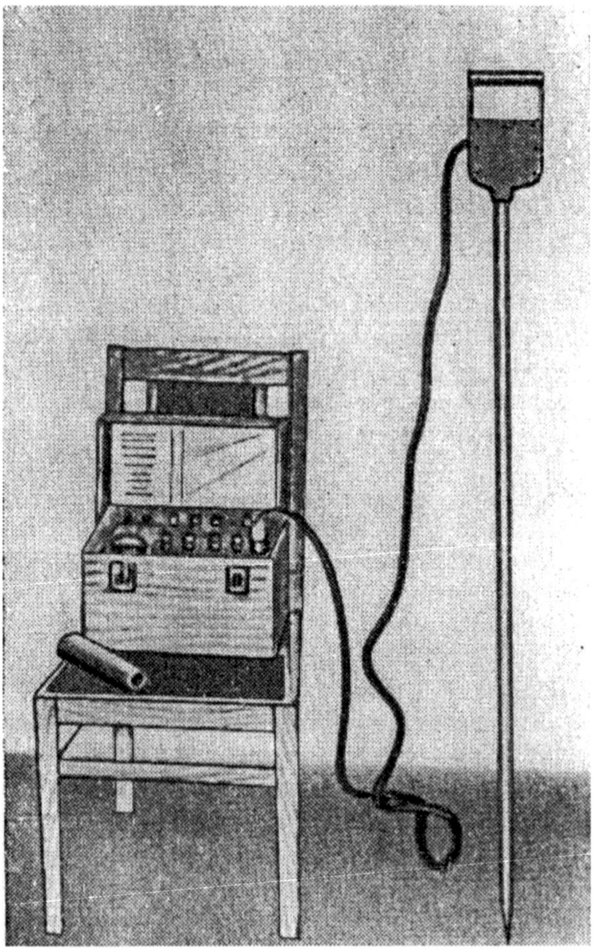

Fig. 3.8 General view of the gamma-ray density gauge with direct-insertion probe. (Ferronsky et al. 1968)

method can be adjusted within limits, the choice of the optimal parameters for the measuring probe and the measurement technology depends in the first instance on the particular problem to be solved.

Constant geometry of the measurement is particularly important. Thus, when measurements are carried out in wells, the well diameter must be as close to the diameter of the measuring probe as possible, since otherwise large and variable errors become unavoidable. Good results can be obtained if the casing of the well consists of thin-walled tubes and the calibration is carried out with these tubes in position. Measurements in uncased wells are better still. However, this is possible only for certain special types of deposits, for example, frozen and loess deposits and also clays of solid consistency. Extensive experiments have shown that the density can be determined with gamma-gamma density gauges to within ± 0.03 g/cm^3.

Calibration measurements are particularly important in this connection. They involve the determination of the scattered gamma-ray intensity as a function of the density of the ground material. This function is determined experimentally under both laboratory and field conditions. Calibration measurements must be carried out with particular care and accuracy for the following reasons. In laboratory calibration measurements the main problem is to develop a model of the ground with given density and uniform density distribution. The size of the model must exceed the possible maximum working depth when the density of the ground is to be measured by determining the scattered gamma-ray intensity. Field calibration is often similar, although one can never be sure that the density distribution at a particular point is uniform. A series of samples must therefore be taken to ensure the calibration graph is reliable. Finally, the technological features of the calibration measurements must be strictly analogous to those of measurements under field conditions.

References

Bulatov BL, Leypunskiy OI (1959) Gamma-ray albedo and buildup factor on reflection. At Energy 7:21–25

Dubinchuk VT (1966) Maximum working depth in studies of ground density by the gamma-gamma method. In: Proc. 10th Conference of Junior Scientists and Engineers at VSEGINGEO, Moscow, pp 24–29

Dyad'kin IG (1955) On the theory of gamma-gamma logging of wells. Izv AN SSSR Ser Geofiz 4:323–331

Emelyanov VA (1962) Gamma rays and neutrons in field soil-melioration studies. Atomizdat, Moscow

Ferronsky VI (1958) Methods of determination the density and moisture content of grounds. In: Mechanized compaction of grounds in construction practice. Gosstroyizdat, Moscow, pp 33–39

Ferronsky VI, Danilin AI, Dubinchuk VT et al (1968) Radioactive investigative methods in engineering geology and hydrogeology. Atomizdat, Moscow

Filippov EM (1962) Applied nuclear geophysics. USSR Academy of Sciences Publishing House, Moscow

Filippov EM, Kuznetsov GA (1959) Determination of the density of rocks and ores, in exposures and excavations, using gamma-ray scattering. Razvedka i Okhrana Nedr 9:11–17

Korablev JIH (1956) A portable radiometric device BK-3. Device Instrum Exp 3:18–21

Makarov RA, Basin YaN (1961) Radioisotope methods of measurement in the construction industry. NIIOMTP, TsBTI, Moscow
Ochkur AP (1957) Density logging. In: Problems in Mining Geophysics. Gosgeoltekhizdat, Moscow, pp 62–68
Verbovenko IG, Fakidov IG (1958) Discussion on gamma-gamma logging. At Energy 4:12–17
Volarovich MP, Churayev NV (1960) Investigation of peat using radioactive isotopes. USSR Academy of Sciences Publishing House, Moscow
Voskoboynikov GM (1957) Theoretical foundations of selective gamma-gamma logging. Izn AN SSSR Ser Geofiz 3:351–362

Chapter 4
Neutron Back-Scattering Method

Abstract The anomalous slowing-down cross-section of hydrogen with respect to neutron back-scattering is the physical basis of the method of soil moisture content determination. The neutron back-scattering method is widely used in geoengineering and hydrogeology to determine the moisture content of soil and rocks under natural conditions. Since the moisture content is determined by this method without taking any samples, the method is used to investigate friable deposits in unsaturated zones and below the groundwater table. The method yields estimates of moisture contents at individual points in the ground and is used by a continuous scan, giving the moisture content as a function of probe position. The sensitivity, working depth of the measuring probe, parameters of the medium and the various factors affecting the neutron slowing-down process are discussed in this chapter.

4.1 Principles and Range of Application

It is known from the physics of the interaction of neutrons with moisture-containing rocks that, when fast neutrons are slowing down to thermal energies, the maximum contribution to this process is due to hydrogen that has an anomalous slowing-down cross-section.

The main forms of interaction of neutrons at their propagation in the medium are inelastic scattering on nuclei and nuclear capture by atoms. If the source of fast neutrons (with energy more than 0.2 MeV) is placed into unbounded media, then their moving is accompanied with the above processes and with the loss of energy. Interaction of fast neutrons with light elements in the form of elastic and non-elastic scattering at the beginning occurs with equal probability. As a result, after reaching the level of thermal energy (~0.025 eV) a neutron comes to the thermal equilibrium with the medium of the atoms. The process of the loss of energy at the neutron's interaction with the medium is called its slowing down. Further motion of a thermal neutron is governed by the diffusion low and is finished by its capture or decay on a proton, electron or neutrino.

The process of neutron's slowing down due to inelastic scattering has probability only at the first moment of motion when its energy is high. The reaction of the inelastic scattering has an energy threshold and is possible in the case when the

V. I. Ferronsky, *Nuclear Geophysics*, Springer Geophysics,
DOI 10.1007/978-3-319-12451-3_4

neutron energy appears to be higher than the energy of the nucleus excitation. For the light nuclei with atomic mass up to 25 this threshold begins after energy of 1 MeV and for the heavier nuclei it has a lower value. So, for the artificial sources of fast neutrons, which are used in practice and have energy about 3–5 MeV, the slowing down process of the fast neutron begins after a few number of acts of elastic interactions. The mean value of neutron's energy loss ΔE due to one act of elastic scattering is constant and equal to:

$$\Delta E = 2AE/(A+1)^2, \tag{4.1}$$

where A is the atomic mass of the element; E is the neutron's energy before its collision.

It is seen from (4.1) that the maximum loss of the neutron's energy at elastic scattering takes place for light nuclei. The maximum value of the energy loss appears to be at the interaction with hydrogen when it reaches half of the initial neutron's energy. It follows from the same (4.1) that at multiple acts of interaction the energy loss can be characterised by a linear logarithmic scale. The mean number of collisions n_m that is necessary for the neutron's slowing down from initial energy E_0 to the mean thermal energy $E_n = 0.025$ eV can be calculated by the formula:

$$n_m = \frac{1}{\xi}\ln\frac{E_0}{E_n}, \tag{4.2}$$

where ξ is the sum of the mean logarithmic loss of the neutron's energy. The value of ξ is found by:

$$\xi = 1 + \ln\frac{(A-1)}{2A}\ln\frac{A-1}{A+1}, \tag{4.3}$$

Radiation capture is the probable reaction of neutrons with medium atoms. As a result of the radiation capture the compound nucleus is created, which occurs in an exciting state. Transition of the exciting nucleus to a steady state is accompanied by emission of gamma-radiation. It is known that the radiation capture reaction of type (n, γ) is probable at any neutron energy and for nuclei of all the elements. A cross-section of the radiation capture reaction for neutrons with the energies $E = 0.001$–10 MeV is proportional to the value of 1/E.

Probability of the radiation capture is increased with a decrease of the neutron's energy and at $E \leq 10$ MeV the elastic scattering is predominated. At the same time the cross-section of the interaction for the neutrons with energy $E < 1$ keV, excluding the resonance regions, is proportional to 1/v, where v is the neutron's velocity.

Reactions of the radiation capture of the slowing down neutrons have broad practical application. This type of interaction is used for analysis of the elementary content of a geological medium based on the ability of the nucleus of each element in emission of γ-radiation of the specific energy. The probability of the neutron's interaction with the nucleus of a medium is characterised by the cross-section of the interaction. The product of the atomic (microscopic) cross-section on the number of atoms in a volume unit is defined by cm^2/cm^3, or cm^{-1}.

The following parameters of a neutron's interaction with the medium are of practical importance. The slowing-down length L_s is characterised by the mean distance between the point of the neutron's yield and the point of its slowing-down up to the thermal energy. The diffusion length L_a is the distance between the point of reaching the thermal energy and point of its capture. The neutron's migration length L_m is its total distance between the yield and capture. These three distances are defined by the ratios:

$$L_s^2 = n_c \cdot S_s \cdot \frac{S_t}{3}; \tag{4.4}$$

$$L_a^2 = S_t \cdot S_s \cdot \frac{S_a}{3}; \tag{4.5}$$

$$L_m^2 = L_s^2 + L_a^2, \tag{4.6}$$

where

$$S_s = \frac{1}{n\sigma_s}; \quad S_t = \frac{1}{n\sigma_s(1-\cos\varphi)}; \quad S_a = \frac{1}{n\sigma_a};$$

S_s is the mean free path distance of the neutron relative to its scattering; S_a is the mean free path distance of the neutron relative to its capture; S_t is the transport distance characterising the neutron's length in the initial direction before its turn on 90°; n is the number of atoms in the volume unit of the medium; n_c is the number of collisions needed for slowing down up to the thermal energy; σ_s is the neutron's effective (microscopic) scattering cross-section; σ_a is the neutron's effective (microscopic) capture cross-section; φ is the mean value of the neutron's angle of scattering.

The most important parameters describing the interaction of neutrons with the most common rock-forming elements are summarised in Table 4.1.

It is clear from Table 4.1 that when neutrons interact with hydrogen, the mean logarithmic energy loss per collision, the number of collisions resulting in slowing down to thermal energy and the thermal-neutron scattering cross-section differ by at least an order of magnitude from the corresponding values for the other rock-forming elements. The thermal-neutron absorption cross-section turns out to be very high only for certain elements. Anomalous absorbers of thermal neutrons include B, Cl, Mn, Fe, K and Li.

In view of the above features of the interaction of neutrons with rocks, two methods are usually employed to determine the moisture content of ground and soil by the neutron back-scattering technique. The first of these methods involves the detection of thermal neutrons and is used whenever the medium under investigation does not contain strong thermal neutron absorbers. The second method involves the

Table 4.1 Parameters characterising the interaction of neutrons with rock-forming elements

Element	Atomic weight	Mean logarithmic energy loss per interaction	No. of collisions per necessary to slow down from $E_0 = 3.5$ MeV to $E_n = 0.025$ eV	Thermal-neutron scattering cross-section σ_s(barn)	Thermal-neutron absorption cross-section for v = 2200 m/s (barn)
H	1	1	19	38	0.332
B	11	0.171	89	4	769
C	12	0.158	112	4.8	0.003
O	16	0.12	154	4.2	0.0002
Na	23	0.085	190	4.0	0.505
Mg	24	0.0811	235	3.6	0.063
Al	27	0.0723	290	1.4	0.23
Si	28	0.0698	297	1.7	0.13
Cl	35	0.0561	343	16	31.6
K	39	0.0504	376	1.5	1.97
Ca	40	0.0492	380	3.0	0.43
Mn	55	0.0359	529	2.3	13.2
Fe	56	0.0353	539	11	2.53

detection of epithermal (0.5–1000 eV) neutrons and can be used to investigate the moisture content of saline soils and grounds. If we place a fast-neutron source in the ground and record the thermal or epithermal neutron flux at some distance from the source, the counting rate as a function of the moisture content will be of the form shown schematically in Fig. 4.1. The curve representing the function $I_{nn} = f(w)$ cuts the vertical axis at a certain distance above the origin. This is so because the mineral fraction of the ground has a definite slowing-down power for fast neutrons. The segment ab of the curve is characterised by an increase in the moderate neutron flax density with increasing moisture content. This situation is observed in practice when the distance between the source and detector is less than the neutron slowing-down length in the medium. When these two distances are comparable, an increase in the moisture content does not give rise to an appreciable change in the neutron

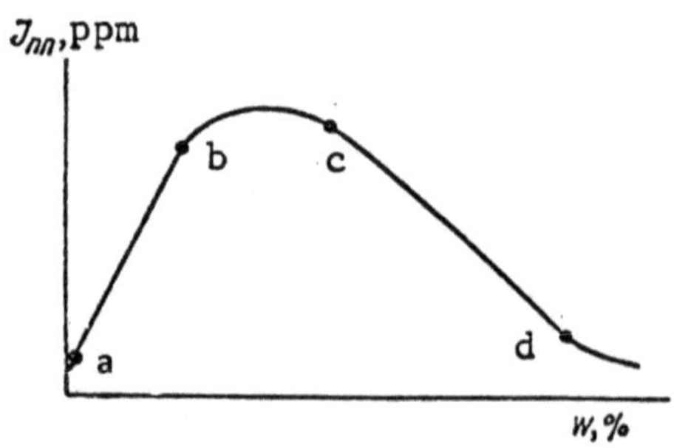

Fig. 4.1 Flux density of slowing-down neutrons as a function of the moisture content at a point located at a given distance from a fast-neutron source

flax density at the detector. This corresponds to the segment bc on the curve of Fig. 4.1 (inversion region). When the neutron slowing-down length is less than the distance between the source and detector, the counting rate falls in inverse proportion to the moisture content (segment cd).

The above considerations are of fundamental importance in the selection of the optimal parameters of the detecting probe and in taking into account the various factors that affect the results of moisture-content measurements under field conditions. The selection of these parameters depends on the conditions of measurement and on the properties of the equipment employed for this purpose (IAEA 1970).

It was found (Kantor 1955, 1958) that for an unbounded uniform hydrogen-bearing medium without strongly captured elements the effect of slowing-down neutrons is well approximated by the expression:

$$N = \frac{Q}{8\pi L_s^3} e^{-r/L_s}, \tag{4.7}$$

where n is the density of the epithermal neutron flux; Q is the activity of the neutron source; r is the distance between the source and a given point.

Distribution of the thermal neutrons in the medium is determined by:

$$N_t = \frac{Q\tau}{8\pi L_s^2 \alpha} e^{-r/L_s}, \tag{4.8}$$

where N_t is the density of the thermal neutron flax; τ is the life time of the neutron.

$$\alpha = 1 - \left(\frac{L_a}{L_s}\right)^2.$$

The expressions (4.7) and (4.8) are valid for qualitative estimation of the epithermal and thermal neutron flax at $r > L_s$. It is seen that at $L_s > r$ $e^{-r/Ls} \rightarrow 1$. So, with an increase of the moisture content the slowing-down length L_s is decreased and the density of the neutron flux N is increased. At $L_s < r$ the role of the exponential term becomes primary.

Neutron moisture gauges consist of a measuring probe and a recording instrument. The probe comprises a fast-neutron source and a detector of thermal or epithermal neutrons. When the fast-neutron source provides an appreciable gamma-ray background and the detector is sensitive to this radiation, a lead shield is inserted between the source and detector in order to minimise the influence of the gamma-rays. The probe elements are located in a suitable container. Both discrete counting devices are rate meters that are used to determine the neutron flux density.

Depending on the nature of the geoengineering or hydrologic problems, one of three possible measuring techniques can be employed. These techniques are not in principle different from the corresponding arrangements in the case of density measurements (Fig. 3.2).

The surface type moisture gauge (Fig. 3.2a) is used to determine the moisture content of the surface layer of the ground and of exposed rocks and excavations. This system can be used to determine the moisture content at individual points and also continuously over a given surface profile. In the last case the probe must be dragged along the ground by a suitable transporter (e.g., a track) that also carries the recording equipment.

The borehole moisture gauge (Fig. 3.2b) is used to determine the moisture content at individual points along the borehole and also continuously, using the neutron-neutron or neutron-gamma logging method.

In the system illustrated in Fig. 3.2c the moisture content is determined by inserting the moisture probe directly into the ground. As in the last system, the measurements can be carried out either at individual points in the ground or continuously by the neutron-neutron logging method. At a small depth (1–1.5 m) and in relatively loose ground, the probe can be pressed manually in, while in other cases it has to be driven in by special equipment.

The moisture content of the uppermost layers of the ground is usually measured by surface-type moisture gauges in the case of land reclamation and agricultural studies. In geoengineering surveys the surface measurements are usually carried out either in mining excavations (e.g., pits and tunnels) or on exposed portions of rocks. It is then necessary to remember that the uppermost layers of soil, ground and rocks are characterised by a considerable variability of the moisture content with time, owing to the continuous variation in the temperature and in the moisture supply due to precipitation.

The above three techniques are widely used in geoengineering surveys carried out on various scales, in searches for suitable building sites, in hydrogeological studies, in geotechnology and in construction operations.

4.2 Optimal Parameters of Measuring Probe

The moisture content of ground and rocks can be determined to a high degree of accuracy by the neutron back-scattering method if the parameters of the measuring probe and the various factors affecting the neutron slowing-down process are suitably taken into account. These parameters and factors include (1) the sensitivity of the method as a function of probe design and conditions of measurements, (2) the maximum working depth as a function of the probe parameters and of the composition of the medium and (3) the effects of the chemical composition, density of the medium and presence of bound moisture. Let us consider these problems in greater detail.

4.2.1 Sensitivity of the Method

The sensitivity of the neutron method in determining the moisture content is defined as the derivative of the measured neutron flux density with respect to the moisture

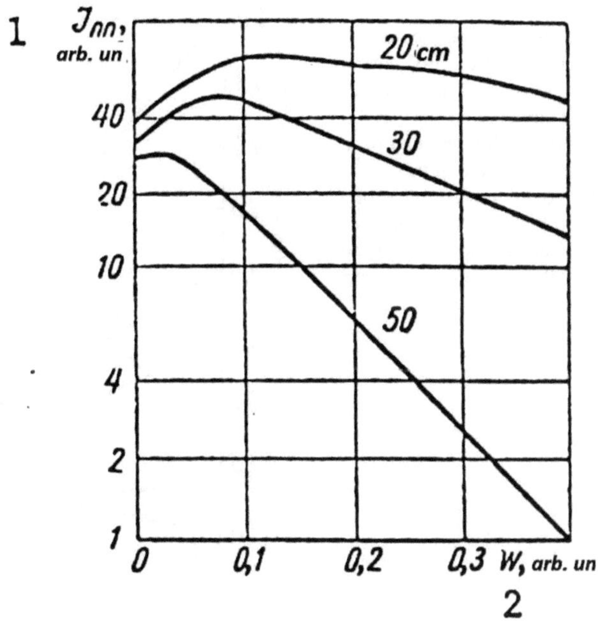

Fig. 4.2 Calculated curves showing the epithermal-neutron flux density (*arbitrary units*) as a function of the moisture content (relative units) for various probe lengths. (Ferronsky et al. 1968)

content of the medium. It follows from Eqs. (4.7), (4.8) and Fig. 4.1 that the characteristic feature of the dependence of the neutron flux density is the presence of the inversion region. This emphasizes the importance of choosing the optimal size for the measuring probe, for which the sensitivity of the method to moisture-content changes is sufficient for the problem in hand. Because of the inversion, the degree of differentiation of the medium with respect to the moisture content increases with increasing distance of the working part of the curve $I_{nn} = f(w)$.

Figure 4.2 shows calculated curves representing the function $I_{nn} = f(w)$ for epithermal neutrons (Filippov 1962). It is clear that the sensitivity of the method (given by the slope of the curve) increases with increasing source-detector distance. Extensive experiments have confirmed this conclusion. Figure 4.3 shows experimental data on the recorded thermal-neutron flux density as a function of the moisture content for different probe lengths. For the sake of convenience, the intensities I_{nn} are referred to as a standard value I_{nn}^{st} corresponding to a volume moisture content w_v of 20 %. These experiments were carried out under field experiments in sandy clay deposits. The measurements were performed by direct insertion of the probe into the ground. The graphs are based on a considerable volume of data obtained with a scintillation detector of thermal neutrons.

It is clear from these graphs that the sensitivity is practically zero when the probe length is 25 cm. This corresponds to the inversion region of the function $I_{nn} = f(w)$. An increase in the probe length leads to a gradual increase in the sensitivity. For the "zero" probe, i.e., for very small source-detector distances (about 5 cm), we have the pre-inversion region of the function $I_{nn} = f(w)$. The sensitivity of the "zero" probe then corresponds to a probe of length 40 cm in the post-inversion region.

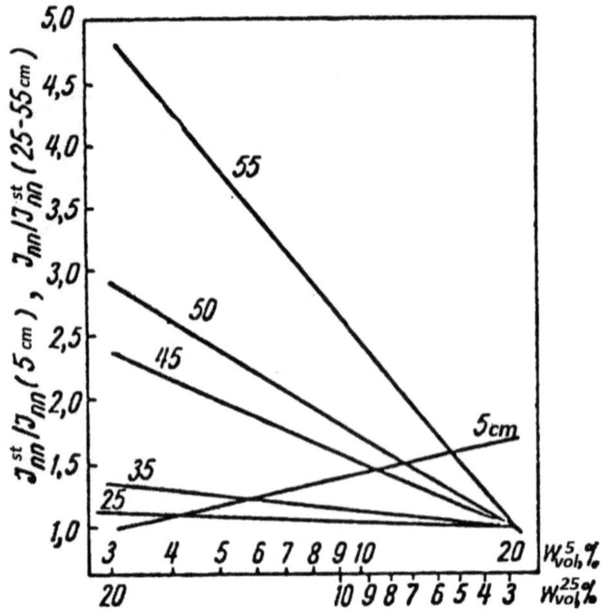

Fig. 4.3 Experimental data showing the recorded thermal-neutron flux density as a function of the volume moisture content for various lengths of the logging probe. *Upper* scale along the horizontal axis is given for a 5-cm long probe; the *lower* scale corresponds to a probe 25–50 cm in length. (Ferronsky et al. 1968)

The "best" length of the measuring probe depends on the particular problem. For example, if one is interested in the moisture content in wells, it is best to use the maximum possible sensitivity, which corresponds to probe lengths of not less than 50 cm. The results of the measurements are then less sensitive to the construction of the well.

Apart from the probe size, the sensitivity of the method is also found to depend on the initial energy of the fast neutrons. However, the radium-beryllium, polonium-beryllium, plutonium-beryllium, americium-beryllium and actinium-beryllium neutron sources have very similar energy spectra with a maximum neutron yield in the region of 3–5 MeV. There is therefore little interest in the effect of the neutron energy on the sensitivity of the method.

In the future, when moisture measurements will be carried out with controlled neutron sources, such as neutron generators, we shall have to return to the question of the effect of the initial neutron energy on sensitivity parameters.

4.2.2 *Maximum Working Depth*

This is a very important concept in moisture-content measurements since it governs the degree to which a particular set of measurements is representative of the given volume. The maximum working depth is directly related to the slowing-down and the diffusion lengths. The slowing-down length ℓ_s is a measure of the mean distance between the point of exit of the neutron and the point at which it first reaches thermal energy. It can be calculated from the formula:

$$\ell_s^2 = \frac{n_c}{3n^2\sigma_s^2(1-\overline{\cos\varphi})} \tag{4.9}$$

where n is the number of atoms per unit volume; n_c is the number of collisions up to the point at which the neutron first reaches thermal energy; σ_s is the effective neutron scattering cross- section and is the mean neutron scattering angle. The values of n_c and σ_s are given in Table 4.1.

The neutron diffusion length ℓ_d is a measure of the distance between the point at which the neutron reaches thermal energy and the point at which it is absorbed. It can be calculated from the formula:

$$\ell_d^2 = \frac{1}{n^2\sigma_s\sigma_a(1-\overline{\cos\varphi})}, \tag{4.10}$$

where σ_a is the neutron absorption cross-section (Table 4.1).

The total distance between the point of axis and the point of absorption, which is called the migration length, can be calculated from the equation:

$$\ell^2 = \ell_s^2 + \ell_d^2. \tag{4.11}$$

In applied nuclear geophysics the maximum working depth is usually estimated in terms of the effective radius r_{eff} (Kantor 1958), which defines the cylindrical volume of the ground in which up to 90 % of the neutrons are scattered and slowed down.

Theoretical calculations based on the data obtained by Filippov (1962) have led to the following approximate formula that is valid for grounds and rocks of low moisture content:

$$r_{eff} = 2.1(1+\Delta)\ell_s, \tag{4.12}$$

where Δ is the relative increase in the maximum working depth as a function of probe size (Fig. 4.4). Table 4.2 gives data on the maximum working depth for a number of media.

The data indicate that the maximum working depth decreases rapidly with increasing moisture content and is not very sensitive to the probe size. There are also some other considerations on the assessment of the maximum working depth of the neutron method.

Van Bavel et al. (1956) estimated the maximum working depth in terms of the "sphere of influence", defined as the volume of the medium around the source of the radiation in which 99 % of all the resulting thermal neutrons are slowed down. The radius of the sphere of influence in pure water has been found by calculation to be 15 cm. It was assumed that for other media the sphere of influence should contain the same amount of water. On this basis, the radius of the sphere of influence for ground materials was found to be given by the empirical formula:

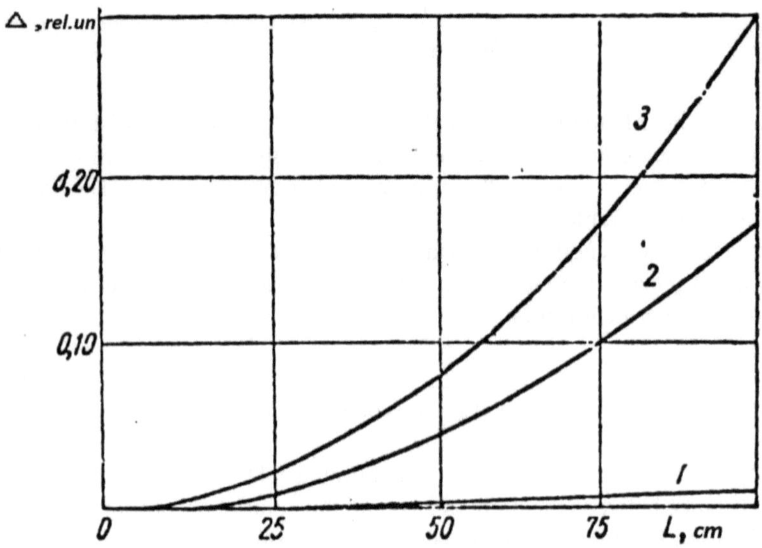

Fig. 4.4 Relative increase in the maximum working depth as a function of the probe length: *1* dry sand, *2* sand with 5 % moisture content, *3* sand with 10 % moisture content

Table 4.2 Maximum working depth in ground and rocks for neutrons with an initial energy of 3–5 MeV

Rocks	Density (g/cm^3)	Slowing down length (cm)	Effective radius for probe length (cm)		
			25	50	7
Sand	1.65	46.0	96.6	97.0	97.2
Sand with 5 % moisture	1.70	25.4	53.9	55.6	58.5
Sand with 10 % moisture	1.75	19.7	42.4	44.5	48.5
Sandstone	2.65	27.5	57.8	58.1	53.6
Limestone	2.72	25.0	52.6	52.8	53.4

$$r = 15\sqrt[3]{\frac{100}{w_v}}, \tag{4.13}$$

where w_v is the moisture content in volumetric %.

Olgaard (1965) showed later that the above procedure was inaccurate because it did not take into account the scattering and slowing-down properties of the ground skeleton. Moreover, not all the thermal neutrons from the sphere of influence reach the detector. He therefore proposed to estimate the maximum working depth by means of a "sphere of contribution", i.e., a sphere in which 95 % of the recorded neutrons were scattered and slowed down.

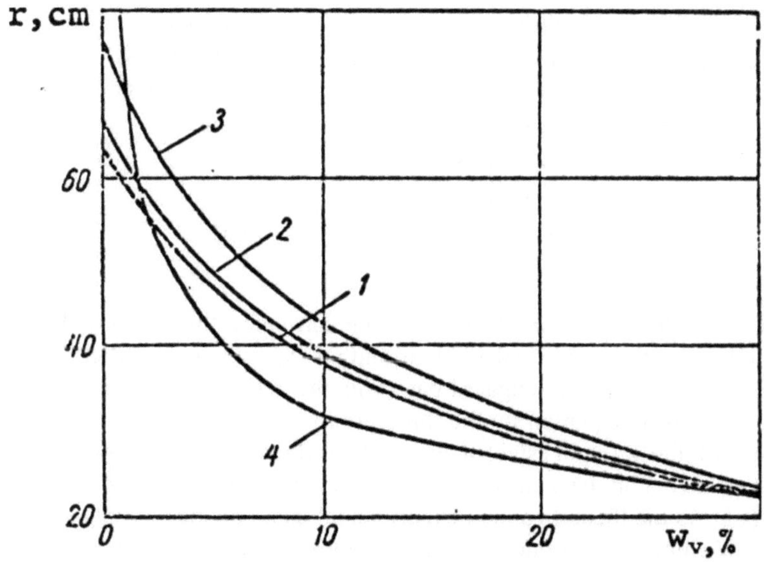

Fig. 4.5 Radius of the "sphere of contribution" as a function of moisture content, calculated for different densities of the ground skeleton, g/cm^3

For a probe in which the source of radiation lies at the centre of a proportional counter (such probes are manufactured by Nuclear Chicago), the radius of the sphere of contribution can be found from the empirical formula:

$$r = \frac{100}{1.4 + 0.1 w_v}, \tag{4.14}$$

where w_v is the moisture content in percent. This formula is valid for moisture up to 35 %.

Figure 4.5 shows the radius of the sphere of contribution as a function of the moisture content, as calculated by Olgaard (1965), for various densities of the ground skeleton ρ_{sk}.

4.2.3 *Effects of Parameters of the Medium*

It is well known that ground materials have a complicated chemical composition. However, variations in the chemical composition have little effect on neutron slowing-down properties. Table 4.3 gives values of the slowing-down cross-section σ_{eff} for different media. The quantities were estimated as the sum of macroscopic neutron scattering cross-sections of the chemical elements multiplied by the mean logarithmic neutron energy loss per collision.

Table 4.3 Slowing-down properties of various ground materials for a bulk density of 1.5 g/cm^3, in the absence of chemically bound water

Medium	σ_{eff} (barn)
Sands and sandy loams	0.0163–0.0169
Clays and loams	0.0165–0.0155
Ditto, 99 %+1 % water	0.0395
Ditto, 95 %+5 % water	0.1315
Ditto, 90 %+10 % water	0.2310
Water	1.54

Very detailed studies of the problem have been carried out by Olgaard (1965), who performed major theoretical calculations and an experimental verification on typical Danish ground material. Olgaard based his theoretical calculations on the solution of the diffusion equation by the three-group method. He considered five types of ground (Table 4.4) and established that the influence of the chemical composition of these media on the measured moisture content was small (Fig. 4.6).

Figure 4.7 shows the theoretical and experimental data obtained by Olgaard on the influence of chlorine, introduced artificially into a medium in the form of a

Table 4.4 Composition of Danish ground materials according to Olgaard's data, wt %

Chemical composition	Type of ground				
	Sandy loam in the region of Blangstedgård	Sandy heavy loam in the region of Risø	Sandy loam in the region of Aadum	Clayey sand in the region Borris	Sand in the region of Northern Yandivand
H	0.292	0.326	0.292	0.173	0.135
Li	~0	~0	~0	~0	~0
B	0.001	0.003	0.0005	0.001	0.0001
C	1.237	0.506	0.475	0.2	0.705
N	0.025	0.055	0.061	0.023	0.052
O	58.8155	53.4998	51.7952	54.651	54.9671
Na	0.65	0.74	0.91	0.5	0.48
Mg	1.27	1.38	0.62	0.27	0.038
Al	2.79	3.75	3.61	1.73	0.86
Si	32.21	34.28	35.21	39.7	40.97
P	0.066	0.058	0.065	0.048	0.051
S	0.0006	0.0012	0.0008	0.001	0.0004
Cl	0.0069	0.037	0.0075	0.007	0.0004
K	1.23	1.73	1.57	1.02	0.82
Ca	4.53	1.55	1.04	0.4	0.33
Ti	0.27	0.32	0.29	0.24	0.17
Mn	0.056	0.034	0.063	0.036	0.031
Fe	1.40	1.76	1.99	1.0	0.39

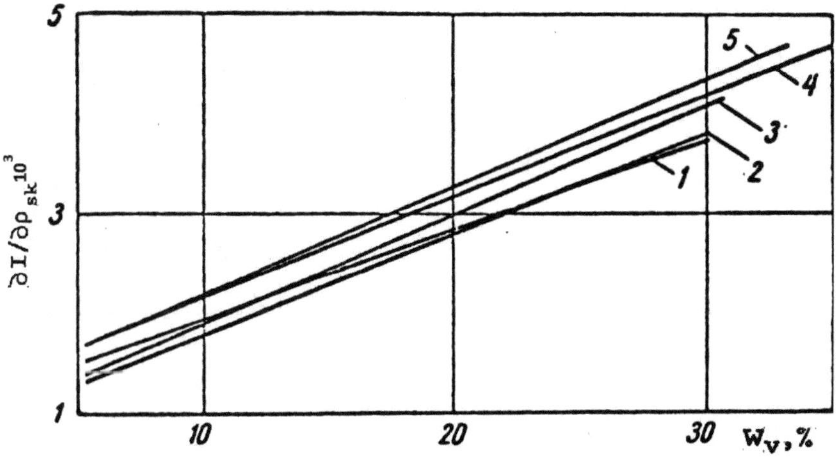

Fig. 4.6 Partial derivative of I with respect to ρ_{sk} as a function of w for various types of Danish ground materials: *1* heavy sandy loam in the region of Risø, *2* sandy clay in the region of Borris, *3* sand in the region of Northern Yandivand, *4* sandy loam in the region of Blangstedgård, *5* sandy loam in the region of Aadum

sodium chlorine solution. It is clear that chlorine has a very substantial effect. Thus, the introduction of only 1 % by weight of sodium chlorine (10 mg/l of water) is equivalent to a 3 % change in the volume moisture content.

The effect of hydrogen on moisture content measurements is very important in practice. Hydrogen is present in various organic compounds, which are particularly abundant in the top layer of soil and in turf covered rocks. Theoretical calculations carried out by Olgaard show that a 0.2 % weight change in concentration of hydrogen due to organic inclusions is equivalent to a 2 % change in the moisture content. However, the experimental data obtained by Churaev and Rode (1966) have not confirmed this conclusion. These workers used thermal neutrons to determine the flux density as a function of hydrogen concentration in various compounds (Fig. 4.8). The hydrogen in organic compounds and in chemically bound water has a scattering power lower by a factor of 1.6–1.7 than that of hydrogen in free water.

Hydrogen present in the hydroxide ions of clay minerals such as kaolinite, montmorillonite and hydrated mica in the form of crystal hydrates has a lower scattering power than hydrogen in free water. Hence it follows that hydrogen in chemical compounds can be regarded as free only when the neutron energy is greater than the energy of bounds holding the hydrogen atoms in the medium. If the neutron energy is less than the bound energy of the hydrogen atoms the collision of a neutron with such an atom cannot be regarded as a two-body problem. Hydrogen then behaves like a particle with a large atomic weight.

The above considerations lead to the following conclusions. In the absence of strong thermal-neutron absorption, variations in the chemical composition of the ground can be neglected without loss of accuracy in the measured moisture content. If the ground material contains elements that give rise to strong thermal-neutron

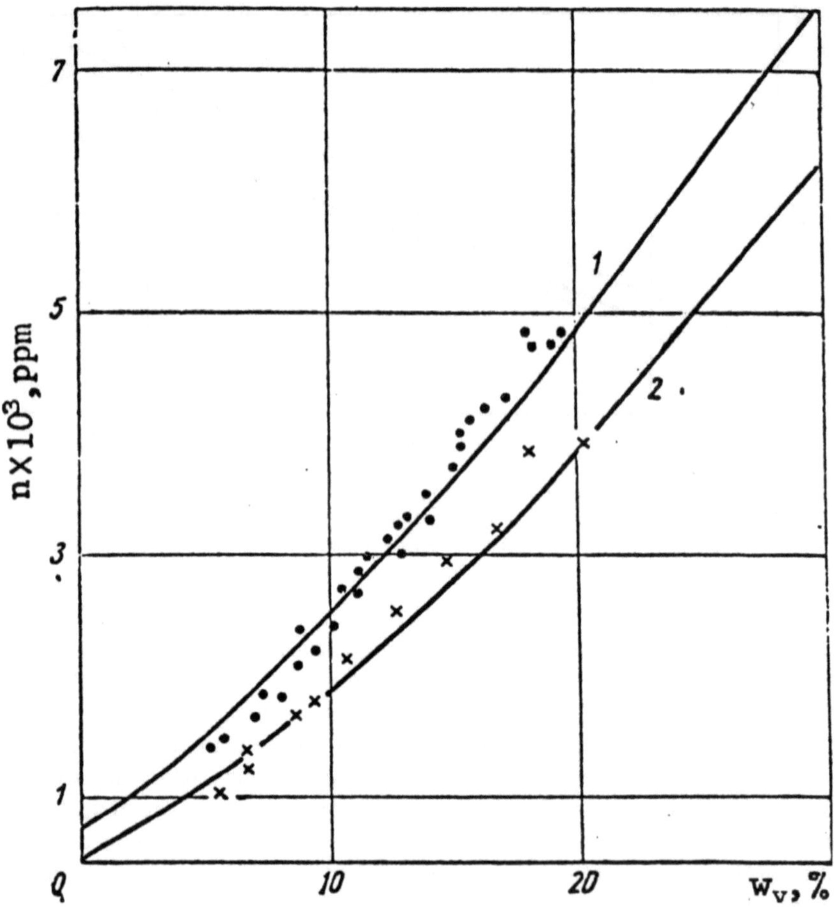

Fig. 4.7 Thermal-neutron counting rate as a function of the volume moisture content in sandy loam: *1* natural ground, *2* natural ground with 1 % by weight of NaCl; +++—experimental data; ---- theoretical data. Density of the skeleton equal to 1.4 g/cm³. (Ferronsky et al. 1968)

absorption, then thermal-neutron measurements of the moisture content cannot ensure the required accuracy. In such cases it is best to work with epithermal neutrons.

Olgaard also carried out an interesting analysis of the effect of ground density on the recorded scattered neutron flux. This problem had been considered earlier by many researchers but no detailed quantitative data were available in the literature. Figure 4.9 shows the thermal-neutron flux as a function of moisture content for different bulk densities of the skeleton of sandy clay. It is clear from the figure that for the most common range of the skeleton, density variation (1.4–1.8 g/cm³) in the error in the measured moisture content may reach 3 % of the absolute value. Density measurements must therefore accompany moisture-content determinations.

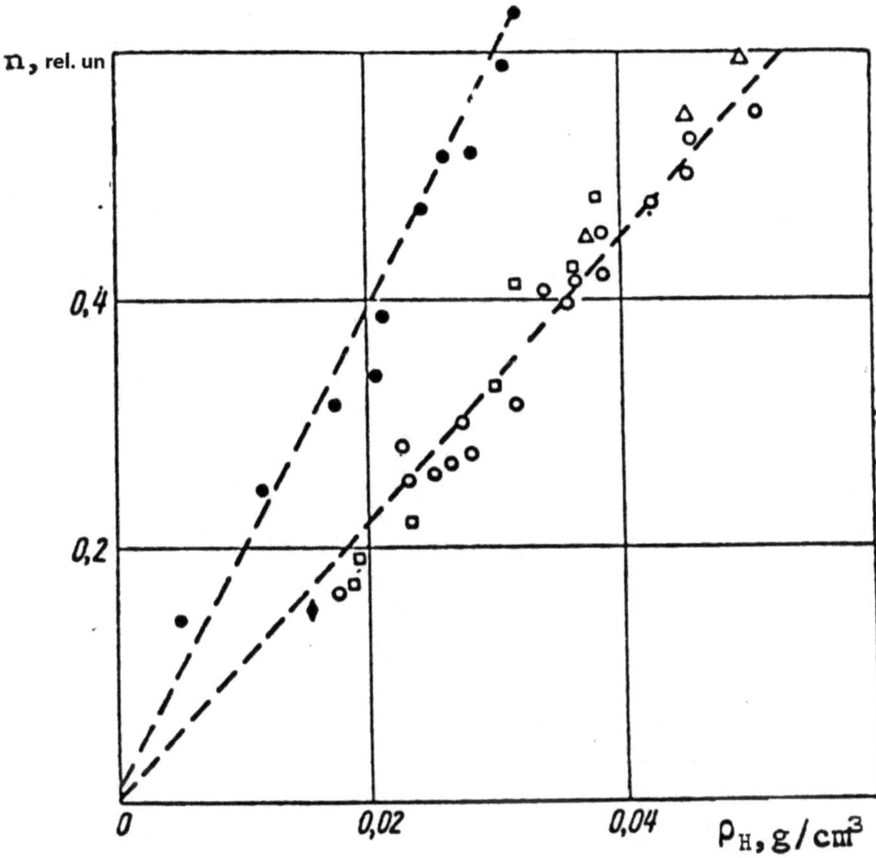

Fig. 4.8 Thermal-neutron counting rate as a function of hydrogen content in various compounds: □ – sucrose + sand; △ – alum + sand; ● – water + sand; ○ – turf; ▰ – sawdust

4.3 Design of Neutron Moisture Gauges

A neutron moisture gauge consists of a measuring probe and a radiometer connected by a cable. The radiometers for measuring the scattered neutron intensity may be scalars or rate meters recording the mean neutron flax over an interval of time with an output to a pointer indicator or a recorder.

The design of neutron moisture gauges differs mainly by the form of the probe. Depending on the problem in hand, they may be of the surface- or depth-type. Surface-type moisture gauges are used to determine the moisture content in the top layer of the soil and in exposed rocks and excavations. The depth types can be used in boreholes. There are also universal moisture gauges with probes that can measure the moisture content both on the surface and in boreholes. The essential element of neutron moisture gauge probes is the geometry of the mutual disposition of the source and detector. Three basic systems are used in existing designs. The first one

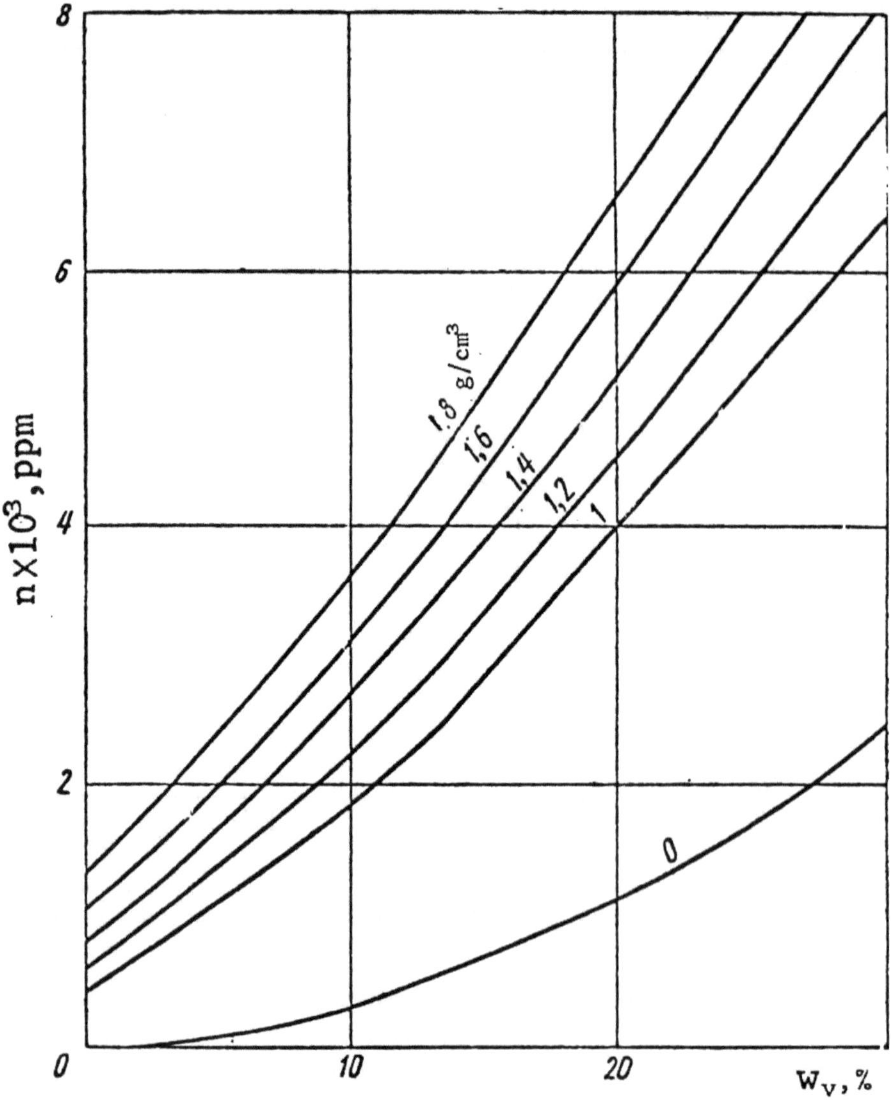

Fig. 4.9 Thermal-neutron counting rate as a function of volume moisture content for sandy loams and various densities of the ground skeleton. (Ferronsky et al. 1968)

is used in probes manufactured by nuclear-Chicago and by Nuclear Enterprises. The neutron source is located near the centre of a proportional gas-filled counter (boron trifluoride). This geometry ensures that the system operates on the linear segment (ab) of the curve shown in Fig. 4.1.

In the second the neutron source is located near the end of the detector. Probes of this geometry are manufactured by Friesich and Hopfner and by the Troksler Company. In the former design the detectors have a small diameter, while in the latter they have a large diameter. The use of a luminescent crystal as the detector in the so-called "zero" probe was used extensively by the authors to investigate the moisture content of friable deposits by pressing the probe directly into the deposit.

When arrangement two is used with a proportional counter, the working region of the $I_{nn} = f(w)$ curve corresponds to the segment ab. If the detector has considerable linear dimensions and the moisture content is high, the inversion region bc may be reached, which is very undesirable.

Arrangement three has the source and detector at a fixed distance from each other and is widely used in moisture-content measurements in large-diameter boreholes and in well logging. The working region for this system is cd (Fig. 4.1). The advantage of this arrangement is that it minimises the influence of interference due to the construction of the well itself. The usual practice in the case of considerable interference is therefore to increase the probe length to the maximum possible value. It is then necessary, however, to increase quite substantially the source activity, which is undesirable from the point of view of radiation hazard.

$^{210}Po + Be$, $^{226}Ra + Be$, and $^{239}Pu + Be$ fast-neutron sources are normally used in different countries. $^{227}Ac + Be$ and $^{241}Am + Be$ sources are used in some other countries. For short and "zero" probes the source activities are usually up to 10^6 fast neutrons/s, while for long probes the activity is up to 10^8 neutrons/s. An output of 10^4 neutron/s is commonly referred to as the specific output.

Proportional counters filled with $^{10}BF_3$ gas are widely used for neutron detection. Proportional counters filled with ^3He are also becoming of interest. The latter counters are more sensitive to thermal and especially epithermal neutrons than the boron trifluoride counters.

Gas-filled counters surrounded with cadmium foil are widely used. In this case the neutrons are detected by recording the capture gammas produced when thermal neutrons are captured by cadmium nuclei, which have a high capture cross-section. The particular advantage of neutron measurements using gamma-ray counters and cadmium filters is the simplicity of the associated electronics. However, neutron-detection efficiency in the presence of a high gamma-ray background is very low.

The most effective and convenient neutron detectors are the scintillation counters. The scintillation detectors of thermal neutrons used most commonly in Russia are type T-1 and T-2 ZnS(B). Te-activated LiI crystals are also used (LiI(Te)). The former have, however, the advantage that they do not respond to gamma radiation. This is very convenient for "zero" probes that do not incorporate screens that absorb gamma rays from the neutron source and from thermal-neutron capture in the ground.

Some of the designs of neutron moisture gauges in Russia were of the NIV-2 type. The NIV-2 gauge includes a probe for measurements directly in the ground or in boreholes, a portable rate meter and a calibration device. The probe incorporates a Pu + Be fast-neutron source producing 10^5 neutrons/s and four SBM-20 gamma-ray counters in a cadmium container. The counters are shielded from the neutron source by a lead screen. Slow neutrons produced in the hydrogen-containing medium around the probe and incident on the detector are captured by the cadmium. This is accompanied by the emission of capture gammas that are recorded indirectly through the capture gamma-rays. The advantage of this method of detection of thermal neutrons is its simplicity. However, it suffers from the following disadvantage: it has very low gamma-ray efficiency and the results are sensitive to the ground density.

To determine the moisture content by the neutron moisture gauge, a hole is drilled in the ground and is cased with a thin-walled aluminium tube 50 mm in diameter.

4.4 Possible Errors in the Moisture Content Measured by the Neutron Method

The possible sources of measurement errors can be divided into three groups. The first group consists in errors due to the inhomogeneities in the physical properties and in the composition of the ground material. The second group includes errors due to the measurement technology, while the third group consists in errors connected with the recording equipment and the radioactive decay of the source.

The factors in the first group include variability in the density of the skeleton and the chemical composition of the medium, the presence of elements with high thermal-neutron capture cross-sections and the presence of chemically bound water. These factors were analysed in the preceding sections and require no further discussion.

Technological errors are connected with disturbances of the groundwater in the borehole drilling and, especially, the presence of cavities, casing pipes and differences in the latter's wall thickness, the presence of groundwater in the borehole, the presence of a gap between the probe and the inner surface of the casing pipe and so on. The errors due to technological factors can be reduced to a minimum of less than 1 % by careful construction of the borehole, special cleaning-up routines for surface measurements and care during the measurements themselves. However, these errors may be considerable if such precautions are not observed. Any differences between the probe diameter and the diameter of the borehole have a particularly large effect on the accuracy of the measurements.

In large boreholes (diameter in excess of 100 mm) there may be considerable cavities and the difference between the borehole diameter and the diameter of the probe is so large that the probe must be held against the borehole wall. It is then practically impossible to determine the moisture content quantitatively.

Errors connected with the counting equipment are governed by statistical fluctuations in the radioactive decay of the neutron source and instabilities of the equipment operation. Experience has shown that statistical fluctuations can be neglected if they are less than 1 %, which, in the case of discrete counting devices, corresponds to 10,000 count/min or more.

The simplest method of checking the stability of the equipment is to verify its reproducibility by using a reference body, which, in our case, can be the shielding container used for transporting the probe (as in the NIV-2 instrument).

Extensive measurements have shown that, when the above factors are properly taken into account, the moisture content can easily be measured to within

± 0.01 g/cm^3 of water, which corresponds to 0.5–1.0% of the absolute moisture content.

4.5 Calibration of Neutron Moisture Gauges

Before a neutron moisture gauge can be used, we must know the function $I_{nn} = f(w)$, which must be established separately for each instrument. This constitutes its calibration. The calibration can be carried out either experimentally or by a combination of calculations and experiments.

The calibration curve can be obtained experimentally, both under laboratory and field conditions, by measuring the intensity of the scattered thermal neutrons for a known moisture content, density of the ground skeleton and chemical composition of the ground.

When the calibration is carried out under laboratory conditions, the model used for the measurements must have a radius in plan that is not less than twice the maximum working depth of the device. It is usually very difficult to prepare such a model with a given uniform moisture content distribution throughout the volume. It is therefore often more convenient to calibrate the device under field conditions corresponding as closely as possible to those in the ground under investigation. Special care must be taken in the case of field calibration because the moisture distribution in the ground is often highly inhomogeneous. Only well tested strata can therefore be used for this purpose.

In the second method, the calibration is based on theoretical formulae and known constants characterising the scattering properties of the ground and rocks and on experimental data on the density of the ground skeleton and the chemical composition of the medium.

References

Churaev NV, Rode LG (1966) Measurement of moisture in peat soil by neutron method. Pochvovedenie 1:96–100

Ferronsky VI, Danilin AI, Dubinchuk VT et al (1968) Radioactive investigative methods in engineering geology and hydrogeology. Atomizdat, Moscow

Filippov EM (1962) Applied nuclear geophysics. USSR Academy of Sciences Public House, Moscow

International Atomic Energy Agency (1970) Neutron moisture gauges. Technical report series N 112, IAEA, Vienna

Kantor SA (1955) Fundamentals of the theory of neutron logging. In: Applied geophysics, vol 3. Gostoptekhizdat, Moscow, pp 3–22

Kantor SA (1958) On the depth study of rocks by the neutron-neutron logging. In: Applied geophysics, vol 13. Gostoptekhizdat, Moscow, pp 111–133

Olgaard PL (1965) On the theory of the neutron method for measuring the water content in soil. Riso Report N 97, Copenhagen

Van Bavel CH, Underwood N, Swanson RW (1956) Soil moisture measurement by neutron moderation. Soil Sci 82:17

Part II
Penetration Logging Techniques

Chapter 5
Penetration Logging Methods and Equipment

Abstract In this chapter a group of field methods for determining the physical and mechanical properties of grounds in situ for geoengineering practice is presented. The solution of a problem is achieved on the basis of combining the soil static penetration test and nuclear logging methods. The first experimental study on the application of the nuclear penetration logging method was started in 1962. The group of methods, which were tested and included in the penetration logging rig, are as follows: natural gamma logging (GL), gamma-gamma logging (GGL), neutron-neutron logging (NNL), static penetration test of the ground with recording of cone resistance and the lateral jacket friction test by electro-sensing strain gauges. These were the objectives: for lithological stratification of the geological section under investigation; for determination of soil density; for determination of soil moisture and identification of the groundwater table; for obtaining the mechanical characteristics (modulus of compressibility and coefficient of the ground friction along the penetration section). The penetration logging rigs and equipment SUGP-10, SPK, SPKT and submerged penetration logging installation PSPK-69 are described.

Geoengineering studies always involve an investigation of the properties and composition of grounds down to a certain depth. A very serious difficulty encountered in this connection is that the lithologic formations found under natural conditions have heterogeneous properties even within the limits of a given genetic type of grounds. So that when various soil parameters are determined from samples the results are often fortuitous and are not really characteristic of the grounds as a whole. It is therefore necessary to take a large number of samples and subject the data obtained from them to a statistical analysis. But in practice the number of samples that can be taken from holes and wells is limited by the amount of effort and the cost of labour. In many cases, when grounds are investigated below the groundwater level and also in the space of water areas, it is difficult to take samples from an undisturbed structure because of the absence of suitable technical means and in media such as sand, marine ooze or clay deposits of fluid consistency it is practically impossible to collect samples. Drainage of ground under planned civil engineering projects is very expensive. An example of this is the Hydropower Station on the Volga River in Russia, where the presence of water-saturated sands under the proposed dam necessitated freezing the sand to a depth of 30 m (Durante et al. 1957).

V. I. Ferronsky, *Nuclear Geophysics*, Springer Geophysics,
DOI 10.1007/978-3-319-12451-3_5

Sampling of the frozen sands was taken in the course of the excavation. The volume of these samples was determined with an Archimedes balance at zero temperature and their weight was measured at temperatures below zero. The porosity and the degree of water saturation were thus found as functions of depth.

The properties and composition of rocks can only be investigated by methods capable to provide estimates of these properties under natural conditions of deposition on the whole depth of the section. The methods of nuclear logging, which are widely used in nuclear geophysics for well studies, are found to satisfy these conditions.

It is well known that up to 60 % of geological and hydrogeological studies for engineering purposes (civil, road, hydraulic and maritime engineering, environmental, agriculture irrigation and so on) are carried out by field methods. More than 70 % of the field work is executed by traditional penetration testing. The tendency for further development and application of field methods for their practical engineering use is explained by the demand of accuracy in quantitative initial physical and mechanical soil data for building civil and industrial constructions on land and in coastal areas.

Efficiency of the field methods was achieved by combining the static penetration ground test and nuclear logging methods. In this case we continuously record physical and mechanical properties through a geological section. The first experimental work on application of the nuclear penetration logging methods was carried out in 1962, to the depth of 25 m, with the SUGP-10 hydraulic equipment mounted on a motorised caterpillar rig of 11 t. This was a really important step in the development of field methods for geoengineering purposes. Further works in the development of equipment and electronics for penetration logging techniques appear to be successful and are described in this book.

5.1 Essence of Penetration Logging Techniques and Conditions of Application

The group of methods, which were tested and included in the penetration logging rig (PLR), are as follows:

1. Natural gamma logging (GL) for lithological stratification of the geological section under investigation;
2. Gamma-gamma logging (GGL) for determination of soil density;
3. Neutron-neutron logging (NNL) for determination of soil moisture and identification of the groundwater table;
4. Static penetration test of the ground with recording of the cone resistance and lateral jacket friction by electro-sensing strain gauges.

All the above methods satisfy the common demands: (a) to provide quantitative information about the studied properties with an acceptable-for-practice accuracy; (b) to be able to superpose emplacement of the probe gauge into the ground and transmit the obtained information; (c) to resolve in continuous recording the obtained information; (d) to apply simultaneously a number or all the methods in one probe.

To the contrary of well logging, in PLR the logging probe is moved continuously on the rod along a geologic section through the studied ground. The parameters under study are recorded during this process.

GL records the natural gamma background of the rocks, which is due to radioactive isotopes such as ^{235}U, ^{238}U, ^{232}Th, ^{40}K and their decay products. These elements emit alpha, beta and gamma radiation. The physical characteristics and the observed distribution of radioactive elements in the Earth's crust form the basis for application of the natural-gamma-logging method for separation of different strata and determining the clay content in different ground types in a section.

GGL is one of the most developed and widely used methods of the nuclear logging technique. In nuclear geophysics it is also referred to as density logging. In principle, GGL involves the recording of rock-scattered radiation from a gamma source, which is carried by a logging probe as it moves along the well. It was shown in Chap. 3, where various physical principles are reviewed, that the scattered gamma radiation is related functionally to the density of the surrounding rocks. So, GGL yields the rock density as a function of depth in the well under investigation.

The intensity of the recorded scattered by medium radiation depends on the length of the measuring probe, the energy of the radiation emitted by the source, the degree of its collimation, the sensitivity of the detector at the energy of the scattered radiation and the angle of incidence of the radiation on the detector. The effect of these parameters has already been analysed in Chap. 3. In addition, well logging is subject to the following specific factors, which have an important and often decisive effect on measurements: the thickness of well casing and the material of the logging probe, diameter of the well, the presence of well casing and of drilling mud, the presence of cavities outside the pipe and the effect of cementation. Since the depth of which studies can be carried out with the aid of scattered gamma radiation is restricted, with the gamma penetration amounting to not much more than 10 cm, these factors give rise to interfering effects around the logging probe. On the other hand, the artificial medium (drilling mud, casing pipes, cementation, etc.) surrounding the probe scatters a substantial proportion of the recorded radiation and introduces considerable errors into the measurements results.

To reduce the effect of the drilling mud and the variable diameters of the well during GGL, specially mounted probes and collimated radiation are used. Another method is to employ amplitude discrimination, so that the most penetrating component can be selected from the scattered-radiation spectrum (Gorsky 1958).

It is considered in practice that, when GGL data are interpreted, the presence of cavities is the most important interfering factor. These cavities produce a positive signal on the logs, which is equivalent to a reduced density in the stratum under investigation. The measurement results are then usually corrected by additional cavitymetry data, or data obtained by some other geophysical logging method.

The NNL method records essentially the scattered neutron radiation originally emitted by a fast-neutron source located in the logging probe and displaced along the well. The method yields a diagram for the moisture content of rocks as a function of depth in the well under investigation.

The physical basis of the interaction of neutrons with various media, the resolution of the neutron method of determining the moisture content (both as far as the sensitivity and maximum working penetration are concerned) and estimates of the optimum parameters of measuring probes were discussed in Chap. 4. The factors affecting the measured parameters by NNL are the construction of the well, the presence of drilling mud, cementation of the region outside the pipe and so on. These factors depend on the technology of well sinking and well construction and, as in the other methods of nuclear logging, they introduce serious difficulties that reduce the accuracy and reliability of the final results.

To reduce the influence of the well construction on neutron-neutron logging it is common to employ the post-inversion region of the dependence of the neutron flax density on the distance from the radiation source. This is done in practice by using logging probes with lengths in excess of 50 cm.

The inelastic interactions between fast neutrons emitted by a neutron source and the capture of the slowed-down neutrons by nuclei of the rocks material are usually employed in neutron-gamma logging. These interactions result in capture gamma radiation, which is recorded by the gamma detector as the logging probe is displaced along the well.

Approximate calculations of the capture gamma radiation in a homogeneous medium from a point source of fast neutrons show that this radiation depends on the slowing down properties of the medium in question, the number of gamma quanta per neutron capture and the density of the medium. Barsukov (1958) showed that the intensity of the capture gamma radiation as a function of the logging-probe length is given by the empirical relationship:

$$I_{n\gamma} = Ae^{-\delta L}, \tag{5.1}$$

where A is a constant that depends on the intensity of the source and the chemical composition of the rock, L is the length of the logging probe and δ is the slope of the curve $\ln I_n\gamma = f(L)$.

The presence of chlorine in the medium under investigation enhances the yield of the capture gamma radiation (about 2.4 gamma quanta per neutron capture). In prospecting geophysics this property is used to determine the position of the water-oil contact, where the water-containing collector is characterised by an enhanced concentration of chlorides and hence an increased intensity of capture gamma radiation.

The dependence of the capture gamma radiation record on the probe length has an inversion region. The point of inversion of this curve for wells filled with water corresponds to probe lengths of 30–40 cm (for sand with 25% moisture). As the well diameter increases, the inversion point is displaced up to 35 cm and the point of inversion corresponds to a probe length of 30 cm.

According to Zolotov (1958), the maximum penetration of the method decreases with increasing moisture content of the medium. Thus, a change of the moisture content from 10 to 25% leads to a 12–15% reduction of radius in the 'sphere of contribution'.

From a practical point of view, the sensitivity of NLG to the moisture parameter is of practical interest. In order to increase the sensitivity of the method, the neutron detector is surrounded by cadmium foil (cross-section $\sigma_{Cd}=2550$ barn). To achieve better differentiation of the rocks by their moisture contents, the counters are also surrounded by a 0.8–1.0 cm lead screen. In this case the soft capture radiation of moist rocks is cut off by the lead filter and the differentiation factor is increased by 30–50 %.

A semi-logarithmic plot of the recorded capture gamma intensity against the moisture content, in the form Inn = f (log *w*), results in two straight lines of different slope. The first line (with the smaller slope) corresponds to moisture contents up to 10 % and the second to moisture contents in excess of 10 %. The accuracy with which the moisture content can be determined in the latter case is about 10 %. Since the capture gamma rays in rocks have high energies (the order of a few MeV), the counters used in NGL usually have a higher sensitivity in the hard region of the spectrum (MS-type counters). The linear dimensions of the counters must be such as to have no disturbing effect on the litholodic differentiation of the profile, for which purpose NGL is used as well.

In addition to the determination of the moisture content of rocks, studies of capture gamma radiation in NGL have shown that this method can be used to estimate quantitatively the amounts of oxygen, carbon and a number of other rock-forming elements. When the gamma-ray spectrum is investigated at energies in excess of 4.5 MeV i.e., when gamma rays due to inelastic scattering of fast neutrons are eliminated, it is possible to determine also the amounts of elements such as Na, Ca, Al and Fe.

Nuclear logging methods originated in prospecting geophysics for investigation of rock sections in deep wells. The development and improvement of these methods has been the subject of many theoretical and experimental studies carried out to estimate the resolution of the methods and to develop ways to interpret logging diagrams. Analysis of the study results has shown that the conditions under which radioactivity logging has been used in wells were far from being optimal. There are many seriously interfering factors connected with the technology of drilling and well construction, which have already been mentioned.

In geoengineering studies the depth of exploration wells is, as a rule, restricted to 25–30 m, whereas a depth of 12–15 m is sufficient for ground exploration for major constructional works (apartment blocks, public and industrial buildings). The well diameter is here sharply reduced, there is frequently no casing, the volume of cavities is smaller and the drilling is carried out without using drilling mud. On the whole, the conditions under which nuclear logging methods can be used are more favourable in the case of geoengineering studies. However, in contrast to studies connected with mineral prospecting, geoengineering work is usually the subject of more stringent requirements, in the sense that the data on the properties and the section of the deposits must be as quantitatively interpreted as possible.

Our research and practical application of the penetration logging techniques started in 1962 in VSEGINGEO (Russia). It involves the use of nuclear logging combined with static penetration, i.e., driving of the logging probe into the ground by a special hydraulic device (Ferronsky 1969). The principle of this is as follows. The logging probe, which is screwed onto the working rod, is forced into the ground to the required depth by special hydraulic installation. The logging diagrams are recorded during this process. The technological problems encountered in connection with the driving of the logging probe into loose deposits were solved relatively simply by analogy with static penetration, which is widely used in engineering geology. Since the diameter of the logging probe is not more than 50–60 mm and the specific penetration resistance of the ground is usually less than 100–250 kg/cm^2, a driving installation developing an axial force of 10 t will usually succeed in driving the logging probe screwed onto the rod to the required depth of 20–25 m.

The method of nuclear logging based on ground penetration does not require preliminary drilling of holes, which means that it is cheaper, self-contained and independent during field investigations. Moreover, it has important advantages from a methodological standpoint. Static driving of the logging probe into the medium under investigation ensures that all the physical possibilities of the nuclear logging methods are utilised, because this eliminates the interference connected with the construction of wells.

Extensive studies have shown that these penetration logging methods hold considerable promise in geoengineering. They can be used in conjunction with the static sounding methods that result in estimates of the mechanical properties of the ground and also in connection with the use of a special ground sampler that is driven under pressure. In this case it is realistic to consider the investigation of the physical and mechanical properties of soils and sections of loose deposits under natural conditions without drilling wells.

The author carried out the following experimental works under field conditions: systematic studies of the parameters of logging probes with respect to the sensitivity and accuracy of nuclear logging by static insertion of the probe and to the effect of various factors on measurement results. The choice of the recording equipment has been reviewed and the conditions and design of apparatus for the static driving of the rod into the ground have been determined. These studies included GL, GGL and NNL, using thermal and epithermal neutrons. The specific penetration resistance and friction of the ground were new mechanical parameters of the ground, which satisfied the nuclear logging methods in the developed techniques.

The experimental field work was carried out under various geological conditions in Moscow, Yaroslavl, Volgograd, Kherson, Krasnodar and the Black Sea (near Novorossysk), as well as in Uzbekistan and was accompanied by the necessary drilling and laboratory works. This enabled us to determine the geoengineering problems that can be solved by penetration logging techniques (PLT) and also to establish rational applications of these methods in conjunction with other geophysical and penetration techniques, which extend the range of application of these procedures in engineering geology.

It should be noted that PLT can be used to investigate deposits at the bottom of reservoirs, which is of great practical importance for the development of new methods for determining the properties and the sections of loose bottom deposits within a water area. A special rig was developed for this purpose, which was used to carry our penetration-logging work at the bottom of reservoirs.

5.2 Experimental Penetration Logging Rig SUGP-10

In the penetration logging method the probe is forced into the ground without prior borehole construction. It has been developed primarily for relatively shallow investigations of friable unconsolidated deposits. The presence of highly consolidated strata, cobbles or pebbles and cemented zones may prevent penetration. Accordingly, it is most readily applicable to engineering geological problems and to hydrogeological investigations of aquicludes, aquitards and shallow sand aquifers and is seldom suitable for the study of poor gravel deposits. Penetration logging methods can be used in both saturated and unsaturated zones.

Figure 5.1 demonstrates a general view of the penetration logging technology for study of friable deposits. The measuring apparatus consists in the following elements: (a) demountable logging probe that is screwed to the end of the first drilling rod; (b) a communication cable that passes through the hollow rod and is used to transfer signals from the probe to the recording apparatus on the surface; (c) ground-based recording equipment that receives signals from the probe and records them in the form of continuous diagrams. Static driving of the logging probe with the column of rods into the ground is provided by a hydraulic device, which develops the force up to 10 t. The device includes a special hydraulic cylinder with a two-way action and an automatic reversing system. This design automates the process of insertion and removal of the rod column with the logging probe and allows the speed of insertion to be adjusted between 0.1 and 4.0 m/min.

The hydraulic cylinder has a straight-through hole with a hydraulic griping chuck for driving and withdrawal of the column of rods. Oil is supplied to the hydraulic system by a pump that is operated by a motor through a distributing box. The in-

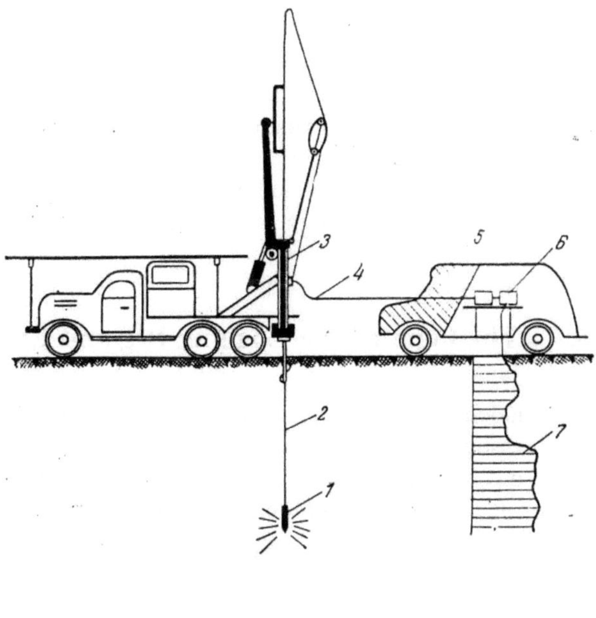

Fig. 5.1 Block diagram of the rig and apparatus station: *1* measuring probe; *2* column of drilling rods; *3* hydraulic device for embedding the rod column; *4* cabin for apparatus equipment; *5* auto base; *6* logging diagram; *7* rod manipulator. (Ferronsky and Gryaznov 1979)

vestigation develops an axial thrust of up to 10 t on the rod column. In this way the probe is driven into the ground to a depth of 25 m.

In the experimental field work, we used the motorised caterpillar rig SUGP-10 developed by the Special Design Bureau of the former USSR Ministry of Geology. The rig, which weighed 11 t, included a hydraulic device for driving smooth drilling rods into the ground with the aid of a special hydraulic device.

The rods themselves (three rods each 8 m in length) are kept in a special cartridge that in its working position is held by a supporting post. The axial load developed while the rod is being driven is balanced by the weight on the machine, which is 11 t.

The logging probes used in this work were of the standard borehole type (NK-60 L Gamma-59 and KU-59). The electronic circuitry for the standard probes was placed in steel pipes of 60 mm in diameter and 4 mm thick. This design of the logging probes was found to withstand considerable mechanical loads.

Probe data were transmitted to the surface apparatus along a coaxial single-core cable and also along a three-core cable. The cable is connected with the probe and is joined by the build-up of rods by plug-in joints.

The surface apparatus was placed in a truck that served as the laboratory. The truck also carried a photographic laboratory for developing the oscillograms. The recording apparatus comprised a logging scintillation radiometer RK-60 L a radiometer incorporating Gamma-59 gas-discharge counters and a KU-59 scintillation radiometer.

Gamma radiation was detected both by gas-discharge counters and by scintillation detectors [NaI(Tl) phosphorus 30×40 mm in size]. Thermal neutrons were detected by scintillation detector types T-1 and T-2 (zinc sulfiole crystals with boric acid). The NLS diagrams were recorded by N-370 records and the EPO-5 electric-prospecting oscillograph.

The gamma-rays were produced by a ^{60}Co source of 8 mg-eqt of radium and a ^{137}Cs source of 15 mg-eqt of radium. The fast-neutron sources were Po-Be and Pu-Be preparations with activities of 0.2–0.3 curie alpha radiation.

In addition to the equipment mentioned above, we used the SPKL apparatus specially developed following our suggestions by the Kiev Design Bureau of Geophysical Instrumentation for a composite penetration-logging station.

The results of the field studies of the application of the above methods for the solution of various geoengineering problems are summarised below.

5.3 The Penetration Logging Rig and Equipment SPK

This station was designed for commercial production to provide geoengineering soil investigations of different purposes. It was the first highly productive machine based on combining the static penetration test of ground and nuclear logging techniques. It is able to obtain lithologic structure, physical and mechanical soil properties by field methods to a depth of 25 m without boreholes and laboratory tests.

The station applies nuclear logging (GL, GGL, NNL) and static penetration with measurement of cone resistance and lateral jacket friction of ground to the logging probe. The above set of methods is able to solve the following tasks:

1. Obtain a diagram of the volumetric soil density along a section by GGL;
2. Obtain a diagram of the volumetric soil moisture along a section by NNL;
3. Obtain a diagram of the ground cone resistance to the probe emplacement along the geological section by the electro-sensing strain gauge for estimation of deformation and rheology properties;
4. Obtain a diagram of the ground friction with lateral jacket surface by the electro-sensing strain gauge for estimation of the steady state property;
5. Make lithological stratification of the sand-argillaceous geologic section under investigation by gamma-logging and penetration test diagrams;
6. Identify the groundwater table position by NNL and GGL data;
7. Estimate the chlorine content in saliferous soils by NNL by recording the thermal and epithermal neutrons;
8. Estimate the degree of the ground uniformity in width and depth with respect to properties;
9. Take undisturbed soil samples from different depths of a section by a special sampler.

Taking into account the above possibilities, the rig SPK was intended to be used for solution of the following geoengineering problems: (a) Engineering geological mapping of different scales; (b) Special combined engineering geological and hydrogeological prospecting for irrigation purposes mapping; (c) Geoengineering prospecting for civil engineering planning and construction of auto and railroads, airfields, hydropower stations and so on.

The rig SPK includes a separate block of the hydraulic system for emplacement of the column of rods with the probe. The block was arranged on the high passing track ZIL–157KG. It includes the following main units.

1. Hydraulic system provides an automatic process of emplacement and withdrawal of the rod column with the probe by a constant velocity up to 6 m/min and ensures other operations including screwing and removal of the anchors;
2. Emplacement mechanism that includes two hydro-cylinders and three-shoe hydro-mechanical gripping chucks;
3. Anchor device, which includes two anchor jacks of original construction, are found in the plane of the hydro-cylinders;
4. A device for raising and laying the rods with mounted communication cable during screwing and unscrewing the column;
5. Mechanised turning protective containers, mounted on the rig frame, in which the two probe heads with radioactive sources are stored;
6. Mechanism of a synchronised drive of the recording tapes. During the rod emplacement the recorder's tapes in the laboratory provide a synchronous record of the measured information in the appropriate scale;
7. The hydraulic pump drive is provided through the gear-box of the car engine;
8. Communication device is used for a two-sided connection between the rig and laboratory;

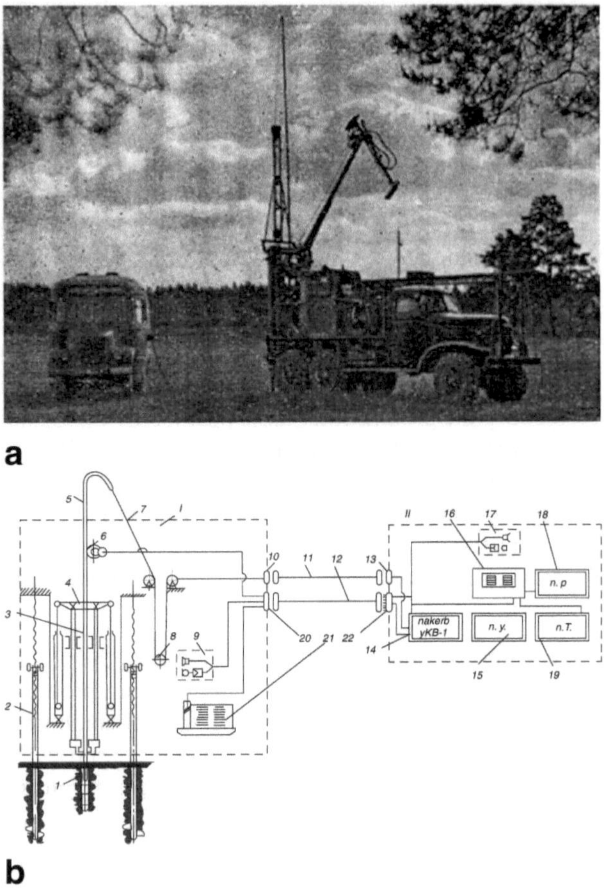

Fig. 5.2 General view (**a**) and schematic drawing of the assembled hydraulic system and apparatus laboratory (**b**): *1* measuring probe; *2* anchor jack; *3* rod column; *4* emplacement mechanism; *5* tip cable; *6* clamp roller; *7* tip cable; *8* pulley block; *9* communication device; *10* joints; *11* connection cable; *12* power supply; *13* control desk; *14* recording block; *15* radiometry desk; *16* electro-sensing strain desk; *17* gasoline engine unit for power supply. (Ferronsky and Gryaznov 1979)

9. Gasoline engine AB-1-0/230 is used for the power supply of all the apparatus and other purposes.

The laboratory for the recording apparatus and initial processing of the obtained information was in the bus GAZ-66. A schematic drawing of the assembled hydraulic system on the track and the laboratory is shown on Fig. 5.2.

The laboratory included the following equipment:

1. Two combined penetration–logging probes, the first of which is intended for measuring the ground cone resistance to the probe emplacement and ground friction with lateral jacket surface and GGL recorded during the probe withdrawal. The second probe is intended for a record of the NNL and GL diagrams;
2. Apparatus desk with recorders for measuring information, power supply and instruments;
3. Everyday equipment for life and rest.

The frontal ground resistance and lateral ground friction gauges are located in the lower part of the first probe where both the cone and lateral strains are originating.

The gauges' cylindrical elastic (steel) elements with glue on the electric tensometers are made from constantan wire. The first elastic element is bound with the probe tip and the second one with the lateral part of the probe. The power supply of the tensometers is provided from a common laboratory desk.

The gamma radiation source of ^{137}Cs of 4 mg-eqt Ra is placed in a removable tip of the probe. The detector, which is a NaI (Tl) crystal with a photoelectric multiplier, is located 40 cm above. The fast neutron Po-Be source of $5 \cdot 10^5$ n/s is placed also in the removable tip of the probe. And about 5 cm above the detector of thermal neutrons the LDNM-2 with a photoelectric multiplier is located. The natural gamma radiation is detected by a (Tl) crystal with a photoelectric multiplier FEU-35, which is placed 1 m above the source.

Further development of SPK led to substantial changes in its mechanical and apparatus design as the SPKT rig. In particular, all the penetration logging equipment and measuring apparatus were mounted on the one track ZIL–157KG where a second cabin for measuring apparatus was designed. In addition, a common one probe with five gauges for the penetration test and nuclear logging was combined. As a result, the multichannel apparatus MAK-5P included five initial transformers and a ground desk with a receiving device and registers. A specific feature of the apparatus was that the signals transformed to the ground desk were transmitted through a radio-channel. The electric impulse from the probe transformer modulates the carrier frequency. The column of the rods was used in the capacity of an aerial. And the aerial of the receiving device was an electrode grounded in 5–10 m from the penetration point. The signal is amplified by the receiving device and moved to the decoder where it is modulated and registered.

The rig SPKT was designed by the Special Design Bureau of the former USSR Ministry of Geology in cooperation with the Institute of Hydrogeology and Engineering Geology (VSEGINGEO).

5.4 The Submerged Penetration Logging Rig PSPK-69 Mounted on the Exploration Catamaran Type Ship "Geologist-1"

Engineering-geological studies on near-shore areas of seas, reservoirs and rivers for different types of hydro-construction design are extremely difficult and expensive. The problems are technological and methodological. This is because drilling works are complicated and the sampling of non-disturbing samples is practically impossible mainly because of saturated sands and silts.

It is only field methods of determination of the grounds' physical and mechanical properties and bottom sediment stratification that provide the best solution to the above problem. The penetration logging techniques and apparatus discussed here are the most developed for that purpose.

A special design and construction of such techniques for practice was done in VSEGINGEO (USSR) in 1963–1970 (Ferronsky 1969; Ferronsky and Gryaznov 1979). Because the apparatus was intended to be submerged on the bottom of a

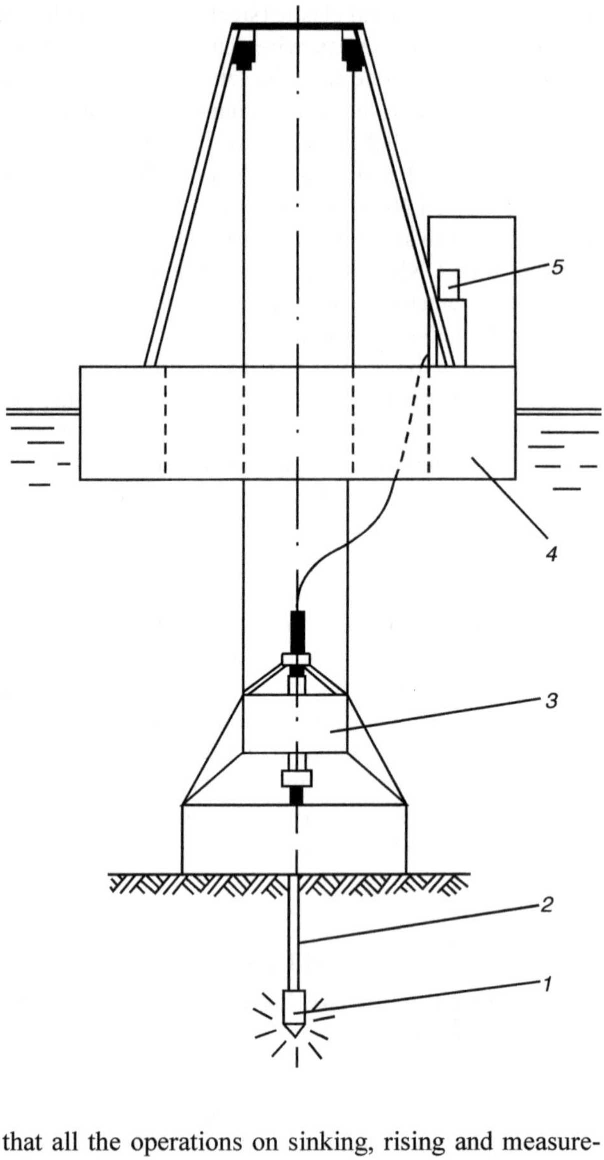

Fig. 5.3 Schematic structure of the experimental submerging installation PSPK: *1* submerging installation; *2* measuring probe; *3* column of penetration rods; *4* operation desk; *5* connection cable; *6* pontoons with equipment. (Ferronsky and Gryaznov 1979)

reservoir it was necessary that all the operations on sinking, rising and measurements be carried out automatically.

The first experimental submerging installation PSPK was mounted onto two pontoons on the Black Sea (Novorossiysk harbor). It included the submerging installation, a set of measuring apparatus and four floating boat-pontoons (Fig. 5.3).

The submerging installation included a hermetic container of machinery oil where aggregates of the hydraulic system, an electric engine with an oil pump and working equipment were mounted. The hydraulic cylinder with a hollow core and gripping chuck for emplacement and rising up of the rod column and the electric

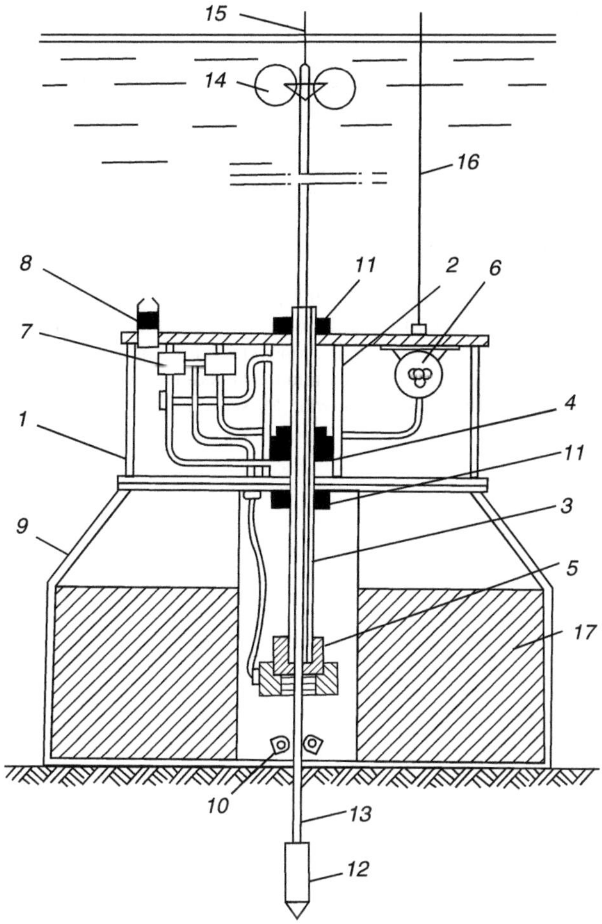

Fig. 5.4 Submerging installation of rig PSPK: *1* hermetic container; *2* hydraulic cylinder; *3* hollow core; *4* piston; *5* gripping chuck; *6* electric engine; *7* hydraulic system; *8* compensator; *9* ballast capacity; *10* non-contact switch over; *12* measuring probe; *13* rod column; *14* a float; *15* apparatus cable; *16* power cable; *17* ballast (10 t). (Ferronsky and Gryaznov 1979)

engine were mounted on the container lids. Figure 5.4 shows a schematic structure of the submerging container.

The measuring apparatus includes a combined measuring probe screwed onto the rod column and a registration desk placed on the floating pontoons.

The measuring probe includes GL, GGL, ground frontal resistance and friction gauges.

The submerging installation PSPK was used for experimental methodical works. This was the first experience of the application of penetration logging techniques for geoenginering purposes at sea. Later on in 1965–1967 the first catamaran type ship for marine geoengineering prospecting was designed and produced. The ship was equipped with the submerging penetration logging installation PSPK, drilling equipment and apparatus for ultrasonic profiling of bottom sediments.

References

Barsukov OA (1958) Model studies of the neutron distribution in water- and oil-bearing collector intersected by a well. In: Exploration and development of mineral deposits. Gostoptekhizdat, Moscow, pp 22–27

Durante VA, Kogan YaL, Ferronsky VI, Nosal SI (1957) Field investigation of density and moisture of soils. In: Papers to the 4th Intern Congress on soil mechanics. Izd. AN SSSR, Moscow, pp 66–78

Ferronsky VI (1969) Penetration logging methods for engineering geological investigation. Nedra, Moscow

Ferronsky VI, Gryaznov TA (1979) Penetration logging. Nedra, Moscow

Gorsky Ya Ya (1958) Scintillation counters and their applications in radiometric equipment. In: Prospecting and industrial geophysics. Gosgeoltekhizdat, Moscow, pp 12–21

Zolotov AV (1958) Effective radiometric radius of wells. In: Prospecting and exploitation of mineral deposits. Gostoptekhizdat, Moscow, pp 144–151

Chapter 6
Theoretical Basis of Penetration Logging Tests

Abstract The theoretical foundation of static penetration testing of the ground by recording of the cone and lateral friction resistance with respect to the modulus of compressibility and coefficient of inner ground friction is discussed. Solutions based on the theory of ultimate equilibrium, emplacement of a spherical probe into an infinite elastic medium, imbedding of spherical sound in elastic media and taking into account creeping in time and a two-dimensional axis-symmetric problem of stress relaxation are considered. The conditions of measurement of the cone ground resistance, friction and normal ground pressure at static penetration are analysed.

Penetration logging is the field method of ground study based on the relationship between the ground resistance to the conic probe loading and ground physical and mechanical properties. Between the other field methods of geoengineering investigation, like ground testing by plate loading, pressiometry and vane shearing, penetration testing is the most popular and extended. Its popularity is explained first of all by high representation of the results characterising the obtained soil properties along the section depth. This merit is especially important if one takes into account the significant dispersion of soil properties even within the same genetic type of sediments. Because of this, the probability in obtaining a random result of a test in a separate point, which is extended on the whole stratum, is always very high.

The static penetration testing as a method of ground study has been known from the end of the 19th century. During 1914–1922 it was widely used by geotechnicians during railway construction in Sweden for determination of clay consistence at its throwing up into the embankment. Since that time the penetration test has been widely used in Europe, America and Asia for different soil studies (Bondarik 1964; Sanglera 1971; Proceedings 1974–1975).

Initially, the sounding test method was used for qualitative estimation of the strength and deforming properties of studied sedimentary grounds. Later on, the quantitative interpretation approaches were developed (Meyerhof 1951, 1956, 1961; Durante et al. 1957; Menzenbach 1959; De Beer 1963; Bondarik 1964; Trofimenkov and Vorobkov 1964; Yaroshenko 1964; Begemann 1965; Schultze and Melzer 1965; Ferronsky 1969; Sanglera 1971). The authors propose the analytical and graphical relationships between the cone resistance of its penetration and the modulus of soil compressibility, specific cohesion force, angle of inner friction,

V. I. Ferronsky, *Nuclear Geophysics*, Springer Geophysics,
DOI 10.1007/978-3-319-12451-3_6

porosity, consistence, bearing capacity of piles and so on. Most of the above relationships, in the form of experimental correlation curves, have been obtained. The attempts to find fundamental solutions were undertaken. This last direction of works is seen as the most important because without solutions like this a justified methodology of experimental data interpretation is not reached.

It is worth noting that dynamical sounding is a simpler form of study than the static one. But the soil processes at dynamical sounding are too complicated from the viewpoint of interpretation of the soil properties. The problem appears to be more complicated for its analytical solution. That is the way the privilege in methodological aspects is given to the static test. In this regards, in our work we discuss only static penetration.

In order to find the relationship between the cone resistance and the studied ground property the physical processes in the soil should be well understood. Note that the measured ground resistance is not a specific ground property but is a complicated function of many changes in a wide range of physical and mechanical ground properties like density, water bearing, granulometric and mineralogic content, structural and water-colloidal bounding between the particles and so on. It means that use of interpolation relations between the ground resistance and one or a number of the above properties is valid only for a given type and definite state of sediments.

6.1 Solutions Based on the Theory of Ultimate Equilibrium

For interpretation of static penetration results the method based on physical prerequisites and solutions of the theory of ultimate equilibrium has been widely used (Meyerhof 1956; De Beer 1963; Yaroshenko 1964). The essence of these prerequisites is as follows.

If the ground is loaded by a local load that is transferred through some plate then because of deformations the last load provides settling. If the load is increased the settling continues. Depending on the ground properties, the character of the settling increase can be expressed by two curves (Fig. 6.1a). Curve 1, starting from some point A, steeply proceeds to a straight line parallel to the ordinate axes. Curve 2 smoothly moves to point A as a straight line having a definite slope to the ordinate axes. In such cases, at the load that corresponds to the settling in point A, ground destruction under the plate occurs. The ultimate load p_u of the ground destruction appears to be the criterion of the ground bearing capacity or ultimate resistance for given loading conditions.

The stage of the wedge-like core I formation proceeds to the process of ground destruction under the loaded plate (Fig. 6.1b). A region between the reference point and point A in Fig. 6.1a corresponds to that stage. The destruction itself is started by separate particles mutually shifted, which leads to occurrence of small-slide planes.

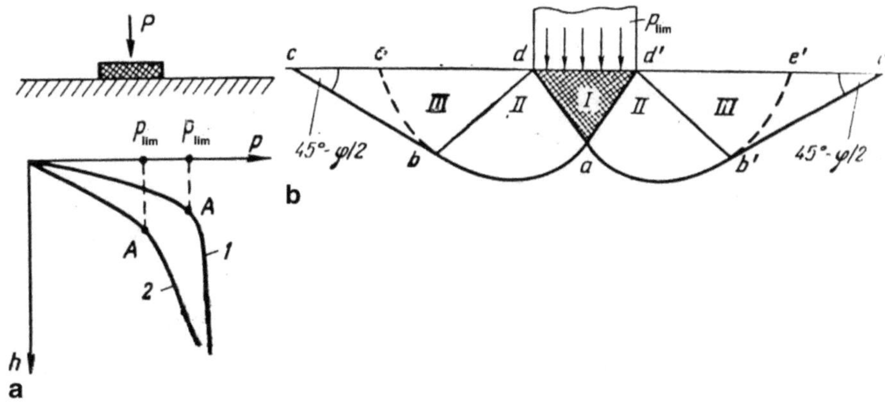

Fig. 6.1 Relationships between the settling of a plate and an outer load at different conditions of ground destruction (**a**); physical processes that appear during ground destruction at its loading (**b**). (Ferronsky 1969)

Their development into the common-slide plane finally leads to bulge ground out onto the surface. If zones of the ultimate strength state enclose the regions II and III, where the ground proceeds into a plastic state, then there will be the phenomenon of a common shift. Here the glide surface *abc*, along which the common shift is occurred, represents a curved region *ab* and linear section bc, which crosses the horizon plane under angle $45° - \varphi/2$. Practically, when the slide planes are large enough and compaction deformations in the regions II and III are absent, then a general shift is observed. As a rule, the horizontal compression in region II occurs to be not enough for the ground in region III to reach the plastic state. This is why zone III within the region *abc* remains in an elastic state and the ground destruction proceeds by local shifts in zone II. In this case the slide surface does not reach the massif surface and ground destruction is accompanied by so-called 'inner bulging'.

Terzaghi (1961), on the basis of Prandtl's (1920) work, gave an approximate solution of the two-dimensional problem of the ultimate equilibrium theory for determining an ultimate vertical uniformly distributed load in the case of ground destruction by common and local shifts. In the case of ground destruction, because of a common shift, the solution has the form (Fig. 6.2):

$$p_u = N_\gamma \frac{\gamma b}{2} + N_p \gamma h + N_c c, \tag{6.1}$$

where p_u is the ultimate load borne by the ground before its destruction; γ is the ground density; h is the depth of load action relative to the surface; b is the width of the load action; c is the specific cohesion force; N_γ, N_p, N_c are the coefficients of the ground bearing capacity depending on the angle of inner friction.

Equation (6.1) has three terms. The first one considers the width of the load; the second takes into account the lateral ground load following from its depth; and the

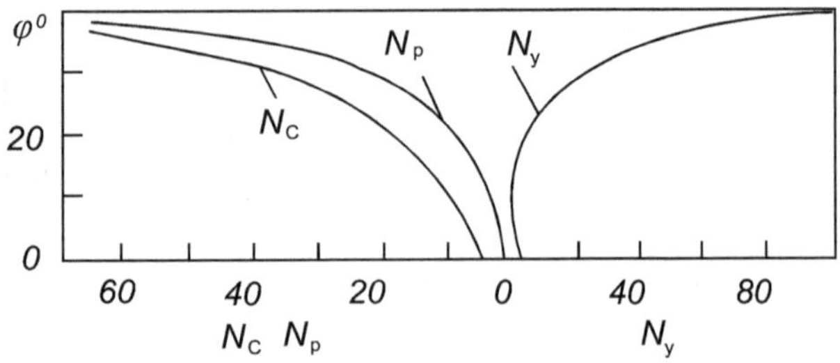

Fig. 6.2 Dependence of the ground bearing capacity coefficients N_γ, N_p, N_c from the angle of the ground inner friction. (Ferronsky 1969)

third comes from the ground cohesion forces. The values of the coefficients N_γ, N_p, N_c for a rough foot are expressed by the following equations:

$$N_\gamma = \frac{1}{2}\operatorname{tg}\varphi\left(\frac{\lambda_\gamma}{\cos^2\varphi} - 1\right), \tag{6.2}$$

$$N_p = \frac{\alpha^2}{2\cos^2(45^\circ + \varphi/2)}, \tag{6.3}$$

$$N_c = \operatorname{Ctg}\varphi\left[\frac{\alpha^2}{2\cos^2(45^\circ + \varphi/2)} - 1\right],$$
$$\lambda_\gamma = \operatorname{tg}^2(45^\circ + \varphi/2), \tag{6.4}$$

where $\alpha = e^{(3/4\pi - \varphi/2)\operatorname{tg}\varphi}$.

For a smooth foot of the loading, the coefficients N_p and N_c have the form:

$$N_p = \alpha_1^2 \operatorname{tg}^2(45^\circ + \varphi/2), \tag{6.5}$$

$$N_c = \operatorname{Ctg}\varphi\left[\alpha_1^2 \operatorname{tg}^2(45^\circ + \varphi/2) - 1\right], \tag{6.6}$$

where $\alpha_1 = e^{1/2\pi \operatorname{tg}\varphi}$.

If the angle of inner ground friction is $\varphi = 0$, then from Eq. (6.1) and the graphs of Fig. 6.2, a rough foot of loading is:

$$p_1 = 5.7c. \tag{6.7}$$

Analogously, for a smooth foot one has:

$$p_1 = 5.14c. \tag{6.8}$$

If the ground destruction takes place because of local shifts, then Eq. (6.1) acquires the form:

$$p_u = N'_\gamma \frac{\gamma b}{2} + N'_p \gamma h + N'_c \frac{2}{3} c. \tag{6.9}$$

The ground bearing capacity coefficients N'_γ, N'_p and N'_c in Eq. (6.9) are determined by formulas (6.2)–(6.6) where the values *c* and *φ* are substituted by values of *c*' and *φ*' using their relations:

$$c' = \frac{2}{3} c, \tag{6.10}$$

$$tg\varphi' = \frac{2}{3} tg\varphi. \tag{6.11}$$

If the angle of inner ground friction is $\varphi = 0$, then from Eq. (6.9) and the graphs of Fig. 6.3, one can find an expression for the ultimate load p_u for a rough foot of loading:

$$p_u = 3.8 \text{ c}. \tag{6.12}$$

Analogously, for a smooth foot one has:

$$p_u = 3.42 \text{ c}. \tag{6.13}$$

It was found that the above solutions are valid when the ratio of the loading depth and width is $h: b \leq 1$. In this case the shift resistance of the ground above the loading plane can be replaced by its load of γh.

Based on the above physical prerequisites and Terzaghi's solutions, Meyerhof developed a method of estimation of the ground bearing capacity based on penetration test data with application to the calculation of shallow foundations and piles. The theoretical solution of the problem was achieved for the elastic wedge (Fig. 6.3).

The figure shows the state of the ground destruction. When the elastic wedge at the surface occurs ($h: b \leq 1$), it is assumed that the slide surface of the common shift is completely developed and the ground is bulging to the outside (Fig. 6.3a). Because of resistance increase at the farther wedge imbedding and development of the tangential strengths in the upper ground mass, the slide surfaces started to close and bulging of the ground occurred (Fig. 6.3b). Finally, starting from a certain depth, the slide surfaces are completely closed and the ground bulge out stops (Fig. 6.3c).

For a smooth wedge (Fig. 6.4) in the area located above the ground destruction surface one can derive zone ACD of the two-dimensional plane-parallel shifts, namely, zone ADE of the radial shifts and zone AEFG of the mixing shifts. The

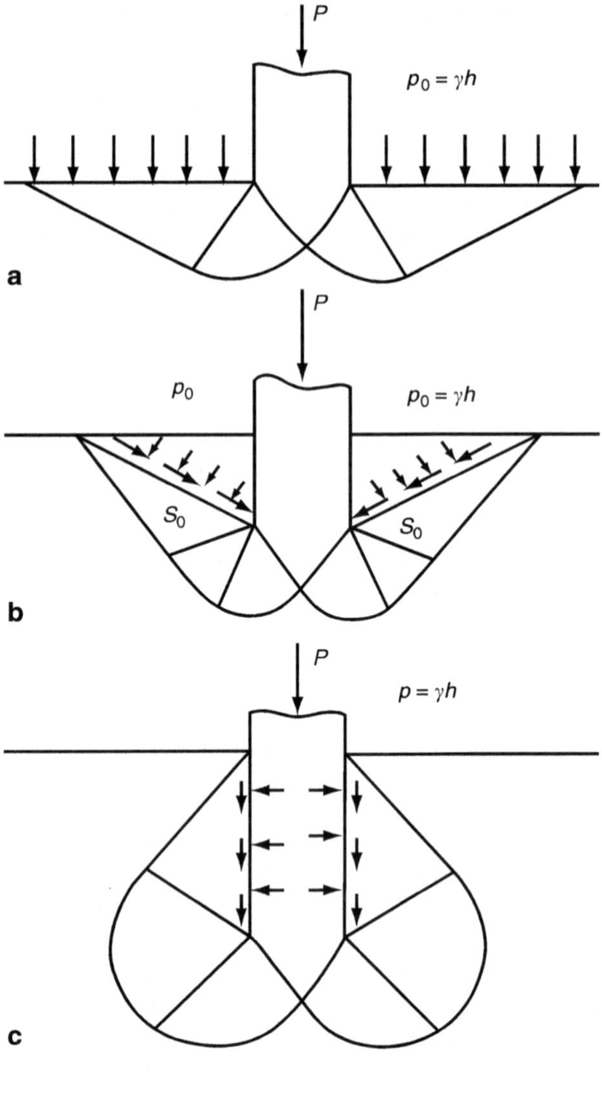

Fig. 6.3 Physical processes taking place at ground destruction while a narrow elastic wedge is forced: **a** Development of a surface slide and ground bulging out; **b** Local ground bulging out onto the surface and partial closing of the slide planes; **c** Complete closing of the plane slides without the ground bulging onto the surface. (Ferronsky 1969)

last one appears when the wedge is located shallow from the ground surface. In the zone AEF of the deep located wedge, plane-parallel ground shifts can be observed. If the wedge roughness is increased then the angle ψ at point A is decreased. For the absolute rough blunt wedge the elastic zone ACD becomes a peculiar continuation of the wedge (Fig. 6.4a). At the same time for the tapered wedge the elastic zone runs together with the wedge.

The scheme represented in Fig. 6.1 and corresponding solutions can be used for calculation of the strains in the ultimate zone of ground equilibrium for the inserted wedge, where the mass of the AFG wedge should be substituted by the equivalent normal p_0 and tangential s_0 strains. The last ones are applied in the plane AF by

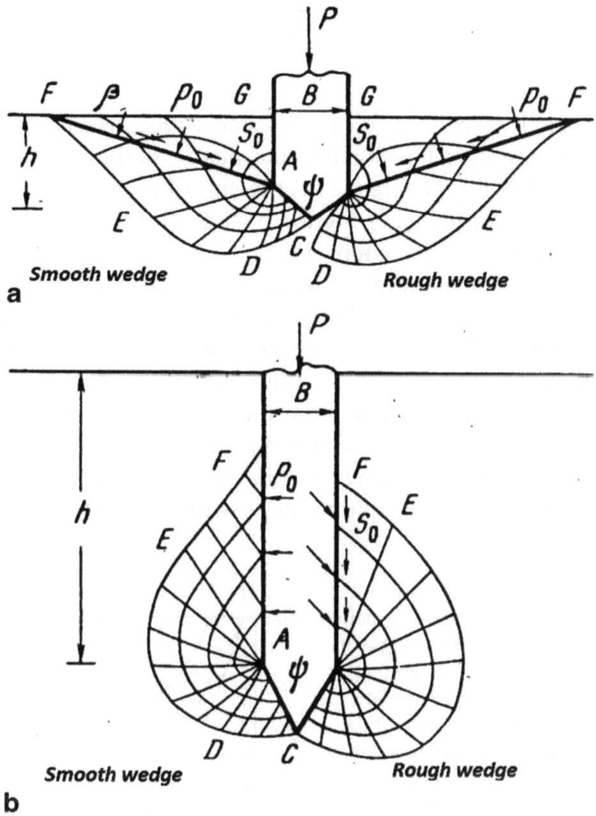

Fig. 6.4 Schematic picture of development of the ultimate ground equilibrium zones. (Ferronsky 1969)

angle β to the horizon. The ground bearing capacity here is determined by formulas (6.1) and (6.9), where bearing capacity coefficients N_γ, N_p and N_c will be dependent on the angle of inner ground friction φ, slope angle of bulging to the horizon β (Fig. 6.4), the ratio of the wedge depth insertion and its width *h*: *b* and the angle of the wedge top. For the ratio $h{:}\ b \leq 1$ the strains p_0 can be taken equal to γh. For the values $h{:}\ b \geq 4$ the value p_0 depends on the angle of inner friction φ:

$$p_0 = k_b \gamma h, \tag{6.14}$$

where k_b is the coefficient of normal ground pressure at the wedge foot equal to 0.5 for sands and 1.0 for clays.

Considering the problem of ground bearing capacity determination by the data of static penetration, Meyerhof assumes that the ground destruction under a wedge loading proceeds analogously to the process of penetration cone insertion. In this connection he extends the principle solution of the problem obtained for the one-dimensional wedge on the penetration cone. The ultimate loading that the ground perceives from the penetration sound and also the ground frontal resistance, can be found from the expression:

$$R = N_{\gamma r}\frac{\gamma d}{2} + N_{pr}p_0 + N_{cr}c, \tag{6.15}$$

where R is the ground cone resistance measured by static penetration; d is the cone diameter; p_0 is the normal strain equal to γh for h: d ≤ 1 and calculated by (6.14) at h: d > 1; $N_{\gamma r}$, N_{pr}, N_{cr} are the coefficients of the ground bearing capacity depending on the angle of the inner friction φ, angle β (Fig. 6.3), the ratio of the cone depth and the angle at the cone top.

For calculation of the bearing capacity coefficients, Meyerhof (1951) used his own solution for the ultimate bearing capacity cylindrical foundation. These coefficients for an absolute rough cone are given in Fig. 6.5. It is interesting to note that from comparison of the bearing capacity coefficients N_{pr} and N_{cr} for the cone and wedge one can see that the values of these parameters for the cone are higher than for the wedge. But the value of $N_{\gamma r}$ for the cone is less than for the wedge.

The method of determination of the bearing capacity based on solutions in the theory of ultimate equilibrium was further developed in the works by Berezantsev

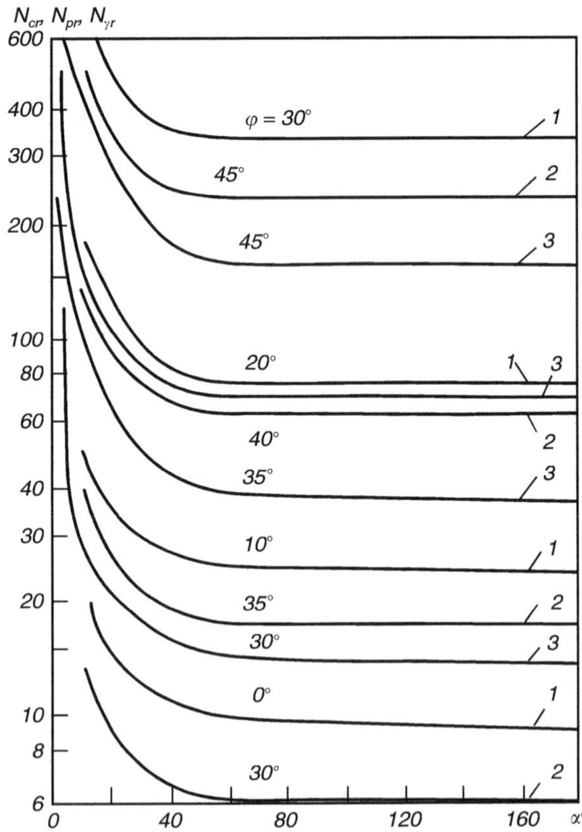

Fig. 6.5 Relationship between the bearing capacity coefficients and the angle of the cone top for determination of the ground frontal resistance at insertion of the rough cone at different values of angle of the ground friction: (*1*) N_{cr}; (*2*) $0.1N_{pr}$; (*3*) $N_{\gamma r}$. (Ferronsky 1969)

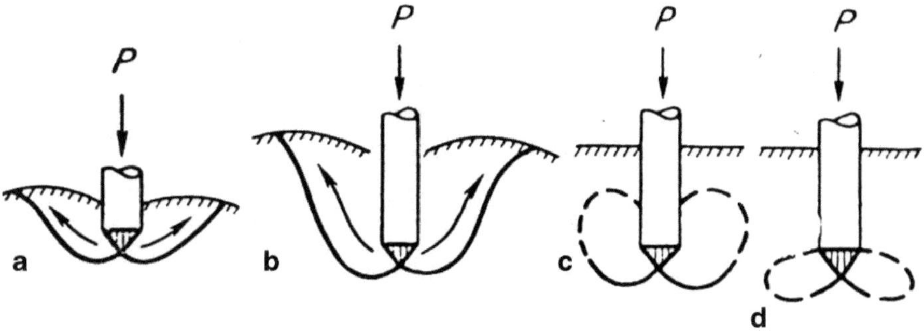

Fig. 6.6 Possible models of sandy ground destruction according to Berezantsev and Yaroshenko (1957). (Ferronsky 1969)

and Yaroshenko (1957). Their theoretical solution of the axial-symmetric problem from the theory and experimental studies on the sand bearing capacity allowed this problem to be considered more accurately from the viewpoint of physical and theoretical positions.

According to Berezantsev and Yaroshenko, the following physical picture can be observed during flat plane (cone) loading. First, the elastic core by moving particles is formed. During further loading of the plate, its packed core slides into the soil masses. The packed core consists in two parts: the elastic that is a continuation of the plate and the non-elastic that is changing depending on the ground properties. In the case of a non-deepening cone (Fig. 6.6a) destruction of dense sand is accompanied by its shift on a slide surface with an angle $45° - \varphi$ to the horizon.

For a sinking cone (Fig. 6.6b) with the value d: h$\leq$3, at the moment of ground destruction a deep shift together with ground compaction over the cone occurs (Fig. 6.6c). This phenomenon is characteristic for dense and intermediate dense sand. Finally, at the cone deepening, which corresponds to d: h>3 for dense and intermediate dense sand, a deep local shift around the packed core is observed, which is accompanied by ground compaction below the cone foot (Fig. 6.6d). This case is also characteristic for a non-deepening cone of a loose sand structure (Fig. 6.6b).

On the basis of its theoretical solutions, Berezantsev (1952) obtained the following formulas for the ultimate sandy ground bearing capacity around the cone (Fig. 6.6):

For the case *a*:

$$p_u = A_k \gamma d, \tag{6.16}$$

where A_k is the coefficient depending on the angle of the ground inner friction.

For the case *b*:

$$p_u = A'_k \gamma d, \tag{6.17}$$

where A'_k is the coefficient depending on the angle of the ground inner friction and the relative plate deepening on d: h.

Table 6.1 Numerical values of the coefficient α

h:*d*	Φ=26°	φ=30°	φ=34°	φ=37°	φ=40°
5	0.75	0.77	0.81	0.83	0.55
10	0.62	0.67	0.73	0.76	0.79
15	0.55	0.61	0.68	0.73	0.77
20	0.49	0.57	0.65	0.71	0.75
25	0.44	0.53	0.63	0.70	0.74

For the case *c*:

$$p_u = A_k \gamma d + B_k \alpha \gamma h, \tag{6.18}$$

where A_k is the coefficient depending on the angle of the ground inner friction; B_k is the coefficient depending on the angle of the ground inner friction and influence of the ground loading above the plate; α is the coefficient taking into account the cone depth (Table 6.1).

The values of A_k and $B_{k,}$ depending on the angle of the ground inner friction, are given in Fig. 6.7. The theoretical solution for determination of the ultimate load in case b has not been obtained. Berezantsev assumed that, because of notable compaction deformations, for this case a combined problem of theories of the ultimate equilibrium and ultimate compaction should be solved.

Analysing possible conditions of application of Berezantsev's axysimmetric problem solution, Yaroshenko (1964) discussed the following three aspects:

1. Applicability of the theoretical solutions obtained for a flat plate to the cone sound;
2. Identity of the deformations occurring at plate loading and conic sound penetration;
3. Necessitate of taking into account the Mohr circle curvilinearity, which is the evidence of the angle of inner friction constancy at different normal pressures.

Taking into account that the packed core of conic form under the flat plate at the moment of the ground destruction is formed, Yaroshenko assumed that the character of deformations and the values of critical loads for the plate and cone should be identical. Analysis of the ground stress state influence on the angle of inner friction value showed that it substantially depends on the normal stresses. The shortage of the above facts leads to error in determination of the ultimate loading by up to 50 %.

Yaroshenko came to the conclusion that the axis-symmetric problem solution of the ultimate equilibrium theory can be used for determination of the angle of ground inner friction, which is the main parameter of sand's bearing capacity. But the correction takes into account the Mohr circle curvilinearity.

Yaroshenko suggests that the solution that is obtained for the depth where the ground loading above the cone influences the shift deformations to be used (Fig. 6.7). Below that depth the solution should be taken without the term that takes into account those deformations. Table 6.2 shows Yaroshenko's data calculated according to the above recommendations for practical application in sandy grounds.

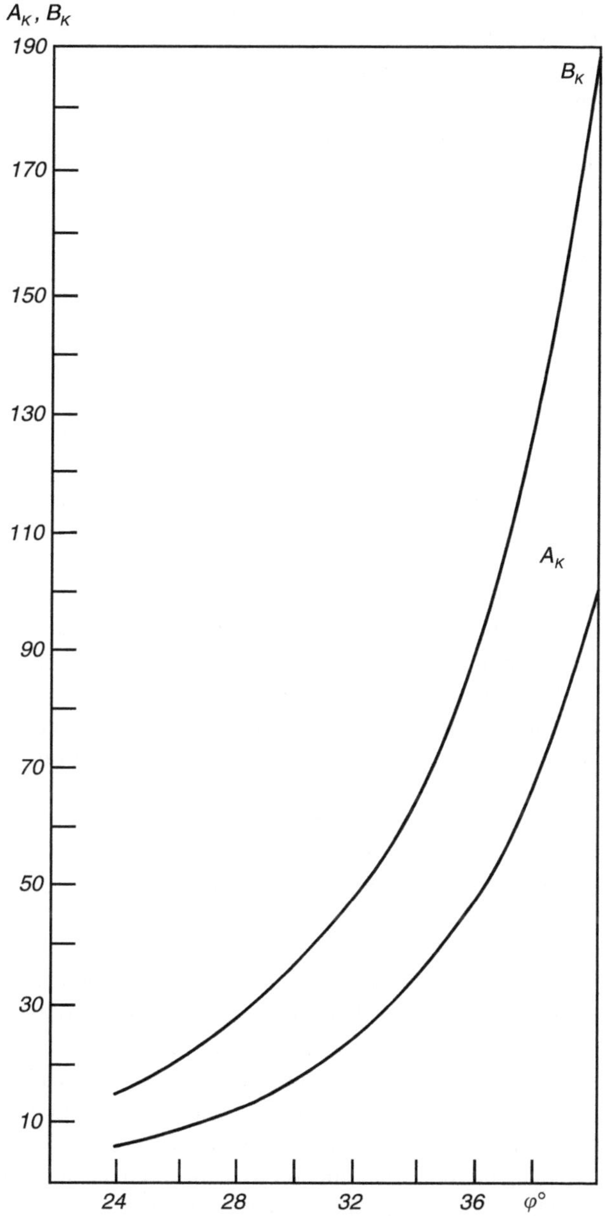

Fig. 6.7 The values of coefficients A_k and B_k depending on the angle of the inner ground friction. (Ferronsky 1969)

In the case when the cone imbedding is accompanied by ground destruction and the predominant deformations are the shift deformation, then the ultimate equilibrium theory solutions appear to be a reliable theoretical basis for determining the strength ground parameters c and φ by means of static penetration. Analysis of experimental data shows that the ground destruction during the cone imbedding is accompanied by shift surface development in the surface ground layer. The resis-

Table 6.2 Values of the angle of the ground inner friction φ depending on the cone resistance R and ground loading γh

Ground loading	Frontal resistance R x 0.1 MPa														
γh x 0.1 MPa	11	16	25	34	48	64	88	115	150	180	230	275	335	400	480
0.08	34	35	36	37	38	39	40	41	42	43	44	45	46	47	48
0.11	33	34	35	36	37	38	39	40	41	42	43	44	45	46	47
0.15	32	33	34	35	36	37	38	39	40	41	42	43	44	45	46
0.21	31	32	33	34	35	36	37	38	39	40	41	42	43	44	45
0.27	30	31	32	33	34	35	36	37	38	39	40	41	42	43	44
0.40	29	30	31	32	33	34	35	36	37	38	39	40	41	42	43
0.55	28	29	30	31	32	33	34	35	36	37	38	39	40	41	42
0.75	27	28	29	30	31	32	33	34	35	36	37	38	39	40	41
1 and >	26	27	28	29	30	31	32	33	34	35	36	37	38	39	40

tance shifts appear to be lower than the compaction deformations. As Yaroshenko correctly noted, for some ground types like loose sands the process of their compaction at sound imbedding starts just from the surface.

An interesting experiment for qualitative estimation of the physical processes in sandy grounds of different density was provided by Gryaznov (Ferronsky and Gryaznov 1979). The experiments were carried out in a laboratory metal box of 1 m high and 0.5 × 1 m in plan. The three layers of sand were placed with different densities of 1.6 g/cm^3 for the upper, 1.3 g/cm^3 for the middle and 1.8 g/cm^3 for the lower layers. The latitudinal cut of the cone of 60 mm diameter was imbedded by a cut along the glass wall of the box (Fig. 6.8). The occurred deformations were fixed by means of colour streaks and a glass network. Each step of.the cone imbedding was photographed.

The described experiments show that the ground destruction process and accompanied displacement deformations are observed only at the ground surface. In this respect the results of the third series of experiments are characteristic (Fig. 6.8c). The bulging out of the ground is observed only within 5–6 colour streaks. Below a certain depth, depending on the density, the cone penetration is accompanied by uniform sand compaction to the sides without any signs of bulging out.

An analogous phenomenon was observed during penetration logging experimental field tests of other ground types, namely, in loams, sandy loams (including loess) and clays. In this case the effects were fixed by visual observation of the excavated penetration holes. For a low-moist loam, especially for a loess loam, estimated by eye, a dense layer around the penetration hole is formed. These observations made the author conclude that during cone penetration below a certain critical depth the compaction deformations are predominate. In this regard, it is suggested that for theoretical consideration of the problem it is useful to turn to the theory of elasticity but taking into account media creeping and relaxation.

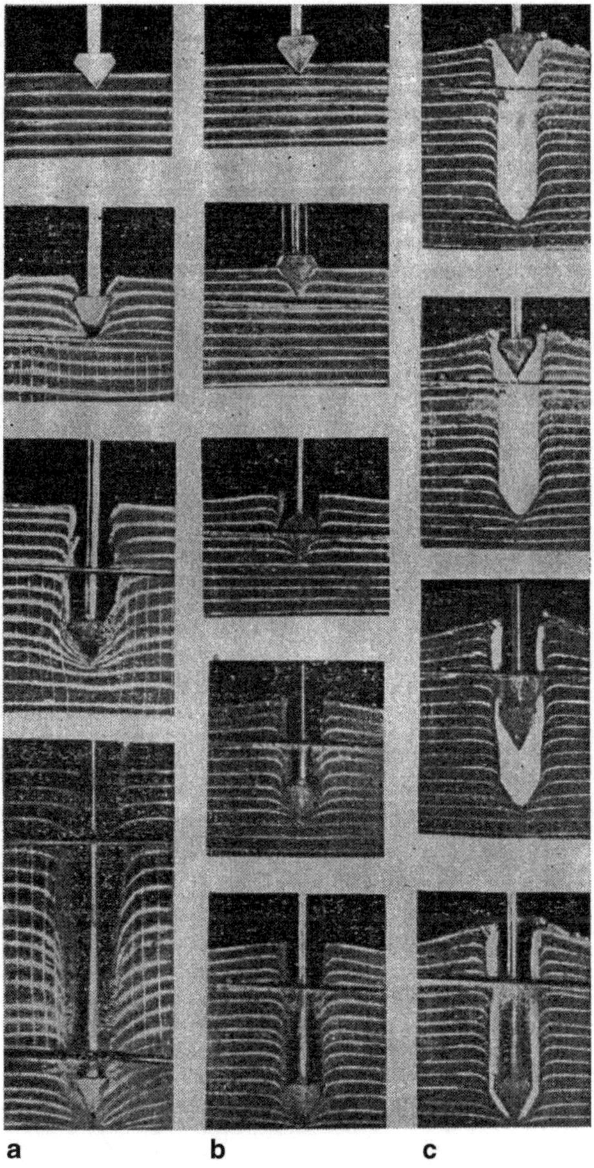

Fig. 6.8 Laboratory experiments on the study of sandy ground deformations at penetration cone imbedding: **a** Three layers of sand of loose and middle density; **b** The dense sand medium; **c** Repetition experiment after the hole was filled by white sand. (Ferronsky 1969)

6.2 Imbedding of Spherical Probe into an Infinite Elastic Medium

Based on laboratory and field experimental data, the following ground processes at static sounding occur. At the sound imbedding from the surface a zone of stressed state is formed in the ground where deformations of compression are developed.

At the same time, by tangential strains and due to ground squeezed-out the shift deformation appears (see Fig. 6.3). Further sound displacement was accompanied by these two competing forms of processes: ground displacement without change of its volume (shift deformations) and ground displacement with change of volume (compaction deformations).

Depending on the physical and mechanical ground properties (porosity, moisture content, water-colloidal and structural bonds), granulometric content and also the depth of penetration one of the above processes will predominate. In the general case, shift deformations accompanied by slide surfaces development and ground bulging will dominate in the surface zone up to a certain critical depth. Here the ground shift resistance exceeds the ground compression resistance. Below a critical depth, the shift resistance becomes higher of the compression resistance because of side ground loading. That is why at further cone displacement the ground compression deformations dominate. The maximum value of these deformations depends on the ground density and is equal approximately to the cone diameter. The surface ground layer as a rule has low-compression resistance and its depth is equal to several diameters.

Thus, interpretation of the cone resistance results with penetration ground tests based on the theory of ultimate equilibrium solutions is limited by the critical depth, i.e., the depth where the shift deformations dominate. The critical depth of sounding is easily determined by a frontal resistance diagram. The resistance starts to decrease with depth up to the point of the critical depth from where it starts to increase. This point shows the depth where the shift deformations are not developed.

Interpretation of the cone resistance below the critical depth, where the compression deformations start to develop, should be considered on the basis of the theory of elasticity solutions. The three classical problems of theory of elasticity are known, which can be considered with respect to application in soil static penetration.

If force P is applied onto the surface of elastic half-space (Fig. 6.9a), then the stresses in the domain below occur, which initiate corresponding deformations. The solution of this problem, with respect to stresses and deformations in the ground

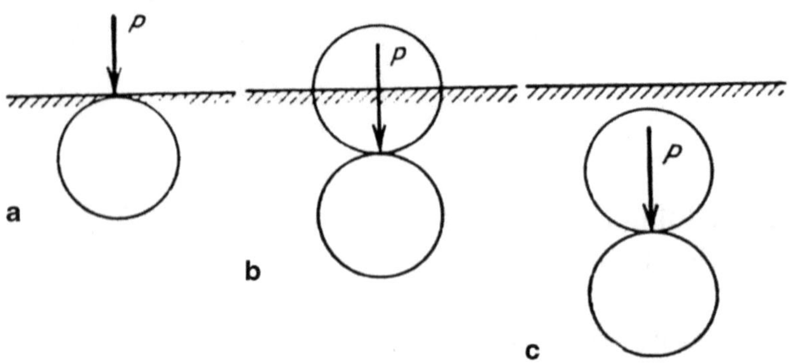

Fig. 6.9 The stress state zones for a loaded cone on an elastic ground surface: **a** Boussinesq problem; **b** Mindlin problem; **c** Kelvin problem. (Ferronsky 1969)

massif from the force action, is known in the theory of elasticity as the Boussinesq problem (Timoshenko 1934).

If the force acts on some depth of the elastic half-space, then two symmetric zones of stress state with opposite signs appear. In the case when the stress state zones go out of the limiting plane (Fig. 6.9b), then the solution of the problem with respect to stresses and deformations is known as the Mindlin (1936) problem. If both stress state zones do not go out of the limiting plane (Fig. 6.9c), then the solution of the problem with respect to stresses and deformations is known as the Kelvin problem (Timoshenko 1934).

For our considered case of sounding below a critical depth the conditions of the stress state correspond to the Kelvin problem. Let us consider this problem with respect to a penetration ground test because in this formulation the problem has not yet been solved. The described below solution of the task about imbedding of a spherical sound into elastic media was published by the author (Ferronsky 1969).

According to Kelvin's solution (Timoshenko 1934), the stress components in the point A(z, r) (Fig. 6.9c) with co-ordinates (z, r) from the force P applied in point B inside the elastic space with co-ordinates (0, 0) in the cylindrical system of reference are:

$$
\begin{aligned}
\sigma_{\text{я}} &= \frac{P}{8\pi(1-\mu)}\left[(1-2\mu)z(r^2+z^2)^{-3/2}+3z^3(r^2+z^2)^{-3/2}\right],\\
\sigma_r &= -\frac{P}{8\pi(1-\mu)}\left[(1-2\mu)z(r^2+z^2)^{-3/2}-3r^2z(r^2+z^2)^{-3/2}\right],\\
\sigma_\theta &= \frac{P}{8\pi(1-\mu)}\left[(1-2\mu)z(r^2+z^2)^{-3/2}\right],\\
\tau_{\text{гя}} &= \frac{P}{8\pi(1-\mu)}\left[(1-2\mu)z(r^2+z^2)^{-3/2}+3rz^2(r^2+z^2)^{-3/2}\right],
\end{aligned}
\tag{6.19}
$$

where σ_z, σ_r, σ_θ, τ_{rz} are the stress components in the considered point; μ is the Poisson's coefficient.

The relative deformations in the same reference system are expressed as:

$$
\begin{aligned}
\varepsilon_z &= \frac{\partial\omega}{\partial z},\\
\varepsilon_r &= \frac{\partial u}{\partial r},\\
\varepsilon_\theta &= \frac{u}{r},\\
\gamma_{rz} &= \frac{\partial u}{\partial r}+\frac{\partial\omega}{\partial z},
\end{aligned}
\tag{6.20}
$$

where ε_z, ε_r, ε_θ, γ_{rz} are relative deformations in the point on corresponding directions; ω, u are the point shifts in the vertical and radial directions.

The relationship between the relative deformations and stresses according to Hooke's law are described by the equations:

$$\begin{aligned}
\varepsilon_z &= \frac{1}{E}\left[\sigma_z - \mu(\sigma_r + \sigma_\theta)\right],\\
\varepsilon_r &= \frac{1}{E}\left[\sigma_r - \mu(\sigma_r + \sigma_\theta)\right],\\
\varepsilon_\theta &= \frac{1}{E}\left[\sigma_\theta - \mu(\sigma_z + \sigma_r)\right],\\
\gamma_{rz} &= \frac{2(1-\mu)}{E}\tau_{rz}.
\end{aligned} \tag{6.21}$$

From Eqs. (6.20) and (6.21) the transfer of point A in the radial direction by action of the force P can be found from:

$$u = \varepsilon_0 r = \frac{r}{E}[\sigma_\theta - \mu(\sigma_z + \sigma_r)]. \tag{6.22}$$

Or, substituting the corresponding stresses from Eq. (6.19), one has:

$$u = -\frac{Pr(1+\mu)}{8\pi E(1-\mu)} z(z^2 + r^2)^{-3/2}. \tag{6.23}$$

For the points located on the plane with co-ordinate z=0

$$u_{z=0} = 0. \tag{6.24}$$

Let us find the transfer of point A along axis z. For this, from expressions (6.20) and (6.21) the equations can be written as:

$$\begin{aligned}
\frac{\partial \omega}{\partial z} &= \varepsilon_z = \frac{1}{E}\left[\sigma_z - \mu(\sigma_r + \sigma_\theta)\right],\\
\frac{\partial \omega}{\partial r} &= \gamma_{rz} - \frac{\partial u}{\partial z} = \frac{2(1+\mu)\tau_{rz}}{E} - \frac{\partial u}{\partial z}.
\end{aligned} \tag{6.25}$$

Substituting from (6.19) and (6.23) the corresponding values of the stress components and displacements *u*, one has:

$$\begin{aligned}
\frac{\partial \omega}{\partial z} &= \frac{P}{8\pi E(1-\mu)}\left[(1-4\mu^2)z(r^2+z^2)^{-3/2}\right] + 3z(z^2 - r^2\mu)(r^2+z^2)^{-3/2},\\
\frac{\partial \omega}{\partial r} &= \frac{P}{8\pi E(1-\mu)}\left[(3-\mu-4\mu^2)r(r^2+z^2)^{-3/2}\right] + 3r(z^2 + r^2\mu)(r^2+z^2)^{-3/2}.
\end{aligned} \tag{6.26}$$

From here, after integration and finding boundary condition arbitrary constants, we obtain:

$$\varpi = \frac{P(1+\mu)}{8\pi E(1-\mu)}\left[(3-4\mu)r(r^2+z^2)^{-3/2}+z^2(r^2+z^2)^{-3/2}\right]. \tag{6.27}$$

For the points lying on the plane with co-ordinate z=0 the displacements are equal to:

$$\varpi_{z=0} = \frac{P(1+\mu)(3-4\mu)}{8\pi Er(1-\mu)}. \tag{6.28}$$

Let us consider now the above solution of Kelvin's problem to our task for a cylindrical sound with a spherical tip of radius r_0 (Fig. 6.10).

In a certain moment of the sound momentary equilibrium the contact plane of elastic media with a spherical tip will have a cercal surface of radius r_0. Let us find the solution of the task with respect to the sagging ω_m of the point M that is placed within the contact plane. The loading on the elementary plane dF, which is formed

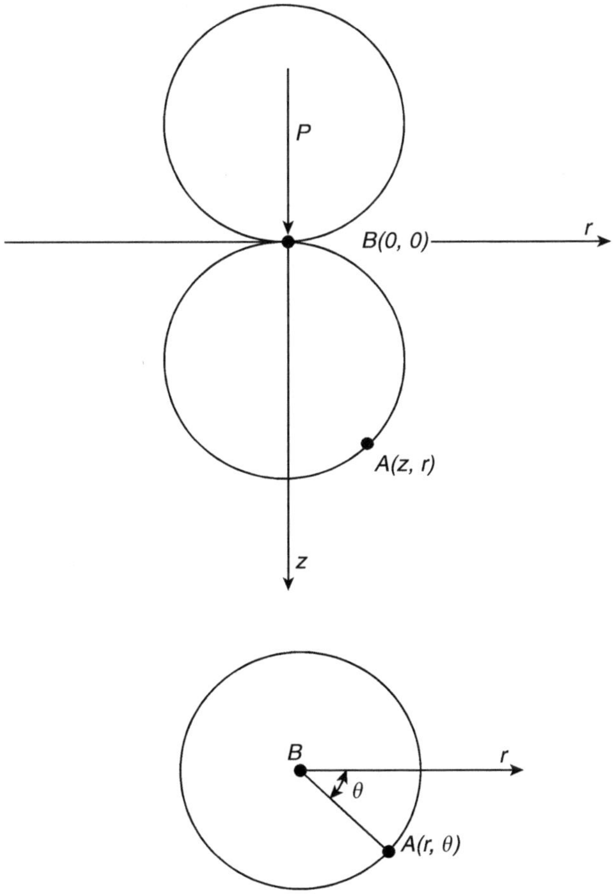

Fig. 6.10 The scheme for the solution of Kelvin's problem with respect to the penetration ground test by sound with a hemispheric tip of radius r_0. (Ferronsky 1969)

by the secant lines mn and m'n' passing through the point M with angle dφ and the arcs of the radiuses s and s+ds, will be equal to:

$$dP + pdF = ps\, d\varphi\, ds.$$

Displacement of the point M by the given elementary force according to expression (6.28) is equal to:

$$d\omega_M = \frac{(1+\mu)(3-4\mu)}{8\pi E(1-\mu)} p\, d\varphi ds.$$

The total displacement of the point M by the loading of the full contact area of the hemisphere is determined by:

$$\int pds = \frac{\pi m^2 n^2}{8} \frac{p_0}{r_0}. \qquad (6.29)$$

If in expression (6.29), under an integral sign, there is an unknown function p, then this equation is integral. It was proved that the pressure p on the contact surface is distributed by the law of hemisphere down onto this surface (Timoshenko 1934). Therefore the diagram of the pressure along the secant mn (see Fig. 6.10) has semicircle shape (dashed line on the figure). From here integral $\int pds$ from Eq. (6.29) will be equal to:

$$\int pds = \frac{\pi m^2 n^2}{8} \frac{p_0}{r_0},$$

where p_0 is the maximum pressure within the loading area; p_0/r_0 is the coefficient of loading.

Because, from geometrical consideration:

$$mn = 2\sqrt{r_0^2 - r^2 \sin^2\varphi},$$

one has:

$$\int pds = \frac{\pi}{2}(r_0^2 - r^2\sin^2\varphi)\frac{p_0}{r_0}. \qquad (6.30)$$

Substituting expression (6.30) into (6.29), we can write:

$$\omega_M = \frac{2(1+\mu)(3-4\mu)}{8\pi E(1-\mu)} \int_0^{\pi/2} \frac{\pi p_0}{2r_0}(r_0^2 - r^2\sin^2\varphi)d\varphi. \qquad (6.31)$$

After integration one has:

$$\omega_M = \frac{(1+\mu)(3-4\mu)p_0\pi}{8\pi\, E(1-\mu)}\left(\frac{r_0}{2} - \frac{r^2}{2r_0}\right). \tag{6.32}$$

At $r=0$ in expression (6.32) we have the deflection deformation $\omega_M = \omega_0$ along the sound axis. Taking into account that in the considered problem $\omega_0 = r_0$, then from (6.32) we can find the expression for the modulus of deformation E:

$$E = \frac{\pi(1+\mu)(3-4\mu)p_0}{16(1-\mu)}. \tag{6.33}$$

In the field ground tests, static penetration provides measurement not of the maximal pressure p_0 on the sound axes but the mean (specific) frontal resistance to the sound emplacement. Because in expression (6.33):

$$p_0 = \frac{3P}{2\pi r_0^2}, \quad \text{and} \quad P = R\pi r_0^2,$$

one obtains:

$$E = \frac{3\pi(1+\mu)(3-4\mu)R}{32(1-\mu)}, \tag{6.34}$$

where R is the specific cone ground resistance to the sound embedding during static penetration.

For practical use of the obtained relationship one circumstance should be taken into account. The stress state of an infinite elastic medium from the action of force according to Kelvin is considered on the plane for which the co-ordinate $z=0$ is the plane of symmetry of the stress distribution. In connection with this only half a loading comes to the medium presented below the point of loading. The second half of the loading acts above that point. This effect can be illustrated, for example, by the action of force on a cercal area. In fact, if in expression (6.19) the force P for the vertical stresses is applied in the centre of the cercal area with radius r_0 then, in this case, the stresses will be described by the expression:

$$\sigma_я = \int_0^{r_0} \frac{2\pi r p}{8\pi(1-\mu)}\left[(1-2\mu)z(r^2+z^2)^{-3/2} + 3z^3(r^2+z^2)^{-3/2}\right]dz.$$

After integration we have:

$$\sigma_\theta = \frac{P}{4(1-\mu)}\left[(1-2\mu)\left(1-\frac{z}{\sqrt{r_0^2+z^2}}\right)+\left(1-\frac{z^3}{\sqrt{(r_0^2+z^2)^3}}\right)\right],$$

where at z=0 one obtains:

$$\sigma_я = \frac{P}{2}. \tag{6.35}$$

Considering real grounds, one should bear in mind that they can receive only small tensile stresses and sandy grounds do not take any at all. In addition, sound imbedding into the ground brakes down its continuity. Thus, the second part of the loading practically initiates the stresses in the zone below the loading plane. In this regards, a coefficient *a* must be put into formula (6.34), which takes into account the developed extra stresses below the sound by the second loading part. The value of the coefficient can theoretically vary from 1 to 2. If one assumes that the entire load is transferred below the symmetry plane then *a*=2 and one obtains:

$$E = \frac{3\pi\alpha(1+\mu)(3-4\mu)R}{32(1-\mu)}. \tag{6.36}$$

The Poisson's coefficient μ is taken, depending on the ground type, from a reference book.

As we can see from expression (6.36), the compressibility modulus value that is determined by the cone ground resistance depends only on the Poisson's coefficient and does not depend on the sound diameter. This is because at static penetration the ground deformations to the sides are equal to the sound radius. Practically we provide penetration ground tests with a given value of the ground deformations. Experimental data on the ground deformations are given in Chap. 7.

Thus, Kelvin' solutions allow obtaining the analytical solution for determining the modulus of ground deformations using the data of penetration ground tests. Note, that the obtained solution is analogous to the solution for determining the same modulus by plate tests. Both solutions are based on the Boussinesq equation solution. From here, the results of the ground deformation modulus obtained by the two methods should be identical. The field works carried out by Gryaznov in VSEGINGEO using penetration and plate tests showed that the values calculated by formula (6.36) are in good agreement for sandy grounds and provide lower values for the plate tests for clays in a plastic state.

Let us note some specific points of the modulus of determination by the penetration tests. The first one is the representation of the measurements, i.e., estimation of the ground volume for which the calculated modulus value is valid. This question is important for comparison of the modulus values determined by penetration and plate methods taking into account that the ground properties are often changed with depth. At present, according to the soil mechanics approach, the ground active zone under the plate test is estimated as a double diameter (for a round shape) or double size (for a square size) plate. From here, comparison of test results can be done only for a geologic section with uniform soil properties in depth not less than double the size of the test plate.

The second specific feature of the penetration test is that static sounding is provided continuously with certain velocity. From here, the compaction deformations are always developed in time faster than the deformation rate at tests. In this case the measured values of the ground resistance will also be higher. The development of compaction deformations can be taken into account by the solution of corresponding tasks from the ground consolidation theory.

Based on the results obtained at the end of the last century by Tsitovich, Ter-Martirosyan, Zaretsky, Abelev and others (Tsitovich and Ter-Martirosyan 1981; Zaretsky 1967), we consider two tasks:

1. The ground is considered as an elastic-creeping medium where deformations are developed in accordance with the creep theory of Boltzmann-Wolterra (Arutiunyan 1952);
2. The above formulation is clarified by the effect of pore water pressure (Zaretsky 1967).

Let us consider the first task.

6.3 Imbedding of Spherical Probe into Elastic-Creeping Media

Let us assume that during imbedding of a cylindrical sound with a spherical tip into ground media, the stress action of the elastic and creep deformations in the ground skeleton are developed (Fig. 6.10). At the same time the ground medium state is characterised by the equation of Maslov–Arutiunyan's hereditary creeping (Arutiunyan 1952), which for one-dimensional compression has the form:

$$\varepsilon_x(t)=\frac{\sigma_x(t)}{E_M(t)}-\int_{\tau_1}^{t}\sigma_x(\tau\tau)(t,\tau)d\tau, \tag{6.37}$$

where $E_M(t)$ is the modulus of instantaneous deformation; $\delta(t,\tau)$ is the creeping deformation at the moment of time t from the unit loading applied at the time moment τ_1, which is defined by the expression:

$$\delta(t.\tau)=\frac{1}{E_M(\tau)}+\frac{1}{E_1(t,\tau)},$$

where $E_l(t)$ is the modulus of the long-time deformation.

For the values E_M and $E_{l,}$ Arutuniyan (1952) proposed the following expressions:

$$E_M(t)=E_M(1-\beta e^{-\alpha\tau}),$$

$$\frac{1}{E_l(\tau)}=\varphi(\tau)\left[1-e^{\eta(t-\tau)}\right], \tag{6.38}$$

where α, β, η are parameters of the ground creeping; $\varphi(\tau)$ is the function of the ground aging that is determined by the expression:

$$\varphi(\tau)=\frac{1}{E_l}+\frac{A_1}{1-\tau}. \tag{6.39}$$

Thus, the considered medium in general will be characterised by four parameters: the modules of instantaneous and long-time deformation, parameter of creeping and coefficient of the transversal deformation.

Applicability of the hereditary creeping theory for describing the stress-deforming ground medium has been proved by a number of researchers (Maslov 1969; Zaretsky 1967; Florin 1961). In order to pass from the elastic-instantaneous to elastic creeping state, Arutuniyan's theorem about identity of these stress states at constancy of the creeping cores of the volumetric contraction and net shift, which is accepted in soil mechanics, is used (Zaretsky 1967). If we define the elastic-instantaneous state by σ_{if}^0 and the elastic-creeping state by $\sigma_{if}(t)$ at any moment of time, then:

$$\sigma_{if}^0=\sigma_{if}(t). \tag{6.40}$$

Now, on the basis of Kelvin's problem solution, provided in the previous paragraph, the stress components and relative deformations in point A (see Fig. 6.9) with co-ordinates z, r from the force P(t) applied in point B with co-ordinates z=0, r=0 of infinity elastic-creeping medium, can be expressed by Eqs. (6.19) and (6.20).

The relationship between the relative deformations and stresses, taking into account the ground skeleton creeping according to the hereditary creeping theory, is written as:

$$\varepsilon_{я}(t)=\frac{\sigma_z(t)-\mu[\sigma_r(t)+\sigma_\theta(t)]}{E_M(t)}-\int_{\tau_1}^{t}\{\sigma_z(\tau)-\mu[\sigma_r(\tau)+\sigma_\theta(\tau)]\}\frac{\partial}{\partial\tau}\delta(t,\tau)d\tau,$$

.... (z, r, θ)

....

$$\gamma_{r,z}(t)=\gamma_{r,z}(t)=2(1+\mu)\left\{\frac{\tau_{r,z}(t)}{E_M(t)}-\int_{\tau_1}^{t}\tau_{z,r}(\tau)\frac{\partial}{\partial\tau}\delta(t,\tau)dz\right\}, \tag{6.41}$$

.... (z, r, θ)

....

Henceforth, the symbols (z, r, θ) mean that the other relations are written by the cyclic change of indexes z, r, θ.

From expressions (6.20) and (6.41) the displacement of point A in the radial direction from the force P(t) can be found as:

$$
\begin{aligned}
u_r = \varepsilon_0(t) r = r\left(\frac{\sigma_0(t) - \mu\,[\sigma_z(t)+\sigma_r(t)]}{E_M(t)}\right. \\
\left. - \int_{\tau_1}^{t} \{\sigma_\theta(\tau) - \mu[\sigma_z(\tau)+\sigma_r(\tau)]\}\frac{\partial}{\partial\tau}\delta\,(t,\tau)d\tau.\right.
\end{aligned}
\tag{6.42}
$$

Or, substituting the values of the stresses components from expression (6.19), we have:

$$
u_r(t) = \frac{(1+\mu)rz(r^2+z^2)^{-3/2}}{8\pi(1-\mu)}\left[-\frac{P(t)}{E_M(t)} + \int_{\tau_1}^{t} P(\tau)\frac{\partial}{\partial\tau}\delta(t,\tau)d\tau\right]. \tag{6.43}
$$

For the points located on the plane z=0, one has:

$$
u_{z=0} = 0. \tag{6.44}
$$

Let us find replacement of the point A along the axes z. From Eqs. (6.20) and (6.41) one has:

$$
\begin{aligned}
\frac{\partial\omega(t)}{\partial z} &= \varepsilon_z(t) = \frac{\sigma_z(t) - \mu[\sigma_r(t)+\sigma_\theta(t)]}{E_M(t)} - \int_{\tau_1}^{t}\{\sigma_z(\tau) - \mu[\sigma_r(\tau)]\}\frac{\partial}{\partial\tau}\delta(t,\tau)d\tau, \\
\frac{\partial\omega(t)}{\partial r} &= \gamma_{rz}(t) - \frac{\partial u}{\partial z} = 2(1+\mu)\left[\frac{\tau_{rz}(t)}{E_M(t)} - \int_{\tau\tau_1}^{t}\tau_{rz}(\tau)\frac{\partial}{\partial\tau}\delta(t,\tau)d\tau\right] - \frac{\partial u}{\partial z}.
\end{aligned}
\tag{6.45}
$$

Substituting the corresponding values of stresses and shifts from Eqs. (6.19) and (6.41) into formula (6.45), we have:

$$
\begin{aligned}
\frac{\partial\omega(t)}{\partial z} &= \left[\frac{(1-4\mu^2)z(r^2+z)^{-3/2} + 3z(z^2-\mu r^2)(r^2+z^2)^{-3/2}}{8\pi(1-\mu)}\right] \\
&\quad \left[\frac{P(t)}{E_M(t)} - \int_{\tau_1}^{t} P(\tau)\frac{\partial}{\partial\tau}\delta(t,\tau)d\tau\right], \\
\frac{\partial\omega(t)}{\partial r} &= \left[\frac{(3-\mu-4\mu^2)r(r^2+z)^{-3/2} + 3r(z^2+r^2)(r^2+z^2)^{-3/2}}{8\pi(1-\mu)}\right] \\
&\quad \left[\frac{P(t)}{E_M(t)} - \int_{\tau_1}^{t} P(\tau)\frac{\partial}{\partial\tau}\delta(t,\tau)d\tau\right].
\end{aligned}
\tag{6.46}
$$

From where after integration:

$$\omega(t)=\frac{1+\mu}{8\pi(1-\mu)}\left[(3-4\mu)(r^2+z^2)^{-1/2}+z^2(r^2+z^2)^{-3/2}\right]$$
$$\left[\frac{P(t)}{E_M(t)}-\int_{\tau_1}^{t}P(\tau)\frac{\partial}{\partial\tau}\,\delta\,(t,\tau)d\tau\right]. \tag{6.47}$$

For the point z=0 the displacements are:

$$\omega(t)_{z=0}=\frac{(1+\mu)(3-4\,\mu)}{8\pi r(1-\mu)}\left[\frac{P(t)}{E_M(t)}-\int_{\tau_1}^{t}P(\tau)\frac{\partial}{\partial\tau}\,\delta(t,\tau)d\tau\right]. \tag{6.48}$$

For the case when the ground skeleton creeping is absent and the force P is constant the above equation will be identical to expression (6.28) for the elastic medium.

We consider now the problem of imbedding of the spherical cylinder-sound with a semispherical tip to the elastic-creeping medium. According to expression (6.48), displacement of point M (Fig. 6.10) from the constant loading effective on the elementary plane should be:

$$d\omega_M(t)=\frac{(1+\mu)(3-4\mu)}{8\pi(1-\mu)}pd\varphi ds\delta(t,\tau_1) \tag{6.49}$$

The total displacement of point M from the loading over the sound contact area is:

$$\omega_M(t)=\frac{(1+\mu)(3-4\mu)}{8\pi(1-\mu)}\delta(t,\tau_1)\int\int_f pd\varphi ds. \tag{6.50}$$

After integration of Eq. (6.50) the following equation is for determining the settling of the points of plane z=0:

$$\omega_M(t)=\frac{\pi p_0}{8}\frac{(1+\mu)(3-4\mu)}{(1-\mu)}\left(\frac{r_0}{2}-\frac{r^2}{4r_0}\right)\delta(t,\tau_1). \tag{6.51}$$

Expression (6.51) is the equation of spherical surface with changing radii:

$$R_{sp}(t)=\frac{1}{\beta(t)}, \tag{6.52}$$

where $\beta(t)=\frac{\omega_0(t)-\omega(t)}{r^2}$, $\omega_0(t)=p_0k_1\frac{r_0}{2}\delta(t,\tau_1)$, $k_1=\frac{\pi(1+\mu)(3-4\mu)}{8(1-\mu)}$.

If the radius of the curved surface (bowl deflection) is too large in comparison with the loaded circle radius (which is as a rule observed) then expression (6.51) can be accounted for as the equation of a spherical surface.

At r = 0 the central deflection is:

$$\omega_0(t) = \frac{\pi p_0}{8}\frac{(1+\mu)(3-4\mu)}{(1-\mu)} r_0 \delta(t, \tau_1). \tag{6.53}$$

If the settle value is measured for a spherical sound-plate at a given loading, then the equation for the modulus of the long-time deformation is determined from Eq. (6.53):

$$E_1 = \frac{\pi p_0 r_0 (1+\mu)(3-4\mu)}{16\left[\omega_0(\infty) - \omega_0(\tau_1)\right](1-\mu)}. \tag{6.54}$$

Using the settle curve change in time ω_0, it is easy to determine the creeping parameter:

$$\eta = -\frac{1}{t}\ln\frac{\omega_0(\infty) - \omega_0(t)}{\omega_0(\infty) - \omega_0(\tau_1)}. \tag{6.55}$$

Let us assume that the stress distribution diagram is represented in the form of a surface. Its ordinates are changing in time with the conservation of spherical distribution law:

$$p(r,t) = p_0\left(t\sqrt{1 - \frac{r^2}{r_0^2}}\right), \tag{6.56}$$

where $p_0(t)$ is the maximum ordinate of the stress distribution diagram on the plane of z = 0.

Then, based on Eq. (6.51), displacement of the point M under the action of such a loading can be written as:

$$\omega_M(r, t) = \frac{\pi(1+\mu)(3-4\mu)}{8(1-\mu)}\left(\frac{r_0}{2} - \frac{r^2}{4r_0}\right)\left[\frac{p_0(t)}{E_M(t)} - \int_{\tau_1}^{t} p_0(\tau)\frac{\partial}{\partial\tau}\delta(t,\tau)d\tau\right]. \tag{6.57}$$

At t = 0 one has:

$$\omega_0(t) = \frac{\pi r_0}{16}\frac{(1+\mu)(3-4\mu)}{(1-\mu)}\left[\frac{p_0(t)}{E_M(t)} - \int_{\tau_1}^{t} p_0(\tau)\frac{\partial}{\partial\tau}\delta(t,\tau)d\tau\right]. \tag{6.58}$$

In the absence of ground skeleton creeping, Eq. (6.58) is identical to Eq. (6.32) for the elastic medium. Equation (6.58) can be used for the solution of a number of tasks on the determination of deformation and rheological characteristics of grounds by the static penetration method and different regimes of testing.

Task 1 The rate of spherical sound loading is given and the value of sound imbedding in time is measured.

In this case the task is reduced to the Eq. (6.57) integration with substitution of the form of loading. We consider first the linear law of loading as $p_0(t) = p_0(1+t)$.

Substituting this expression into Eq. (6.57), after integration we obtain:

$$\omega_0(t) = \left[\frac{\pi r_0 (1+\mu)(3-4\mu)}{(1-\mu)}\right]\left\{\frac{p_0(1+t)}{E_M} + \frac{p_0}{E_l}\left[1 - e^{-\eta t}\left(1 - t + \frac{1}{\eta}\right)\right]\right\}. \quad (6.59)$$

By initial values of the pressure $p_0(t)$ and settling $\omega_0(t)$ the instantaneous modulus of deformation can be found by:

$$E_M = \frac{\pi p_0 r_0 (1+\mu)(3-4\mu)}{16\omega_0(\tau_1)(1-\mu)}. \quad (6.60)$$

Taking into account that the creeping parameter is small and the value $e^{-\eta t}$ can be equated to unity, by the settling value $\omega_0(t)$ from Eq. (6.59) the modulus of long-time deformation and creeping parameter η can be found. In this case Eq. (6.59) has the form:

$$\omega_0(t) = \pi r_0 p_0 \frac{(1+\mu)(3-4\mu)}{16(1-\mu)}\left[\frac{(1+t)}{E_M} + \frac{1}{E_l}\left(t - \frac{1}{\eta}\right)\right]. \quad (6.61)$$

Task 2 The rate of spherical sound loading is given and the cone resistance at its imbedding is measured.

In this case the task is reduced to the Eq. (6.57) integration relative to $p_0(t)$. For this, the equation will be transformed into the linear differential equation of second-order with a variable coefficient. This is reached by two times differentiation by *t*. The obtained equation is:

$$p_0''(t) + p_0'(t)\left\{\eta\left[1 + \frac{\varphi(t)}{m(t)} + \frac{m'(t)}{m(t)}\right]\right\} = \frac{1}{km(t)}\left[\omega_0''(t) + \eta\omega_0'(t)\right], \quad (6.62)$$

where:

$$m(t) = \frac{1}{E_M}$$

$$k = \frac{\omega r_0 (1+\mu)(3-4\mu)}{16(1-\mu)}.$$

The initial conditions for the Eq. (6.62) solution are:

$$p_0(\tau_1) = \frac{\omega_0(\tau_1)}{km(\tau_1)}$$
$$p_0'(\tau_1) = \frac{\omega_0'(\tau_1)}{k} - \eta p_0(\tau_1)\varphi(\tau_1). \tag{6.63}$$

The general solution of Eq. (6.62) with initial conditions (6.63) can be written as:

$$p_0(t) = \frac{p_0'(\tau_1)}{k}\int_{\tau_1}^{t} e^{-\int_{\eta}^{\tau} A(x)dx} dx + \int_{\tau_1}^{t} e^{-\int_{\tau_1}^{\tau} A(x)dx} d\tau \int_{\tau_1}^{t} e^{-\int_{\tau_1}^{\tau} A(\xi)d\xi} B(x)dx + p_0(\tau_1), \tag{6.64}$$

Where

$$A(x) = \eta\left[1 + \frac{\varphi(x)}{m(x)} + \frac{m'(x)}{m(x)}\right],$$
$$B(x) = \frac{1}{m(x)k}\left[\omega_0''(x) + \eta\omega_0'(x)\right].$$

In a particular case when the ground aging and the instantaneous ground modulus changes are absent, one has:

$$p_0(t) = \frac{p_0'(\tau_1)}{k}\int_{\tau_1}^{t} e^{-\int_{\tau_1}^{\tau}\eta\left(1+\frac{E_M}{E_l}\right)dx} d\tau + \frac{E_M}{k}\int_{\tau_1}^{t} e^{-\int_{\tau_1}^{\tau}\eta\left(1+\frac{E_M}{E_l}\right)dx}$$
$$\times d\tau\int_{\tau_1}^{t} e^{-\int_{\tau_1}^{x}\eta\left(1+\frac{E_M}{E_l}\right)d\xi}\left[\omega_0''(x) + \eta\omega_0'(x) + p_0(\tau_1)\right]. \tag{6.65}$$

At the constant value of sound imbedding we have:

$$\omega'(t) = \omega_0 = \text{const},$$
$$\omega_0(t) = \omega_0(1+t),$$
$$\omega''(t) = 0,$$
$$p_0(\tau_1) = \frac{\omega_0(\tau_1)E_M}{k},$$
$$p_0'(\tau_1) = \frac{\omega_0'(\tau_1)}{k} - \eta\frac{\omega_0(\tau_1)E_M}{kE_l}. \tag{6.66}$$

Substituting the value of Eq. (6.66) into the integral Eq. (6.65), we obtain an expression for determining the value of the maximal ordinate of the cone resistance diagram:

$$p_0(t)=\left[p_0^{'}(\tau_1)+\frac{E_M}{1+E_M/E_1}\right]\frac{1-e^{-\eta(1+E_M/E)t_1}}{k\eta(1+E_M/E_1)}+p_0(\tau_1). \tag{6.67}$$

At $t=0$, one has:

$$p_0(\tau_1)=\frac{\omega_0(\tau_1)E_M}{k}. \tag{6.68}$$

For $t>5/\eta(1+E_M/E_1)$, with high accuracy, this formula is valid:

$$p_0(t)=\left[p_0^{'}(\tau_1)+\frac{E_M(t)}{1+E_M/E_1}\right]\frac{1}{k\eta(1+E_M/E_1)}+p_0(\tau_1). \tag{6.69}$$

Task 3 Initial imbedding of a spherical sound-plate is given and the relaxing stresses in time are measured.

In this case the space relaxation problem is solved. Let the spherical sound-plate into the elastic-creeping medium on the plane level $z=0$ with imbedding $\omega_0(\tau_1)$ be displaced. Thereafter, the sound is fixed in some position (Fig. 6.11). Consider the task in a general form when the system of the sound initial imbedding position has a finite rigidity. The spherical sound imbedding along axes *z* will be defined by Eq. (6.57). Imbedding of the fixed system (rods and dynamometer) is determined depending on the system rigidity:

$$l(t)=\frac{P(t)}{C_p^{'}}, \tag{6.70}$$

where $C_p^{'}$ is the initial deformation; $l(t)$ is the system elongation under action of the force $P(t)$.

For a spherical diagram of the contact stresses distribution, we have:

$$l(t)=\frac{p_0(t)}{C_p}, \tag{6.71}$$

where $C_p=3C_p^{'}/2\pi r_0$ is the reduced system fixation rigidity.

The condition of imbedding equality on the contact of ground—spherical sound in the process of stresses damping is written in the form:

Fig. 6.11 The calculating scheme for determining stress relaxation in the system dynamometer—sound—ground for the elastic-creeping medium. (Ferronsky 1969)

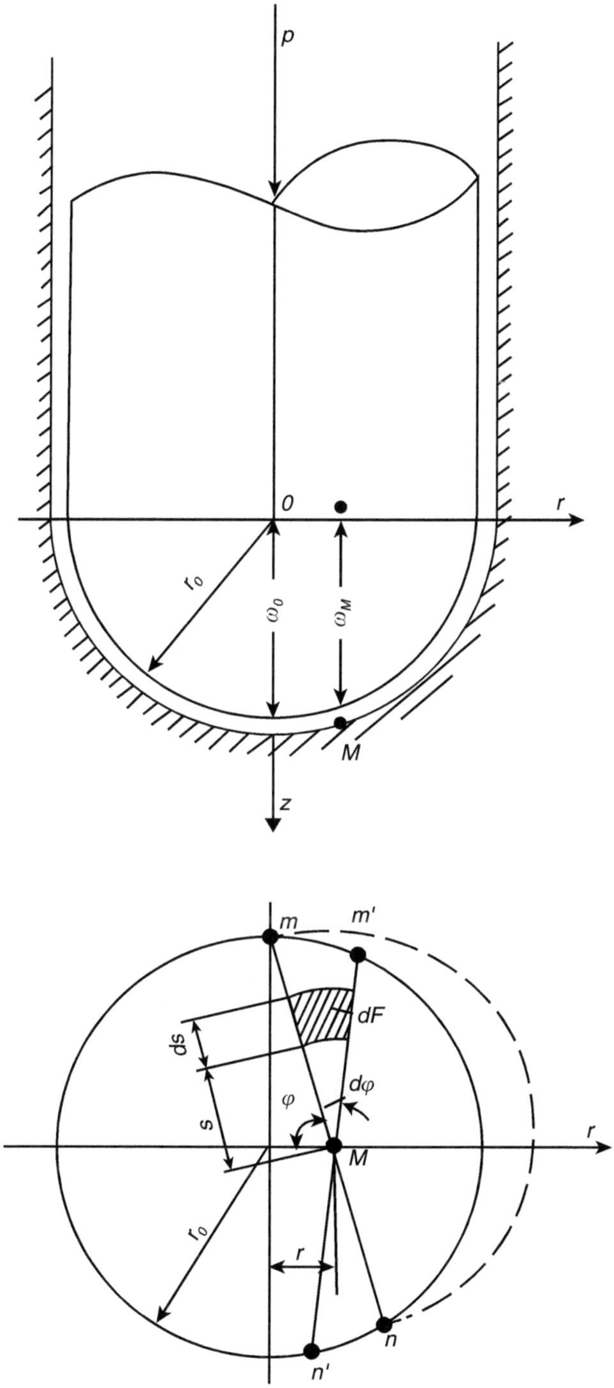

$$\omega_0(t)-\omega_0(\tau_1)=-1(t)+1(\tau_1). \tag{6.72}$$

Substituting the values $\omega_0(t)$ and l(t) from (6.57) and (6.71) into (6.72), we obtain:

$$p_0(t)k\left\{\frac{p_0(t)}{E_M(t)}--\int_{\tau_1}^{t}p_0(\tau)\frac{\partial}{\partial\tau}\delta(t,\tau)d\tau\right\}-\omega_0(\tau_1)=-\frac{p_0(t)}{C_p}+1(\tau_1).$$

After some transformations the last expression can be written in the form of the integral equation:

$$p_0(t)=\frac{\Delta(\tau_1)}{1+kn(t)}+C_pk\int_{\tau_1}^{t}p(\tau)\frac{\partial}{\partial\tau}\frac{\delta(t,\tau)}{1+kn(t)}d\tau, \tag{6.73}$$

where:

$$\begin{gathered}\text{Л}=\frac{\pi\kappa_0(1+\mu)(3-4\mu)}{16(1-\mu)},\\ n(t)=\frac{C_p}{E_M(t)},\\ \Delta(\tau_1)-\omega_0(\tau_1)+l(\tau_1).\end{gathered} \tag{6.74}$$

Differentiating Eq. (6.73) twice by *t* and reducing it to a differential equation of second-order with a variable coefficient, one has:

$$p_0^{"}(t)+p_0(t)\left\{\left[1+\frac{C_pk\varphi(t)}{1+kn(t)}\right]+\frac{kn'(t)}{1+kn(t)}\right\}=0 \tag{6.75}$$

And taking the initial conditions:

$$\begin{gathered}p_0(\tau_1)=\frac{\Delta(\tau_1)C_p}{1+kn(\tau_1)},\\ p_0^{'}(\tau_1)=-\frac{p_0(\tau_1)kn\varphi(\tau_1)}{1+kn(\tau_1)},\end{gathered} \tag{6.76}$$

the general solution of Eq. (6.75) with initial conditions is written as:

$$p_0(t)=p_0(\tau_1)\left\{1-\frac{\eta kC_p\varphi(t)}{1+kn(t)}\int_{\tau_1}^{t} e^{-\int_{\tau_1}^{\tau}\left(\eta\left[1+\frac{kC_p\varphi(x)}{1+kn(x)}\right]+\frac{kn'\varphi(x)}{1+kn(x)}\right)dx} d\tau\right\}. \tag{6.77}$$

Expression (6.77) determines the stress changes in time on the ground—spherical sound contact and takes into account the system fixation rigidity, change of the instantaneous modulus of ground deformation, aging and ground skeleton creeping.

In a particular case when the instantaneous modulus of ground deformation is constant, expression (6.77) is simplifying and has a form:

$$p_0(t)=p_0(\tau_1)\left\{1-\frac{kC_pC_0}{1+kn+C_pC_0}\left[1-e^{-\eta\left(1+\frac{kC_pC_0}{1+kn}\right)t}\right]\right\}, \tag{6.78}$$

where $C_0=1/E_1$.

In the case when the system is rigid, where fixing the initial deformations substantially exceeds the ground rigidity, i.e., at $n=C_p/E_M\rightarrow\infty$, the condition of pure stress relaxations in the ground massif is:

$$p_0(t)=p_0(\tau_1)\left\{1-\frac{E_M}{E_1+E_M}\left[1-e^{-\eta(1+E_M/E_1)t}\right]\right\}. \tag{6.79}$$

In the above considered cases, in order to pass to the cone resistance the relationship $p_0=3R/2$ can be used.

Let us determine the parameters of ground deformability and creeping. The expression (6.77) can be written for the condition R(t) as:

$$R_0(t)=R_0(\tau_1)\left\{1-\frac{kC_pC_0}{1+kn+C_pC_0}\left[1-e^{-\eta\left(1+\frac{kC_pC_0}{1+kn}\right)t}\right]\right\}. \tag{6.80}$$

From this equation of the initial value of $R(\tau_1)$ the modulus of instantaneous deformation is:

$$E_M=\frac{3kC_pR(\tau_1)}{2\Delta(\tau_1)C_p\pi r_0^2-3R(\tau_1)}, \tag{6.81}$$

where k, C_p, $\Delta(\tau_1)$ are determined by Eqs. (6.71) and (6.74).

The modulus of long-time deformation by the stabilizing value of $R(\infty)$ is determined by:

$$E_1 = \frac{C_p\left[k - 1 - \dfrac{R(\infty)}{R(\tau_1)}\right]}{(1+kn)\left[1 - \dfrac{R(\infty)}{R(\tau_1)}\right]}. \tag{6.82}$$

By the measuring data of damping in time of the value R(t) the ground creeping parameter is determined:

$$\eta = -\frac{1+kn}{t\left[1 - k(n + C_pC_0)\right]} \ln\left\{1 - \frac{\left[1 - \dfrac{R(t)}{R(\tau_1)}\right][1 + kn + C_pC_0]}{kC_pC_0}\right\}. \tag{6.83}$$

Thus, the theoretical solution of the problem on imbedding of spherical sound into an elastic-creeping medium allows determining of deformation and rheological ground parameters by the measured data on the cone ground resistance to sound imbedding at static penetration. The above obtained formulae allow calculating the modulus of instantaneous and long-time deformation and the creeping parameter at different regimes of investigation.

Sometimes, in engineering geological practice, there are cases when mechanical properties of the ground in flowing consistency should be obtained. This is an actual problem for marine design and construction. The grounds of flowing consistency do not possess elastic characteristics and at small stresses run as a viscose medium. Calculation of such a ground is provided by the model of Bingham—Shvedov's viscose-plastic body or by Newton's viscose body (Maslov 1969). In the last case the relationship between the deformation rate and stresses is written in the form:

$$T = \eta_t v, \tag{6.84}$$

where τ is the tangential stress, 10^5 Pa; η_t is the viscosity flowing coefficient; v is the shift deformations, c^{-1}.

For determination of the viscosity flowing coefficient of loose grounds at tests by imbedding of spherical sound, the Stokes solution on imbedding of a heavy sphere into a viscose medium can be used:

$$\eta_c = \frac{\gamma_s - \gamma}{18v} d^2, \tag{6.85}$$

where γ_s is the bulk density of the sphere; γ is the ground density; d is the sphere diameter; v is the sphere imbedding velocity.

6.4 Two-Dimensional Axis-Symmetric Problem of Relaxation Stress

In addition to the cone ground resistance, the penetration logging complex includes the friction jacket gauge, which measures ground friction along the jacket surface.

The ground friction parameter has been measured at static penetration by a number of researchers. One group of them used the results of measurements for estimation of the pile bearing capacity (Faierstein and Makarov 1964) while the other group tried the parameter to be used for estimation of characteristics of ground shift resistance. An interesting experiment was carried out by Begemann (1965) where a well correlation between the friction value and specific rotation shift resistance at the vane gauge test was found. On that basis, an empirical relationship for determining the shift parameters (c and φ) was proposed.

Let us consider in more detail the possibility of measured parameters at the study of loose sediments.

As was noted earlier, with the imbedding of the cone sound below a critical depth the process of preferential lateral compaction occurs. It is known from the theory of ground compaction that in this case two forms of deformation occur, namely, elastic or reversible and non-elastic or residual. Development of non-elastic deformations results in dislocation of separate particles relative to each other. This process is close to the process of the ground shift and some researchers by mistake accept it as the beginning of the ground destruction. In this case one should follow the classical definition of shift and compaction conceptions.

The shift deformations are deformations under outer forces that do not lead to a volume of media change. The compaction deformations are those that lead to change in form and volume of the media.

Cone ground resistance gauge 1 (Fig. 6.12) is fixing the ground resistance R to its compaction, which occurs because of elastic and non-elastic deformations. By friction gauge 2, the shift resistance T along the lateral surface of the jacket at normal ground pressure σ is measured. The normal pressure is the pressure of normal ground resistance that occurs at its compaction with the elastic deformations. If the elastic deformations are absent then the value of normal pressure will be equal to zero. The ultimate resistance of ground shift fixing by the friction gauge for the ground without cohesion is written as:

$$T = \sigma \, \mathrm{tg}\, \varphi_t \tag{6.86}$$

where $\mathrm{tg}\varphi_t$ is the coefficient of ground friction along the lateral jacket surface.

From expression (6.86) it is easy to find the angle of friction φ_t when the values T and σ are measured. Because the material of the gauge jacket at penetration tests is the same then it is possible to find the transition coefficients for different ground types and to determine the angle of inner friction using the value of φ_t.

Determination of the friction coefficient $\mathrm{tg}\varphi_t$ by measuring of the friction parameter T and the normal pressure σ is also of great interest for estimation of the pile bearing capacity. As was said above, the registered friction T represents the

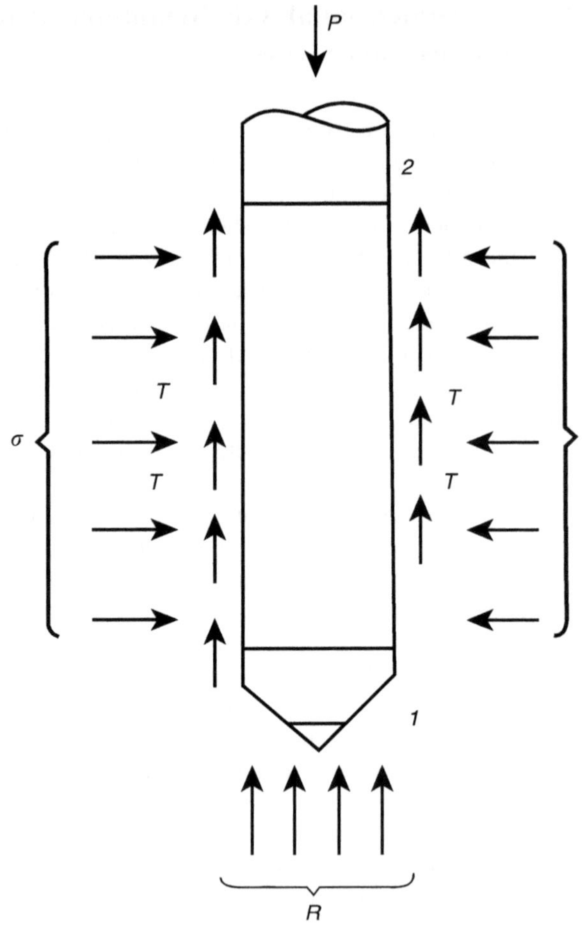

Fig. 6.12 Scheme of forces acting along the friction gauge: (*1*) is the cone ground resistance gauge; (*2*) is the friction resistance on the ground along the jacket surface. (Ferronsky 1969)

instantaneous ultimate resistance to the ground shift of the gauge lateral surface at the normal pressure σ. The last value will be changing in time owing to the relaxation stress and the pore pressure drop in water bearing grounds. In this regards, the most objective characteristic for the pile bearing capacity with respect to friction is the friction coefficient $\mathrm{tg}\varphi_t$, which is easily determined on the basis of experimental data. At the same time it should be borne in mind that the $\mathrm{tg}\varphi_t$ value substantially depends on the roughness of the gauge jacket and pile material, which can be taken into account by corresponding correction coefficients obtained by calibration work.

Measuring of the normal pressure is also of interest in connection with determination of the compaction thickness border for calculation of construction settling. At present, the border of the compaction thickness is considered the ground depth where extra stresses by the mass construction are accounted for by 20 % of the natural pressure of the upper ground masses. This criterion of the compaction thickness is accepted conventionally from observational data. More justified recommendations have not been proposed. A more reasonable idea follows from the principle of ultimate equilibrium of the construction and ground thickness work at a given settling. Such a method of calculation for a foundation based on the principle of ultimate deformations has been considered (Kuzmin and Ferronsky 1963). But at that time, the field method for recording continuous diagrams of ground property was absent. Now such methods are available and the problem of foundation calculation based on ultimate deformations can be reanimated.

It is obvious that elastic ground deformations are unessential for normal work of a construction. The normal pressure measurement at static penetration can be the direct method of the pressure diagram recording along the geologic section, which initiates elastic ground deformation. In this case the border of compaction thickness locates on the depth where extra pressure initiated by the outer loading will be equal to the normal pressure initiated by elastic resistance at the sound imbedding. In addition, the normal pressure σ is the pressure of elastic ground resistance and it can be used for determination of the real modulus of the ground elasticity, which is needed for the foundation calculation at short-periodic loading in seismic active regions, action of explosive and impact loadings.

Let us present a theoretical solution for the axial-symmetric problem of relaxation stress adaptable to the conditions of normal ground pressure measurement at static penetration. The solution is given on the physical basis presented in the previous section where the grounds are considered as an elastic-creeping medium described by the equation of hereditary creeping.

The problem solution can be used for calculation of pile bearing capacity using the data on the friction and normal pressure measurements at static penetration. The solution can also be applied for determination of the ground deformation modulus and rheology parameters.

Consider the stress-deformable state of the ground around an elastic cylinder (Fig. 6.13). The ground deformation properties will be characterised by the

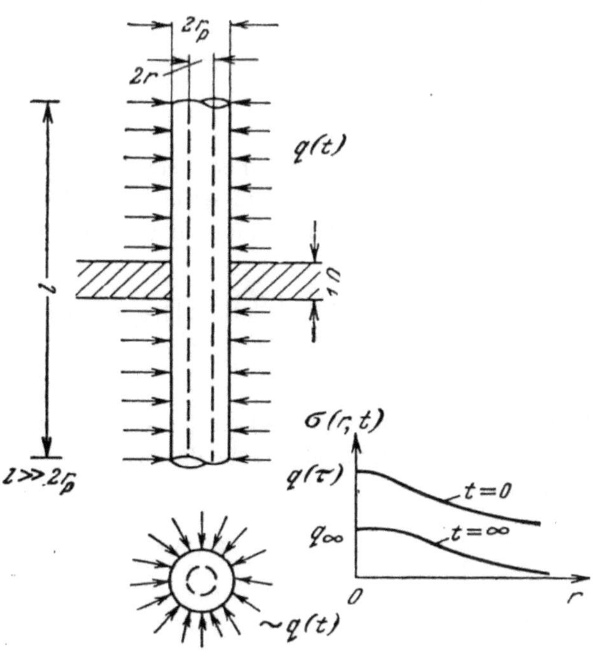

Fig. 6.13 Calculating scheme of relaxation stress from elastic ground resistance. (Ferronsky 1969)

following parameters: $E_{i,}$ is the instantaneous deformation modulus; E_l is the modulus of long-term deformation; μ is the coefficient of transversal compression and η is the creeping parameter.

The problem is solved in a general case when the material of the displaced cylinder has a finite rigidity and the ground skeleton is satisfied to the hereditary creeping theory. We consider sandy, frozen and non water-bearing clay grounds. In this regards, the initial values of the pore pressure and the filtration phenomenon are excluded. The super stresses in the ground massif around the cylinder holes are considered in the framework of a two-dimensional problem. In the process of super stresses damping in ground, the condition of equality of ground and cylinder displacement is kept. In addition, it is assumed that the appeared plastic zone around the sound is determined from the condition of ultimate equilibrium.

During displacement of the elastic cylinder in the ground massif, extra stresses occur. Because the problem is axial-symmetric, then tangential stresses are absent. First consider the task of the sound and ground deformations under the contact stresses changing in time. The stresses from the ground masses are not taken into account because they cannot relax. The stress components in the surrounding ground massif can be found from the theory of elasticity relationships (Bezukhov 1968):

$$\sigma(r)=\sigma(\theta)=\pm q\left(\frac{r_p}{r}\right)^2, \tag{6.87}$$

Where σ (r) and (θ) are the normal stresses in the ground in the distance r; r_p is the radius of the cylindric sound; q is the contact stresses.

Expression (6.87) can be used for a description of the elastic-creeping medium stressed state on the basis of Arutiunian's theorem of a constant Poisson's coefficient. We then have:

$$\sigma(r)=\sigma(\theta)=\pm q(t)\left(\frac{r_p}{r}\right)^2. \tag{6.88}$$

This equation means that the stress state of the elastic-creeping medium induced by outer forces is identical to stresses occurring for the elastic-instantaneous problem. The radial relative deformations in this field are written as follows (Arutiunian 1952):

$$\varepsilon_r(t)=\frac{\sigma_\theta(t)-\mu\sigma_r(t)}{E_M(t)}-\int_{\tau_1}^{t}[\sigma_\theta(t)-\mu\sigma_r(t)]\frac{\partial}{\partial\tau}\delta(t,\tau)d\tau. \tag{6.89}$$

Taking into account Eq. (6.89) and bearing in mind the known relationship between relative deformation $\varepsilon_r(t)$ and displacement value $u_r(t)$ in the radial direction, namely:

$$\varepsilon_r(t)=\frac{\partial}{\partial t}u_r(t)$$

we obtain an expression for displacement of the sound surface, which is in contact with the ground in the form:

$$u_r(t)=(1+\mu)r_p\left[\frac{q(t)}{E_M(t)}-\int_{\tau_1}^{t}[\sigma_\theta(t)-\mu\sigma_r(t)]\frac{\partial}{\partial\tau}\delta(t,\tau)d\tau\right]. \tag{6.90}$$

Radial displacement of the contact with the ground elastic sound surface initiated by the corresponding stresses is defined by the expression:

$$u_p(t)=q(t)\frac{1+\mu}{E_p}r_p, \tag{6.91}$$

where E_p and μ_p are the elasticity modulus and Poisson's coefficient for the sound material.

We consider now the law of the contact stresses damping in time, which occur by the dislacement of elastic sound of radius r_p into the ground. The condition of consistency on the contact surface sound—ground is written as:

$$u(t)-u(\tau_1)=u_p(\tau_1)-u_p(t). \tag{6.92}$$

Substituting the values of u(t) and $u_p(t)$ into (6.92), one has:

$$q(t)=\left[\frac{1+\mu_p}{E_p}+\frac{1+\mu}{E_M(t)}\right]=\int_{\tau_1}^{t}(1+\mu)q(\tau)\frac{\partial}{\partial\tau}\delta(t,\tau)d\tau. \tag{6.93}$$

After some transformations, an integral equation is obtained in the form:

$$q(t)=\frac{E_p\Delta}{(1+\mu_p)[1+\omega n(t)]}+\frac{\omega E_p}{1+\omega n(t)}=\int_{\tau_1}^{t}q(\tau)\frac{\partial}{\partial\tau}\delta(t,\tau)d\tau, \tag{6.94}$$

where

$$\omega=\frac{1+\mu}{1+\mu_p},\quad n(t)=\frac{E_p}{E_M(t)},\quad \Delta=u_p(\tau_1)+u(\tau_1).$$

Thus, the problem of determination of the contact stresses q(t) is reduced to solution of the integral Eq. (6.94) with the core:

$$k(t,\tau)=\frac{1}{1+\omega n(t)}\frac{\partial}{\partial\tau}\left[\frac{1}{E_M(\tau)}+\frac{1}{E_l(t,\tau)}\right]$$

with the free term:

$$f(t)=\frac{E_p\Delta}{(1+\mu_p)[1+\omega n(t)]}$$

and with the parameter:

$$\lambda_0=\Omega E_p.$$

The integral equation after some transformation can be rewritten as:

$$q''(t)+q'(t)\left\{\eta\left[1+\frac{\omega\varphi(t)E_p}{1+\omega n(t)}\right]+\frac{\omega n'(t)}{1+\omega n(t)}\right\}=0. \tag{6.95}$$

So, the solution of the integral Eq. (6.94) I is reduced to the solution of the homogeneous linear differential equation of second-order (6.95) with a constant coefficient and the following initial conditions:

$$\begin{aligned} q(\tau_1)&=f(E_p,\mu_p,E_M(\tau_1)\varphi,c),\\ q^{'}(\tau_1)&=-\frac{q(\tau_1)\eta\omega\varphi(\tau_1)}{1+\omega n(\tau_1)}. \end{aligned} \tag{6.96}$$

The general solution of Eq. (6.95) is written in the form:

$$q(t)=q(\tau_1)\left\{1-\frac{\eta\omega E_p\varphi(\tau_1)}{1+\omega n(\tau_1)}\int_{\tau_1}^{t}e^{-\int_{\tau_1}^{t}\left[\eta\left(1+\frac{\omega E_p\varphi(x)}{1+\omega n(x)}\right)+\left(\frac{\omega n'(x)}{1+\omega n(x)}\right)\right]dx}d\tau\right\}. \tag{6.97}$$

Expression (6.97) determines the rule of the stress changes in time on the contact cylinder—ground including the creeping and aging on the ground. The components of stresses in the surrounding massif can be determined by the above written relations (6.88). For the task, we accept that the instantaneous modulus of ground deformation is constant, i.e., $E_M(\tau)=E=\text{const.}$ and the ground aging is absent, i.e., $\varphi(\tau)=1/E_l=\text{const.}$ In this case expression (6.97) acquires the form:

$$q(t)=q(\tau_1)\left\{1-\frac{\omega E_p./E_l}{1+\omega n E_p./E_l}\left[1-e^{-\eta\left(1+\frac{\omega E_p/E_l}{1+\omega n}\right)t}\right]\right\}. \tag{6.98}$$

In the case of higher cylinder rigidity in comparison with the ground, i.e., at $n = E_p/E_1 \to \infty$ the condition of pure relaxation of the ground massif occurs:

$$q(t)=q(\tau_1)\left\{1-\frac{E_p./E_1}{1+E_p./E_1}\left[1-e^{-\eta(1+E_M/E_1)t}\right]\right\}. \tag{6.99}$$

In this case the stresses are determined by the expression:

$$q(t)_{=\infty}=q(\tau_1)\frac{1}{1+E_p./E_1} \tag{6.100}$$

Let us determine initial contact pressure, assuming that in the surrounding ground massif around the sound a plastic zone of radius r_p has formed. We employ for this the equation of ultimate equilibrium in the form (Sokolovsky 1969):

$$\sigma_r - m\sigma_\theta = n, \ (\sigma_r = \sigma_1 > \sigma_\theta = \sigma_3), \tag{6.101}$$

where:

$$m=\mathrm{tg}\left(\frac{\pi}{4}+\frac{\varphi}{2}\right)\mathrm{ctg}\left(\frac{\pi}{4}-\frac{\varphi}{2}\right),$$
$$n=2\mathrm{ctg}\left(\frac{\pi}{4}+\frac{\varphi}{2}\right). \tag{6.102}$$

Substituting Eq. (6.101) into the differential equation of equilibrium and assuming that $\tau_{rz}=0$ and $\sigma_z=\mathrm{const}$, one has:

$$\sigma_k - \sigma_\theta + r\frac{d\sigma_k}{dr}=0, \tag{6.103}$$

After integration we obtain:

$$\sigma_k=\left[q(\tau_1)+\frac{n}{m+1}\right]\left(\frac{r}{r_r}\right)^{\frac{1-n}{n}}-\frac{n}{m-1}=0. \tag{6.104}$$

Substituting this solution into Eq. (6.88) and assuming that at the contact of elastic and plastic zones the normal stresses are equal to one another, then for $r=r_n$ we obtain:

$$q(\tau_1)=\left[\frac{n}{m-1}+\frac{n}{m+1}\right]\left(\frac{r_r}{r}\right)^{\frac{1-m}{m}}-\frac{n}{m-1}. \tag{6.105}$$

Thus, initial contact normal pressure on the cylindrical sound surface depends on the sound radius r_r, radius of the plastic zone r_p and the strength parameters m and n. In the particular case when $r_r \approx r_n$, i.e., when the plastic zone radius is small, we have:

$$q(\tau_1) = \frac{n}{m+1}, \tag{6.106}$$

And at $m=0$, i.e., when $\varphi=0$, we obtain:

$$q(\tau_1) = c, \tag{6.107}$$

where c is the ground cohesion force.

The plastic zone radius around the sound is also in a first approximation and can be determined from the condition of equality of the sound transversal section area. The ground ring area has formed by the ground displacement aside from the sound, i.e., we have:

$$r_n = r_r\sqrt{2} - u_{r_r}, \tag{6.108}$$

where u_r is the shift of the ground cylindrical surface at $r = r_r$ determined from the condition (6.90), i.e.,

$$u_{r_r} = \frac{1-\mu}{E_M(\tau_1)} r_r\, q(\tau_1). \tag{6.109}$$

Substituting value r_n from (6.108) into (6.105), we obtain the relationship between $q(\tau_1)$, r_r, $E_M(\tau_1)$, μ, m, n.

For determining the ground deformation parameters the contact stresses around the cylindrical sound should be measured by the normal pressure gauge. The rigidity Cp of the cylindrical gauge of normal pressure is determined from the expression:

$$C_p = \frac{r_K q}{u_r}, \tag{6.110}$$

where q is where the stress occurs around the relaxometer; r_r is the relaxometer radius; u_r is the radial displacement of the outer surface.

On the basis of the above considered solution it is possible to write the expression for determining the ground contact stresses around the relaxometer:

$$q(t) = q(\tau_1)\left\{1 - \frac{\omega C_p . C_0}{1+\omega_1 n_1 + \omega_1 C_{p1}}\left[1 - e^{-\eta\left(1+\frac{\omega C_p C_{01}}{1+\omega_1 n_1}\right)t}\right]\right\}, \tag{6.111}$$

where:

$$n_1=\frac{C_p}{E_M},\omega_1=1+\mu,C_0=_0=\frac{1}{E_Д}$$

By the value and rate of contact stress damping it is easy to determine the ground deformation and rheological parameters. In fact, if the relaxation radius $r_{p,}$ rigidity C_p and initial values of the contact stresses q (τ_1) are known, then it is easy to de-terssmine the modulus of instantaneous deformation:

$$E_M=q(\tau_1)\frac{r_r(1+\mu)}{u(\tau_1)}. \tag{6.112}$$

If in addition the stabilising values of the contact stresses $q(\infty)$ are known, then from (6.111) the modulus of long-term deformation E_1 can be determined by substituting there the value:

$$E_1=\frac{E_M}{q(\tau_1)-q(\infty)}\left[\frac{\omega q(\tau_1)}{}-1\right]. \tag{6.113}$$

With known values of E_M and E_1 and the curve of the contact stress damping, the parameter of ground creeping can be determined, using (6.111):

$$\eta=-\frac{1+\omega n}{t(1+\omega n+\omega E_{M_1}/E_1)}\ln\left\{1+\frac{[q(t)-q(\tau)_1][1+\omega nE_{M_1}/E_1]}{\omega E_{M_1}/E_1(\tau_1)}\right\}. \tag{6.114}$$

Thus, the problem about initial contact pressure on a cylindrical sound surface at its displacement in the elastic-creeping medium including formation of the plastic zones of flow is completely solved. By field measuring of contact stresses by normal pressure gauges it is possible to determine the ground deformation and rheological characteristics based on rigorous theoretical solutions. If the ground does not develop the creeping property then the expression (6.111) can be used for determining the elastic properties.

6.5 Conditions for Measuring Ground Parameters by Static Penetration

The conditions for measurement of frontal ground resistance, friction and normal ground pressure at static penetration are of great importance for quantitative interpretation of mechanical ground properties in a geological section and for use of these data for lithological strata separation. Some time ago the cone ground resistance at a penetration test was measured on the surface by a number of different dynamometric apparatus. Such a method is simple with respect to technical implementation and service. But it was noticed that during sound displacement the penetration rod is subjected to great loading. In this regards, the recorder of the cone resistance fixes only 30 % of its real value and 70 % comes from the rod resistance. The attempt to use casing tubes decreases the rod influence but it is not sufficient to solve the problem.

The only way to obtain accuracy is application of the measuring gauge mounted in the sounding probe. In this case one can be sure that the recorded parameters are real resistance that we use for interpretation.

We have not discussed the problem of rod stability during loading at penetration. This is a separate problem of mechanics. But the conclusion from our practice is that for qualitative interpretation of penetration test data, the gauges should be placed in the measuring probe.

References

Arutiunyan NH (1952) Some problems of the creep theory. Gostekhteorizdat, Moscow

Bondarik GK (1964) Dynamic and static penetration ground test in engineering geology. Nedra, Moscow

Begemann HK (1965) The friction jacket cone as an aid in determining the soil profile. In: Proc. 6th ICSMFE, Montreal 1:17–20

Berezantsev VG (1952) Axial-symmetric problem of the ultimate equilibrium theory for the quick medium. Gostekhizdat, Moscow

Berezantsev VG, Yaroshenko VA (1957) Bearing capacity of the deep sandy foundation. In: Papers to the 4th Intern. Congress on Soil Mechanics. Izd. AN SSSR, Moscow, pp 143–152

Bezukhov NI (1968) Fundamentals of the theory of elasticity, plasticity and creep. Vyshaya Shkola, Moscow

De Beer EE (1963) The scale effect in the transposition of the results of cone analysis. Geotech 13:39–40

Durante VA, Kogan YaL, Ferronsky VI, Nosal SI (1957) Field investigation of density and moisture of soils. In: Papers to the 4th Intern. Congress on Soil Mechanics. Izd. AN SSSR, Moscow, pp 66–78

Faierstein BD, Makarov PN (1964) Rig C-832 for the ground sounding. In: Installations for for the ground sounding and drilling. CBTI Gosstroy USSR, Moscow, pp 3–24

Ferronsky VI (1969) Penetration logging methods for engineering geological investigation. Nedra, Moscow

Ferronsky VI, Gryaznov TA (1979) Penetration logging. Nedra, Moscow

Florin VA (1961) Fundamentals of soil mechanics. Gosstroyizdat, Moscow

Kuzmin PG, Ferronsky VI (1963) Foundation project on the ultimate state. Rosvuzizdat, Moscow

Maslov NN (1969) Steady state and deformation of support wall. Energiya, Moscow

Menzenbach E (1959) Die Anwendbarkeit von Sonder zur Prüfung der Festigkeitseigenschaften des baugrundes. Forschugsberichte des Landes Nordrheit-Westfalen, N 713. Westdeutcher, Opladen, pp 215–220

Meyerhof GG (1951) The ultimate bearing capacity of foundation. Geotech 2:301–332

Meyerhof GG (1956) Penetration tests and bearing capacity of cohesionless soils. J. Soil mech. Found Div. Proc. ASCE 82:1–19

Meyerhof GG (1961) The ultimate bearing capacity of wedge-shaped foundations. In: Proc. 5th ICSMFE, Paris 2:105–109

Mindlin RD (1936) Force at a point in the interior of a semi-infinite solid. Physics 7:195–202
Prandtl L (1920) Über die HÓ"rte plastischer körper. In: Nachr Ges Wiss Göttingen, Math Phys, pp 75–85
Proceedings (1974–1975) Proceedings of the European Symposium on Penetration Testing. ESOPT, Stockholm
Sanglera G (1971) Investigation of soils by penetration test (transl. from French). Stroyizdat, Moscow
Schultze E, Melzer K (1965) The determination of the density and the modulus of compres- sibility of non-cohesive soils by soundings. In: Proc. 4th ICSMFE, Montreal, pp 354–358
Sokolovsky VV (1969) The theory of plasticity, 3rd edn. Vyshaya Shkola, Moscow
Terzaghi K (1961) Theory of soil mechanics (transl. from Germany). Gosstroyizdat, Moscow
Timoshenko SP (1934) Theory of elasticity (transl. from English). Gostekhizdat, Leningrad
Trofimenkov YuG, Vorobkov LN (1964) Field methods of the grounds study. Stroyizdat, Moscow
Tsitovich NA, Ter-Martirosyan ZG (1981) Fundamentals of applied geomechanics in construction. Vischaya Shkola, Moscow
Yaroshenko VA (1964) Interpretation of the ground static penetration results. In: Design of complicated foundations. Fundamentproject, Moscow 99:14–24
Zaretsky YuK (1967) Theory of the grounds consolidation. Nauka, Moscow

Chapter 7
Experimental Studies and Interpretation of Penetration Logging Data

Abstract Penetration logging data, which are obtained at field works in the form of a complex of logging diagrams, hold extensive information on the physical and mechanical properties of studied sediments, the lithological structure and content of a section, variability of properties and contents in-depth and spread of strata. Some knowledge and experience are needed for transfer of that information into the form of quantitative and qualitative data used for the calculation and design of different building constructions and geological mapping. Good knowledge of the physical and theoretical fundamentals of methods is needed for the understanding of different effects affecting the interaction of nuclear radiation with soil and rocks. And also the purpose of that information and its quantity should be clear. In this chapter different aspects of interpretation of penetration logging results and some questions of conditions for application of the methods are discussed. In particular, attention is paid to the physical processes in the ground at probe displacement and the effects of probe parameters (diameter, displacement rate, position of the friction gauge, probe dimension and so on). Finally, an estimate is made of the applicability of the theoretical solutions presented in Chap. 6 on using cone resistance and lateral friction for determination of soil mechanical properties.

7.1 Density, Moisture, Porosity, Groundwater Level

The GGL method is analysed from the standpoint of its application in geoengineering studies to obtain quantitative logging diagrams of the density distribution across deposits and to analyse lithologic sections and the persistence of individual lithologic-genetic rock types by the compactness of the deposits. The most interesting problems in this regard, which are examined in the first instance, are the sensitivity of the method to changes in density and the accuracy of the measurements. Studies of the effect of probe length and of the source energy on the sensitivity of the technique were carried out in experimental areas by the following means. The experimental areas were chosen on the basis of exploratory drilling in such a way that at least two types of ground were included in the section through the deposits (as a rule, sand and clay) with different values of absolute densities and in the immediate neighbourhood of each other (approximately 1 m apart). The depth to

V. I. Ferronsky, *Nuclear Geophysics*, Springer Geophysics,
DOI 10.1007/978-3-319-12451-3_7

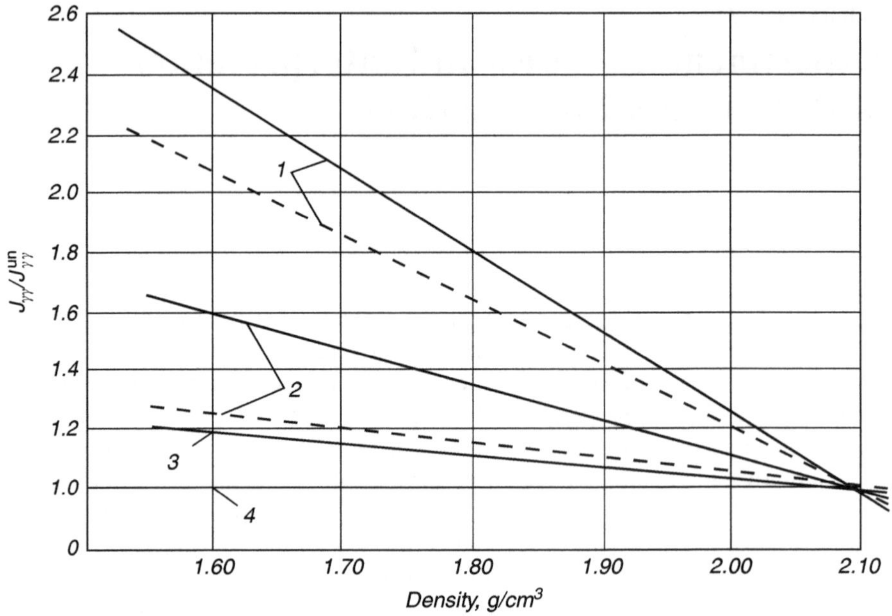

Fig. 7.1 Sensitivity of the GGL method with respect to changes in the ground density as a function of probe length and gamma energy ———— ^{137}Cs, – – – – – – – ^{60}Co. (Ferronsky and Gryaznov 1979)

which the probes were driven did not exceed 8 m and the distance from the gamma source to the phosphor varied between 20 and 50 cm. The sources were ^{60}Co and ^{137}Cs. When the probe length was changed the measurements were carried out in the same hole by repeated insertion of the rod carrying the probe and also by driving in the probe at new points. Ground samples were taken in each case at the end of the measurements.

Figure 7.1 shows the experimental results for the sensitivity of the GGL method as a function of the probe length. Data were obtained in an experimental area near the town Istra, in the Moscow region, for a section of sandy loam deposits with densities between 1.6 and 2.1 g/cm^3. The intensity of the recorded radiation was expressed in relative units to facilitate comparison of the results, with $I_{\gamma\gamma}^{sb}$ representing the intensity of the radiation in moraine loam having a density of 2.1 g/cm^3.

It is clear from the analysis of the experimental data that for a probe length of 20 cm the method has practically zero sensitivity. As the probe length increases, the sensitivity begins to increase. For a probe length of 30 cm, the change in density by 0.1 g/cm^3 leads to a 4 % reduction in the recorded gamma radiation. For a probe length of 40 cm, a change in the density by the same amount reduces the recorded intensity by 12 %, whereas for a 50 cm probe the reduction is 27 %.

When the optimal length of the logging probe is chosen it must be remembered that the sensitivity should ensure that the density is determined to within 0.2–0.3 g/cm^3

in geoengineering works. The sensitivity necessary for this purpose is achieved for a probe length of 40 cm or more. Further increase in probe length does lead to a higher sensitivity and requires the use of stronger sources, which is undesirable from a radiation hazard point of view.

Since in penetration logging the measuring probe is in direct contact with the medium under investigation, the maximum working depth is not a decisive factor for obtaining quantitative data on the density of the ground (which is the case for casing wells). This feature of the method enabled us to use ^{137}Cs as the source of gamma rays. As a result of the use of softer radiation it was possible to increase the sensitivity of the method and avoid complications like source decay corrections in long-term work (half-life of ^{137}Cs is 33 years). Comparative studies using cobalt and cesium sources were carried out in a number of experimental areas. It was found that the sensitivity of the method with respect to density is more than 20 % higher in the case of ^{137}Cs as compared with ^{60}Co.

Analysis of the results of these investigations has shown that for GGL with ^{137}Cs source, mean density of the ground material of 1.8 g/cm³ and density changes of up to 0.5 g/cm³ with probe length L, the accuracy α for a relative measurement error of 3 % was as follows:

L, cm	20	30	40	50
α, g/cm³	1.00	0.9	0.03	0.02

We note that the maximum working depth of the method is also reached for a logging probe length of 40 cm or more (as was shown in Chap. 3). The maximum working depth under the conditions of penetration logging is of interest from two points of view. First, when loose Quaternary deposits are investigated, the ground samples to be analysed must be representative to characterise the particular geologic formation by the selected parameter. Since the properties of the samples are highly variable, even for the same geologic-genetic types of rock, there is a tendency in geoengineering practice to take large samples so that they are more representative for the deposits under investigation. It is important to note that diagrams obtained by penetration GGL can be used to investigate in detail the rock variability as far as density is concerned, throughout the depth of the investigated section. In this sense, penetration GGL is more convenient than other methods. Since the effective radius of investigation in sandy and clayey soils is on average 10–12 cm, it follows that, for the optimal probe length of 40–50 cm, the volume of the ground under investigation is equal to the volume of a hollow cylinder 40–50 cm long and 10–12 cm in radius, i.e., to about 15,000–20,000 cm³. Such a sample will be much more representative of the medium under investigation than the samples obtained by other methods used at present. Secondly, the penetration GGL method suffers from (perhaps the only one) disadvantage that the medium around the probe becomes compacted. Existing data obtained by a number of practical specialists have shown that the radius of this compacted zone depends on the type of the ground, the density of its structure and the presence of structural connections. It is equal to 2–3 times of the probe diameter.

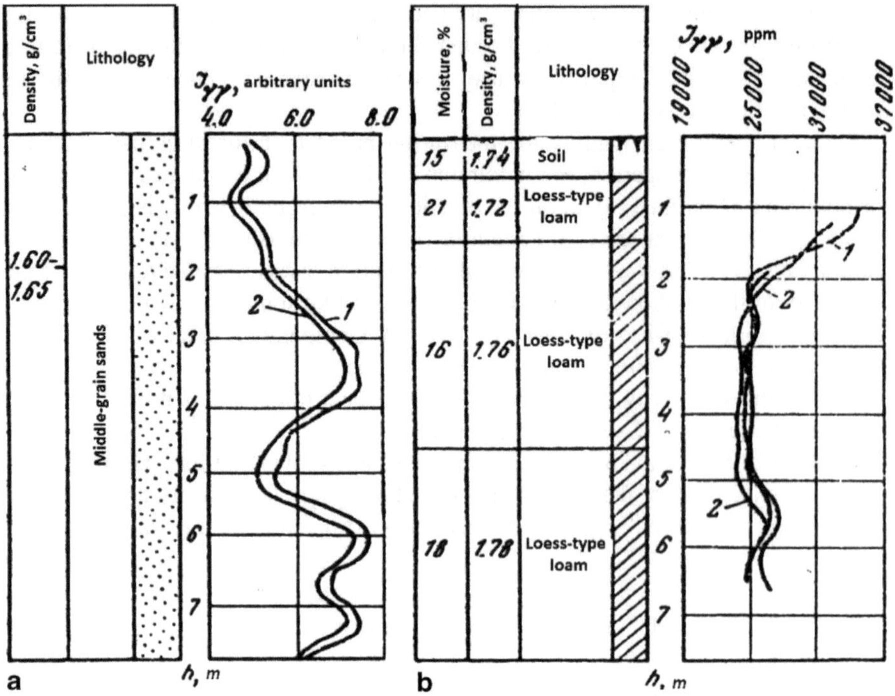

Fig. 7.2 Effect of lateral compaction of sandy soil on intensity of the recorded scattered gamma and neutron radiation: *1* probe driven into undisturbed ground for the first time; *2* second insertion of the probe into the hole filled with sand. (Ferronsky and Gryaznov 1979)

A number of experiments were carried out to verify the effect of such a lateral compaction of the ground on the density determinations. Penetration GGL diagrams were recorded in grain sand medium. After the probe was withdrawn, the well was filled with sand (taken from an adjacent hole), which was then lightly rammed. The probe was then again pressed in and the diagram was recorded once more. In the second operation the sand was compacted in the axial direction and produced an interference effect (Fig. 7.2a). It is clear from these data that the error produced as a result of lateral compaction does not exceed 4 % of the initially recorded gamma intensities.

Another series of experiments was carried out on an area covered with loess-type loams. A well was produced by pushing in a hollow cylinder 50 mm in diameter to a depth of 8 m with the resulting core being removed. GGL and NNG diagrams were then recorded in the well by inserting into it a probe of 60 mm in diameter. No lateral compaction was produced during this process. One metre away from the experimental well a similar diagram was recorded with later compaction of the ground. The lateral compaction has practically no effect in this case. Evidently the interference produced by lateral compaction lies within the limits of experimental error. An analogous experiment was carried out with recording of the NNL diagrams (Fig. 7.2b).

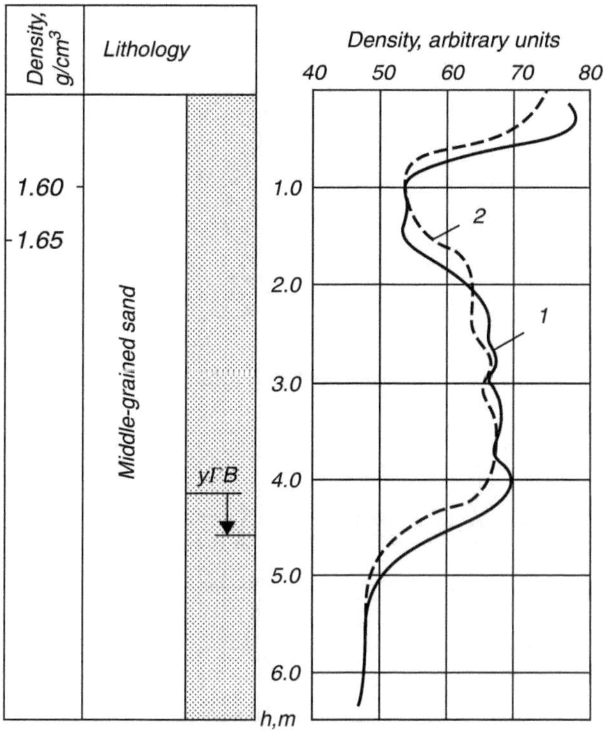

Fig. 7.3 Effect of the speed of probe insertion on variation of the recorded intensity: *1* 2 m/min; *2* 6.5 m/min. (Ferronsky and Gryaznov 1979)

Since it is possible to use highly efficient installations for GGL, which are capable of driving logging probes into the ground at the rate of 6 m/min or more, let us estimate the effect of the rate of insertion of the probe on the final data. This type of estimation was carried out in a number of experimental areas. Figure 7.3 shows the GGL diagrams obtained in sections of sand deposits with various speeds of insertion (v) of the logging probe. When the necessary constant time of the measuring instrument is suitably chosen, the recorded radiation intensity is independent of v.

The GGL diagrams were calibrated in units of density by taking direct specimens from holes in the neighbourhood of the region where the soundings were carried out. This automatically excluded the effect of lateral compaction of the ground when the probe was driven in since the GGL diagrams were thus referred to the density of undisturbed ground. An analogous procedure for calibration of the NNL diagrams was used.

Let us now compare the results of some experimental estimates of the accuracy of the measured density and moisture and the differentiating capability of both methods as used to investigate sections of loose deposits. The GGL and NNL data were compared with standard laboratory determinations of the same parameters by direct weighting of the soil samples. The samples were taken from holes (four from each) in steps of 20 cm, using 200 cm^3 rings.

Figure 7.4 shows the comparing results for density measurements. The points represent data obtained by direct weighting of the specimens taken from holes. The

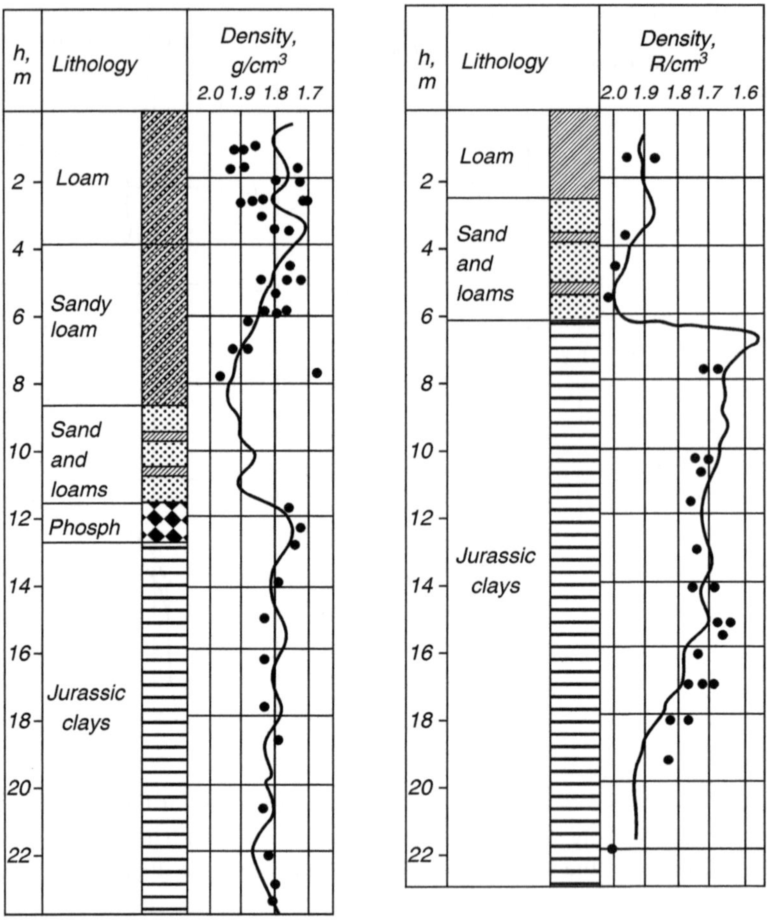

Fig. 7.4 Comparative data on the accuracy of density measurements by the GGL method and by the standard cutting-ring method for different types of deposits. (Ferronsky and Gryaznov 1979)

results were obtained under very inconvenient conditions, in the region of glacial deposits, whose properties are very variable in depth. It is clear from the graphs that, on the whole, there is a satisfactory agreement between the GGL data and the results of the sample measurements. The overall character of the variation of density with depth is also the same. Moreover, the following interesting phenomenon was observed. A spread in the values of the density, even within a given type of deposit, was noted throughout with the deviations of individual determinations from the mean density occasionally reaching 0.1–0.2 g/cm^3. The GGL diagrams, on the other hand, take an automatic average over these data. The result is a reasonably smooth curve of density as a function of depth, where each point represents an average over a volume of 15,000–20,000 cm^3. This is the great advantage of GGL as compared with direct weighting. It is also important to note a further interesting phenomenon. Analysis of an enormous volume of experimental data showed that the density of

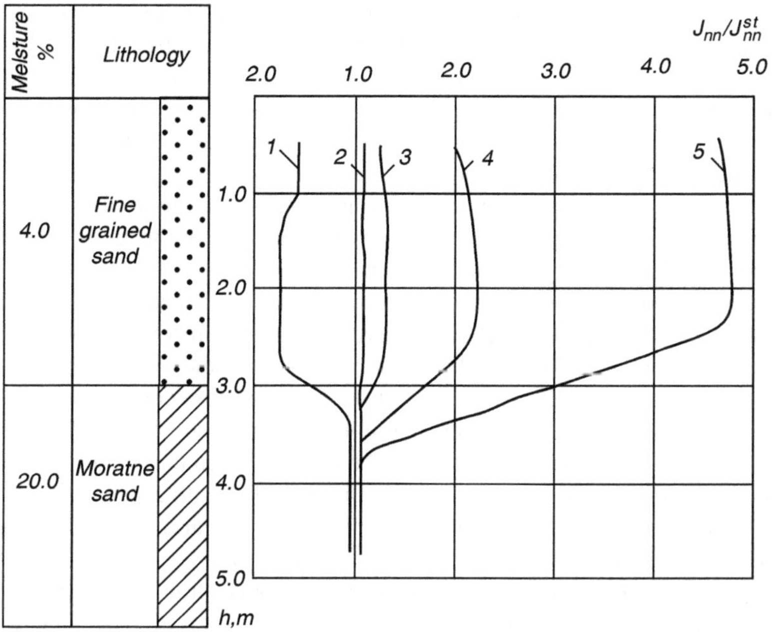

Fig. 7.5 Sensitivity of the NNL method with respect to changes in the moisture content for various probe lengths. (Ferronsky and Gryaznov 1979)

sands and loams below the groundwater level shows practically no fluctuations and persists in depth within individual lithologic-genetic soil types.

In addition to the above properties, GGL diagrams show a considerable differentiating capability with regard to the lithology of a section and the persistence of individual types of deposit as far as the density is concerned. Thus, for example, it is clear from the figures that the diagrams clearly indicate transmissions from sandy deposits to moraine loams by density change. When a sufficient number of GGL diagrams are available for a given density-depth profile, it is possible to follow the thickness of the individual strata.

We carried out experiments under field conditions designed to provide an estimate of the effect of the probe length on the sensitivity of the NNL method. The experimental area was chosen on the basis of preliminary borehole exploration, which was aimed at finding locations with a sufficiently broad range of variation in moisture content. Sandy-argillaceous sections in glacial deposits usually satisfy these conditions. Logging probes of various lengths were then driven into the ground at points separated by about 1 m, so that the final data were comparable. Since there were no mineralised waters or rock-forming elements with anomalous neutron capture cross-sections on the experimental areas, the scattered neutral radiation was recorded by detecting thermal neutrons.

Figure 7.5 shows the results obtained in a section of sand and moraine loam, using probe lengths of 5–50 cm and for a moisture variation range from 4 to 20%.

For sensitivity estimating, the experimental results are given in relative units. The standard intensity $I_{\gamma\gamma}^{sb}$ was taken to be the density of the neutron flux for moraine loam. It is clear from the graph that when the moisture content w_0 changed up to 16 % the recorded thermal-neutron flux density for the 5-cm long probe increased by a factor of 1.7. For the 25-cm probe the sensitivity was practically, zero, which corresponds to the inversion region on the relationship under investigation. Further increase in the probe length resulted in an increase of sensitivity. In this case, we have a reciprocal dependence of I_{nn} on w, i.e., the intensity of the recorded radiation decreases with increasing moisture content.

Differentiation of the moisture content in the unsaturated zone is of great interest. Analysis of the experimental data showed that the 5-cm NNL probe has a completely satisfactory sensitivity to the moisture content in the unsaturated zone. The dependence of the recorded neutron flux I_{nn} on the moisture content w is linear for moisture variations up to 20 %. A change in the moisture content by 3 % gives rise to more than a 10 % change in the recorded intensity. When the 5-cm probe is used, the practical sensitivity of the NNL method is such that with an instrumental error of 2–3 % the moisture can be determined to less than 1 % and this accuracy can be substantially improved when longer probes are employed.

Analysis of experimental data shows that in the unsaturated zone (w = 3–20 %) and for instrumental errors of the ordered 3 %, the accuracy with which the moisture content can be determined is as follows (function of the probe length):

L, cm	5	25	35	45	50
α, g/cm^3	1	5	2	0.7	0.6

When the possible error in the measured moisture content is estimated, the most important factor that must be examined is the effect of the lateral compaction of the ground when the probe is driven into it. We carried out a number of experiments to investigate this effect and to estimate the errors in the measured moisture applying the above described GGL procedure.

Let us now consider the results of field experiments on the accuracy of quantitative determinations of the moisture content by NNL, in which the NNL data were compared with standard analysis of specimens by drying. The ground specimens for thermostatic drying were taken from boreholes in the region where the sounding was carried out. Figure 7.6 shows the results for a number of types of glacial deposits. The points indicate data obtained by thermostatic drying.

Analysis of these and other experimental data shows that the results under comparison are very close in absolute magnitude. There is also a satisfactory agreement between the overall variations of the moisture content with depth. At the same time, there is a spread of the individual values of the moisture content obtained by the weighing method with these deviations occasionally reaching 5–7 % of the absolute value of moisture. This phenomenon is explained by the fact that "point" determinations represent average moisture for sample volumes that do not exceed 10–20 cm^3. In the NNL diagram, on the other hand, each point represents on average a sphere

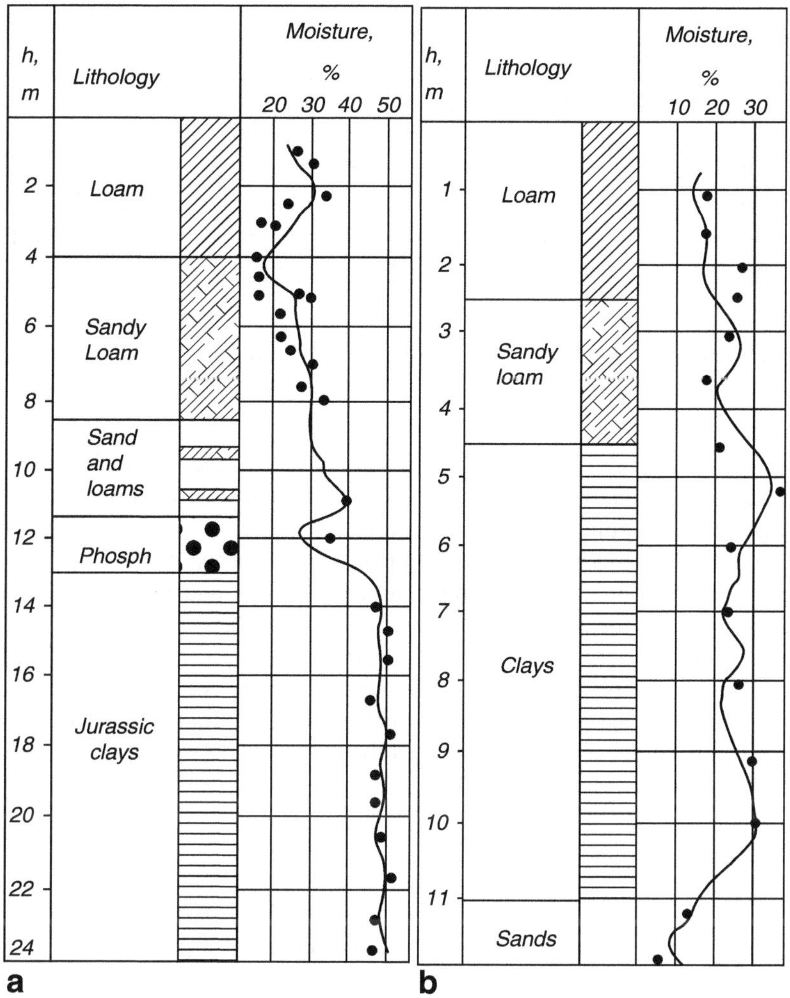

Fig. 7.6 Comparative data on the accuracy of moisture measurements by the NNL method and by the standard cutting-ring and drying method for different types of deposits. (Ferronsky and Gryaznov 1979)

having a radius of about 20 cm or about 30,000 cm^3 in volume. The NNL diagram therefore does not show point variations in the moisture content and its dependence on depth turns out to be reasonably smooth.

The diagrams also illustrate the differentiating capability of the NNL method with respect to the lithology of the section and the persistence of the individual types of deposit as far as the moisture content is concerned. The groundwater level is very clearly defined in all cases characterised by a rapid increase in the moisture content.

For experimental data obtained with probes of various lengths, it was found that the 5-cm and 40-cm probes give practically the same accuracy. The differentiating capability with respect to lithology and the persistence of moisture content of rocks

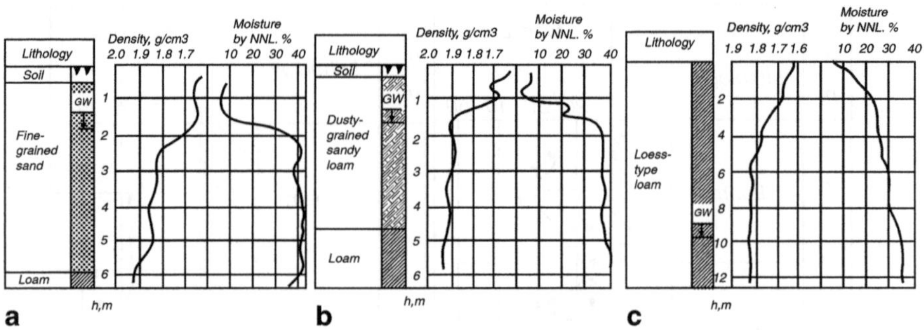

Fig. 7.7 Comparative results of groundwater level identification by GGL and NNL diagrams **a** in sand layer, **b** in sand-loam, **c** in loess stratum. (Ferronsky and Gryaznov 1979)

is higher for the 5-cm probe because, owing to the reduced probe length, there is a substantial reduction in the effective volume over which the moisture content is averaged for each point under investigation.

Theoretical consideration of the comparative estimation of the data having different representation characterising non-uniform soils and rocks with respect to nuclear logging is discussed in detail by Czubek (1976).

An important parameter of ground in geoengineering is porosity, which defines the total volume of pores in the ground volume. Porosity is expressed in per cents or relative units. For the unsaturated zone, porosity is determined by calculation using known values of density and moisture of the ground:

$$n = 1 - \frac{\gamma - \varpi}{\gamma_b},$$

where n is the porosity; γ is the density; γ_b is the bulk density; ω is the moisture.

For the saturated zone the porosity is determined as:

$$n = \frac{\gamma_b - \gamma}{\gamma_b - 1}.$$

The position of the groundwater level in the sedimentary soil is determined by a distinct increase of the registered thermal neutron flux. But only in sandy and sandy loam soils is the clear border of the water table observed (Fig. 7.7a and 7.7b). Because clear changes of moisture content are also characteristic for the stratum changes in lithology structure (for example, change of sandy layer by loam strata), having only the NNL diagram is not enough for identification of the groundwater level. Both NNL and GGL diagrams are useful for solving the problem. Its final solution is found by means of determining the saturation of the soils, i.e., volume $\omega = n$.

In the loam ground a distinct border of the groundwater table is not observed. This is because of the high level of the capillary border. In this connection a smooth increase of moisture and density on the NNL and GGL diagrams with depth is observed. Only by calculation can the level of complete saturation of soil be found. A good illustration to the above is the groundwater level position in a loess stratum presented in Fig. 7.7c. The diagram was obtained in the Tashkent region (Uzbekistan) during engineering geological mapping for irrigation purposes. Here the groundwater level was determined at a depth of 6.8 m by the water saturation coefficient G = 0.86 and the capillary zone level occurs in the depth of 4 m.

7.2 Influence of Sounding Parameters on Ground Resistance and Friction

A number of methodical questions should be studied for interpretation of cone resistance and lateral friction diagrams with respect to determination of soil mechanical properties. Inasmuch as the successful solution of technical problems of construction and design of the combined probe with the gauges for measurement of the ground cone resistance R and its friction T, the possibility has appeared to investigate methodical questions. The questions are the influence of the sound parameters on the cone ground resistance, influence of the penetration rate on the measuring R and T parameters, design of sound and some others.

The researches dealing with experimental studies on physical processes in ground with sound imbedding up to now have not come to a common opinion related to a picture of the real processes. For this reason, there are many models for the interpretation and calculation of the deformation and strength parameters of ground obtained by its sounding. It seems the majority of researchers are doing treatment correctly because compactions and shifts at sounding always take place. It was noted in Chap. 6 that during probe imbedding there are always two mutually competing processes, which are particle motion without volume change (deformation of particle shift) and particle motion with volume change (deformation of compaction or discomposure). The questions are which and when of the above processes prevail or whether both are equal in value. The essence of the studied process is not in motion of the particles but in changes of the volume of the deforming ground.

It follows from the results of most researchers that in the general case the probe displacement into ground from the surface is mainly accompanied by shift deformations up to some depth. Below that depth, deformations of compaction are observed (Fig. 6.9). That depth depends on the sound diameter, type of the ground and its density and moisture. It is called the critical depth of sounding.

In practicing static sounding by different methods and because of technical reasons, diameters of sounds from 30 to 110 mm and more are used. In order to improve the method of sounding results, interpretation of the influence of the sound diameter on the front resistance and lateral friction should be estimated. In addition, the value of total force, which the installation engine should develop for displacement of the probe to a certain depth, depends on the sound diameter. At the same time, the diameter of the sound influences the engine power of the installation.

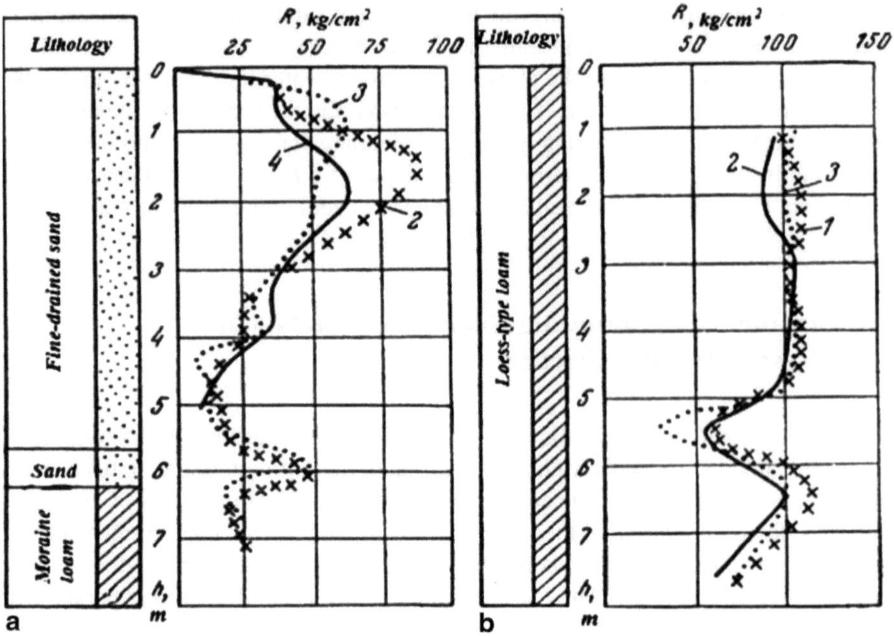

Fig. 7.8 Influence of sound diameter change on its cone resistance R: **a** for sand and loam sediments in the Moscow region; **b** for sandy-loam loess near the South Dnepr region; *1* d=60 mm; *2* d=100; *3* d=120 mm. (Ferronsky and Gryaznov 1979)

Finally, static sounding is widely used for estimation of the bearing strength of hanging piles.

On the basis of physical phenomena in ground during displacement of sound and also on the theoretical solution of the problem, it can be accepted that for the prevailing shift deformations the cone resistance to sound displacement depends on its diameter. This relationship follows from Eqs. (6.15–6.18). If displacement of the sound is accompanied by prevailing compaction of the ground, then, according to theoretical solutions (6.36), (6.81) and (6.82), the sound diameter cannot influence the cone resistance.

The influence of the probe diameter on the cone resistance is studied in different geological conditions. The geological section near Istra town, Moscow region, is represented by fine and moist sands with a density of 1.7–1.95 g/cm³ and moisture from 4 to 20 % and also by moraine sandy loam with a density of 2.1 g/cm³ and moisture of 20 %. Here the diagrams of the cone resistance in a number of near-placed points were recorded with a constant rate of displacement equal to 1 m/min with diameters of sound of 80, 100 and 120 mm. Figure 7.8a shows the obtained results, where it is seen that the cone resistance practically remains unchanged at different sound diameters. Some small decrease of the cone resistance for a sound diameter of 100 mm and an increase of this parameter for a sound diameter of 80 mm within depths from 1 to 2 m were observed. Such irregular deviations in a non-saturated zone of glacial sediments for all the physical and mechanical properties are very often observed. Hence, such deviations in ground properties like the above should be accepted as natural phenomena.

Analogous studies with sound diameters of 60, 80 and 100 mm were carried out on another experimental site near the village of Novo-Troitsk in the South Ukraine region. The geological section was represented by rather well uniform physical and mechanical properties on depths of sandy-loam loess sediments. Their density changed from 1.7 to 1.94 g/cm^3 and moisture from 16 to 23 %. The sounding tests were carried out up to 7 m and the rate displacement was 1 m/min. The results of the tests are shown in Fig. 7.8b, were one can see that an increase of diameter does not practically effect the value of recorded frontal resistance. Some deviations in the registered parameter reflect natural variations of the property.

It was found in other experiments, realised near the villages of Kolomenskoe and Shchemilovo in the Moscow region, that the measurements of frontal sound resistance in sand, sandy loam and loam sediments with sound diameters of 60, 80, and 100 mm showed that there is some tendency for resistance to decrease with diameter increase. Such a tendency is observed for sand with a density of 1.9–2.0 g/cm^3 for which the resistance is ~8 MPa and more.

The opposite results were noted in Kerisel's experiments (Kerisel 1964). His tests were carried out in an artificial metal box with a diameter of 6.4 and 10.2 m depth. The sand was packed with a given density. Then still piles of 42–320 mm were emplaced into the sand using a 200 t jack. In these experiments the conditions of the pile edge work were identical to sound emplacement. During the pile displacement the cone resistance was recorded by means of special hydraulic gauges. The total resistance was measured on the pile head. Figure 7.9 shows the results of the above tests for four pile diameters of 42, 110, 216 and 320 mm and for three sand bulk density values of 1.58, 1.68 and 1.75 g/cm^3.

It is seen from the graphs that in the sand of 1.58 and 1.68 g/cm^3 densities the influence of the pile diameters on their frontal resistance is not observed. The results of this test part coincide with our data. At the maximal sand density of 1.75 g/cm^3 some increase of the cone resistance for a diameter of 216 mm is observed. In these conditions the piles of diameter 42 and 110 mm are effected by the same sand resistance.

The results of Kerisel's tests on artificially packed sand completely coincide with our experiments in natural conditions. Some deviation of the frontal resistance for a pile of 216 mm diameter with a sand density of 1.75 g/cm^3 is explained by a random local variation of the sand density and moisture in artificial and natural conditions. In any case, there is no reason that these facts could be explained by a basic regularity.

Kerisel's results allow estimation of the sounding critical depth location. As noted earlier, we define this depth at the border above which the shift deformations are prevailed. It was found on the basis of the ultimate equilibrium theory that up to the critical depth the cone resistance depends on the sound diameter. It is seen in Fig. 7.9 that the upper part of the diagram R = f(h) is characterised by increasing the resistance with the depth. This relationship is especially clear for the sand density of 1.75 g/ cm^3 (see Fig. 7.9c). Here the critical depth is increased from 1 to 3 m for the pile diameter change from 42 to 216 mm. For the diameter of 320 mm the critical

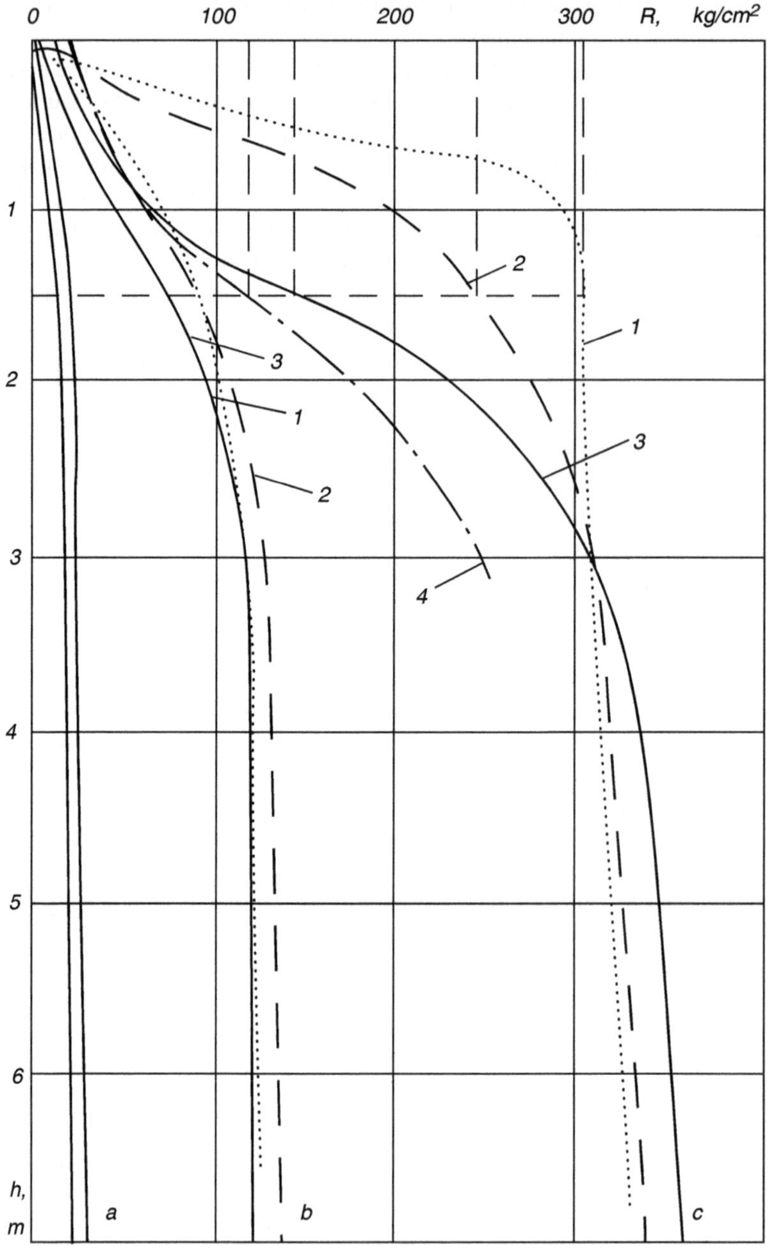

Fig. 7.9 Results of Kerisel's experiments on the influence of pile diameter on sand cone resistance to pile displacement at different sand densities **a** 1.58 g/cm^3, **b** 1.68 g/cm^3, **c** 1.75 g/cm^3. Size diameters: *1* 42, *2* 110, *3* 216, *4* 320 mm

depth was not reached because of the limited power of the jack. The cone resistance on the same depth is also sharply changing depending on the pile diameter. For example, at the depth of 1.5 m the piles with diameters of 42, 110, 216 and 320 mm are tested with the cone resistance correspondingly equal to 30, 24, 14 and 12 MPa. Further displacement below the critical depth is accompanied mainly by compaction deformations and the cone resistance does not depend on the pile diameter.

Thus, it follows from the experimental tests of relationship between the sound diameter and its cone resistance of ground that there is a critical depth of sounding that really reflects physical processes in sediments tested by displacement of measuring sound. The critical depth is an important parameter that determines the field of application of individual methods for interpretation of the sounding test results on the basis of the theories of ultimate equilibrium, elasticity, elastic and creep medium. This is especially important for calculation of the pile bearing capacity.

For static ground sounding, because of the small diameter of measuring sound used in practicc (not morc than 100 mm), the critical depth is insignificant and placed just at the surface. This is why the main form of deformations accompanying sound displacement is accepted to be deformations of compaction and why at static sounding the results of cone ground resistance measurements should be used for determination of the modulus of momentary and long-term deformations, the creeping coefficient and for plotting the diagram of these parameters at the section depth based on the elasticity creeping theory solutions (see Chap. 6). It follows from here that in practice sounds of different diameters can be used and their results should be comparable. But the measuring gauge of the frontal resistance should be placed just at the sound tip.

Let us discuss the influence of the sound tip angle on the measuring parameter. While considering physical processes in the ground during conic sound displacement it is seen that the cone angle effects the value of the measuring parameter. This effect is developed together with the ground breaking down by the local and common shifts in the region of strains under the conic sound. As was discussed earlier, during the standard plate load test on a surface, massif ground starts to brake down after a compact core under the plate has formed. The angle at the core tip should depend on physical and mechanical ground properties. It changes during the process of imbedding and provides optimal conditions for the ground breaking down at minimal load value. It is certain that if the core angle is kept constant then it will not be optimal for different types of ground.

It is obvious that below the critical depth of sounding for different types of sediments, there also exists an optimal angle at which the ground is breaking down and lateral compaction and its compression will take place with minimal power of the rig. But from a practical viewpoint, for small cone angles (15–30°), the penetration hole often becomes curved because different solid objects (boulders, gravels and so on) appear. In order to estimate the influence of the cone angle a number of test series in different type sediments were carried out. One of such tests in a glacial sandy and loam geological section was executed near the town of Podolsk in the Moscow Region. In that test, cone angles from 30 to 180° at a constant displacement rate were used. Figure 7.10a demonstrates the test results. The recorded diagrams show that the cone angle weakly effects the cone ground resistance. Nevertheless, the effect is uniquely traced. It was shown that the value of the cone resistance is lower at 60° of the cone angle as compared with others. It seems, in this case, the ground particles do the shortest way at their compaction motion.

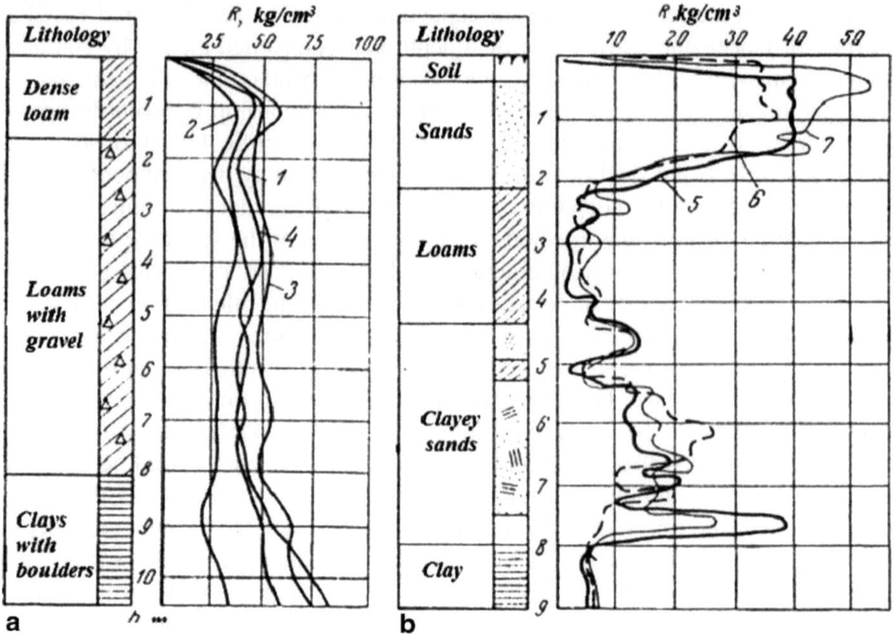

Fig. 7.10 Test results of ground cone resistance change as a function of the cone angle: *1* 30°; *2* 60°; *3* 90°; *4* 180°; *5* 15°. (Ferronsky and Gryaznov 1979)

Figure 7.10b shows the results of another series of tests near the village of Kolomenskoe in the Moscow Region. Here the geologic section was presented also by sand and loam sediments and the cone angle was changed from 15° to 180°. Any significant influence of the cone angle on the measuring cone resistance in these tests was not observed. At the same time, in dense sand the influence of the cone angle is notable. This effect was seen by testing dense, intermediate and loose sands in glacial sand and loam sections near Moscow. Here the cone resistance for the cone angle of 15° increases but the difference compared with larger angles is diminished for increased sound diameter.

Taking into account the above experimental data the cut-off cone angle for the penetration logging installations SPK, SPKT and PSPK-69 was accepted to be 60°. The shape of such an angle is the closest to a spherical mode. This form is also practically more convenient with respect to avoiding the rod curving from solid subjects.

The problem of the relationship between the sound and rod diameters has both methodological and practical meaning. From a methodological viewpoint two principle schemes can be used. In the first (Fig. 7.11), the cone base is directly connected with the rod. A number of researches used this scheme trying to provide some portion of the shifted ground to be moved to the space between the penetrating hole and rod. But the practice shows the effect like this only for the loam and silt of marine sediments in fluid consistence is observed. The lows of viscous liquid but not elasticity and plasticity are more applicable for such a ground. In addition,

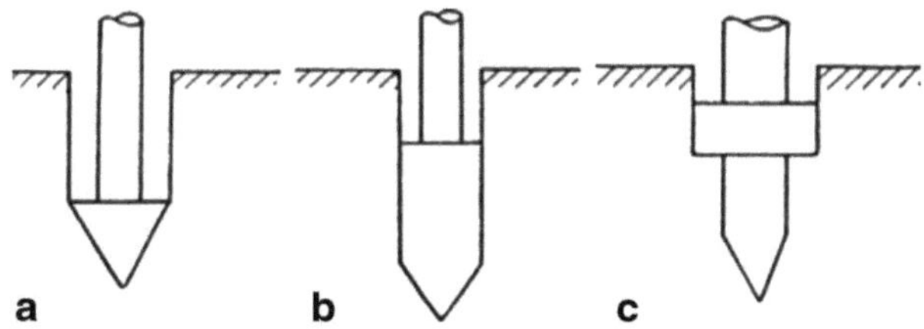

Fig. 7.11 Possible schemes for measurement of frontal ground resistance at different probe and cone diameters. (Ferronsky and Gryaznov 1979)

scheme (a) has negative effects. During displacement of such a sound the elastic deformations capture the rod that then needs to be withdrawn back. During dynamical sounding in such cases the cone is left in the ground. And finally, scheme (a) does not permit to combine it with a simultaneous measurement of the lateral friction.

The second scheme (b) of the cone resistance measurement includes a cylindrical element between the cone and rod. This scheme excludes possible shift deformations and protects compaction of the squeezed cone ground. The cylindrical element is used as the lateral friction gauge.

The optimal ratio between the cone and rod diameters is also insured by complete use of outer hydraulic pressure for displacement of the probe to the needed depth. It is known from the theory and practice of static penetration that in some types of sediments their elastic deformations are able to capture and keep not only the cone but the whole rod. Because of the large rod surface, in this case, most of the engine's energy is spent on overcoming the rod friction. In order to exclude that friction, the difference in the cone and rod diameters should be the possible ground elastic deformations.

Experimental studies of the above effects for different cone diameters at a rod diameter equal to 50 mm in glacial sediments were carried out. These results are shown in Fig. 7.12. It is seen that for displacement to the same depth at minimal outer power the cone diameter should be 80 mm. The ratio of diameters is $d=80/50=1.6$. But taking into account different sediment properties and durability of the existing rod standards, the penetration logging rig SPK was equipped by rods with a diameter of 63.5 mm and probe diameter $d=80$ mm.

One more scheme of diameter difference for the cone and rod is possible. The calibration ring belt can be taken off with changing diameter. But this scheme has not been tested.

The designed and used in practice penetration logging rigs SPK, SPKT and PSPK with their technical parameters displacement rate up to 8 m/sec can be developed. There is technical possibility for further increase of that value. From the point of view of productivity there is reason to do this. But because the ground deformations develop in time, the quality of obtained information demands the probe rate should not effect the measured parameter.

Experimental studies of this problem were carried out in the Kherson Region (Ukraine) in a loess geological section, where ground density was from 1.7 to 1.9 g/cm^3 and the moisture was 16–23 %. The probe was of 80 mm in diameter and its displacement rates were 0.1; 0.25; 0.5; 0.75; 1.2 and 3.5 m/min. The distance between the sounding points was 1 m (Fig. 7.13).

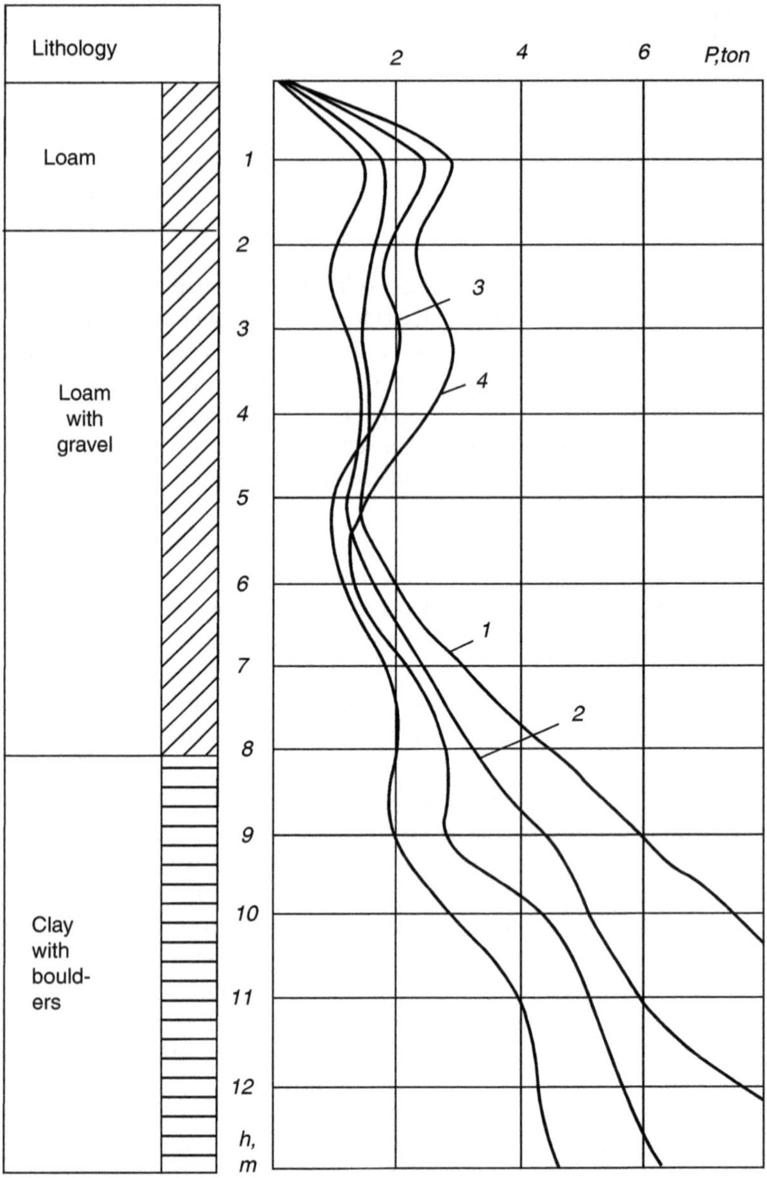

Fig. 7.12 Test results on the influence of probe and rod ratio on common outer displacement power: *1* d_p=60 mm; *2* d_p=70 mm; *3* d_p=80 mm; *4* d_p=90 mm. (Ferronsky and Gryaznov 1979)

It follows from the above data that displacement rates of the probe within 0.1–3.5 m/min in the studied loess sediments effect only slightly the measuring cone resistance. But in the case of more dense sediments, within the upper 2 m with gypsum inclusions, the higher rate of the probe displacement leads to increase of ground resistance. An analogous picture in denser sandy sections was observed. The

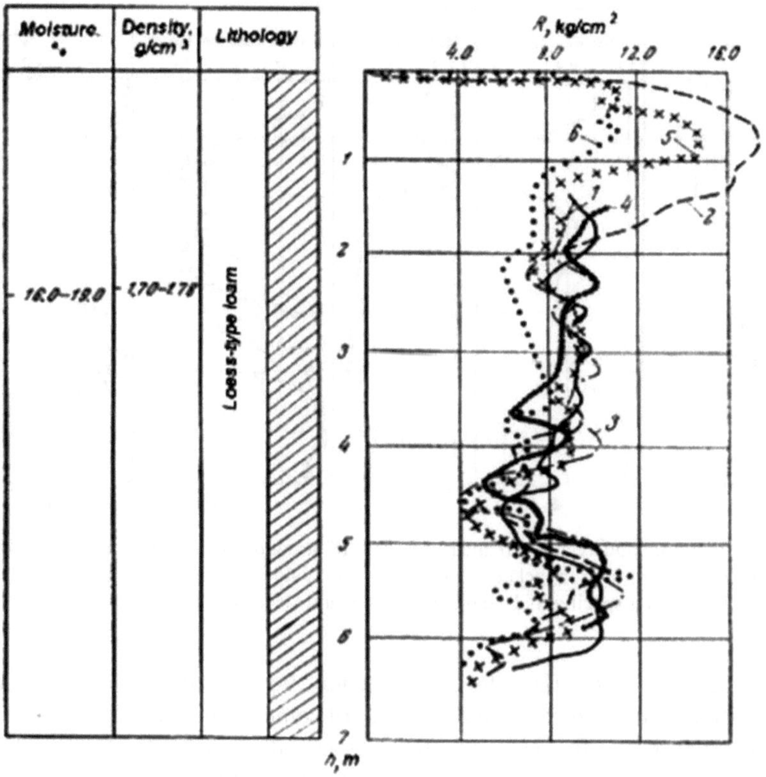

Fig. 7.13 Test relationship between ground cone resistance and displacement rate of the probe: *1* 3.5 m/min; *2* 1.2 m/min; *3* 0.75 m/min; *4* 0.5 m/min; *5* 0.25 m/min; *6* 0.1 m/min. (Ferronsky and Gryaznov 1979)

conclusion is as follows. For the works in unknown sandy and loam sections where dense sediment layers are possible the displacement rate of the probe should be low. The stability against the rod column curvature in dense sediments is also the effect that limits the rate of probe displacement. In general the effect of the displacement on the ground resistance has theoretical solutions (see Chap. 6).

Recently, the lateral ground friction parameter in static penetration has been used. The gauge of ground friction is placed in the probe of rigs SPK, SPKT, PSPK and S-832 designed by the Bashkortostan Research Institute (Faierstain and Makarov 1964; Begemann 1965). Since researchers began quantitative interpretation of static penetration, it became clear that separate measurements of the normal and lateral ground resistance were necessary. It was proposed to use the parameter for estimation of the pile bearing capacity for obtaining the ground share resistance and for lithologic separation of a sedimentary section. It is obvious that for development of methodology interpretation the influencing factors on the ground friction should be studied. Such factors are the position of the friction jacket on the sound, its optimal diameter, displacement rate and the cone angle. The ratio of cone and lateral resistance is also of interest.

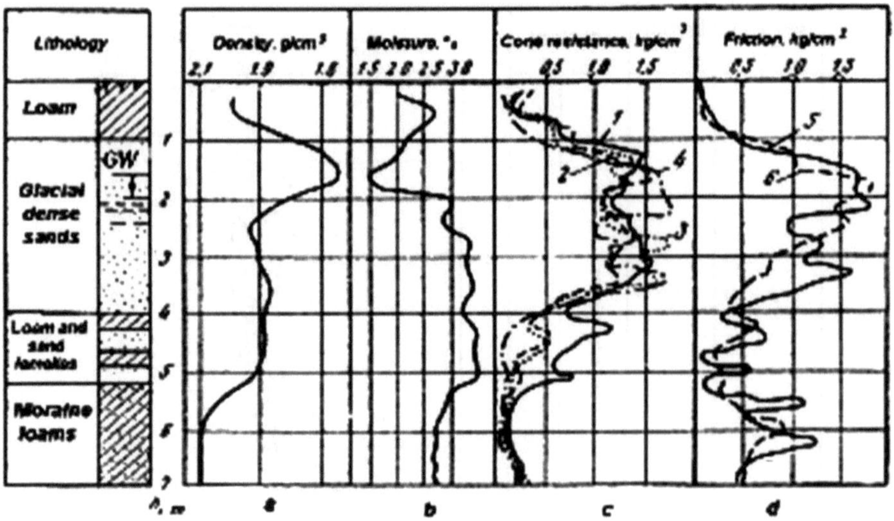

Fig. 7.14 Experimental results on measurements of ground friction as function of the space between the cone and friction jacket: *1* 15 cm; *2* 30 cm; *3* 45 cm; *4* 60 cm (*5* and *6* the values during the probe rise at a spacing of 15 cm). (Ferronsky and Gryaznov 1979)

Some researchers (Kerisel 1961; Yaroshenko 1964) noted that a constant value of cone resistance in uniform sediments lower of some depth due to the unloading effect above the cone is fixed (see Fig. 6.4). As a result, a zone with decreased density above the cone should appear. It was decided to check the presence of the unloaded ground effect by experimental work. The influence of the relaxation strength on the friction measurement should also be studied. The field work was carried out in a sand and loam glacial sediment section in the Moscow region. The results of these experiments together with density and moisture profiles are presented in Fig. 7.14.

The friction jacket was made as a steel cylinder with cone diameter placing at 15, 30, 45 and 60 cm from the cone tip. The size of the jacket itself was 200 mm, diameter—62 mm and surface—400 cm^2. The probe displacement rate was 1.5 m/min.

It is seen from the data that at the spaces between the cone tip and middle jacket equal to 30, 45 and 60 cm the friction diagrams for sand and sandy loam sediments practically coincide. The deviations within each layer do not exceed 3 % of the area. At the space equal to 15 cm for the depth in the section between 4 and 5.5 m, where sandy and sandy loam layers are interstratified, the friction values are growing. It is possible that the density change was effected. An analogous picture is observed at the probe rise (see Fig. 7.14, curves 5 and 6). Here two sand layers at the depths of 5.7 and 6.3 m appear and the friction strains are averaging. A better coincidence of the ground friction diagrams at the probe displacement and its raising in saturated sediments is observed. In dry holes and low-moisture sediments the above regularity is not kept.

The experiments have shown that the assumed unloading effect of the shifting ground is absent. It should be characterised by lower density values and hence by the friction decrease fixed by the gauge placed above the cone. This phenomenon does

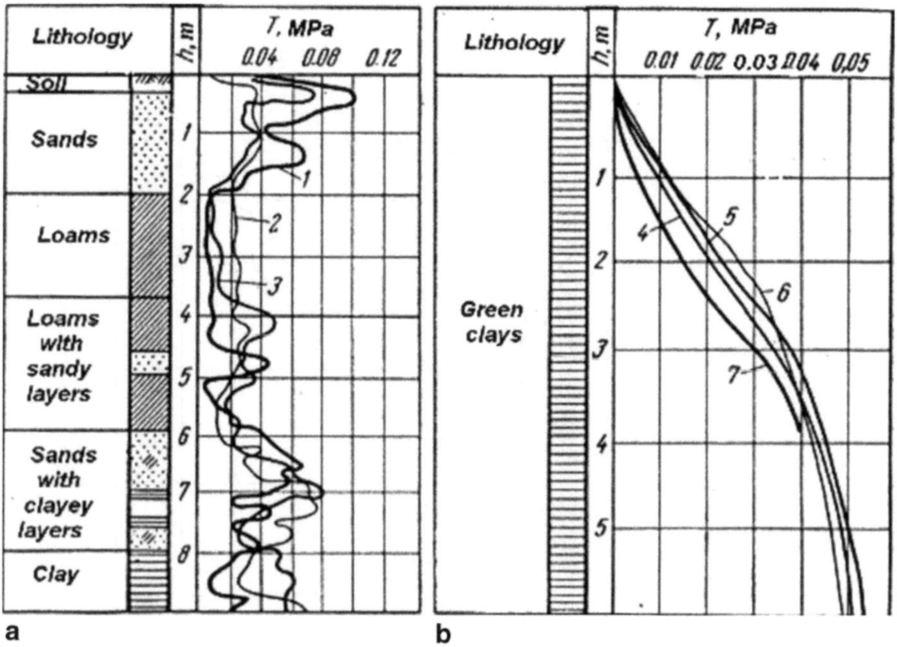

Fig. 7.15 Relationship between the ground friction and diameter of a friction jacket **a** the author's data for diameters: *1* 60 mm, *2* 80 mm, *3* 100 mm, *4* 45 mm, *5* 75 mm, *6* 110 mm, *7* square pile with perimeter 125 cm; **b** are Kerisel's data. (Ferronsky and Gryaznov 1979)

not occur neither in sand nor in sandy loam. A decrease of the ground stresses during their relaxation at ground compaction under cone displacement is also not observed. It is possible that the value of the stress relaxation is lower than the measured parameter fluctuations because of its natural variation within the studied sediments in the section. These effects are not fixed since they lay within the measuring errors.

The practical conclusion is that the friction jacket can be placed in any place within the measuring probe and its size should be not minimised by design reason. We place the friction jacket just above the cone gauge and its surface is determined by means of sensitivity of the recording apparatus.

The effect of probe diameter on the friction parameter was studied in the sand and loam sediments section near Kolomenskoe Village in the Moscow Region. Three jacket diameters of 60, 80 and 100 mm at a constant rate of probe displacement were tested. The results of one from four series of experiments carried out in different sections are shown in Fig. 7.15a.

It appears that for the experimental conditions the friction jacket's diameter does not effect the measured friction parameter value neither in sandy nor in loam sediments of different density. Analogous results were obtained by Kerisel (1964). Figure 7.15b demonstrates his data. The experiments were carried out in uniform green loam near Paris. The ground moisture was equal to 29–31.5 % and the plasticity parameters were $W_f=74–96\,\%$, $W_p=28–34\,\%$, $A=W_f-W_p=28–34\,\%$. The tests were carried out by use of penetrating cone diameters of 45, 75 and 110 mm and

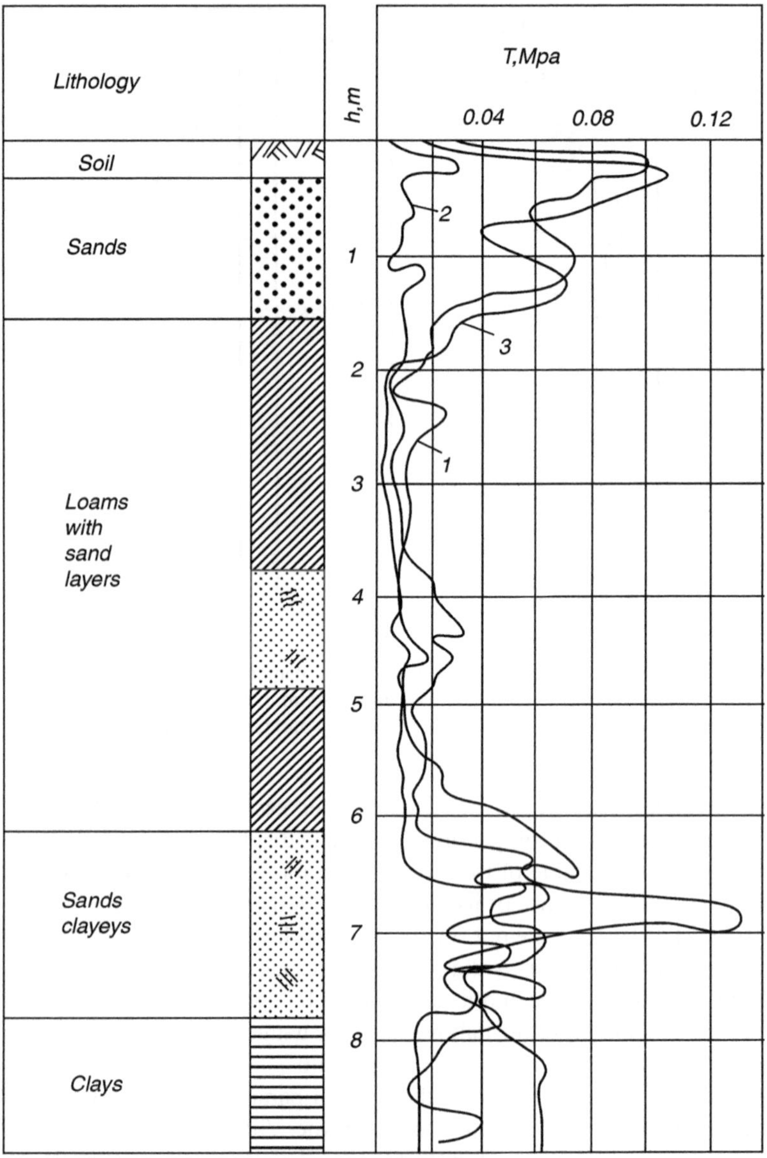

Fig. 7.16 Ground friction dependence on tip cone angles: *1* 15°, *2* 60°, *3* 180°. (Ferronsky and Gryaznov 1979)

also by a hollow square pile of 125 cm on the perimeter. The conclusion is important for interpretation of pile bearing capacity based on penetration logging ground tests and for interpretation of cone resistance measurements.

Field tests of the cone angle influence on the friction parameter were carried out in the same Moscow region. Four series of experiments with three tip cone angles of 15°, 60° and 180° and three jacket diameters in the sandy and loam sediments showed that friction resistance change is negligible. Figure 7.16 shows these results.

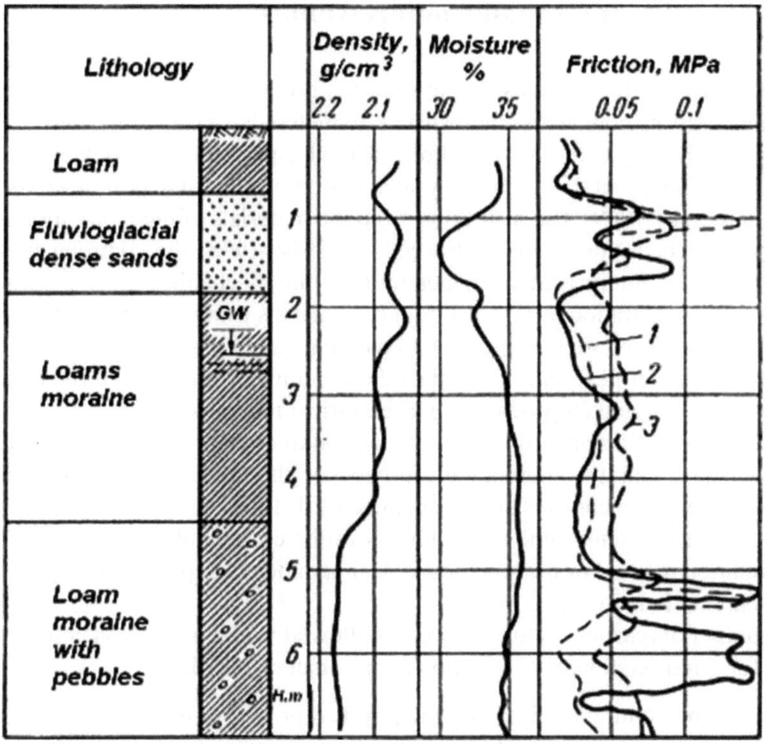

Fig. 7.17 Relationship of ground friction resistance from the probe displacement rate: *1* $v = 0.45$ m/min; *2* $v = 1.65$ m/min; *3* $v = 3.3$ m/min. (Ferronsky and Gryaznov 1979)

It worth noting that, as was shown in the previous paragraph, there was not a fixed systematic change in cone resistance depending on the tip cone angle. Evidently, for significant absolute values of cone resistance, which need to be applied to the probe displacement, its insignificant variations related to the cone angle are not distinguished from other effects.

The obtained results show that the physical phenomenon of the ground and sound interaction is one more argument in favour of large cone angles for the probe design.

Experimental tests of the ground friction reaction from the probe displacement rate in sand and loam sediments of glacial genesis in the Moscow Region were carried out. The friction jacket was placed just at the cone resistance cone gauge with 15 cm size from the cone tip to the middle of the jacket. The probe was displaced with a rate from 0.45 to 3.3 m/min. Figure 7.17 shows the test results.

It is seen from the figure that friction values for most non-homogeneous in density and moisture sediments from 2 to 5 m in depth, the displacement rates 0.45 and 1.65 m/min are practically the same. At the probe rate of 3.3 m/min the friction value is somewhat increased. But in the sand and sandy-loam sediments below 5 m this regularity breaks. In general, we observe that at the probe displacement rate of 2 m/min the value of the friction parameter is not changed and does not bring in notable errors.

7.3 Modulus of Ground Compressibility

The existence of the relationship between the ground resistance to the probe displacement and its compaction and share parameters has been confirmed by all the researchers dealing with theoretical or practical aspects of the problem (Bondarik et al. 1967; Sanglera 1971; Terzagy and Pek 1958; Trofimenkov and Vorobkov 1974; Ferronsky 1969). But as follows from the materials of the International Symposium on Penetration Testing (Proceedings 1974–1975) the common principles and methods of interpretation for determining the mechanical ground parameters are based on penetration test results. Some specialists using penetration test data for quantitative interpretation of the ground parameters are absent.

On the basis of penetration test data the modulus of ground deformation is determined by applying empirical or semi-empirical relations obtained by comparing with the results of a standard laboratory method of compression or field plate tests. Table 7.1 demonstrates the relationship between the modulus of ground deformation and penetration cone resistance obtained by different researchers. The proposed equations for the same type of sediments differ considerably. The choice of the interpretation method of the soil mechanical parameters depends on the school that the researcher belongs to.

A number of reasons, as to why it is not allowed to elaborate upon common principles for interpretation of penetration results, can be stipulated:

1. There is no common opinion on the physical processes that occur in the ground at probe displacement. The multiple experiments carried out by different researchers with different content, properties and state grounds, brought about manifold ideas about the occurring processes. In connection with the observed and accepted physical premises researchers obtain corresponding theoretical solutions for interpretation of their experimental data (see Chap. 6).
2. The attempts in analysis and comparison of experimental data obtained by authors from different countries encounter methodological and technical difficulties. This is because researchers use penetration rigs equipped by gauges with different technical parameters (type and design of the probe, gap between the rod and penetration hole, probe diameter and cone angle, friction jacket, displacement rate). As a result, the experimental records appear to be non-comparable.
3. The natural conditions of the sedimentation process led to exceptional heterogeneity in properties and structure of the grounds. In this connection. the obtained experimental relations, strictly spiking, are valid only for the tested sediments. In this regards, the interpolation relation appears to be the perspective direction for the problem solution.
4. The stressed and deformable ground state in nature, strictly speaking, cannot be characterised by the modulus of deformation, angle of inner friction and a coefficient of cohesion as a constant value. These parameters are variable values depending on the action of outer conditions on the ground. This statement is proved by the different results of their determination by compression, field gauge plate and pressiometer gauge tests in laboratory and field conditions.

Table 7.1 Relationship between the modulus of deformation E and specific cone ground resistance R found by different authors

Author	E and R relationship laboratory compression test	Ground type
E. Paproth (1943)	$E = \frac{3}{2}\mu kR$ [a]	Sand
	$E \approx 7R$	Sand (at flat tip)
A. Buismann (1944)	$E = \alpha R$ [b]	Sand
	$E = 1.5\frac{R}{p_n}$ [d]	Sand
E. Schultze and K. Meltzer (1965)	$E = vR^{0.522}$ [c]	Sand
L. Parez and M. Bachelier (1965)	$E = 2.3\frac{R}{\alpha_0}$	Loam $0.3 < \alpha_0 < 1$
J. Schmertmann (1970)	$E = 2R$	Sand $1.08 < \alpha_0 < 2.5$
G. Meyerhof (1951)	$E = 1.9R$	Sand
	Field plate test and pressiometer test	
G. Sanglerat (1971)	$E = 1.1R$	Sand
	$E = 2.4R$	Loam
V. Ferronsky (1969)	$E = \frac{3\pi(1+\mu)(3-4\mu)R}{16(1-\mu)}$ [e]	Different types
State Standard (1982) of Russian	$E = 3R$	Sand
	$E = 7R$	Loam

[a] μ and k are the coefficients depending on the cone angle, scattering of the stresses and probe size;
[b] α is the coefficient depending on the ground type; for sand $\alpha = 1.5$;
[c] v is the coefficient depending on the ground type;
[d] p_n is the natural ground pressure on the study depth;
[e] μ is the Poisson's coefficient

In order to obtain comparable testing results from applying different techniques, the similarity conditions in the methods of study should be satisfied.

Coming back to analysis of the relationship between the deformation modulus and the cone ground resistance to the probe displacement we can note that the relationship between the two parameters is accepted to be linear (see Table 7.1).

The difference in the coefficient of proportionality, depending on the methodology and techniques of tests and also from some other factors, can be explained by the non-comparability of the experimental data obtained by different researchers. Despite this, in many countries penetration data are accepted for quantitative interpretation of the ground mechanical properties.

In Chap. 6 we presented fundamentals, methodological justification and theoretical solutions for determination of the ground deformation modulus (without and with relaxation stresses). The formulas for the deformation modulus based on ground cone resistance are obtained with the assumption that a spherical probe is displaced in elastic media. The results obtained by penetration tests and the deformation modulus calculated by formula (6.36) with field standard plate tests were compared with sand and loam sediments in the Moscow region, on landslides in the Crimea and the Black Sea shore.

Comparative tests were carried out on alluvial sands in the Lower Volga River basin, on sandy loam and loess sandy loam sediments in the Karshy Steppe in Uzbekistan, on moraine sediments in the Moscow region and on Black Sea bottom sediments. Experimental sections sufficiently uniform in ground properties were selected. The size of the test plate was 5000 cm^2. Table 7.2 presents a part of the results obtained in three regions.

It is seen from the table that the ratios of the deformation modulus obtained by standard plate tests and calculated by formula (7.36) based on the cone resistance records for alluvial sands of the Lower Volga River Region are changed from 0.57 to 2.58, for loess of Karshy Steppe from 1.68 to 4.5, for sandy loam of the same region from 0.79 to 0.89, for sands of the Moscow region from 1.12 to 2.1, for sandy loam from 0.5 to 17.0 and for loam of the same region from 0.37 to 1.15. The changes appear to be non-regular by sign. We come to the conclusion that the method of penetration tests based on theoretical solutions satisfies the conditions for foundation design for civil engineering purposes.

7.4 Ground Shear and Rheology Parameters

Let us consider the experience of researchers from different countries on interpretation of the penetration tests for determination of ground shearing and rheology characteristics.

According to Lowsberg et al. (Proceedings 1974–1975), for determining the shearing parameters for non-cohesion soils Belgium researchers use de Beer's method. For this purpose, applying Terzagi's bearing capacity coefficient and the ground cone resistance of penetration tests, the diagram of friction angle φ changing in depth is plotted and its ultimate value is found.

According to other methods, the ultimate value of the inner friction angle φ for non-cohesive soils is determined on the basis of Meyerhof's ground bearing capacity coefficient (see Fig. 6.5).

The cohesion of bonded soils *c* in non-draining test conditions (instantaneous cohesion) is determined by the value of the ground cone resistance R on the empirical formula:

$$c = \frac{R}{A}, \tag{7.1}$$

where A is the coefficient depending on the sound design; A = 15–20 for the cone with a friction jacket and A = 10–15 for the Danish cone without a friction jacket.

According to Heinen (Proceedings 1974–1975) in the Netherlands for determination of the instantaneous cohesion, expression (7.1) with A = 15 is used. This value is obtained by comparison of penetration tests with laboratory three-dimensional

Table 7.2 Comparative results of deformation modulus determination

Region and type of sediments	E_{plate} MPa	E_{sound} MPa	E_{plate} E_{sound}	Region and type of sediments	E_{plate} MPa	E_{sound} MPa	$E_{plate}E_{sound}$
Lower volga river, alluvial sands	83.5	78.0	1.07	Karshy Steppe, sandy loam	15.1	9.0	1.68
	22.6	39.6	0.57		56.6	12.6	4.5
	92.0	84.0	1.10		28.2	9.0	3.12
	24.0	27.0	0.89				
	107.5	78.0	1.38	The same sandy loam	42.3	48.0	0.89
	34.5	36.0	0.96		47.0	57.0	0.82
	60.5	40.8	1.48		52.2	66.0	0.79
	44.0	39.0	1.51				
	53.8	48.0	1.11	Moscow region sands	16.2	14.4	1.12
	64.5	39.0	1.65		6.6	4.2	1.57
	30.7	36.0	0.85		5.1	4.5	1.13
	21.5	30.0	0.72		8.6	4.8	1.8
	74.5	28.8	2.58		4.7	3.0	1.55
	64.5	75.0	0.86		36.0	17.5	2.1
	57.0	36.0	1.58				

Table 7.2 (continued)

Region and type of sediments	E_{plate} MPa	E_{sound} MPa	E_{plate} E_{sound}	Region and type of sediments	E_{plate} MPa	E_{sound} MPa	$E_{plate}E_{sound}$
	40.3	18.0	2.24	The same sandy loam	8.6	9.0	0.96
	97.0	129.0	0.75		17.1	12.6	1.35
	57.0	108.0	0.53		4.1	6.0	0.66
	84.0	54.0	1.55		7.7	7.2	1.08
	53.7	26.4	2.04		15.1	30.0	0.5
	74.5	54.0	1.38		56.6	33.0	1.7
	97.0	138.0	0.70		28.2	33.0	0.85
	74.5	60.0	1.25		17.1	15.0	1.1
	60.5	54.0	1.11				
	51.0	78.0	0.65	The same loam	4.1	6.3	0.65
	91.0	78.0	1.16		7.7	7.0	1.20
	68.0	93.0	0.73		4.7	4.9	0.96
	97.0	81.0	1.19		2.8	3.5	0.80
	64.5	42.0	1.53		2.8	2.4	1.15
	80.7	111.0	0.73		5.6	11.2	0.47
	64.5	30.0	2.14		5.0	13.3	0.37
	84.0	142.0	0.59		7.0	10.5	0.66
	71.5	60.0	1.20		9.4	16.8	0.56

Table 7.3 Values of the specific cohesion force for weak loam sediments (Begemann 1965)

Depth from the surface, (m)	Cohesion, 0.1 MPa (at $\varphi = 0$)		Depth from the surface, (m)	Cohesion, 0.1 MPa (at $\varphi = 0$)	
	Measured by cone jacket	Measured by vane sound test		Measured by cone jacket	Measured by vane sound test
0.5	0.3	0.4	4.5	0.1	0.11
1.5	0.1	0.1	5.5	0.1	0.11
2.5	0.15	0.07	6.5	0.175	0.18
3.0	0.07	0.05	7.5	0.2	0.21
3.5	0.15	0.05	8.5	0.13	0.15

compression tests and by the field rotation cut-off test. In Greece (Proceedings 1974–1975) the coefficient A for water-bearing soils is taken to be 15 to 18.

In Italy (Proceedings 1974–1975) for plastic loam sediments the expression for instantaneous cohesion takes the form:

$$c = \frac{R - p_n}{15}. \tag{7.2}$$

where p_n is the natural pressure on the study depth. For the loam sediments in a heavy-plastic state:

$$c = \frac{R - p_n}{25}. \tag{7.3}$$

It is assumed here that there is no relationship between *c* and R for loam ground in the flow consistency.

It worth noting that interpretation of the shearing parameters by ground frontal resistance is mainly provided for by measurements done by a Danish penetrometer. In this case the cone displacement is accompanied by partial ground extrusion into the gap between the rod and the hole side. From the viewpoint of methodology, the searing conditions are not always held. For example, displacement of the cone in loose sand should be accompanied only by its compaction without shifting and bulging out. But in such a case the relationship between the shearing and conel resistance should be kept. This relationship appears to be a consequence of the physical nature of the dispersed media compaction because of the porosity decrease.

A closer picture of shifts modelling based on lateral friction measurements by the friction jacket was given by Begemann (1965). According to his data regarding the study of loam sediments in the Netherlands, for which the inner friction angle is equal to zero, the shearing parameter record by the friction jacket allows to obtain the value of the specific cohesion force. These data correlate well with the results of the cohesion force value measured by the vane sound test (Table 7.3).

It is seen from the above data that convergence in the mean for the two methods is high. On the basis of his experimental results Tomlinson (1957) came to the same conclusion. He found that the correction coefficient should be introduced when the cohesion force of the loam sediments is interpreted based on the lateral friction measurements. The correction must take into account the consistency of the loam ground in the form:

$$c = k_c T. \tag{7.4}$$

According to Tomlinson the value of k_c is changing in the following limits:

Consistence	Flowing	Plastic	Tightly plastic	Solid state
Coefficient k_c	1–1.2	1.2–2.0	2–4.0	4.0

For the studied sandy sediments, for which the cohesion is accepted to be equal to zero, the relationship between the measured jacket lateral shift resistance and the angle of inner ground friction can be expressed by the equation:

$$T = k_c \sigma \operatorname{tg} \varphi, \tag{7.5}$$

where k_φ is the coefficient taking into account the difference in friction values of the ground with metal and the ground itself; σ is the normal elastic reaction of the ground on the friction jacket, which can be determined by the gauge of cone ground resistance.

It worth noting that the formulas (7.4) and (7.5), which are recommended for determination of tangential and normal determinations by the corresponding gauges, appear to be empirical. Formula (7.5) has reliable physical justification and represents the known Coulomb's equation. The existing correlation between the shearing and cohesion by formula (7.4) seems to be explained by developing of water–colloidal cohesion forces between the loamy particles and material and the friction gauge at its displacement. Our experiments carried out on Juristic loams and marine sandy–loams of plastic consistence have shown that there is good correlation between the above two parameters. The k_c coefficient in (7.4) for the Juristic loam is equal to 1.5 and for the sandy–loams is equal to 2. Our experience shows that both parameters of inner friction and normal pressure measurements are needed for obtaining a reliable correlation (7.5).

One more advantage of the ground penetration tests is their possibility in obtaining the rheological parameters, namely, the modulus of instantaneous and long-term deformations, coefficient of creeping and changes of porous pressure.

There are two technologic methods of penetration tests widely used in different countries (Proceedings 1974–1975). The first is real static, which provides measurement of ground resistance at the stopping of probe displacement through 10, 20, 30 cm. The second one provides a continuous record of the measured parameters. The choice of technology method depends mainly on the rig design. The first option allows the ground resistance to be measured by a manometer, the second one does it by means of strain gauges or electric tensometers. A combination of measurements of sounding parameters at stopping and continuous probe displacement allows determination of the ground rheological parameters.

The penetration logging rigs described in Chap. 5 allow to provide both real static and continuous measurements of three parameters, namely, the ground cone resistance and lateral and normal elastic resistances. The gauge test of the last one is described in the next paragraph. Thus, there is a possibility in application of the theoretical solutions presented in Chap. 6 for determining rheological parameters by equations (6.81–6.83) and (6.112–6.114).

A combined sound for determining cone, lateral resistance and pore water pressure has been used in Norway (Proceedings 1974–7975). The pore pressure gauge is presented by a special piezometer with a cell and string-acoustic element that records the pore pressure during continuous probe displacement with a given velocity. Combination of sounding parameters and pore pressure in time is an important experimental base for study of ground rheological parameters.

7.5 Normal Pressure

Separate gauges for recording the ground normal pressure and lateral friction were designed for the SPK rig (Ferronsky 1969). The theoretical basis for determination of normal pressure was discussed in Chap. 6.

The normal pressure gauge, which was used in our first field tests, was a thin-walled cylinder with an outer diameter of 62 mm and height of 70 mm. The cylinder was placed just above the cone (Fig. 7.18). The normal pressure, induced by the ground elastic reaction at the probe insertion, was measured by electrical tensometers sticking across the inner cylinder side. The tensometer for the temperature compensation was stuck along the same cylinder side. These two electric tensometers formed the two bridge arms of the measuring gauge. The two other arms were placed on the common control desk of the rig. In this design the frontal cone resistance was not fixed by the normal pressure gauge (see Fig. 6.12) because the last one was separated from the cone by a rubber ring seal (Fig. 7.18a).

Some errors in measurements, which can occur because of friction forces over the outer cylinder surface and compensating tensometers, were taken into account by calibration of the gauges. Together with the normal pressure recording, the ground friction by the separate gauge is recorded.

Figure 7.18b shows the diagrams of the normal pressure and ground friction recorded by the above described sound on the moraine-sand and loam sedimentary section in the Moscow region. Here is seen that in the sandy sediments at depths from 0 to 1.5 m, from 4.5 to 5 m and from 6 to 7 m together with an increase of normal pressure the friction resistance is sharply increased. In the loam sediments a notable increase of the normal pressure does not lead to a friction resistance increase. This is seen especially clear for the Jurassic loam in the depth of 7 m where increase in normal pressure leads even to decrease of friction resistance.

This fact proves well the above presented hypothesis about the mechanism of the physical phenomenon that occurs in the ground during probe insertion. Further experience in use of normal pressure and ground friction will help in development

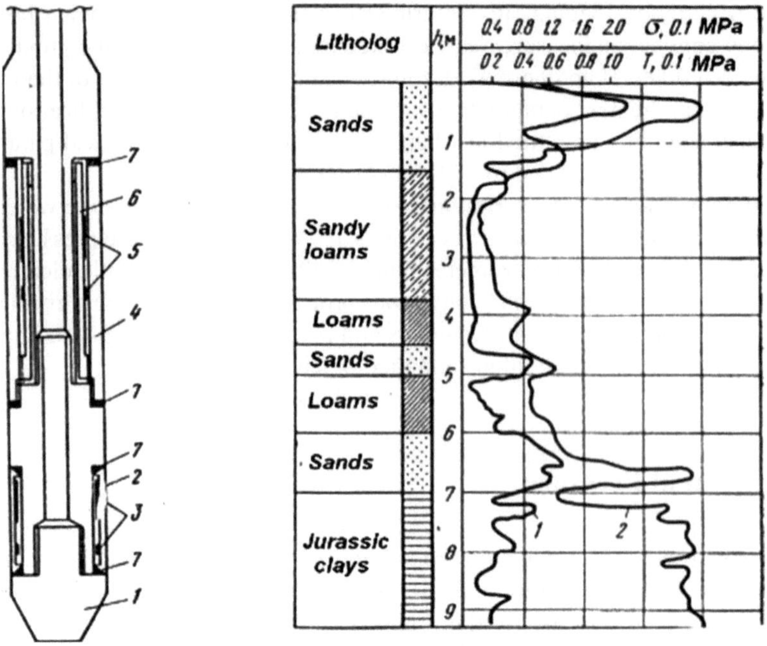

Fig. 7.18 Probe design with ground normal pressure and friction gauges **a** *1* sound cone; *2* thin-walled cylinder; *3* electric tensometer of the normal pressure gauge; *4* jacket of the friction gauge; *5* electric tensometer of the friction gauge; *6* working element of the friction gauge; *7* rubber ring seal. **b** Diagrams of ground friction (*1*) and normal pressure (*2*) recorded in a section of sandy and loam marine sediments. (Ferronsky and Gryaznov 1979)

of reliable interpretation of penetration data in design and calculation of bearing capacity and building foundations.

The experimental results have shown that the existing design of the normal pressure gauges allows the pressure parameter to be recorded while stopping the probe penetration. This is because of disturbance from continuous ground compaction on the normal pressure reaction.

7.6 Lithology Stratification

Up to now, stratification of different lithogenetic type sediments in a geologic section and their geoengineering description have been based on visual core and outcrop studies and laboratory analysis. Estimation of the grounds geoengineering properties for calculation of the building foundation and determination sediment type according to existing classifications by the granulometric content and plastic properties is provided. This is the generally accepted situation resulting from the real possibility of the existing methods of investigation. But the above methodology is quite subjective.

The main purpose of sediment stratification in a geological section and differentiation of the ground stratum is separation from the ground massif of such elements that are possible to be characterised by some close mean values of physical and mechanical properties. Existing methods allow determination of physical and mechanical ground properties only for a separate core. Therefore the characteristic of a separated stratum obtained by analysis of one or few samples can be recognised as legitimate if the single-valued property of the stratum is proved. For regional studies, the observed fluctuations in quantitative values of individual properties have no importance and the separated strata can be characterised by the obtained mean values. But in geoengineering investigations, which are used for calculation of durability and the steady state of engineering constructions, mean values of the ground properties are not acceptable. As a rule, the basis of most engineering constructions appears to be friable Quaternary sediments, all the properties of which are non-uniform and their fluctuations are out of the acceptable ranges to use for foundation design.

The modern penetration logging methods for geoengineering studies open new possibilities in obtaining quantitative data of ground physical and mechanical properties in the form of continuous diagrams along a geological section and stratifying it with respect to properties. Lithological stratification of a geological section with respect to the soil properties is a new quality of the discussed methods, which satisfies the highest demands of any construction designer. For this purpose the information of any one method as well as the total complex of methods can be used.

First of all, the data of natural radioactivity in soil should be used for separation of individual ground types and estimation of loam particle content. Gamma-logging records the natural radioactivity of soils by penetration. The radiation from radioactive elements such as uranium and thorium series and potassium-40 and their decay products are detected. The contents of other radioactive elements, like samarium, lanthanum, rhenium etc., in rocks are less than 1 % and in nuclear logging interpretation are not taken into account. Although alpha, beta and gamma radiation are emitted during the course of decay of many radioactive elements, only gamma radiation is of interest and is the main radiation that is measured.

Uranium has three isotopes, ^{238}U, ^{235}U and ^{234}U, the natural abundances of which are approximately 99.27, 0.72 and 0.01 % respectively. During the spontaneous decay of uranium a whole series of daughter products is formed (radium, radon, bismuth, actinium, etc.) and in the course of decay, uranium and its products emit gamma radiation, the energy of which has a wide range. Owing to the high chemical activity of uranium, it forms a large number of simple and complex compounds and is found in numerous mineral forms in nature.

Natural thorium consists of a single long-lived isotope, ^{232}Th ($T_{1/2}=1.41 \cdot 10^{10}$ years). Like uranium, thorium forms a series of daughter products (radium-228, thorium-228, thallium, polonium, etc.), which also emits gamma radiation of various energies during the process of radioactive decay. Thorium is an element with a wide distribution in the Earth and its abundance by weight is estimated to be four times greater than uranium.

A characteristic of the cations U^{4+} and Th^{4+} is their relatively large size. The radii of these cations measure 0.97 Å and 1.02 Å respectively. For this reason, it is

Table 7.4 The main physical characteristics of natural radioactive isotopes

Isotope	Symbol	Half-life (year)	Particles emitted	Ground-state energy of α and β particles (MeV)	Ground-state energy of gamma radiation (MeV)
Uranium-238	$^{238}_{92}U$	$4.51 \cdot 10^9$	α	4.20; 4.15	0.030; 0.043
Uranium-235	$^{235}_{92}U$	$7.13 \cdot 10^8$	α	4.20; 4.37; 4.40	0.143; 0.185; 0.204
Uranium-234	$^{234}_{92}U$	$2.47 \cdot 10^5$	α	4.77; 4.72	0.053; 0.58 0.117; 0.48
Thorium-232	$^{232}_{90}Th$	$1.41 \cdot 10^{10}$	α	4.01; 3.95	0.042; 0.055
Thorium-230	$^{230}_{90}Th$	$8.0 \cdot 10^4$	α	4.68; 4.62	0.068–0.253
Thorium-228	$^{228}_{90}Th$	1.91	α	5.43; 5.34	0.084–0.214
Thoriun-227	$^{227}_{90}Th$	18.2 d	α	5.72–6.04	0.05–0.237
Radium-228	$^{286}_{88}Ra$	6.7	β	0.05	0.31; 0.005
Radium-226	$^{226}_{88}Ra$	1620	α	4.78; 4.60	0.186; 0.26 0.42; 0.61
Bismus-212	$^{212}_{83}Bi$	60.6 min	α 66.3%	2.2	0.040–1.620
Bismuth-214	$^{214}_{83}Bi$	19.7 min	β>99%	3.26	14 γ-lines
		19.9 min	α–0.04%	5.51; 5.45	0.609 –2.445
Polonium-210	$^{210}_{84}Po$	138.4 d	α	5.305	0.803
Protactin-ium-231	$^{231}_{91}Pa$	$3.43 \cdot 10^4$	α	4.73–5.06	0.027; 0.29
		$3.25 \cdot 10^4$			
Actinium-227	$^{227}_{89}Ac$	21.7	β>98.8%	0.046	0.070; 0.166
		22.0	α–1.2%	4.94; 4.86	0.19
Thallium-208	$^{208}_{81}Tl$	3.1 min	β	1.8	0.511–2.614
Potassium-40	$^{40}_{19}K$	$1.26 \cdot 10^9$	β 89%	1.314	1.46
			K-capture		
			11%		

difficult for them to enter the structure of the usual rock-forming minerals, particularly in the accessory minerals of granites.

Natural potassium has three isotopes: ^{39}K, ^{40}K, and ^{41}K. Of these, only ^{40}K possesses natural gamma radioactivity and its abundance in nature is 0.0119% of all potassium. During decay, ^{40}K forms two daughter isotopes, ^{40}Ca and ^{40}Ar, with the emission of beta and gamma radiation. Potassium is an extremely common element and occurs in a large number of minerals, including varieties of feldspars, micas and clays. The potassium content of these rock-forming minerals can vary from a few tenths of a per cent to several per cent by weight.

Table 7.4 shows the main physical characteristics of uranium, thorium and potassium, together with those of the most important daughter isotopes (Lederer et al. 1967).

Direct and indirect studies of the distribution of the radioactive elements in the Earth's shells shows that the Earth's crust is enriched with these elements. In turn,

Table 7.5 Natural radioactivity of magmatic rocks (per 1 gram of rock)

Rock	Ra (10^{-12} g)	U (10^{-16} g)	Th (10^{-16} g)	Th/U	^{40}K (g)
Acid	1.34	4.0	13.4	3.3	0.026
Intermediate	0.51	1.4	4.4	3.2	0.020
Basic	0.38	1.1	4.0	3.6	0.14
Ultrabasic	0.2	0.6	2.0	3.3	0.004

Table 7.6 Natural radioactivity of some sedimentary rocks (per 1 gram of rock)

Rock	Ra (10^{-12}g)	U (10^{-16} g)	Th (10^{-16} g)	Th/U	^{40}K (g)
Limestone	0.5	1.5	0.5	0.33	0.001
Dolomite	0.11	0.3	–	–	–
Quartzite	0.54	1.6	–	–	–
Sandstone	0–1.5	>4	5.0	1.2	0.004
Shale	1.00	3.0	10.0	3.3	0.008
Loam	1.3	4.3	13.0	3.0	0.011

through analysis of distribution, for example, uranium in the Earth's core and upper mantel, it was found to progressively decrease with depth. The uranium content in amphibole faces and granitoids is accounted by unit grams per ton. In the pyroxene-granulated faces and basalts it decreases up to tens of grams per ton. And in eclogite rocks of the mantel origin it drops to hundreds of parts per ton (Alekseev 1972). Up to now, there is no definition of the mechanism of the planet's crust and upper zone enrichment by radioactive elements. The only undisputable idea is that the observed picture of chemical differentiation of the Earth as a whole appears to be an effect of its thermodynamical conditions of its formation and evolution. In this regards, we refer to our recent work that discusses this problem (Ferronsky and Ferronsky 2013).

The summery value of radioactivity for different type of rocks notably differs. Table 7.5 shows contents of radioactive elements in magmatic rocks reduced to the radium equivalent (Eq. Ra). As follows from the table, within eruptive rocks the highest radioactivity is contained in acid rocks and the lowest in ultrabasic rocks.

Tables 7.6 and 7.7 demonstrate radioactive element contents in sedimentary rocks compiled from Baranov and Khristianov (1963).

Table 7.7 Total radioactivity in some sedimentary rocks (per 1 g of rock)

Rock	Mean value and range of change (10^{-16} eq. Ra)	Rock	Mean value and range of change (10^{-16} eq. Ra)
Anhydride	0.5	Argillaceous sandstone	2–20
Coal	1.0	Argillaceous limestone	2–20
Rock salt	2.0	Carbonaceous loam	3–25
Dolomite	0.5–10	Shale	4–30
Limestone	0.5–12	Potassium salt	10–45
Sandstone	1–15	Deep-water loam	10–60

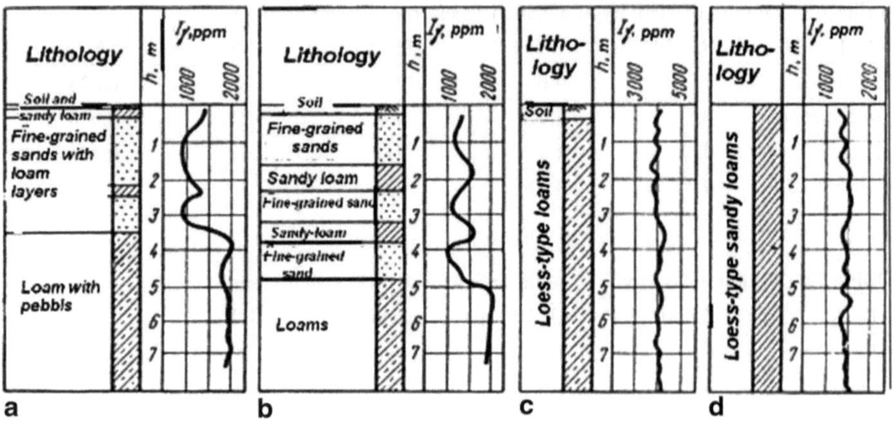

Fig. 7.19 Comparative data of GL and standard borehole methods for resolution of sediment stratification in sections: **a** sandy-loam, **b** interbedded sands, **c** loss sandy-loams in South Ukraine, **d** loss sandy-loams in Uzbekistan. (Ferronsky and Gryaznov 1979)

It is clearly shown that the homogeneous clay-free organic-genic and guart zone sediments (anhydrites, coals, limestone, dolomite, pure quartz sands, and sandstones) are the least radioactive. Sandy or carbonaceous clayey and argillaceous sandstone rocks are characterised by an increase in radioactivity, particularly with increasing clay content, although the range of their radioactivity is variable. All the sedimentary rocks and shale have the highest and comparatively most constant radioactivity.

The natural relationship between a rock type and its radioactive content forms the physical basis for detecting individual zones within a geological section by the gamma logging method and for analysing a geological profile. Interrelation between the radioactivity level of a stratum of homogeneous composition and its clay content is of particular interest in hydrogeological and engineering geological studies. Investigations of friable Quaternary deposits have established that the variation of radioactivity of sands depends entirely upon their clay content because the clays have a high sorption capacity for radioactive elements migrating through the formations in groundwater. The permeability of a stratum is also known to depend, to a considerable extent, on its clay content. Therefore, by identifying the amount of clay present in a stratum through the GL method, a qualitative and sometimes quantitative estimate of permeability for the layer becomes possible if sufficient other parameters such as rock type and degree of fracturing are known.

An experimental test of the GL method resolution for stratification of a geological section was done for different types of sediments: glacial, losses, sands and loams of different genesis. Figure 7.19 shows a number of geological profiles plotted by borehole data and with GL recorded in the same sections.

It is seen in Fig. 7.19a and 7.19b that passing from the microgranular sands to the glacial loams the natural radioactivity increases by a factor of 2. Here a sandy loam interlayer of a foot thick 0.4 m at 2.5 m in depth from the surface is clearly seen on the GL diagram with local higher radioactivity. In Fig. 7.19c the geological profile of loamy loss from Kahovka (South Ukraine) is presented. The section is characterised by rather uniform granulometric contents, which is explained by its natural history of creation. The absolute value of radioactivity of such sediments is two times higher than for the glacial loamy sediments. The same picture (Fig. 7.19d) in sandy loam sediments from Karshy (South Uzbekistan) is observed. Here a steady state level of gamma radiation uniform sedimentary thickness, with respect to clay fraction, is demonstrated. To the contrary of Ukraine losses, the absolute values of natural radiation for sediments are close to the glacial loamy soils.

It was found by experimental tests that for stratification of a geological profile by gamma logging the elementary calibration works of taking samples with respect to their gamma-radiation should be provided. More justified calibration by measuring of granulometry and gamma radiation of the samples should be done. And the most reliable calibration of lithology stratification can be reached by application of a total nuclear and penetration complex of methods.

On the diagrams of density and moisture, obtained by the GGL and NNL methods, clear borders of the property changes, as a rule, coincide with the borders of separate lithology-genetic sediment type. The sands in the geologic section have lower values than loams. The moisture of sands in the non-saturated zone appears to be also lower than loam. And in the saturated zone it is higher.

Taking into account that density values often depend on the moisture values, then a more informative parameter for interpretation of the lithology stratification is the porosity. This parameter is defined by calculation and in the case of full water-saturated soil, porosity is determined directly by a moisture diagram. The porosity diagram of a given section of sediments most represents the objective history of consolidation of the sand and loam sediments.

Analysis of penetration diagrams shows that the loam sediments have lower normal resistance values than the sands. Some researchers (Proceedings 1974–1975) have demonstrated tables of quantitative values of cone resistance figures for sands and loams. But it worth noting that such figures have a wide range of variation in different geographic regions. More justified recommendations for lithology stratification of sand and loam sediments were proposed by Begemann (1965) who found that a good criterion for the above task is the value of ratio between the cone resistance and friction of the soil. For soils having high moisture content up to complete saturation the above ratio is increased with decrease of loam particles content. This regularity is explained by different effects of compaction and shift phenomena in sands and loams. Based on analysis of experimental data, Begemann obtained a nomogram, which allows determination not only on the type of soil but also the value of loam particles for water-saturated sediments (Fig. 7.20).

Our application of Begemann's nomogram in different regions shows that the proposed relationship works well. First of all the ratios between ground cone resistance and friction for identification of soil type were checked. The studies were carried out in the Moscow region, Kazakhstan and Ukraine. The sediments were presented by different genetic soils of Jurassic-Quaternary ages. All studies were performed by ~5000 borehole meters (Table 7.8).

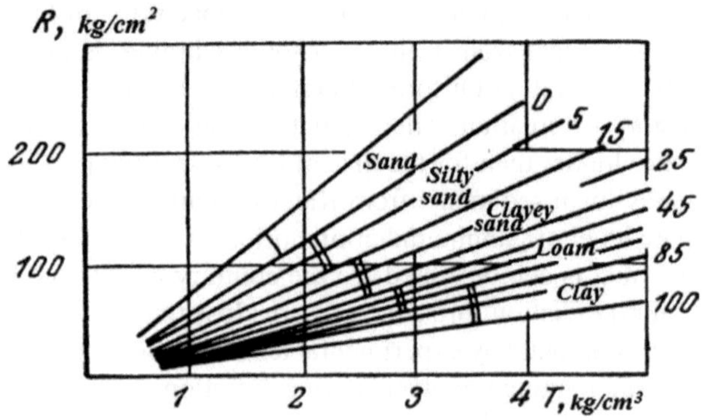

Fig. 7.20 Begeman's nomogram for soil type identification and loam particle content determined by means of ground cone resistance and friction. Figures mean the loam particles <0.016 mm content (%). (Ferronsky and Gryaznov 1979)

Table 7.8 Mean static values of frontal resistance R and friction T of sandy loam sediments studied for checking Begeman's nomogram

Soil	R (MPa)				T (MPa)			
	Max	Min	R (mean)	σ	Max	Min	T (mean)	Clay
Clay	6.2	0.5	2.1	10.3	0.40	0.03	0.13	0.054
Loam	10.5	1.2	3.8	0.64	0.40	0.04	0.14	0.075
Sandy loam	12.5	2.0	5.6	2.9	0.36	0.04	0.11	0.07
Sand	22.5	3.0	12.2	3.3	0.40	0.025	0.14	0.07

It is seen from the table that, in general, the cone resistance values are increased in passing from clay to sand. It means that an increase of the R values results because of a decrease of clay particles determined by the R mean values. But the minimal and maximal R values for different soil types are overlapping within significant ranges. This is not allowed to solve uniquely the problem of the soil type determined by the given parameter.

Analysis of the friction quantitative values shows that both the extreme and mean values of that parameter for different soil types are practically the same. In this regards, it cannot be used for soil type identification.

Further analysis of experimental data was provided from the viewpoint of resolution of the cone resistance and friction ratio with respect to soil type identification, as was proposed by Begemann. At the same time, it was taken into account that the cone diameter in our probe was 62 and 80 mm but in Begemnn's it was 35.6 mm. In addition, the penetration rate in our rig was not less than 2 m/min and in Begemnn's it was 0.5 m/min. Finally, by our classification the clay particle diameter is equal to less than 0.005 mm. And the studies were carried out in saturated and non-saturated soils.

Figure 7.21a shows experimental data of the ratio between the frontal resistance and friction R/T and percentage of clay fraction C_c. The mean values R/T and clay particle content for the sediments of different genesis and age are taken in this case. For example, the mean values of the studied points are placed on the plot within the

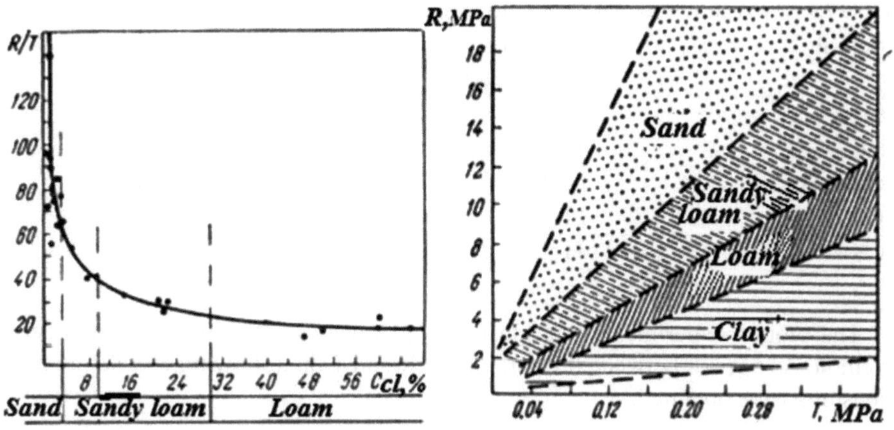

Fig. 7.21 Experimental relationship between R/T and clay particle content (**a**) and relationship between R/T and R/T (**b**). (Ferronsky and Gryaznov 1979)

clay, loam, sandy loam and sands intervals and are represented accordingly by the 133, 56, 26 and 227 measurements in the soil layers.

Analysis of the presented data shows that a definite relationship between the R/T ratio and the clay particle content is observed. In the sand and sandy loam region the curve has a sharp slope to the side of low values of R/T and asymptotic flattening out in the loam and clay region. The obtained experimental relationship is used in practice for estimation of clay particles within the 3–30 % range. In this interval a certain determining error varies from ±2 % of the absolute value at 3–15 % of clay particles content to 4 % at 15–30 % of C_c. The value of error is sharply increased at 30 % and more clay particles and determination is practically impossible.

A correlation formula was derived on the basis of numerous determinations in the form:

$$C_c = 2.698 \cdot 10^4 \left(\frac{R}{T}\right)^{-2.15}. \quad (7.6)$$

A study on the relationship between R/T and density–moisture was carried out. It was found that the values R and T for a given ground type are changing with the coefficient of proportionality of density and are low depending on moisture content. In this case, the observing changes of R and T values from moisture content are in-line with the constant coefficient for both parameters. In soils having constant values of clay particles the ratio R/T is constant.

Figure 7.21b shows the experimental relationship between the cone soil resistance and friction and ground type. The experimental points of the graph form four clear outlined zones corresponding to different ground types from existing classification based on clay particle content. The region between the ordinate axes and the sands zone can be referred to as the values of R and T for gravel and pebble sediments.

Table 7.9 The value ranges of R/T for different ground types

Grounds	R/T values	
	By H. Begemann	By the author
Sands	55–72	57–137
Silt sands	45–55	30–57 (sandy loams)
Clay sands	24–45	30–57 (sandy loams)
Loams	22–29	22–30
Clays	14–22	5–22

The plotted experimental data in Fig. 7.21b were treated by the least squares method. As a result between the parameters R and T the following correlation relationships were obtained:

$$\text{for clays } R = 4 + 13.1\,T\ (r = 0.68 \pm 0.08,\ \text{reliability}\ \mu = 8);$$

$$\text{for loams } R = 4.4 + 26.3\,T\ (r = 0.95 \pm 0.02,\ \text{reliability}\ \mu = 42);$$

$$\text{for sandy loams } R = 17.6 + 32.1\,T\ (r = 0.97 \pm 0.01,\ \text{reliability}\ \mu = 83);$$

$$\text{for sands } R = 11 + 77.1\,T\ (r = 0.89 \pm 0.03,\ \text{reliability}\ \mu = 28).$$

The above results show that practically 90 % of determining results of R/T ratios coincide with determination done by granulometry content and plasticity state. Thus, Begemann's idea of using cone resistance and ground friction at static penetration for determination of lithology strata is completely proved by our experience. Some numerical discrepancies in R/T ratios for separate ground types that follow from Begemann's data and our experiments (Table 7.9) can be explained by differences in classification of the ground types.

The most justified physical possibility for the lithology stratification of a friable geologic section appears to be if the ground cone and friction resistance measurements are carried out by the scheme shown in Fig. 6.12 (Chap. 6). As follows from Coulomb's law, in sandy soils, with growing normal pressure with depth, its shift resistance along the jacket surface is also proportionally increased. In clay soils, for which coefficients of friction have very small values, this effect does not occur. At the same time, clay soils have significant elastic properties and long relaxation time relative to the sands. These properties create a good physical basis for experimental and theoretical use of experimental data for lithology stratification. The results of experimental works in this direction are presented below.

Let us consider a complex of penetration logging diagrams obtained in the Moscow region by the rig SPK and its resolution with respect to the lithology stratification of the geological section. Figure 7.22 shows a complex of the GL, GGL, NNL, R and T diagrams recorded up to 38 m in depth during the experimental study. The geological section was composed of friable Quaternary sediments of glacial and Jurassic ages.

By the GL diagram within 0–5 and 10–11.5 m depths the loam layers having γ-radiation of 12–14 μR/h were identified. The sandy loam layer in the depth of

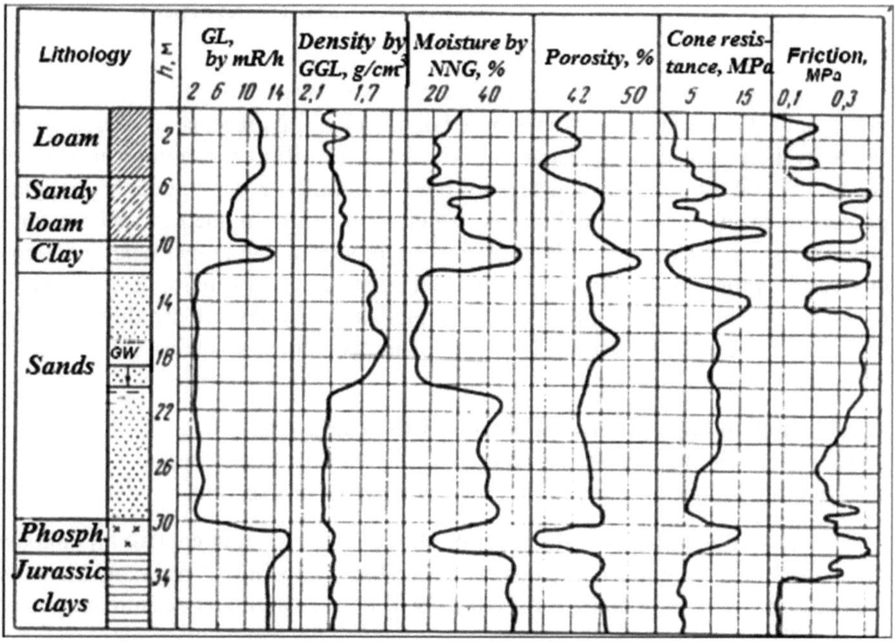

Fig. 7.22 Resolution of the nuclear and penetration logging methods with respect to lithology stratification of a glacial sediment section in the Moscow region. (Ferronsky and Gryaznov 1979)

5–10 m having γ-radiation of 6–8 μR/h was derived. By borehole sampling, that interval of sediment was qualified as stratified sand and loams. Within the depths of 10–30.5 m the sands having γ-radiation of 3–4 μR/h were found. And below 32 m the strata of Jurassic clays having γ-radiation of 12–14 μR/h were determined. Note, that here, as well as in the other regions with glacial sediments, the sands, sandy loams and loams are well correlated with the natural γ-radiation, which in this case has a proportion of 1:2:3. Within the depth of 30.5–32 m the phosphorite layer by γ-radiation of 20 μR/h is uniquely separated.

The bulk density (GGL) and moisture content (NNL) have practically identical resolution with respect to lithology stratification. By these methods the border between the loam and sandy loam is clearly tracing a depth of 11.5 m and the phosphorite layer at 32 m is fixed. In addition, in the non-saturated zone at the depth of 2.5–5.0 m there is a denser layer, which by these parameters is classified as sandy loam. On the basis of these parameters a diagram of the porosity was calculated and plotted. The diagram is of interest from the viewpoint of sediment profile formation history study as well as its practical use for building's foundation design.

By the diagrams of R and T resistance the sand, sandy loam, loam and clay strata with their borders were separated. In addition, some details within the separate type of soils were recognised. For example, two intercalations at depths of 5–7 and 7.5–9.5, which by R/T values are defined as sands, were recognised.

7.7 Application of Statistical and Probability Methods

Interpretation of nuclear and penetration data with respect to lithology stratification of a friable sediment section is based on the physical presence of characteristic limits in the physical and mechanical properties for different lithological types of sediments. Taking into account that we have a complex of continuously changing properties in the form of nuclear and penetration diagrams then the search for a characteristic combination of these properties for separate sediment types evidently ensures the problem solution. Such a problem consideration is possible on the basis of the statistical and probability methods application. And there are three particular tasks that can be solved on the basis of mathematical statistics and theory of probability:

1. the search of borders between separate sediment layers;
2. identification of soil types based on existing classification criteria;
3. estimation of limits of sediment property changes in a section.

The identification criteria for the separate above tasks are sediment natural radioactivity, frontal and lateral friction resistance and their ratio, bulk density and moisture content. Each of these criteria for a given soil type is changing in definite ranges. In general, the less amount of studies the more narrow rates of changes for each criterion. But this does not exclude the presence of specific natural conditions where the amplitude of the property variation can be quite wide.

In order to study the task related to soil type identification and strata borders, a discriminator was used. It determined the ranges of change of each property in a given region for pairs: clay—loam, loam—sandy loam, sandy loam—sand. At the same time the selective dispersions σ_i^2 for each indicator was calculated by:

$$\sigma_i^2 = \sum \frac{(x_i - \overline{x}_i)^2}{n-1}, \tag{7.7}$$

where n is the number of measurements; $\overline{x}_i$ is the mean value of the indicator.

Further, the free dispersion by each indicator for every pair was determined by the formula:

$$s_i^2 = \frac{(n_i - 1)\sigma_{1i}^2 - (n_2 - 1)\sigma_{2i}^2}{n_1 + n_2 - 2}, \tag{7.8}$$

where σ_{1i}^2 and σ_{2i}^2 are the selective dispersions of the first and second selections.

Finally, the discriminator threshold for each soil pair was calculated by the formula:

$$\alpha = \frac{1}{2}\sum_{i=1}^{m} \frac{(\overline{x}_{1i})^2 - (\overline{x}_{2i})^2}{s_i^2}, \tag{7.9}$$

where $\overline{x}_{1i}$and $\overline{x}_{2i}$ are the mean values of the first and second selective criteria.

Practical determinations of the soil types were reduced to finding the mean values of criteria by a separated soil layer in the geological section and by the discriminator calculation z_0 by the formula:

$$z_0 = \sum_{i-1}^{m} \alpha x_i \,, \tag{7.10}$$

where *a* is the value $(\overline{x}_{1i} - \overline{x}_{2i})/s_i^2$.

After drawing the discriminator on the appropriate diagram of the discriminators distribution with the specified thresholds the soil type is determined.

The practical use of the above method shows good results. The statistical treatment of large factual data on separate regions with a compilation of nomograms should be done for a wider application of the method.

The problem of studying the variability of ground properties and accumulation and use of geoengineering information has been developed in special works from a number of researchers (Bondarik 1971; Komarov 1972).

References

Alekseev FA (1972) Application of radioactive and stable isotopes in geological investigations. Nuclear Geophysics, Atomizdat, Moscow, pp 177–193

Baranov VI, Khristianov LA (1963) Radioactivity of the ocean sediments. Chemistry of the Earth's crust. Moscow, Academizdat, pp 401–409

Begemann HK (1965) The friction jacket cone as an aid in determining the soil profile. In: Proceedings 6th ICSMFE, Montreal 1:17–20

Bondarik GK (1971) Fundamentals of the theory of engineering geological rock properties. Nedra, Moscow

Bondarik GK, Komarov IS, Ferronsky VI (1967) Field methods of engineering geological studies. Nedra, Moscow

Buisman AS (1944) Drondmechanika, 2nd edn. Waltman, Delft

Czubek JA (1976) Comparison of nuclear well logging data with results of core analysis. Nuclear techniques in geochemistry and geophysics. IAEA, Vienna, pp 93–106

Faiestein BD, Makarov RM (1964) Rig C-832 for the ground sounding. Installations for for the ground sounding and drilling. CBTI Gosstroy USSR, Moscow, pp 3–24

Ferronsky VI (1969) Penetration logging methods for engineering geological investigation. Nedra, Moscow

Ferronsky VI, Ferronsky SV (2013) Formation of the solar system. Springer, Dordrecht

Ferronsky VI, Gryaznov TA (1979) Penetration logging. Nedra, Moscow

Kerisel J (1961) Foundations profondes en milieu sableux: variation de la force portante limite en function de ladensite, de la profondeur, du diameter et de la vitesse d'enfoncement. In: Proceedings of the 6th ICSMFE, vol. 2, Paris, pp 73–79

Kerisel J (1964) Deep foundations basic experimental facts. In: Proceedings of the 5th ICSMFE, pp 8–31

Komarov IS (1972) Accumulation and assessment of information in engineering geological investigations. Nedra, Moscow

Lederer CM, Hollander JM, Perlman I (1967) Table of isotopes, 6th edn. Wiley, New York

Meyerhof G G (1951) The ultimate bearing capacity of foundation. Geotechniqueb 2:301–332

Paproth E (1943) Der Prüfstab Künzel, ein Gerät für Bauuntersuchungen. Die Bautechnik, pp 21–38

Parez L, Bachelier M (1965) Contribution a a l'etyde la compressibilite des sols a l'aide du penetrometre a cone. In: Proceedings of the 6th ICSMFE, Vol. II, Montreal, pp 3–7

Proceedings (1974–1975) In: Proceedings of the European Symposium on Penetration Testing. ESOPT, Stockholm

Sanglera G (1971) Investigation of soils by penetration test (transl. from French). Stroyizdat, Moscow

Schmertmann JH (1970) Static cone to compute static settlement over sand. Proc ASCE 96:1011–1043

Schultze, Melzer K (1965) The determination of the density and the modulus of compressibility of non-cohesive soils by soundings. In: Proceedings 4th ICSMFE, Montreal, pp 354–358

State Standard (1982) Mining rocks: Methods of field test by penetration logging, GOST 25260-82, approved on 1982, May 17

Terzagy K, Pek R (1958) Soil mechanics in engineering practice (Transl. from German). Gosstroyizdat, Moscow

Tomlinson MI (1957) The adhesion of piles driven in clay soils. In: Proceedings of the 4th ICSMFE, London, vol 2, pp 66–71

Trofimenkov YuG, Vorobkov LN (1964) Field methods of the grounds study. Stroyizdat, Moscow

Yaroshenko VA (1964) Interpretation of the ground static penetration results. Design of complicated foundations. Fundamentproject, Moscow, 99:14–24

Chapter 8
Application of Penetration Logging Techniques for Geoengineering Exploration

Abstract Specific geological and geographical conditions in Belorussia, Ukraine, Moldova, Volga River basin, Central Asia and Kazakhstan for application of the penetration logging technique are discussed. The practical application of penetration logging methods in engineering geological and hydrogeological mapping for irrigation and drainage projects in the Alma-Ata (Kazakhstan), Kizil-Orda (Uzbekistan), Saratov (Russia), Kiev (Ukraine), Neman River (Belorussia) regions is analysed. The results of geoengineering studies of glacial sediments in the Moscow region and bottom sediments in the Black sea off-shore area are presented (Ferronsky, Penetration logging methods for engineering geological investigation, 1969; Ferronsky and Gryaznov, Penetration logging, 1979).

8.1 Geological and Geographical Conditions for Application of Penetration Logging

Despite the short history of their development, the penetration logging techniques have already found practical application for solving a broad number of engineering geological problems. The problems are as follows: engineering geological and irrigation hydrogeological mapping at different scales, exploration for civil and industrial engineering, exploration of landslide areas, engineering geological exploration for marine and power hydraulic engineering, environmental investigation. At present, there is no doubt about the high informative ability and technical and economical efficiency in application of the penetration logging techniques in solving the above listed problems. The possibility of penetration logging methods is determined mainly by geological conditions, which enable ground sounding, availability of the rig to get to the work area and the climatic conditions for exploitation of the apparatus.

Geological conditions where people provide active economic activities are favourable for use of penetration logging. This is, as a rule, thick massifs of sandy-clay sediments of different ages and formation conditions. In southern regions, the loamy loess also allows sounding. In general, the areas with favourable geological conditions for penetration logging are those sandy-clay sediments that can be penetrated by a probe of 80 mm in diameter up to 25 m depth by loading up to 20 t.

V. I. Ferronsky, *Nuclear Geophysics*, Springer Geophysics,
DOI 10.1007/978-3-319-12451-3_8

Table 8.1 Mean values of depth and confidence intervals of the possible sounding depth for a number of regions within Eastern Europe

Region	Mean value of sounding depth, m	Confidence intervals of sounding depth with 90% of probability, m
Northern Europe	7.6	1.8–13.4
Belorussia	9.8	2.5–17.1
North of Volga river region	10.1	3.0–25.4
South of Volga river region	16.3	6.2–23.5
Southern Europe	14.2	3.8–22.6
Ukraine and Moldova	15.1	6.6–23.6
River valleys of Central Asia and Kazakhstan	7.9	1.9–13.9
Foothills and inter-mountain valleys	8.8	0.6–20.4
Alluvial and wind-borne plains	23	20.0–25.0

Analysis of the possible depth sounding in different geographical and geological conditions, based on the experience of commercial organisations using the penetration logging station SPK on the territories of Belorussia, Ukraine, Moldova, Volga River basin, Central Asia and Kazakhstan, was performed (Table 8.1).

One may note that for the European part of the region the possible depth of sounding is increased from the north to the south, which is related to the regularity of expansion of moraine and fluvioglacial sediments with inclusion of boulder and pebble material. In the north of the Central European region and in Belorussia the possible depths of sounding are varied from 5 to 19 m. In the south direction the depths are increased to 10–20 m. Analogous regularities in the depth sounding remain for the Volga River region. Here the sediments that limit the sounding depth are the moraine and Quaternary fluvioglacial sands with gravel or dense Cretaceous sands and Jurassic clays with fossil fauna and veins of phosphorites. In favourably geological conditions the maximum sounding depth reaches up to 40 m.

For the regions not undergone by glaciation (South of Ukraine, Moldova, South of Volga basin), the prevailing possible sounding depths vary within 15–20 m and in the river valleys 10–15 m. The sounding depth in these regions is limited by the layers of cemented carbonates and gypsum, dense alluvial sands with gravel, pebble and tightly plastic clays.

For Kazakhstan and Central Asia the prevailing possible sounding depths in alluvial-eolian and loess soils ranged within 20–25 m. In the foothills, intermountain valleys and large river valleys the sounding depth did not exceed 10 m. Here the sounding depth was limited by rocky and semi-rocky massifs. For the areas of Kazakhstan and Central Asia the mean sounding depth increases from the river valleys to the alluvial-eolian planes.

Some areas, limited by geological conditions for the application of the penetration logging techniques, are represented by local permafrost sediments, rubble rocks, salt-marsh areas and dun sands.

Finally, the rocky and semi-rocky areas on the surface covered by colluvial, colluvial-delluvial deposits and permafrost regions pertain to the areas where the penetration logging method is not applicable due to the geological conditions. But even in this case there are local places for application of the method. Later on it will be shown that penetration logging can be used in combination with a rotary in order to make a hole.

The important factor affecting the sounding depth is the ultimate bearing capacity of the upper ground layer with respect to the crew anchors of the penetration logging rig. The rig's device is capable of accepting the axial force that the rig's hydraulic system for the sound displacement develops. This is because the rig's own weight is too small to take the total (about 20 t) force. In this regards, in constructions of existing facilities, so-called anchor-piles are designed. If the upper ground layer is too loose (swamp, peat, salt-marsh and so on) then even anchor-piles do not accept such a reactive force of sounding.

From the point of view of geographical conditions, the most unfavourable situation for possible transport is near marine salt-marsh deserts, delta salt-marsh planes, non-conductive bands and depressions, sandy dunes, ridges and quicksands, low swamps and quagmires and sliding slopes. The above listed unfavourable types of locality, as a rule, do not cover a studied region as a whole but occupy only a small area.

Climatic conditions can also affect planning of field works. From our experience the penetration logging methods and rigs by their design and technical parameters guarantee reliable exploitation of the apparatus at an air temperature from +5 to +45 °C.

The geological and geographical conditions for application of use of underwater penetration logging techniques on the shelf of inner and outer seas are different.

For the polar basin shelf a wide extension from the shore of friable and clay sediments is characteristic. The majority of these elapse under the edge of steady ice. The far-east basin shelf is covered also by friable sediments but has a steeper shore slope. The presence of a considerable amount of coarse-detrital material is characteristic of the area.

The southern seas shelf is characterised by an extension of friable non-consolidated sediments enclosing shells. Our experience in the study of the Black Sea shelf by the PSPK station and ship "Geologist-1" showed that sea-bottom sediments represented by tightly plastic loams with pebble layers are sounding well up to 25–40 m depth. In the polar and far-east basins, coarse detrital and rocky materials are widely extended and limit the areas of sounding.

8.2 Practical Applications

Penetration logging stations SPK and PSPK were manufactured by the experimental plant of geological exploration instrumentation and equipment of the former USSR Ministry of Geology. From 1968 to 1976 it produced 40 stations that were used in the Ministry's geo-exploring organisations for geoengineering and hydrogeological mapping of large areas for planning irrigation in the southern regions of the country. Later on, penetration logging stations were used for engineering geological exploration in organisations connected to the Ministry of Irrigation and other corporations for civil engineering; water power design, agricultural needs. Technical and economical interest of the new techniques at that time was quickly developing in order to satisfy the growing economy demands of society.

We have analysed the work of 32 SPK stations that were manufactured during 1968–1975. Two of them were destroyed in accidents and three were not used for technical reasons. All other rigs have been working successfully for a long time. It can be stated that during the analysed period the efficiency of the work was continuously growing. So, to the end of 1969, the total metric penetration holes of the penetration logging works of all the stations was determined by 7064 running meters (run. m.). To the end of 1975 it increased up to 81,243 run. m., i.e., grew by more than 10 times. The first period of familiarisation with the new techniques had some difficulties related to teaching specialists, obtaining experience, adjustment of the hydro-mechanical systems and electronic equipment. With better operating experience and planning of works the efficiency of each penetration logging rig improved. For example, the mean efficiency of one working station in 1971 was 1636 run. m. and to the end of 1975 it increased to 3437 run. m. But the efficiency of individual stations in different organisations was notably different. In general, the situation with planning and use of the new techniques started to improve after the State Standard "Mining rocks: Methods of field test by penetration logging" was approved in 1982 (State Standard 1982). This USSR official governmental document establishes the order of planning, operation of test, treatment and calculation of the ground physical and mechanical parameters and their use for design and building construction.

It worth noting that the problems solved by penetration logging techniques continuously increase. The techniques have been widely used for engineering geological and hydrogeological mapping for planning and design of irrigation systems and industrial objects like the Kama automobile plant, enterprises in Neftekamsk and Sterlitamak, the Almalik mining metallurgy complex, engineering geological prospecting for industrial and civil engineering construction in Tolyatti, Kalinin, Zagorsk, Klimovsk, Ufa, Almalik, Leningrad (Sankt-Petersburg), Riga and so on.

The USSR Ministry of Geology used the penetration logging techniques mainly for engineering geological and hydrogeological mapping of wide areas intended for irrigation in Belorussia, Ukraine, Volga River valley, Kazakhstan and Central Asia. Application of the new techniques allowed the quality of geological information to be improved. By doing so, in order to increase the reliability and accuracy of engineering geological and hydrogeological field information, the boring works decreased by 25–30 % as well as the volume of borehole and laboratory tests being reduced.

Experience of the penetration logging techniques application has shown that their use in geoengineering prospecting with respect to technical and economical efficiency is effective. This is shown by a decrease of the work time duration, increase of obtained information and objectivity of lithology identification and correlation, improvement of the sediments physical and mechanical properties reliability especially for loose and water bearing soils and increase of the pile bearing capacity estimation.

It worth noting that because of the high productivity of the penetration logging techniques their efficiency use depends first of all on existence of the front and volume of work. Especially high efficiency of use the penetration logging techniques for geoengineering prospecting of marine bottom sediments where the boring holes for that purpose is expensive an technically low effective work.

8.3 Engineering Geological and Hydrogeological Mapping

The first manufactured penetration logging rigs were used for obtaining initial field materials at engineering geological and hydrogeological mapping for the planning of irrigation lands. Such lands were located in the Alma-Ata (Kazakhstan), Kizil-Orda (Uzbekistan), Saratov (Russia), Kiev (Ukraine), Neman River (Belorussia) regions. This was the first time it was decided to substitute part of the boring works by penetration logging geophysics. After an initial successful experience of high productivity and qualified information and a sharp decrease of boring and laboratory works, the geophysics was accepted without any objection. Let us consider a number of examples demonstrating application of penetration logging results.

8.3.1 Study for Irrigation Land Projects

During 1968–1969, penetration logging studies in the Kizil-Orda irrigating area (Uzbekistan) were carried out. The work was the field component of a common project for engineering geological and hydrogeological mapping provided by the USSR Ministry of Irrigation. The field materials were intended for design and construction of hydraulic and engineering structures. In this connection it was specified in the project to provide a rare drill net of boreholes with detailed studies of the key points. The penetration logging studies were focused on such key points. It was planned to obtain the data on the soil physical and mechanical properties and geological structure of the studied sites. One of the sites was located on the right bank of the Sir-Darya River and represented by a narrow strip of 130 km long and 20 km width. Its natural border on the south and west-south was the Sir-Darya River and on the north and north-east were the canal Kara-Usak and Koksuyskoe depression. The second site was represented by two sections within the Abay national farm where 150 points of the penetration logging studies were planned to be carried out.

From a geological viewpoint, the first site was created mainly by alluvial, partly by eolian sandy and argillaceous sediments. The alluvial soils are underlaid by basic waterproof clays. The alluvial sediments in the lower part of the section are represented mainly by sands with layers and lenses of clays. In the upper part of the section the alluvial sediments were created by the material carried out by the Sir-Darya River in the form of loams and sandy loams alternating by layers and lenses of fine-grain sands. The groundwater table was located up to 10 m from the surface.

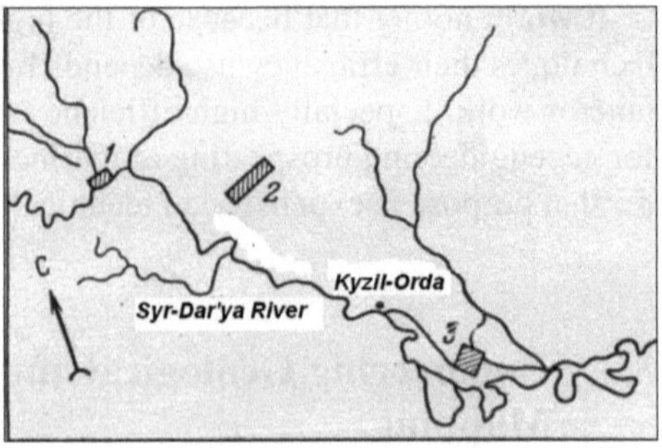

Fig. 8.1 Schematic view of the Kizil-Orda irrigating area with key sites of the penetration logging studies. (Ferronsky and Gryaznov 1979)

The penetration logging works were carried out on three separate sites (Fig. 8.1). At site 1 the studies were executed in nine points with a 25–30 m distance between the points. At site 2 the studies were realized in 27 points at a ~200 m distance between the points. On the right-hand canal, penetration logging was completed in 23 points at 500 and 1000 m between the loggings. And at site 3, 17 soundings were carried out at a 180–200 m distance between the points.

Analysis of the penetration logging diagram from the the Kizil-Orda irrigating area to obtain the main criteria for separation of sediments by means of their properties. It appeared that with the cone resistance logging diagram at its values of less than 2.0 MPa, the loam and clay sediments separate. The sand layers are separated by the same parameter at a value equal to 7.0–8.0 MPa and at a value of 2.0–7.0 MPa the sandy loam sediments are determined. The lateral jacket friction parameter indicates the clay and loam sediments by a value <0.15 MPa; the sand sediments are determined by 0.3 MPa and sandy loam by intermediate values from 0.15 to 0.3 MPa. The diagrams of the natural gamma-logging identify the sands by a value of 600–800 counts/min; the clays by 1200 counts/min; and sandy loam from 800 to 1200 counts/min.

It seems that the diagrams of GGL and NNL are strongly differentiated. The higher values of moisture are characteristic of clay grounds. The groundwater table on the NNL diagrams is identified by a sharp increase of the measured scattering neutron radiation, which corresponds to an increase of water content. The diagrams of GGL in this case show a decrease of the measured gamma radiation, which corresponds to an increase of the scattered mass. The NNL and GGL diagrams below the water table in the section have, as a rule, weakly differentiated character.

At the developed site of work, alluvial and eolian sediments mixed up by water flows caused strong variability in the ground content and properties within the section in width and depth. This explains some difficulties in use of GL for interpretation of lithology of the section's upper part. In such cases stratification of soil thicknesses has conventional character. The other methods of penetration logging diagrams helped. Only complex information obtained by five penetration logging methods provides a reliable basis for interpretation of ground lithology.

Figure 8.2 represents the results of a penetration logging study along the profile between boreholes 117 and 127 at site 2, Alexandrovsky. The starting point was located 1.5 km north-west from the Alexandrovsky settlement. The profile is more than 5 km long and includes 27 soundings with penetration logging records up to 10–17 m in depth.

It was known from the data of borehole 117 at zero point that the upper part of the section up to 3 m depth is represented by clays and loams. Below these are fine sands with thin layers of clay ground. From the data of borehole 127 (at picket 60), the whole section up to 15 m is structured by sands. It appears by penetration logging data that the borehole data give very preliminary and even non-correct information about the lithology and soil properties, as shown in Fig. 8.2.

It is seen from Fig. 8.2 that the upper part of the section up to 6–8 m consists of loose clay grounds and below those are sands. By combined interpretation of the penetration logging diagrams the upper part of the section includes clay grounds with thin sand intercalations within a non-saturated zone. The thickness of the upper layer is 1–2 m, seldomly 2–5 m and below which are loams. At zero picket, the upper layer of loams is underlaid by sandy loams 2.5 m thick, which are characterised by $R=3.0–5.0$ MPa and by lateral friction $T=0.4–0.5$ MPa. Between sandy loams a loam intercalation occurred of 0.25 m thick with a higher moisture content of $W=28\,\%$ and a lower value of $R=2.5$ MPa. In point 2 up to 0.8 m the loams are identified by values of $R\approx 1.0$ MPa and up to 1.75 m the sandy loams by $R=3.0$ MPa and $T=0.4$ MPa. In points 4–26 the loams are generally identified at a depth of about 1 m by values of $R\leq 1.0$ MPa and $T=0.1–0.3$ MPa depending on the ground moisture. In point 28, an intercalation of sandy loam 0.5 m thick with characteristic values of $R=2.5$ MPa and $W=13\,\%$ within the loam layer was found. In point 30, below the pack of clay sediments, sand with clay fraction are deposited. They are characterised by lower values of gamma background: $I_\gamma=1500$ counts/min and resistance $R=8$ MPa. In points 32–38, the sandy loams are identified by lower values of gamma radiation: $I_\gamma=1000–1200$ counts/min (for loams here $I_\gamma=1500$ counts/min) and higher values of $R>2.5$ MPa. In points 40–46, within packs of clay sediments, sand deposits were found.

It worth noting that the performed interpretation of the penetration logging records with respect to the sediment separation of clays, loams, sandy loams and sands is conventional because the values of natural gamma radiation for the sandy loams are lower and the values of cone resistance for loams are higher relative to the general rule. The data of granular content of sediments should be used for determining their standard type.

In the lower pack of the clay water bearing sediments, mainly the silt loam deposits form the upper part of the thickness. The dense sands underlie them. Like in the previous case, separation of silt loam, loam and sandy loam was executed conventionally on the basis of their properties. The silt loams were separated by their maximal moisture (porosity), minimal volumetric mass, the same cone resistance as for normal loams and lower natural radioactivity. The last condition is accepted as characteristic for silt loam but not for clay.

The silt loams are tracing throughout in all the points of penetration logging and have a thickness from 0.5 to 2.0 m. Their characteristics are: $R\approx 1.0–1.5$ MPa, $T\approx 2.0$ MPa, $I_\gamma=1500$ counts/min in the lower part and $I_\gamma<1500$ counts/min in the upper part of the section, $W\approx 40\,\%$, $\gamma=1.93–1.95$ g/cm^3.

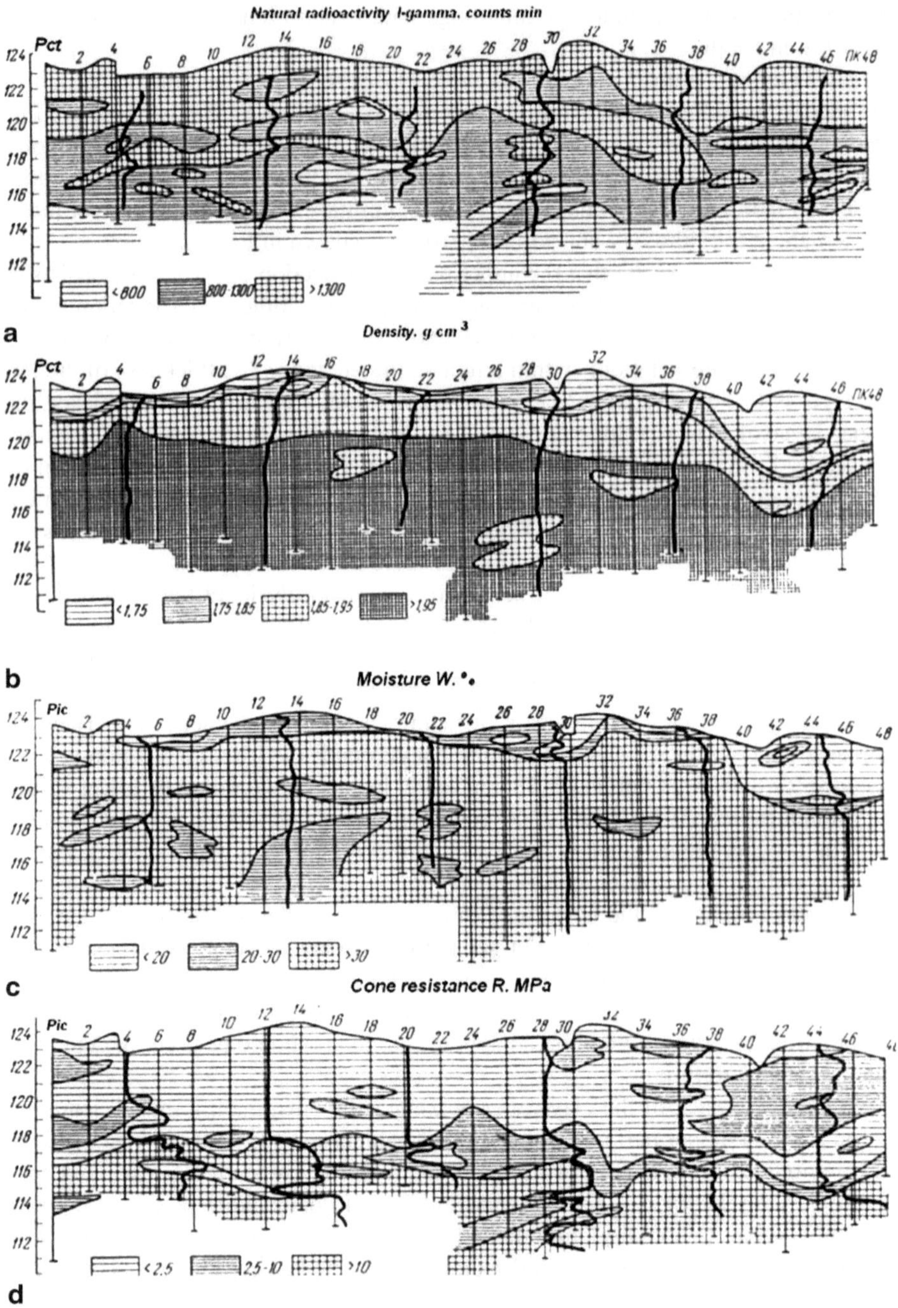

Fig. 8.2 Distribution of properties and geologic section at site 2 Alexandrovsky, obtained by penetration logging studies: **a** natural GL; **b** GGL; **c** NNL; **d** RL; **e** TL; **f** geologic section: *1*—loam; *2*—sandy loam; *3*—sand; *4*—silty clay. (Ferronsky and Gryaznov 1979)

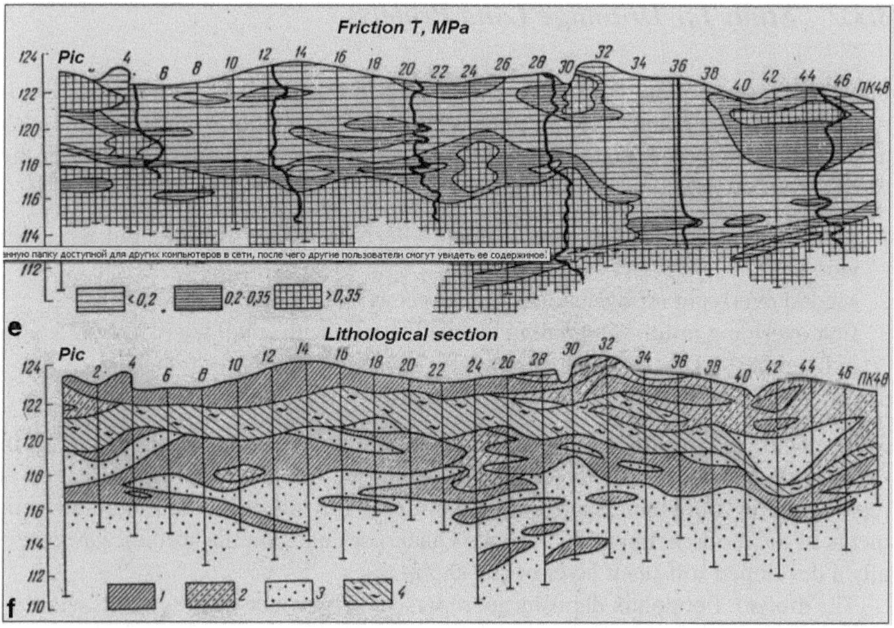

Fig. 8.2 (continued)

Below in the section that underlies dense sands, the loams overlie. They have the same cone resistance and lateral friction as the silt loams. But their density is higher and their moisture is lower ($\gamma = 1.98$ g/cm^3, W$\approx$35%). The loams below have lateral friction T>0.2 MPa. Between the loams are intercalations of sandy loams and sands occur. The sandy loams are characterised by intermediate values of natural gamma radiation ($I_\gamma = 800$–1200 counts/min) and also by the values R<7.5 MPa and T<0.3 MPa.

The underlying dense sands have R=12.0–15.0 MPa, T=0.4–0.5 MPa and minimal $I_\gamma = 600$–800 counts/min. Between the sands small intercalations of clay sediments occur. In point 38, clays at a depth of 14.0–16.5 m were found.

The groundwater table identified by NNL and GGL diagrams coincides with the cover of the silt loams.

In order to control the reliability of lithology stratification of the sediments based on penetration logging data, two screw conveyer boreholes were made in points 20 and 42. The results of compared granular content data proved our criteria for identification of clay, loam, sandy loam and sand by penetration logging data. Both independent methods give compatible results.

The carried out penetration logging works at the Kizil-Orda irrigating area have shown the possibility to provide a complete complex of engineering geological studies for design of an irrigation area and construction of hydraulic installations. The results were available for geologists and engineers for practical use. The results of these studies contributed to improve the quality of the irrigation project and to reduce the preparation of project documentation.

8.3.2 Study for Drainage Land Projects

In 1974, penetration logging studies were carried out at the Morino region (Belorussia) on an area planned for a drainage project. The intention was to accomplish air-landscape mapping along the profile of the right-handed bank floodplain of the Neman River where four sites were traced:

1. mildly-undulating fluvial-glacial plain with hills of 2–4 m in height and with wood;
2. second overlying mildly-undulating terrace with blowing dunes;
3. first overlying mildly-undulating wooded terrace with small wood hills of 5–7 m in height;
4. low-level floodplain with swampy complexes.

Along the traced profile, a series of penetration logging soundings up to 22.5 m in depth with 250–1000 m between them was carried out. The geological section is represented by floodplain sand and clay alluvial, fluvial-glacial and moraine sediments of the modern, upper and middle Quaternary age. On the surface was generally a developed soil-plant layer of 0.3–0.5 m thick.

The modern floodplain deposits are represented by water-bearing sands with below 2 m, silt of 4.5 m thickness, underlaid by glacial-like origin clays of 2.1 m thick. Within the depths of 6.6–7 m sand intercalation was found. The moraine sediments are located below 7 m. The main parameters of the above listed sediments determined by the penetration logging study are shown in Table 8.2.

The first and second mildly-undulating terraces are composed of alluvial sediments of Upper Quaternary age. Sediments are sands with intercalations of sandy loams 0.25 m thick. The penetrated sand thickness changes from 1.7 to 10.5 m. The alluvial sands were uniquely identified by cone resistance, lateral friction and natural gamma radiation (Table 8.3).

The fluvial glacial deposits of the middle quaternary age are traced in the upper part of the section in the first three penetration logging diagrams. Their lithologic stratification does not keep either in depth or in extent.

Under the soil and plant layer up to 0.8–1.5 m the sands are deposited. Deeper in the section, loam and clay deposits are located. The thickness of loam varies from 3.4 to 6.5 m and clays from 2.5 to 4.5 m. The sandy loams were identified in one of the soundings within the depth interval of 4.9–7.5 m. The ground properties are shown in Table 8.4.

Table 8.2 Physical and mechanical properties of floodplain deposits in the Morino region by penetration logging studies

Grounds	Gamma radiation I_γ, convent. unit	Frontal resistance R, MPa	Lateral friction T, 0.1 MPa	Density γ, g/cm^3	Moisture W, %
Sands	20–25	>6.7	>1.5	1.85–1.94	>40
Clays	>55	1.2–1.3	0.8–1.4	1.85–1.95	>40
Sands (intercal.)	–	>10.6	>1.8	–	–

Table 8.3 Physical and mechanical properties of alluvial deposits in the Morino region by penetration logging studies

Grounds	Gamma radiation I_γ, convent. unit	Frontal resistance R, MPa	Lateral friction T, 0.1 MPa	Density γ, g/cm³	Moisture W, %
Sands					
Low-moisture	20–30	6.6–15.6	1.4–3.1	1.63–1.68	5–11
Water-bearing	20–30	6.6–15.6	1.4–3.1	1.9–1.95	50

Table 8.4 Physical and mechanical properties of fluvial-glacial deposits in the Morino region by penetration logging studies

Grounds	Gamma radiation I_γ, convent. unit	Frontal resistance R, MPa	Lateral friction T, 0.1 MPa	Density γ, g/cm³	Moisture W, %
Sands	20–30	>10	>2.4	<2	>30
Sandy loams	–	>16.0	>5	<2	>30
Loams	45–55	0.4–4	0.65–1.8	<2	>30
Clays	>55	0.4–4	1.8	–	–

The glacial middle-quaternary deposits were identified by four penetration logging diagrams at a depth of 5.5–10.0 m, which are underlaid by floodplain fluvial-glacial sediments. These were mainly the loam (I_γ=40–50 convent. units), excluding one sounding where at a depth interval of 5.6–7.6 m, sandy loam with a clay lens was found. Glacial sediments have higher densities more than 2.0 g/cm³ and a moisture lower than the fluvial-glacial and floodplain deposits (W=25–39%). The cone resistance and lateral friction diagrams have often a saw tooth shape because of pebble and gravel material. Figure 8.3 shows such diagrams obtained at one point in the Morino region. The penetration logging diagrams were controlled by standard borehole sampling and laboratory measurements.

The studies in the Morino region improve the quality of engineering geological materials and enable a shorter time period for the drainage area project.

8.3.3 Geoengineering Studies in a Region of Glacial Sediments

A geoengineering study of glacial soils has serious complexities because of variegated depositing of different types of sediments and their properties. In 1964, penetration logging studies at the Istra River site (Moscow region) were carried out. The purpose of the work was development of ground properties interpretation from penetration logging records and lithology plotting.

The site was located in the south end of Istra town. The site size was 200 by 400 m. From a geomorphological viewpoint it relates to the Istra River watershed and partly to a stream of the Peschaney floodplain. Four components of the glacial complex sediments are involved in the geological structure of the site: (1) the

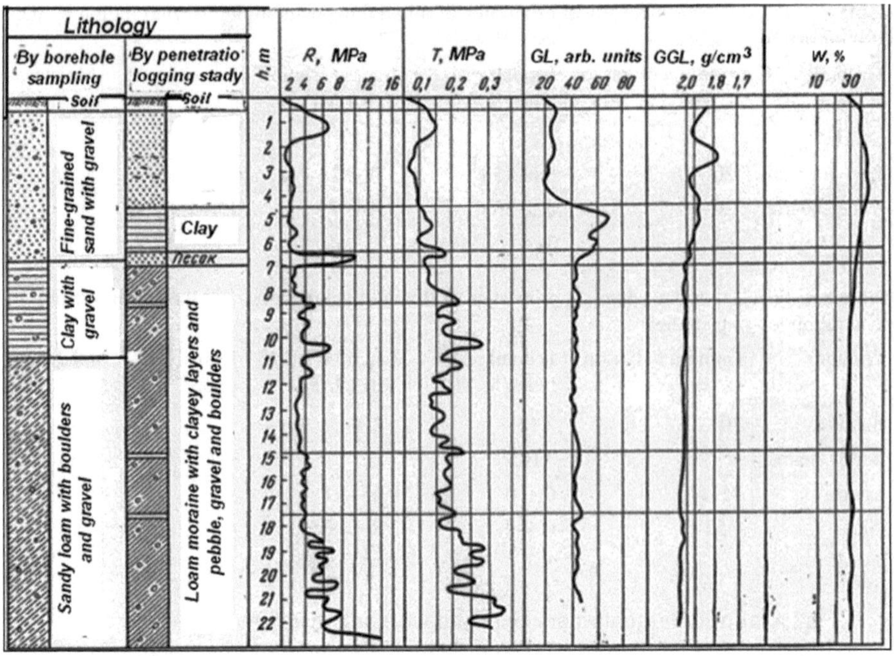

Fig. 8.3 Interpretation of penetration logging diagrams with respect to the ground properties and lithology stratification in the Morino region (Belorussia) controlled by borehole data. (Ferronsky and Gryaznov 1979)

modern covering soil layer; (2) middle Quaternary fluvial glacial sediments; (3) middle Quaternary moraine deposits; (4) low Quaternary under-moraine sands. In the eastern part of the site there was a small area with modern alluvial deposits from the Peschaney stream. The thickness of the Quaternary deposits was 15–20 m and they are underlaid by sandy and clayey sediments from the Jurassic system. From the data of the Moscow regional company of engineering geological prospecting, the property parameters based on the borehole sample study are the following.

The modern covering soils lie on the fluvial glacial deposits in the form of loam with sandy laminas of 1.8–2.3 m thick. The loams have brown colour with heavy to plastic consistency and include plant remains.

The physical properties vary in the following limits: bulk density is within 2.70–2.73 g/cm^3; volumetric density 1.89–2.0 g/cm^3; porosity 39.8–43.2 %; moisture 17.4–25.5 %; plasticity value 8–13; consistency 0.10–0.57.

Middle Quaternary fluvial glacial sediments are spread everywhere and lie from the surface of 1.8–2.3 m under the covering soils in a layer 2–15 m thick. The fluvial glacial deposits are represented by sands of different grain sizes from fine to gravel. The section prevails by fine soils with laminas and lenses of sandy loams and loams.

The physical properties are as follows: bulk density 2.70 g/cm^3; volumetric density 1.60–2.0 g/cm^3; porosity 38.5 %; moisture 5.0–20 %; plasticity value 3–7. In one of the boreholes, loam lamina of brown colour 0.5 m thick was found within the fluvial glacial sediments.

Table 8.5 Physical and mechanical properties of sands from the Istra plot (Moscow region)

Sands	Fractions content				
	>2 mm	2–0.5 mm	0.5–0.25 mm	0.25–0.1 mm	<0.1 mm
Dusty		0–0.5	0.2–7.3	62.9–66.9	25.2–36.9
Fine		0–6	2–35	53–92	7–15
Middle grain	0–1	25–40	19–51	18–36	1–4
Coarse	1–12.5	66–67.5	13–25	5–8	0.5–1
Sands	*Bulk density g/cm*^{3}	*Volumetric density, g/cm*^{3}	*Volumetric moisture, %*	*Porosity, %*	*Angle of natural slope, grad*
Dusty	2.66	1.64–1.78	11–15	41.6–44.7	31
Fine	2.66	1.65–1.70	3.7–15	37.6–44	33
Middle grain	2.64–2.65	1.58–1.64	2.5–3.6	30.6–42.5	30–32
Coarse	2.64	1.62–1.69	3.2–38	37.8–41.0	32–33

Sands are different shades of yellow, partly clayey, flaky, middle dense and dense, dry and water-bearing of 0.6–7.5 m thickness. The properties and granular content are shown in Table 8.5.

Middle Quaternary moraine sediments are spread everywhere and located on the surface and at a depth of 2–10 m under fluvial glacial or partly in alluvial deposits. The thickness of the moraine sediments is 4–6 m. They are presented by brown loams with laminas of sands including pebble, gravel and separate boulders.

The moraine loams contain 2.5–4.0 % of gravel and pebble, 53–55 % of sandy grains, 27–28 % dusty particles and 14.7 % of clayey particles. The physical and mechanical properties of the moraine loams are: bulk density within 2.70 g/cm^3; volumetric density 2.04–2.18 g/cm^3; porosity 31.5–32.2 %; moisture 9.0–12.0 %; plasticity value 8–12; consistency 0.15–0.35, cohesion 0.02 MPa, angle of inner friction 28–34°, modulus of deformation 23 MPa.

In 1963, penetration logging soundings were carried out near the boreholes, with excavations made next door during the study. The distance between the boreholes and excavations was 3–70 m. The sediment sample tests for their comparison with penetration logging data followed as follows. The samples for sand's density, moisture and granular content were taken each 20 cm in depth by metal rings of 200 g/cm^3 volume. Clay sediment samples were treated by the same scheme and in addition a monolith of 15 × 15 × 15 cm size for all other tests in the laboratory was taken.

For determination of the clay content by GL data and further lithology stratification an empirical relationship following from the relationship $I_\gamma = f(C_\gamma)$ is:

$$C_c = \frac{I_\gamma - I_{\gamma\min}}{I_{\gamma\min} - I_{\gamma\min}}, \tag{8.1}$$

where C_c is the relative content of the clay particles in the studied sediments; I_γ is the natural gamma radiation in the section; $I_{\gamma\,\min}$ is the gamma-radiation of pure sands of the section; $I_{\gamma\,\max}$ is the gamma radiation of pure clays in the section.

Table 8.6 Natural gamma radiation values for glacial sediments in the Moscow region

Sediments	Particle content of <0.005 mm, %	Natural gamma radiation, counts/min
Pure sands	2–3	700–1000
Clayey sands	3–8	1000–1300
Gravely sands	2–8	up to 1500
Sandy loams	8–17	1300–1800
Loams	15–24	1600–2200

Relationship (8.1) can be found by experimental tests in a given geological area. For example, it was found for glacial sediments in the Moscow region that the natural radioactivity for moraine sands, sandy loams and loams has relative values of 1.0: 1.5: 2.0. Table 8.6 shows quantitative data for this region. After testing it was derived from (8.1) that:

$$C_c = 0.019\left(I_\gamma - 1000\right), \ \%. \tag{8.2}$$

It is obvious that if the gamma radiation of sands is less than 1000 counts/min, then formula (8.2) provides the clay particle content equal to zero. In case the magnitude error is 10% then the clay content is determined with an accuracy of 25%. This accuracy allows single-valid stratification of the section and separation of their sands, sandy loams, loams and clays. Quantitative determination of density, moisture, porosity and water table was done by GGL and NNL methods by means of calibration curves.

The most precise values of the ground characteristics of a section can be obtained by plotting the property diagrams obtained in separate points of the profile. Figure 8.5a, b, c demonstrates the profile properties for volumetric density, moisture and cone resistance. The common interpretation of the properties was done by plotting the section lithology compared with the profile plotted by borehole sampling data (Fig. 8.5d, e).

Comparing the two engineering geological profiles in Fig. 8.4, one can see that drawing compiled on the basis of drilling data between points 1–6 is limited in depth by the first two metres. This is because of difficulties in the sampling of sands below the groundwater table. The profile compiled by penetration logging data coincides in general with the borehole version. But it gives more detailed geoengineering information related to the borders between separate lithologic layers and their genetic types. In addition, variation analysis was used for establishing a unique value of the layer type using the quantitative data of a parameter. The variation graph P=f (M) is compiled for each layer of sands, sandy loams and loams by:

$$P = \frac{n_i}{\sum n_i},$$

where P is the frequency of yield or probability of the given case; n_i is the frequency of each physical or mechanical parameter (density, moisture, porosity and so on); M is the physical or mechanical parameter.

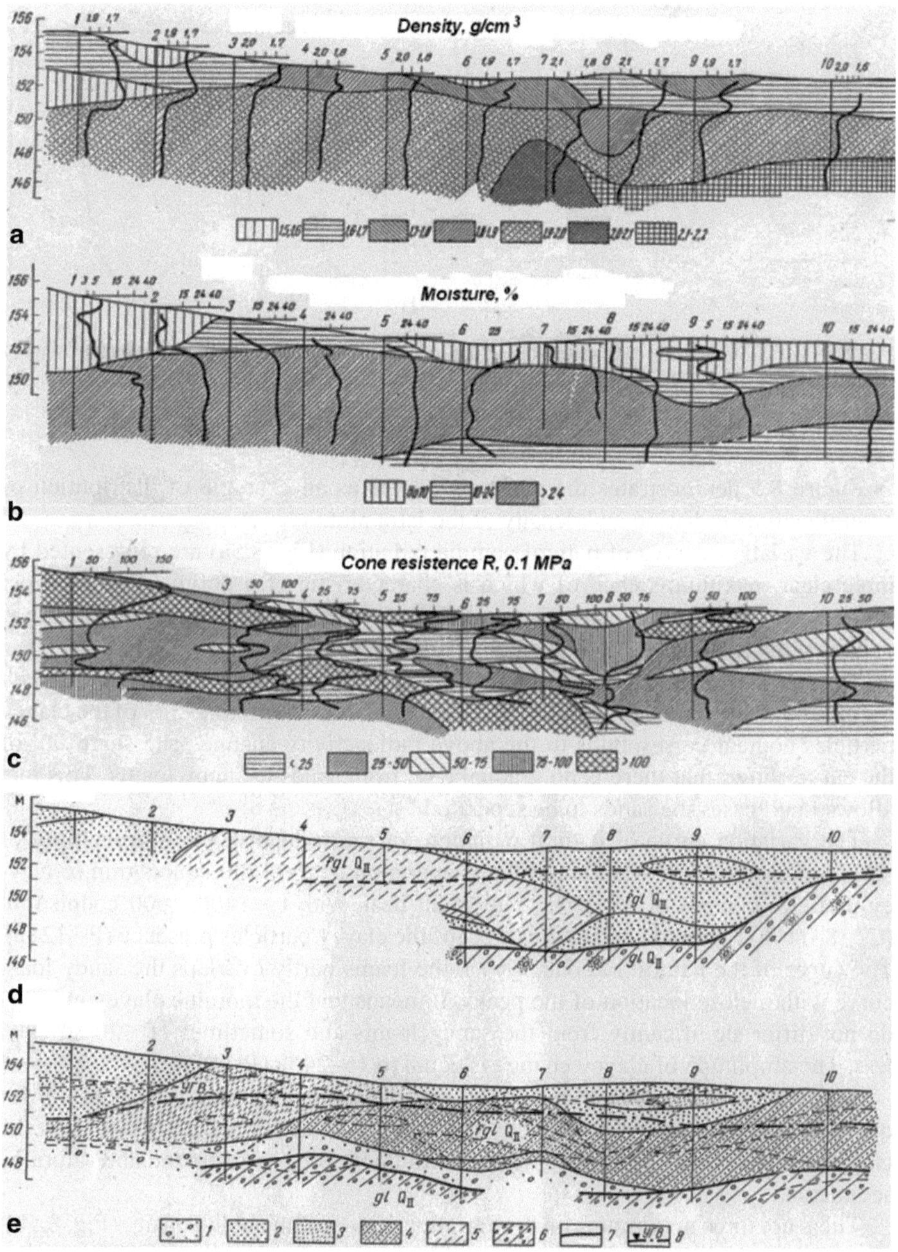

Fig. 8.4 The profiles of the main properties obtained at the Istra site (Moscow glacial region) for: **a** GGL; **b** NNL; **c** R; **d** profile plotted by borehole sampling data; **e** profile plotted by penetration-logging diagrams: (*1*) sands with gravel; (*2*) sands; (*3*) sandy loams; (*4*) non-separated sands and sandy loams; (*5*) loams; (*6*) loams with gravel; (*7*) stratigraphic borders; (*8*) groundwater table. (Ferronsky and Gryaznov 1979)

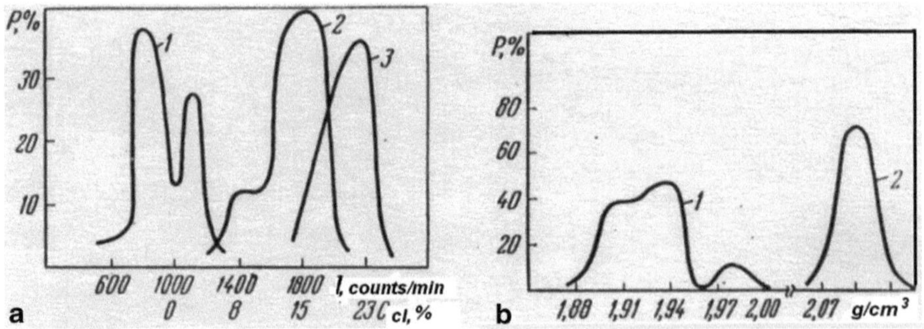

Fig. 8.5 Variation curves for distribution of natural gamma radiation (**a**) and volumetric density (**b**) at the Istra site (Moscow glacial region): (*1*) sands; (*2*) sandy loams; (*3*) loams. (Ferronsky and Gryaznov 1979)

Figure 8.5 demonstrates the variation curves as an example of distribution of natural gamma radiation and density of the Istra plot.

The variation curves of natural gamma radiation (Fig. 8.5a) are represented by three clear maximums each of which is characteristic of a definite ground type: sands, sandy loams and loams. The presence of two maximums in the sand curve is explained by both the method and data treatment error.

The sands are characterised by natural radiation from 70 to 1100 counts/min with an approximately equal probability of 1000 counts/min. Zero to 2–3 % of the clayey particles content corresponds to the above radioactivity change. The sharp fall of the curve shows that there is no gradual pass from sands to sandy loams. This fact allows in all cases the sands to be separated.

The variation curve of natural radiation for sandy loams with a probability of 40 % proves the conclusion about their separation by $I_\gamma = 1800$ counts/min or clayey particles content of 13–17 %. The small peak with $I_\gamma = 1400$–1600 counts/min ($C_c = 8$–11 %) can be related to the effect of the clayey particles presence (P = 12 %). The curve of the natural radioactivity in the loams partly overlaps the sandy loam curve with a close location of the peaks. It means that the moraine clayey grounds do not differ significantly from the sandy loams and sometimes (P < 20 %) even less. The amplitude of clayey change is equal to 15–24 % (P = 10 %).

It follows from the above discussion that the GL data can be used not only for qualitative separation of the lithology but also for quantitative determination of the clay particles content. This conclusion is valid both for non-saturated and saturated sediments.

There are three maximums on the variation curves of the bulk density (Fig. 8.5b). The sands and sandy loams are presented by two maximums. One is in the density interval 1.90–194 g/cm³ and the other is in the interval 1.97–2.00 g/cm³. The first one characterises fine sands (1.90–192 g/cm³) and sandy loams (1.92–1.94 g/cm³). At the accuracy of the method equal to 0.03 g/cm³ they cannot be separated by the parameter value. The second peak characterises the middle-grained sands that have a mean value of density equal to 1.97 g/cm³. The low probability of these sands (P = 11 %) is explained by their content limit of presence in the section ($n_i = 6$). That is why the common frequency was taken for all the sands. The loams in the section with a bulk density of 2.07–2.13 g/cm^3 are determined by a clear maximum on the variation curve. Thus, by the GGL data, the moraine sands and middle-grained sands are single-valued determined. The fine-grained sands and sandy loams of 1.92±0.03 g/cm^3 are characterised as a single lithology layer.

The variation curves of the porosity in their form are analogous to the bulk density curves that were discussed above. The loams are characterised by porosity $n_m = 29\,\%$ and middle-grained sands by $n_m = 41\,\%$. Thus, the sands, sandy loams and loams can also be separated by porosity.

The variation curves of specific cone resistance are represented by a number of maximums, which indicate that they are characteristic for different ground types. The maximal values of the frontal resistances from 1.5 to 4.0 MPa ($P = 20\,\%$) in the section are characteristic of fine-grained sands and sandy loams. The low value for the sandy loams is even lower (1.0–1.2 MPa). For the loams this border on the graph is shifted to the right (2.2 MPa). The mean value of the parameter is close to those ground types at 3.0 MPa.

Applying the penetration logging methods, the fine-grained sands are single-valued identified by R-diagrams, the middle-grained sands identified by R, GGL and NNL, the sandy loams by GL and the loams by GGL, NNL and GL data. The lateral friction in this work was not recorded because it was not necessary.

The main conclusion from this study of glacial sediments is that they are extremely inhomogeneous in depth and cross-section with respect to all the properties. It means that the mean values of the physical and mechanical properties of glacial grounds for foundation calculation and engineering purposes are not acceptable. The penetration logging diagrams can be used as a good alternative to the borehole sampling methodology.

8.3.4 Prospecting for Building Construction

Penetration logging studies in Tashkent (Uzbekistan) and Moscow for designing and calculation of the grounds bearing capacity in civil engineering are described below.

The penetration logging studies in Tashkent and its suburbs were carried out because of city reconstruction after a destructive earthquake in 1966. The purpose of the work was obtaining qualitative data on the physical and mechanical characteristics of the friable grounds and their content up to 25 m in depth, evaluation of the sediment properties variability and groundwater table location.

The site of the work was a flat slightly hilly plain disturbed by the valleys of ancient canals, dry washes, pits and artificial dams. The Chirchik River was responsible for the geomorphology of the area formation. The formation of five terraces was connected by the river activities. Because the penetration logging studies were performed only up to 25 m within Quaternary sediments, a geological description of the older rocks is not presented.

The first floodplain terrace is developed outside of the city and is in the form of a 1 km zone along the river bed. It is formed by well-rolled alluvial pebble and gravel more than 300 m thick. In the upper part of the section the pebble is often interstratified with sands. Sands and sandy loams cover the area from the surface.

The second floodplain terrace covers the south-east suburb of the city. Loess loams and clays interstratifying by sands, gravel and pebble compose the upper part of the section. They are underlined by well-rolled alluvial pebble 300 m thick.

The third terrace is to the south-east of the work site. It is composed of a thick layer of pebbles (300–350 m) overlaid by 3–15 m of loess loams, sandy loams and sands. All the area of work is covered by a discontinuous layer of about 2 m of modern deposits.

The forth terrace is traced in the south-west direction and is built by alluvial middle and heavy loams. This is where the central part of the city is located. In the lower part of the loams, laminas sandy loams and sands are indicated. Their thickness is equal to 10–12 m in the south-east part and exceeds 40 m in the north-west direction.

The fifth terrace serves as a basis for the north-west part of the city. It is composed of proluvial and alluvial loess dusty loam sediments and infrequent carbonates and laminas of sandy loams and sands of the Tashkent complex. The upper horizon (10–12 m within the city) has undergone changes from human activities and represents overset deposits by loess and modified loess grounds. The loess loams are mainly underlaid by rocky loess and sandy gravel sediments.

The penetration logging studies were carried out in 30 points on four profiles (Fig. 8.6). On the main profile, passing through the centre of the city, the distance between the sounding points was 1–1.5 km. On the others it was 200–300 m. The depth of sounding was 20–23 m.

Let us consider some results of the work carried out at Tashkent's work site. Figure 8.7 shows diagrams of the distribution of moisture, bulk density, cone resistance and lateral friction in the main profile. The diagrams of property changes are presented in the form of the profile of each property along the section. In some sounding points of the section the lamellas and lenses of sands and sandy loams appear with their specific value of the property.

In the profile of moisture distribution (Fig. 8.7a), a notable growth of value from 5 to 15 % in the upper zone and from 45 to 50 % in the saturated zone is observed. Three characteristic rather clear zones can be separated with respect to moisture content above the saturated thickness. The first of them occurs from the surface to a depth of 1–5 m with a moisture value from 5 to 25 %. Its formation is determined by surface soil evaporation. The second zone extends within the depth limits of 5–15 m with relatively stable moisture values changing from 25 to 35 %. And the third zone, having a depth interval of 1–4 m, borders with the saturated zone where the moisture value growth is up to 35–45 %. This is the zone of capillary moisture.

The common regularity of the moisture distribution is its growing with depth and its significant variation within separate zones. The moisture regime in a non-saturated zone is completely controlled by the seepage of water from irrigation canals. Its inhomogeneous distribution within separate zones is determined by multiplicity of the sand and sandy loam lamellas and lenses.

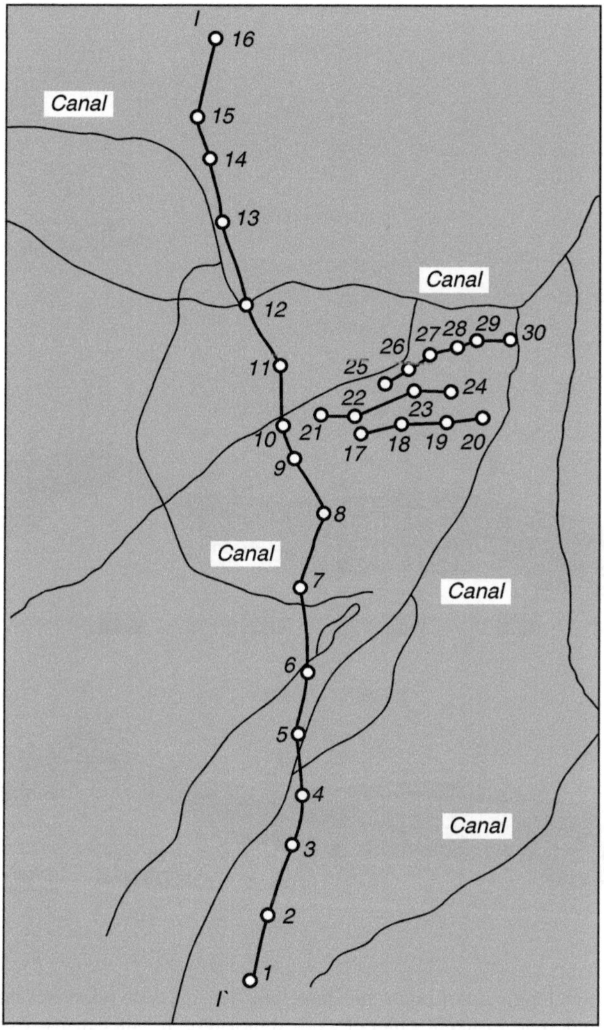

Fig. 8.6 Schematic plan of penetration logging profiles and points on Tashkent's site of penetration logging work. (Ferronsky and Gryaznov 1979)

The values of the bulk density in the section (Fig. 8.7b) vary within 1.35–1.87 g/cm^3. In a number of points (9, 10, 11) the more loose filled sediments with values of 1.19–1.27 g/cm^3 are distinguished. In the upper zone, up to a depth of 2.5–5.0 m, sediments with a density of 1.6 g/cm^3 occur. The grounds below the groundwater table have a characteristic density of 1.80–187 g/cm^3. The general regularity of the bulk density distribution is its growth with depth because of moisture growth. The porosity of the sediments slightly drops and sometimes rises depending on the local conditions.

By frontal resistance (Fig. 8.7c), a number of zones are distinguished in the section. The upper zone's sediments of 1–5 m thickness have R values up to 4 MPa,

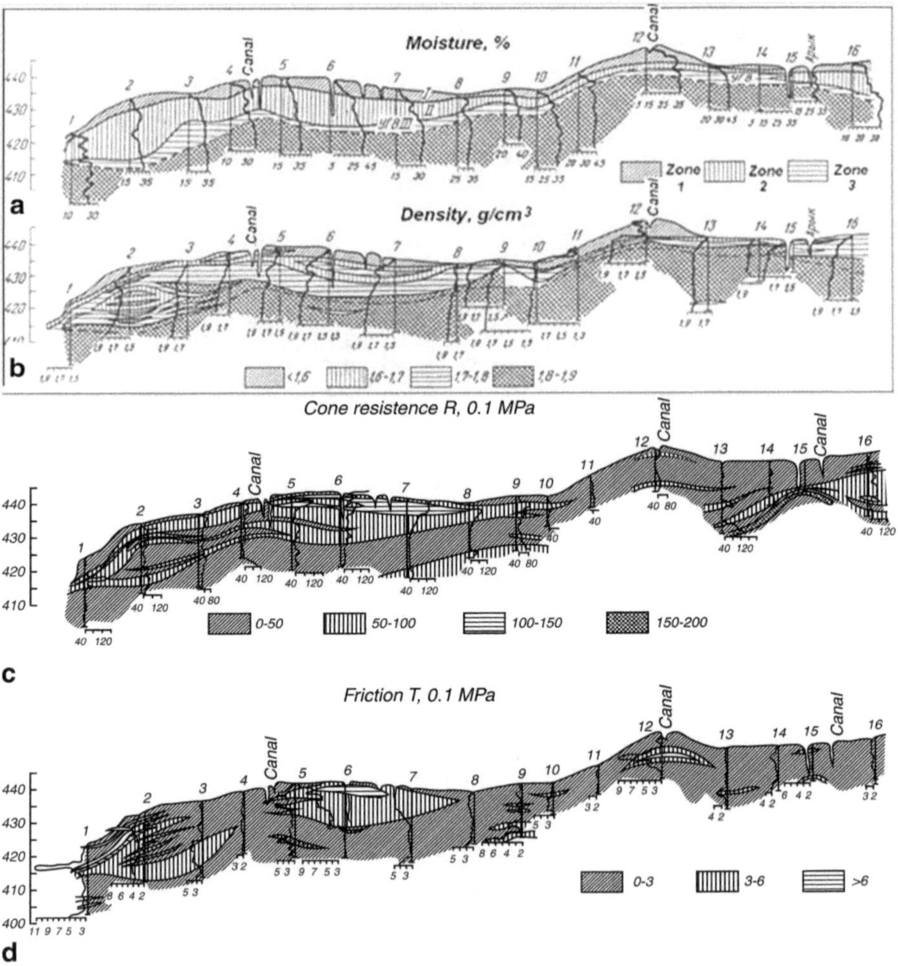

Fig. 8.7 Penetration logging profiles of moisture, bulk density, cone resistance and lateral friction along the main section on Tashkent's plot of work: **a** moisture; **b** bulk density; **c** cone resistance; **d** lateral friction. (Ferronsky and Gryaznov 1979)

which grow in the lower zone up to 6–10 MPa. Here, within the point 6–8, the anomalous zone with R = 16 MPa occurs. In the capillary water zone the resistance decreases to 2–5 MPa and below the depth of 18–20 m the values again rise. Diagrams of cone resistance values in loess sediments often correspond very well to moisture content distribution. Increase of the moisture leads to a decrease of sediment resistance capacity with respect to the outer loading. This regularity is characteristic for the section as a whole. The occurrence of sandy and gravel lenses breaks that regularity.

An analogous picture is observed in the profile of the lateral friction distribution (Fig. 8.7d). The changes in the sediment friction measurements, as a rule, correspond to the cone resistance. In the unsaturated zone the values of friction do not exceed 0.2–0.3 MPa and sometimes they reach 0.9 MPa and more. Such high values were not observed in other regions. In the zone of capillary water and below the groundwater table the specific friction is determined by the value of 0.2–0.35 MPa but below the depth of 18–20 m it rises again.

The subsidence of loess sediments at their wetting is an important problem. Penetration logging methods have the possibility to overcome subsiding loess sediments. The ratio between the cone resistance and lateral friction forms the basis for solving the problem. Those parts of the studied section where the water content increases without porosity change but is accompanied by a cone resistance and lateral friction decrease should be delivered to subsided. This effect was proved by the results of penetration logging studies at the Tashkent and Karshy (Uzbekistan) regions. The limited information, which we have about subsidence of loess sediments in these areas, shows the above said regularities are indeed valid. It is interesting to note that in the points 5, 6, 7 and 8 of Fig. 8.7 up to 6 m of depth, where in the upper studied layers the sediments with maximal values of cone resistance and friction were separated, the subsidence properties do not occur. Here, within the studied site are numerous large-sized canals, from which the water systematically percolates. Long-time saturation of the loess sediments and the outer loading from near existing constructions led to compaction of the grounds in those zones up to a non-subsiding state. It is assumed that special studies of loess sediment subsidence based on penetration logging allow obtaining not only the methodology of zone separation but also quality criteria of ground subsidence. The changes in porosity of the testing of the penetration ground in the section should be the main parameter for the tests interpretation. It worth noting that penetration logging data provide information for a preliminary assessment of loess ground subsidence. According to studies in the Dnepr River region the following values of the physical properties are characteristic for loess subsiding sediments: moisture $<18\,\%$; bulk density <1.9 g/cm^3; porosity $>42\,\%$. All these parameters can be obtained by penetration logging techniques.

The conclusions from the research and practical application, carried out in the region of loess sediments distribution by penetration logging techniques, are as follows. Contrary to existing opinion, a substantial variability of the properties and content of the grounds in a section in depth and space extension are characteristic for that type of grounds. This property is determined by heterogeneity of the mineral and petrography content of the grounds and regime of the moisture content in the unsaturated zone. At the presence of groundwater in a section the durability of the loess sediments is decreased and reaches its minimal values in the capillary horizon. The compact layers do not follow this regularity. Like the glacial sediments, the loess grounds cannot be characterised by a mean value of any property. A detailed picture of the section in depth and width should be presented for geoengineering consideration.

8.3.5 Study of a Landslide Slope

Stratification of lithology, structure of a slide and its surface of sliding are important and time-consuming problems of engineering geology. A landslide study by conventional methods does not give complete answers to all questions. One question relates to friable sediments like water-bearing sands, sandy loams and silt grounds for which the core yield from a borehole is not steady and not sufficiently representative.

In 1967, a landslide study by penetration logging techniques was carried out by our colleague Yu Yurkshtovich (Ferronsky and Gryaznov 1979).

The study object was an old stabilised slide on the right-hand bank of the Moscow River. The aim of the work was determination of the slide structure, bed fixation and recognising breakdowns and crushing zones. The full complex of the penetration logging methods was used in this work and 10 soundings were done to depths from 15 to 28 m along a profile perpendicular to the river plane.

The works were carried out on a second sliding amphitheatre of 440 m long and 200–250 m width. The two ridges of sliding heaps were located at an angle to the Moscow River. Within the working area the remainders of non-separated first and second floodplain terraces were separated.

The studied site is composed of Quaternary, Cretaceous, Jurassic and Carboniferous sediments. The sliding slope sediments are underlined by limestone rocks of the Carboniferous system at a depth of 8–55 m from the river cut.

Jurassic deposits in the upper part are represented by fine-grained grey sands of 8–11 m thickness, lower are micaceous black clays of up to 30 m thickness and on the bottom are glauconite grey sands of 5–7 m thickness.

The Cretaceous deposits are represented mainly by sands of different colour, grain and clay particle content.

Quaternary sediments consist of glacial loams with inclusions of pebble, gravel, boulders and sand laminas. All the deposits are covered by blanketing loams.

As a result of the studies, the profiles of penetration logging properties and lithologic sections were plotted (Fig. 8.8), which characterise the sliding slope structure.

The gamma-logging diagrams (Fig. 8.8a) are clearly fixing the natural radiation of sediments depending on their lithologic content. The most characteristic values of the sediments gamma radiation in the section are given in Table 8.7. It follows from Table 8.7 that each type of sediment is characterised by a definite interval of gamma radiation values with the most characteristic values of separate deposits. Even a negligible portion of the sandy fraction to clayey sediments has notable changes in the natural gamma background.

Specific values of the frontal resistance R diagrams and lateral friction (Fig. 8.8b) distinctly repel different types of sediments and fracturing zones. For example, characteristic values for the sand are $R \approx 5.0$–6.0 MPa and $T \approx 0.1$ MPa. In the Jurassic clays R and T are increasing up to 8–10 and 0.2–0.25 MPa correspondingly. The clayey sands and sandy-dusty clays have intermediate values.

The bulk density and moisture values were determined by GGL and NNL diagrams (not demonstrated here) and used for correlation and refining of the slide slope structure.

Thus, applying the penetration logging methods we separated with a high degree of reliability the lithology section of sediments, distinguished the fractured blocks and traced the shifted blocks of the sliding slope.

In general form, the structure of the sliding slope in a section obtained by a penetration logging study (Fig. 8.8) is as follows. The formation takes place in three stages. The found three large blocks of rocks shifted from their basic plateau prove this idea. The first block occurs between the sounding points 18

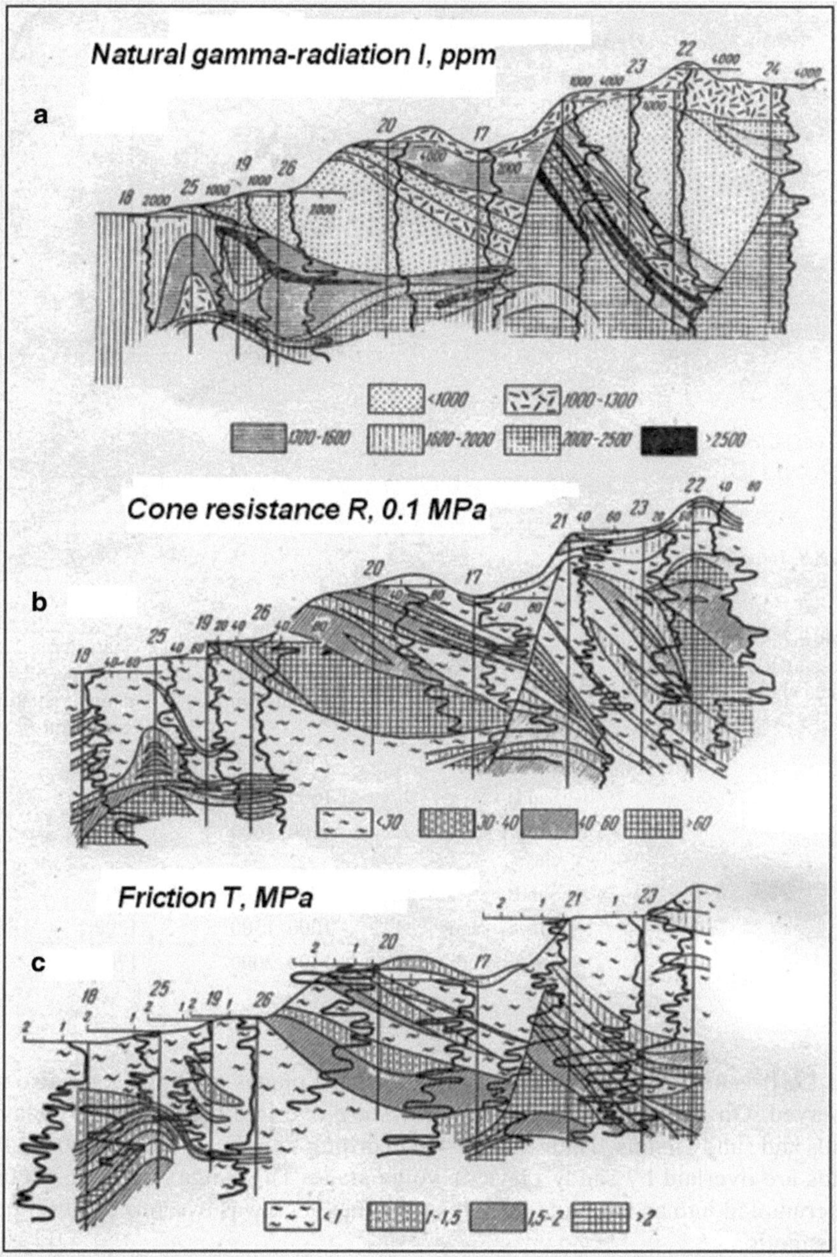

Fig. 8.8 Profiles of GL (**a**), R (**b**), T (**c**) and geological section (**d**) plotted by penetration logging records on the sliding slope of the Moscow River bank: (*1*) limestone; (*2*) clay with disturbed structure; (*3*) sandy clay with disturbed structure; (*4*) heavy sandy clay; (*5*) sands; (*6*) clayey sands; (*7*) lake-boggy clays; (*8*) silicate inclusions; (*9*) loose zones; (*10*) phosphate lens. (Ferronsky and Gryaznov 1979)

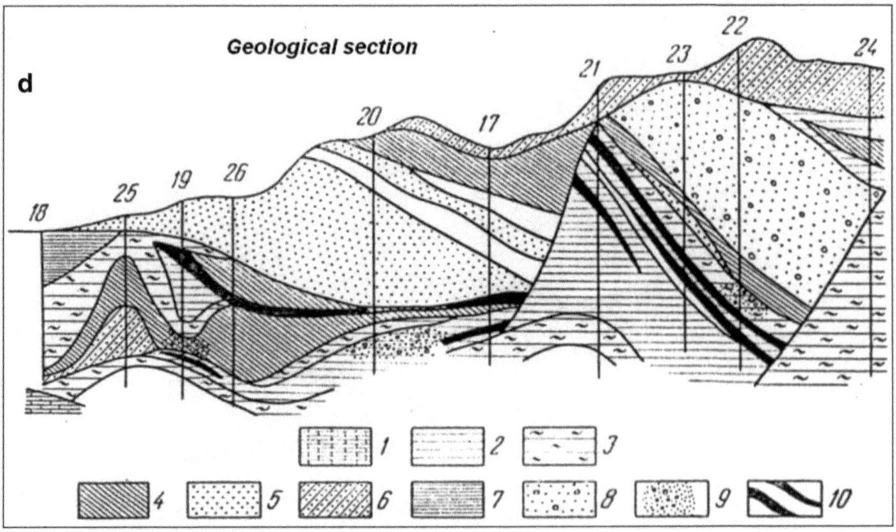

Fig. 8.8 (continued)

Table 8.7 Gamma radiation values of sediments in the studied sliding slope section

Sediments	Age	Changing limits of gamma radiation (counts/min)	Characteristic values of gamma radiation (counts/min)
Pure clays	$J_{3\ ox}$	2000–2500	2000
Sandy clays	$J_{3\ ox}$	1600–2000	1700
Highly sandy clays	$J_{3\ ox\ v}$	1300–1600	1400
Sands	$C_{r\,i}$	800–1200	1900
Clayey sands	Q	1000–1300	1200
Lake-boggy clays	Q	1600–2000	1700

and 19. It is not clearly separated and well traced planes of shifting are also not observed. On the non-shifted Oxford clays are overlaid the Quaternary clayey sands and sandy loams. Their thickness, according to point 25, is 4.5–5.0 m. The sands are overlaid by sandy clays of Volga stage. The Volga and Oxford clays are crumpled into an anticline fold. It seems the block was overturned during the slide motion.

The second block, located between the 25 and 21 sounding points and in the lower part, comprises the Oxford stage clays overlying the Volga sandy clays with laminas of phosphates. The block is very clearly separated by gamma radiation of the GL diagrams. The R and T values sharply increase at the phosphates laminas, which are 40 cm thick.

The Jurassic clays are overlaid by fine sands covered by clayey sands with laminas of loam and clay. The lithology evidences that the block was overturned at

45° with respect to the basic slope side, which was oriented at 75° to the horizon surface. The plane of shifting here is clearly fixed by the sharp rock changing. In sounding point 17, the fractured zone of rocks, by a decrease of R and T values within the block, is fixed.

In the points 20, 26, 19 and 25 the slide plane shift is fixed by increasing gamma radiation on the GL diagrams. This phenomenon is explained by an increase of clayey content along the sliding plane (tectonic clayey type in igneous rocks).

Increase of gamma radiation in the shift zone possibly happened as result of intensive migration of the natural radioactive elements along that zone together with water. In points 20–25 just over the zone the fractured segment of rocks is observed by R and T diagrams.

The third block is between the points 21 and 24. It is represented in the lower part by sandy clay laminas and pure Jurassic clays. In the upper part of pure clays a layer of sediments with higher gamma radiation (2500–3000 counts/min) and lower values of moisture occurs. The values of gamma radiation here are lower than for the phosphates but higher than for the clays. The higher radiation can be explained by monolithic fauna inclusions. The Cretaceous sands are underlying the Jurassic clays and overlaid by laminas of clays and loams. The clayey sands comprise the upper layer of the block.

Location of the Jurassic clay layers evidences that the block was turned to the plateau side at 65–70°. In the northern part it is bounded by the shift plain that is fixed by the sharp change of sediment properties. The assumed slope of the plain shift is 70–75° to the river side. The lower part of the plain shift has not been fixed because of its deeper location.

It follows from analysis of the obtained data that the studied slide slope is rather old and all the occurred processes were smoothed. But the applied complex of methods appears to be effective for restoration of those processes after a long time.

The experience of studying sliding processes was used for an investigation in Noviy Afon on the Black Sea shore and in the Tashkent region (Uzbekistan), where modern slide processes were studied by the same techniques.

8.3.6 Study of Bottom Marine Sediments at Novorossiysk Port

Engineering geological studies of bottom deposits of seas, rivers and lakes are an intensively developing area of activity in connection with ports and hydraulic construction and active utilisation of coastal areas. The use of boat borehole techniques for the study of bottom sediments was always laborious and expensive work. Non-disturbed bottom sediment sampling up to now is an unsolved technical problem. That is why the developed submerged penetration logging installation PSPK was the first serious step in solving important problems in engineering geological studies of marine bottom sediments.

The first experimental studies of the installation prototype PSPK started in December 1965 at the Black Sea at Novorossiysk port. In 1966, at cabotage pier, the first experimental works on determining bottom sediment properties were carried out (Ferronsky 1969).

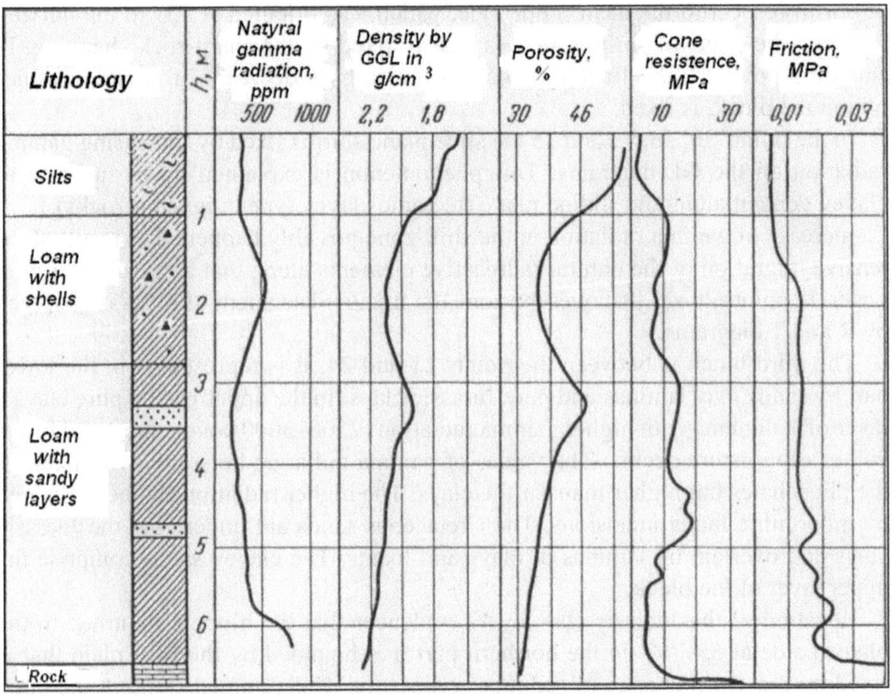

Fig. 8.9 Characteristic diagrams of GL, GGL, porosity, R and T of bottom sediments at Novorossiysk port, Black Sea. (Ferronsky and Gryaznov 1979)

The geological section was represented by a silt layer of 1–2 m thickness. Underlying were gravel and pebble deposits of up to 2 m. Next was a 1–1.5 m layer of silt with pebble and 3–4 m loam underlaid by basic dense clay thickness. There were not any available reliable data about the physical and mechanical characteristics of the studied sediments. This was the first serious difficulty in the experimental work. We did not have the possibility to compare quantitative experimental data and estimate their accuracy and reliability. In this regards, the data obtained by on-land works were the only available material.

The profile of seven sounding points with 4–5 m between the points was carried out during the experimental tests at cabotage pier. Natural gamma logging, gamma-gamma logging, cone resistance and lateral friction were recorded. Figure 8.9 shows the results in the form of Gl, GGL, R and T diagrams at each point. Because the logging points are located close to each other the corresponding diagram values had similar character and ordinate properties. But the layers thicknesses were different: 0.5–0.7 m.

It follows from the obtained data that the cone resistance values of the gravel and pebble deposits vary from 3 to 10 MPa. For the loam sediments it is 1–2 MPa. The silts have shown this parameter to be equal to zero. The loams are separated by GL values up to 1500 counts/min and gravel–pebble deposits with 200–400 counts/min.

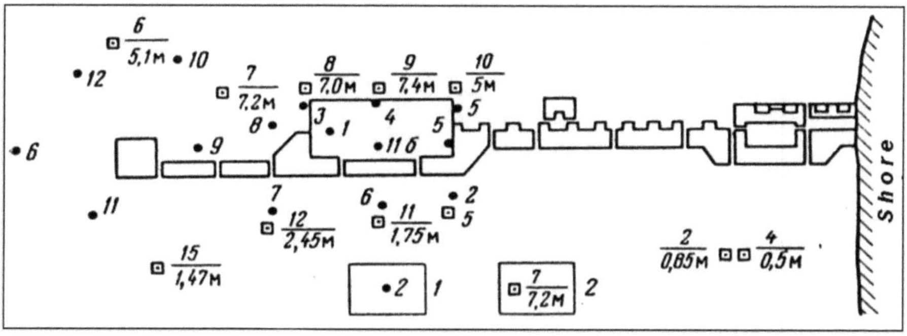

Fig. 8.10 The plan of the oil jetty site and penetration logging points where the bottom sediments were studied in Novorossiysk: (*1*) boreholes; (*2*) penetration logging points. (Ferronsky and Gryaznov 1979)

The peat layer is separated by GGL data with low density and is characterised by gamma-scattering radiation of 12,000–13,000 counts/min rate. The loam and gravel–pebble deposits have this value equal to 7,000–9,000 counts/min.

8.3.7 *Study of Novorossiysk Oil Jetty Structures*

A new area of study was at the constructing oil jetty structures in Novorossiysk (Black Sea). The experimental works were combined with an important practical task. The jetty was the marine hydraulic construction for shipping oil from the pipeline. The upper armoured concrete construction was founded on a stone bed overlaid by concrete blocks. Soon after the construction works were finished, during a heavy storm, the construction of the jetty was seriously damaged because of large (about 1 m) foundation subsidence. The officials of the port invited our team to study the bottom sediments by penetration logging techniques.

The submerged PSPK installation was used for this study. Figure 8.10 shows the plan of the penetration logging points performed around the jetty in 1967.

In order to have the most reliable interpretation of the ground properties the sounding points were selected at the stone bank of the foundation at both sides of the jetty. Figure 8.11 shows the property profiles of natural gamma radiation (GL), bulk density (GGL), porosity (GGL and NNL), cone resistance (R) and friction (T) along the point 6–10 (see Fig. 8.10).

It was found by the studies that non-correspondence in the thickness of loose sandy-clayey sediments and also incorrect values of the physical and mechanical properties were obtained by drilling works and sampling. A loose silt layer of deposits, underlying the upper clay layer, was omitted during sampling. Taking this fact into account and also the determined low values of the deformation modulus and angle of inner friction of sediments in the foundation, the conclusion was that changes of the sandy-clayey thickness in foundation and its low bearing capacity were the cause of the high jetty deformation during the storm.

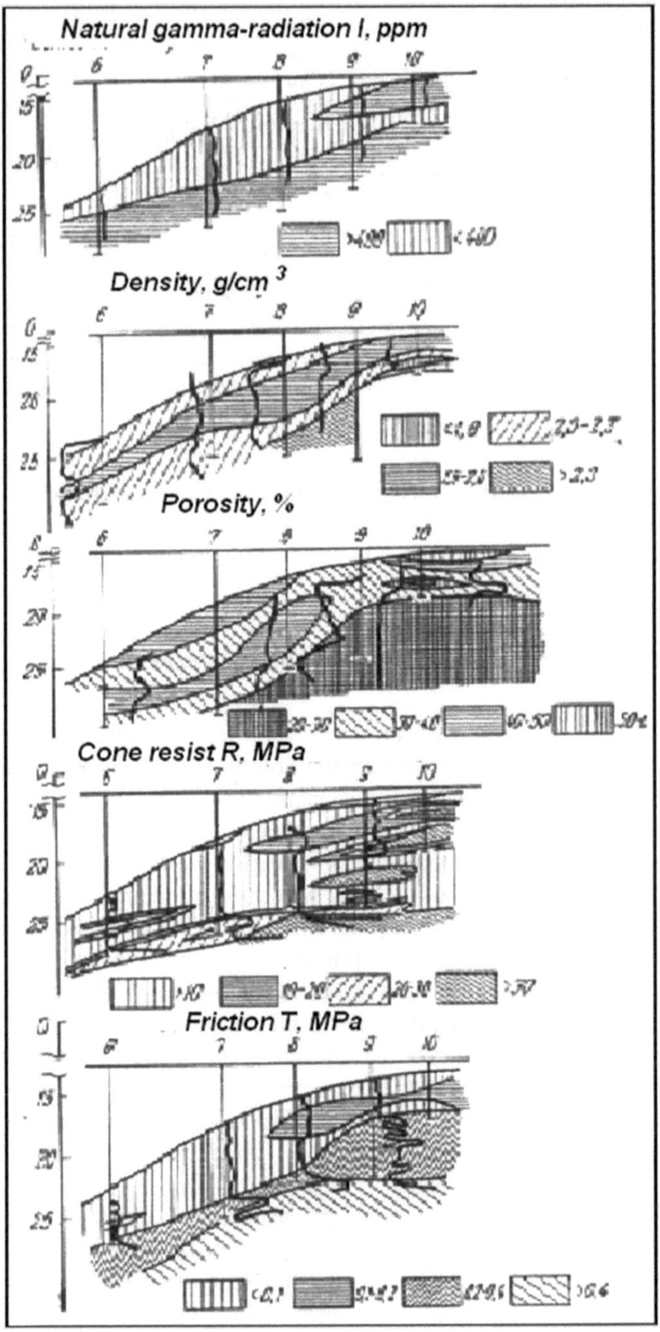

Fig. 8.11 Property profiles of GL, GGL, porosity, R and T obtained at the oil jetty structures in Novorossiysk. (Ferronsky and Gryaznov 1979)

The first experience of application of the penetration logging techniques for engineering geological exploration of marine bottom sediments has shown their technical and economical advantage in comparison with the standard borehole and sampling methodology. Further works in Novorossiysk and Odessa regions were used for the design and production of the improved PSPK-69 apparatus, which was installed on the special investigative ship "Geologist-1".

8.4 Combined Application of Penetration Logging and Traditional Geophysical Methods

Depending on the aims and solving of problems, the specifications of the engineering geological and hydro-geological studies and degree of accuracy and reliability of the ground properties are different. For example, in irrigation and drainage study, main attention is paid to water-physical properties and filtration parameters of large areas. In civil and industrial engineering studies, more serious attention is paid to the physical and mechanical properties and especially to the bearing capacity properties.

Depending on the requirements of the volume and quality of the initial information, the use of appropriate field and laboratory works are planned. The optimal complexity of the study methods in engineering geology and hydrogeology for planning works is a separate important theoretical and practical problem. Underestimation of the methods complexity role leads to difficulties in use of the field materials at project compilation. For example, very often the available voluminous geophysical materials cannot be used because of the absence of basic data for quantitative interpretation. And vice versa, detailed materials with reliable quantitative data in separate points cannot be extrapolated over the area without the corresponding geophysical materials.

The use of penetration logging methods foresees their complexity with other geophysical, drilling and laboratory works. On this basis, the maximal effectiveness of the study is achieved. In some cases, when the penetration logging complex ensures a problem solution, then it uses no extra studies from other methods.

In the general case, when the works have a real character like engineering geological or hydrogeological mapping, their implementation has the following priorities. First, the geophysical studies start, followed by penetration logging and finally, drilling and sampling finish the work. Such an order seems to be reasonable by logistic connection and interconnection of further solutions. In fact, the geographical real methods cover study of considerable areas with minimal cost and time. The obtained qualitative information ensures understanding characteristic sites with identical properties. The detailed quantitative information about the structure, content and properties of deposits in the section provides penetration logging techniques, which is the basis for quantitative interpretation of geophysical information. The drilling and laboratory studies ensure obtaining lacking qualitative information, which is used for calibration of the instrumental response to penetration logging and geophysical works.

Table 8.8 Geo-electric parameters of some sediments obtained by Yu Panda

Sediments	Electric resistance (Ω m)	Velocity of longitudinal waves (m/s)
Dry sands	50–150	400–600
Loose loams	15–30	450–600
Wet sandy loams and sands	30–50	1200–1800
Water bearing clays	3–15	1800–2000
Limestone	300–800	2500–2700

For complexity with penetration logging techniques in engineering geological and hydrogeological investigations, the most popular forms of geophysical methods are vertical electrical sounding (VES) and the induced potentials (IP) study. Both of the methods are distinguished by simplicity in application and high productivity in field work. The physical basis of the methods is the difference in electrical resistance for different types of rocks. The ambiguity in interpretation of the measuring results depends on their relation not only with the rock type but also with moisture, mineralisation, texture and structure of the natural formations and so on. In a specific geological region, the influence of those effects can be taken into account with the drilling basis or penetration logging studies. As a result, the possibility of quantitative or semi-quantitative estimation of lithology, salinity, ground mineralisation, groundwater table and filtration properties occurs. Such a possibility of VES and IP data interpretation depends on the specific geological and geographical conditions of the studied area. The advantage of the methods is in the possibility of obtaining a large sediment properties picture in depth and width.

From the other geophysical methods, which found application in complexity with penetration logging, is seismic prospecting on the basis of wave refraction. The physical basis of the method is the presence of longitudinal seismic waves, a number of refraction borders like the groundwater table, cover of the stone rocks and so on. All the researchers found that in sandy-gravel sediments the groundwater table is fixed uniquely with an accuracy of ~10%. But in the case of an increase of clay, the contrast of seismic wave propagation decreases. Moreover, sometimes with the appearance of natural screens (like perch water), the water table is not separated. Separation of other borders in the sediment thickness needs more complicated seismic apparatus to be applied.

The example of successful use of the geophysical and penetration logging techniques is the work of Yu Panda who performed hydrogeological mapping near Orgeev town in Moldova (VSEGINGEO 1973). In order to separate the lithological structure of a section, groundwater table position and mineralisation he applied penetration logging, electrical sounding and seismic techniques.

By the obtained values of the specific electric resistances and velocities of the longitudinal seismic waves Panda separated for different lithlogic elements a number of geo-electric borders (Table 8.8).

Modern deposits are characterised by an extreme variation of the above parameters in depth and width. This is especially typical for silts and heavy loams, for which the electrical resistance varies from 0 to 20 Ω m. The deposits with anomalous resistance, as a rule, have higher salinity in unsaturated zones and higher mineralization of groundwater, which reaches in separate areas up to 5 g/l.

Electrical prospecting by the VES method was carried out through 500 m on profiles with 3–4 km of density and a symmetrical position at 2000 m length on a watershed and 500 m on the floodplain of the river. Panda states that during work on the same stations the VES penetration logging measurements were carried out. Interpretation of the obtained results was done highly effectively. His conclusion was that the complexity of the penetration logging and traditional geophysical techniques during engineering geological and hydrogeological investigations in the Moldova areas can decrease expensive borehole works by four times.

A number of years V. Lisichko applying complexity of the penetration logging and laboratory determination of the ground parameters (deforming modulus, filtration and subsidence coefficients and so on during hydro-irrigation mapping in a number of southern Ukraine regions on the total area of 10,000 km^2, where only in 1975 22,700 m of penetration logging soundings were carried out (VSEGINGEO 1973).

His object of investigation was Ukrainian Quaternary loess deposits represented by watersheds of eolian-delluvial sandy and clayey deposits overlaid by eolian-delluvial loams and clays. Here, the penetration logging methods were used for lithology section separation, identification of the groundwater table and the capillary zone and the determination of the depth of infiltration at pouring tests.

Very interesting results were obtained by Lisichko in determination of the ground bearing capacity by penetration logging tests and some laboratory measurements. His conclusion was that penetration logging techniques are highly effective methods in both technical and economical aspects.

In the study of bottom sediments on marine areas, the complexity of penetration logging methods with geophysical and drilling works was in the construction of the ship “Geologist-1”. In addition to the emerged apparatus PSPK-69 the ship is equipped by drill UGB-50 and seismic-acoustic apparatus “Ground”.

The technology of the works is, first of all, providing a seismic-acoustic profile along the drafted range in the area. After this, the penetration logging and drilling works are carried out. The available experience of works in the near-shore areas proves this technology.

References

Ferronsky VI (1969) Penetration logging methods for engineering geological investigation. Nedra, Moscow

Ferronsky VI, Gryaznov TA (1979) Penetration logging. Nedra, Moscow

State Standard (1982) Mining rocks: methods of field test by penetration logging. GOST 25260-82 (approved on 1982 May 17)

VSEGINGEO (1973) Annual annals. Nedra, Moscow

Part III
Natural Isotopes in Environmental Studies

Chapter 9
Stable Isotopes in Study of the Global Hydrological Cycle

Abstract The famous Finnish geochemist Rankama wrote in 1954 that isotope geology is a branch of science in which geologic phenomena are studied by investigating stable and unstable isotopes of individual elements, in particular by determining the variations in their abundance. Isotope geology deals with both stable and radioactive isotopes. In this chapter the distribution, hydrogeochemistry and geophysics of naturally occurring stable hydrogen and oxygen isotopes in the atmosphere, oceans, surface and groundwater in brief form are discussed. Using the isotope variation, occurring under natural effects, the origin, dynamics and resident time of water in hydrosphere reservoirs are analysed.

9.1 Separation of Hydrogen and Oxygen Isotopes at Phase Transition of Water

Isotope geology is concerned with both stable and radioactive isotopes. Whereas the possibilities of radioactive isotopes in geologic studies have always been readily appreciated, it was first considered that stable isotopes were encountered in nature in constant ratios and were therefore of minor importance.

However, soon after the development of precise methods of determination of isotopic composition, especially the mass-spectrometric method, it was found that the isotopic composition of many elements was not in fact constant. The experience gained during the forties and the early fifties indicated that the isotope mixtures of oxygen, sulphur, carbon, hydrogen and other atoms (especially light atoms) in natural processes do not remain constant and partial separation of isotopes (fractionation) occasionally reached up to 5 % in either direction. Now it is established that the partial fractionation of hydrogen in natural objects reaches up to 100 %.

In general, light elements show a greater degree of fractionation than heavy elements. This is readily understood if only because for light elements the isotopic mass ratios are higher than heavy elements. For example, masses of isotopes $m_H^{1}/m_H^{2} = 2$ (100 % difference), $m_C^{13}/m_C^{12} = 1.08$ (8 % difference), $m_S^{34}/m_S^{32} = 1.06$ (6 % difference) and so on and this tendency continuous as the atomic weights of the elements increase. Rankama proposed, as a working hypothesis, that the primary isotopic composition of all elements is a natural constant (Rankama 1954). But sub-

V. I. Ferronsky, *Nuclear Geophysics*, Springer Geophysics,
DOI 10.1007/978-3-319-12451-3_9

sequent physical and chemical processes occurring during the geologic history of the Earth have modified this composition. Quoting other authors, Rankama lists the following processes as being responsible for natural fractionation: solution, melting, crystallisation, isotope-exchange reactions, ion migration, gravitational and biochemical separation, phase transitions and so on. These fractionation reactions can, of course, also be produced under laboratory conditions. In nature, the separation occurs on a much smaller scale than under laboratory conditions but the time factor is much greater and therefore the final effect is also much greater. Hence it is clear that, by investigating isotope separation, it is at least in principle possible to judge the conditions governing certain geological processes.

Rankama also pointed out that differences in the abundance of isotopes clearly reflect the energy relationship in different compounds. For example, it is well known that the energy and, consequently, the thermodynamic properties of vibrating and rotating molecules, depend on the masses of the constituent atoms and the atoms of the lighter isotopes, having a higher mobility, have greater kinetic capabilities as compared with the heavier isotopes and are therefore more inclined to migration than to accumulation. The above properties provide the foundations on which geologic and, in particular, hydrologic studies involving the use of stable isotopes can be based.

The first attempt to use isotope methods in hydrogeology was made by Harpaz et al. (1963). Using the data that had become available by the time of the First Symposium on Radioisotopes in Hydrology in Tokyo (IAEA 1963), attempts were made to classify the methods for the determination of various hydrogeological parameters (distribution and origin of underground water, physical and chemical properties of water and water-bearing media, water flow direction and velocity and so on).

In measurements of basic parameters related to the underground water dynamics in porous grounds, for example, the distribution of porosity and of permeability tensors, the isotope methods are more advantageous because they are more rapid and accurate. In measurements of flow velocity and in establishing the origin of the water (meteoric, relict, juvenile, etc.) isotope methods provide the most direct means of establishing these parameters. The isotopes that are of particular interest in hydrogeology can be divided into three groups: (1) stable water tracers, namely D and ^{18}O; (2) radioactive ^{3}H and ^{14}C (used to establish the age of underground water); and (3) radioactive isotopes specially introduced into a water-bearing system.

For a quantitative estimation of the isotopic fractionation effect in a physicochemical process the fractionation factor α is used (Brodsky 1957):

$$\alpha = \frac{N/(1-N)}{n/(1-n)} = \frac{R_1}{R_2}, \tag{9.1}$$

where

$N/(1-N)=R_1$ is the atomic part of the isotope in the enriched fraction (component)
$n/(1-n)=R_2$ is the atomic part of the isotope in the depleted fraction (compound).

In the case of a minimal content of the component being enriched, which is the case, for example, during the fractionation of a natural mixture of isotopic species of water, one might assume that $(1-\mathrm{N})\approx(1-n)=1$ and hence:

$$\alpha=\frac{\mathrm{N}}{\mathrm{n}}. \tag{9.2}$$

If the process of fractionation is successively repeated k times, then the total isotopic fractionation factor will be α^k.

During the fractionation of isotopes in a liquid-vapour system the content of the isotopic components in liquid is N and $1-\mathrm{N}$ and their vapour pressures in a pure form will be p_1^0 and p_2^0. According to Raule's law for ideal liquid mixtures (Rabinovich 1968), the vapour pressure of each component above the mixture may be expressed as:

$$\rho_1=\mathrm{N}\rho_1^0;\ \rho_2=(1-\mathrm{N})\rho_2^0. \tag{9.3}$$

Then

$$\frac{p_1}{p_2}=\frac{p_1^0}{p_2^0}\frac{\mathrm{N}}{1-\mathrm{N}}. \tag{9.4}$$

The partial pressures of the isotopic species of molecules of liquid in the vapour phase are proportional to the ratio of their molar ratios n and $(1-\mathrm{n})$. Hence, Eq. (9.1) may be rewritten in the form:

$$\frac{\mathrm{n}}{1-\mathrm{n}}=\frac{\mathrm{N}}{1-\mathrm{N}}\cdot\frac{p_1^0}{p_2^0}.$$

or

$$\alpha=\frac{p_2^0}{p_1^0}. \tag{9.5}$$

The process of equilibrium isotopic fractionation occurring due to the energy state difference of the isotopic molecules is sometimes called the thermodynamical isotopic effect.

In a series of physicochemical processes occurring at non-equilibrium conditions isotopic fractionation can be the result of kinetic effects.

The isotopic exchange in the system under equilibrium conditions may be described both by the equilibrium constant K and the fractionation factor α.

For such a reaction as $\mathrm{Ax}+\mathrm{Bx}^*=\mathrm{Ax}^*+\mathrm{Bx}$, where there is only one atom of the exchangeable isotopes x and x^* in each molecule of the reacting substances, the equilibrium constant is equal to the fractionation factor (Brodsky 1957):

$$\alpha = \left(\frac{X^*}{X}\right)_{Ax} : \left(\frac{X^*}{X}\right)_{Bx} = \frac{[Ax^*]}{[Ax]} : \frac{[Bx]^*}{[Bx]} = K. \tag{9.6}$$

In more complicated reactions, where several atoms of exchangeable isotopes in each molecule are involved, $\alpha \neq K$.

The relationship between α and K for isotopic exchange reactions was studied by Brodsky (1957), who demonstrated theoretically that, in general:

$$\alpha = (\kappa/\kappa_o)^{1/ab}, \tag{9.7}$$

where

K_0 is the limiting value of the constant; a and b are the stoichiometric coefficients of the reaction.

In order to discuss application of the natural D and ^{18}O in hydrology and hydrogeology their separation in the global water cycle needs to be considered.

9.2 Isotopic Composition of Ocean Water

On average there are 320 molecules of HDO, 420 molecules of $H_2{}^{17}O$ and about 2000 molecules of $H_2{}^{18}O$ for each 10^6 molecules of $H_2{}^{16}O$ in natural waters on the Earth. The ratio of isotopic abundances of deuterium to protium is D/H = 0.000155 (0.0150 atom. %) and that of oxygen $^{18}O/^{16}O = 0.002$ (0.2 atom. %).

The differences in the limits of deuterium variation are one or more orders of magnitude higher than those of heavy oxygen. This may be explained by the more effective separation of D and H in natural processes than that of any other pair of stable isotopes due to the greater difference in their atomic weights. For this reason, D may be considered as one of the most interesting isotopes from a geochemical viewpoint. Since the end of the 1950s, researchers studying the isotopic composition of natural waters have preferred to determine both isotopes of water hydrogen and oxygen simultaneously.

Of all natural waters the ocean, which is a unique reservoir, remains most constant regarding its isotopic and other physicochemical properties. Several authors (Craig and Gordon 1965; Epstein and Mayeda 1953; Friedman et al. 1964) have shown that ocean water at a depth of more than 500 m is homogeneous in isotopic composition. This allowed Craig (1961) to propose it as a standard for reporting concentrations of D and ^{18}O content in natural water. The standard of the ocean water (SMOW) has the following values of D and ^{18}O isotope ratios: D/H = $(155.76 \pm 0.08) \cdot 10^{-6}$ (Hagemann et al. 1970) and $^{18}O/^{16}O = (2005.20 \pm 0.45) \cdot 10^{-6}$ (Baertschi 1976). Fluctuations in the ratios of the absolute values for hydrogen and oxygen isotopes arise from the difficulties incurred in the precise preparation of the synthetic isotope mixtures for mass spectrometer calibration.

The variation in the relative content of D in a deep ocean layer is about 4‰ and that of ^{18}O is of the order of 0.3‰. In the surface ocean layer the regional variations, depending upon water temperature, are 35‰ for deuterium and about 3‰ for oxygen-18. The lowered content of deuterium in the surface ocean layer occurs in those regions where ice melting water affects isotopic composition. In high latitudes, where the surface layer of the ocean is freezing, the isotopic fractionation factor in this case is ~1.0180 for deuterium and ~1.0030 for oxygen-18. In the equatorial region of the ocean, where intensive evaporation of water takes place, there is an enrichment of the surface layer with heavy isotopes.

On the basis of investigations carried out by Epstein, Friedman and Craig with co-authors we can compose a sufficiently detailed picture of D and ^{18}O distribution in ocean waters. The choice of samples, which have been taken from the most characteristic points of individual basins according to oceanographic data in common use, combined with the high precision of measurement and the coincidence of experimental data with principal conclusions that have been made by different authors, suggests that the results obtained are reliable.

Figures 9.1 and 9.2 show isolines of deuterium distribution in the profile and in the surface layers of the Atlantic and Arctic Oceans plotted by Redfield and Friedman (1964). Three layers are commonly distinguished according to the conditions of the distribution of individual isotope concentrations and the observed picture of the ocean waters mixing with depth. The surface layer (down to ~500 m) is characterised by the greatest ranges of local and regional variations of isotopic composition. The deep layer (below ~1000 m) is distinguished by thorough mixing of water and uniform isotopic composition in the whole ocean. The intermediate, or

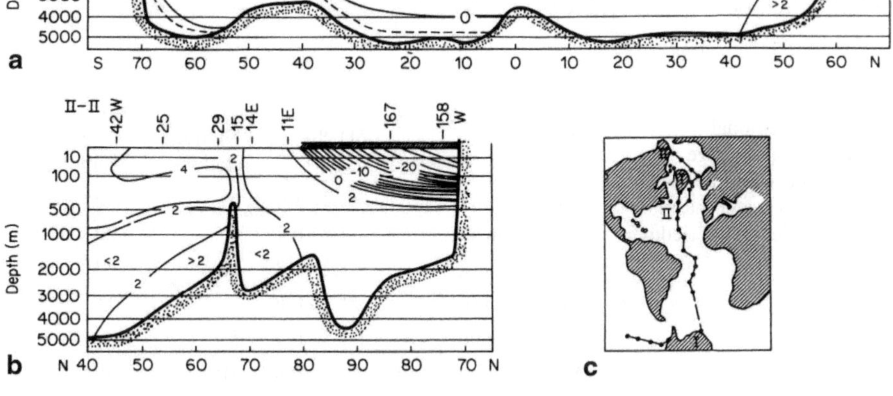

Fig. 9.1 Distribution of deuterium concentrations (in δD‰) in a vertical section of the Atlantic Ocean and Arctic basin waters: **a** section from the Antarctic through the Atlantic Ocean to Greenland. **b** section from the North Atlantic across the Norwegian Sea and the Arctic Ocean to Alaska. **c** schematic plan of sections. (Redfield and Friedman 1964; Ferronsky and Polyakov 2012)

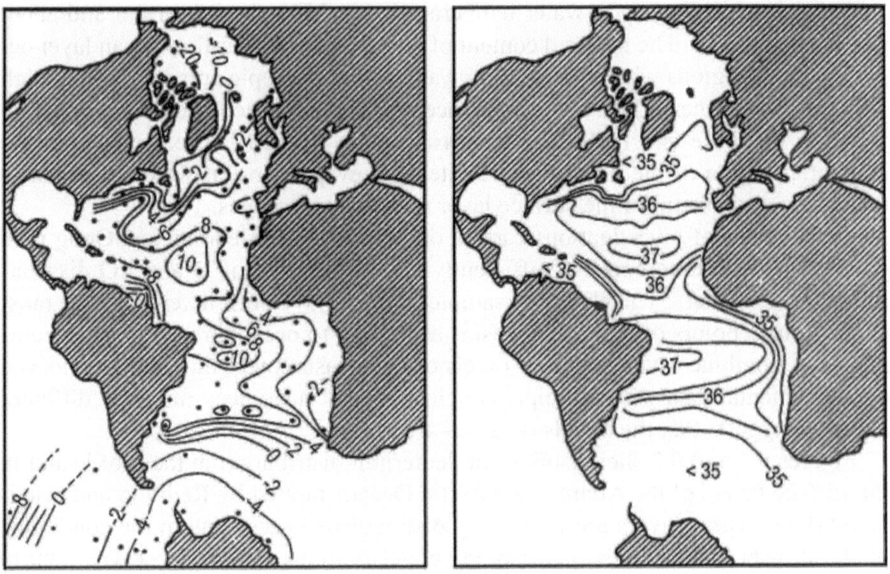

a b

Fig. 9.2 Distribution of deuterium concentrations (in δD ‰) (**a**) and salinity (in ‰). (**b**) on the surface waters of the Atlantic Ocean. (Redfield and Friedman 1964); Ferronsky and Polyakov 2012)

mixing layer (from ~500 to ~1000 m) is characterised by some intermediate parameters. Sometimes in the oceanic water profile a bottom layer may be distinguished, which has characteristic regional peculiarities in some basins.

The deep ocean layer, including approximately three-quarters of its total mass, can be practically considered representative of all the Earth's hydrosphere. On the basis of experimental data Redfield and Friedman give the following average deuterium δ-values (relative to the SMOW standard) for the major deep water masses in individual oceans (see Table 9.1).

Table 9.1 Regional distribution of average deuterium concentrations in deep oceanic water masses. (After Redfield and Friedman 1964)

Ocean	δD, ‰	Number of stations	Number of samples
Arctic ocean	+2.2±1.0	6	22
Norwegian sea	+2.2±0.7	6	22
North Atlantic	+1.2±0.8	18	44
South Atlantic	−1.3±0.6	7	11
Pacific ocean	−1.4±0.4	5	8
Antarctic intermediate	−0.9±0.8	10	19
Antarctic circumpolar	−1.7±0.8	6	16

Table 9.2 Regional distribution of average oxygen-18 concentrations in deep oceanic water masses. (Craig and Gordon 1965)

Ocean	$\delta^{18}O$, ‰	Salinity, ‰
North Atlantic	±0.12	34.93
Antarctic (bottom water)	−0.45	34.65
Indian Ocean	−0.18	34.71
Pacific Ocean:		
Antarctic basin (55–65° S)	−0.21	34.700
Shousern region (22–40° S)	−0.17	34.707
Equatorial basin (6° S –30 N)	−0.17	34.692
Northern region (44–54° N)	−0.17	34.700
Antarctic circumpolar	−0.3–0.2	34.69

The picture of oxygen-18 δ-values for deep ocean waters according to the data of Craig and Gordon (1965) is given in Table 9.2.

From the analysis of these data the following principles of D and ^{18}O variations in deep ocean waters were obtained.

1. The ocean waters in the northern hemisphere are heavier in deuterium and oxygen-18 than those of the southern hemisphere.
2. The waters of the Arctic Ocean are highly uniform in isotopic composition and differ little from those of the Norwegian Sea but differ significantly from the North Atlantic deep waters.
3. The waters of the Atlantic Ocean are mixing preferentially northward and have considerable variations even in the subtropical latitudes of the northern hemisphere.
4. The deep waters of the Indian and Pacific Oceans are highly homogeneous in isotopic composition but have some variations in high southern latitudes. The isotopic composition of water becomes heavier in the direction of low latitudes.
5. The Pacific and Atlantic ocean waters and circumpolar Antarctic waters are approximately similar in isotopic composition.
6. In deep Pacific trenches water is enriched in ^{18}O by about 0.2‰ relative to the bulk of deep waters in this ocean.
7. For deep ocean waters as a whole there is a relationship between D and ^{18}O content, which is sufficiently close to linear: $\delta D = n\delta^{18}O$. The value of n is approximately equal to 10, with some regional deviations.

It appears from the information on heavy isotope distribution that waters of the ocean display a natural tendency towards constant isotopic composition. The main factors governing the observed D and ^{18}O variations are those processes that occur in surface ocean waters and have a regional character. In surface waters the constant enrichment or depletion of water with heavy isotopes takes place due to evaporation of water and exchange with the atmosphere. Further redistribution of isotopes in the ocean proceeds as a result of mixing of surface and deep waters on global and regional scales.

The isotopic composition of the surface ocean waters undergoes considerable variations with latitude. But for equatorial and temperate latitudes the linear relationship between D and ^{18}O contents in the form $\delta D = n\delta^{18}O$ holds. The variation of n for a given region depends on the ratio of the amounts of evaporation and precipitation and is equal to 7.5 for the North Pacific, 6.5 for the North Atlantic and 6 for the Red Sea.

The factors governing regional isotopic variations in the surface ocean waters are: the ratio between evaporation and precipitation; the continental runoff and in a number of cases the submarine discharge of groundwater runoff; the freezing of seawater; the contribution of water from Polar and Circumpolar glaciers and also ice melt-water; the mixing of surface and deep ocean waters. From all the enumerated factors the main one is evaporation from the ocean surface and its ratio with precipitation in a given region. According to Craig and Gordon (1965) the values of the ratio of evaporation to precipitation amount to 0.67 for the equatorial belt, 2.0 (maximum value) for the subtropical trade winds region and 0.5 for high latitudes (above 40°).

The isotopic composition of precipitation varies with latitude in accordance with principles given in the next section. The primary precipitation, condensing from vapour, has the maximum content of heavy isotopes, since the standard vapour pressure for heavy molecules of water is lower than that for the light ones and, thus, they are condensing preferentially before the light molecules. Practically, the first precipitation has an isotopic composition close to that of the seawater in a given region. Consequently, during the movement of vapour away from the evaporation region they are depleted in heavy isotopes. Therefore, in those latitudes where the balance of evaporating and precipitating moisture is observed, the isotopic composition of surface waters remains constant and is close to the mean ocean isotopic composition. In those regions where evaporation exceeds precipitation, enrichment D and ^{18}O of the surface waters occurs. In the Polar Regions where precipitation exceeds evaporation and the incoming water vapour has the lowest D and ^{18}O composition, the surface waters are considerably depleted. Craig and Gordon (1965) gave the following regional limits for oxygen-18 variations in precipitation: from 0 to −5‰ for tropical and subtropical regions, from −5 to −15‰ for temperate latitudes and below −15‰ for the Polar Regions.

It is commonly known that the classical approach for experimental oceanographic studies is based on the relationship between the temperature and salinity of ocean water. The logical question arises whether or not data on isotopic composition of seawater contain any new information. It appears they do for the following reasons. Salinity characterises seawater in terms of a solution, whereas deuterium and oxygen-18 composition characterises just water, which is a solvent. During changes of salinity and heavy isotope composition in surface ocean waters, due to evaporation and precipitation, the salinity depends on the isotopic ratios of water in precipitation and evaporation. Therefore, the data of isotopic composition and salinity of seawater complement each other, characterising the solution from different sides: from the side of components dissolved in it and from the side of water, the solvent.

Experimental data obtained from a number of coastal regions (Redfield and Friedman 1964) representing the relationship between deuterium content and salinity

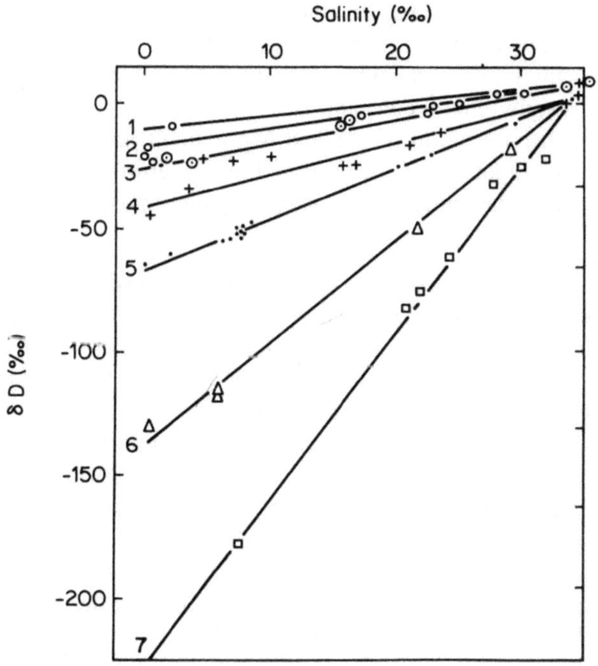

Fig. 9.3 Relationship between D content and salinity for various coastal waters: (*1*) Gulf of Venezuela; (*2*) Georgia Coast; (*3*) Albemarle, N.C.; (*4*) Chesapeake Bay, Va.; (*5*) Baltic Sea; (*6*) Labrador, Hamilton Inlet; (*7*) Greenland Fjords. (After Redfield and Friedman 1964; Ferronsky and Polyakov 2012)

are shown in Fig. 9.3. From this figure it follows that the experimental points fall on lines representing different proportions of fresh and salt water with gradients representative of given regions. All the lines meet in a point, which is characteristic of the relationship between deuterium content and salinity for North Atlantic deep waters or their modifications at lower latitudes and shallower depth.

The deuterium content, corresponding to zero salinity, is characteristic of fresh water and varies with latitude over a wide range, attaining extreme values in the fjords of Greenland. These data are close to those for the deuterium content in continental surface waters of corresponding regions.

The isotopic and chemical composition of the ocean water has more than likely remained the same at least during the last 250–300 million years. This is confirmed by many facts summarised during paleotemperature studies and based on the analysis of the isotopic composition of oxygen in the shells of modern and ancient mollusks (Bowen 1966) and also of oxygen and hydrogen isotopes in cherts of different ages (Kolodny and Epstein 1976).

Arguments in favour of the stability of the isotopic and chemical composition of the ocean water were considered by Lowenstam (Bowen 1966). They are:

1. The crystalline form of the carbonate of castle brachiopoda remained the same from the Mississippian time. This supports the theory that the crystallochemical process of Mg and Sr accumulation has remained stable during the last 250 million years, the period that corresponds to the time interval of the studied samples.

2. Modern organisms, living in ocean waters of variable chemical composition, have different ratios of Sr/Ca and Mg/Ca. This fact demonstrates the absence of any homeostatical mechanism due to which the definite Sr/Ca and Mg/Ca ratios might remain constant.
3. Differences in the Sr and Mg content found in fossil species of mollusks are analogous to those observed in modern ones. The interdependence between Sr and Mg content and oxygen isotopic ratios in both fossil and modern mollusks is very close.

Bowen notes that since the isotopic composition of oxygen, the concentrations of $SrCO_3$ and Mg CO_3 and the values of the Sr/Ca and Mg/Ca ratios in those shells of the fossil mollusks studied by Lowestam always varied within the limits, which are characteristic for modern samples, it seems probable that the concentration of ^{18}O in ocean waters has remained constant.

Glacial epochs may have played an important role in varying the isotopic composition of ocean waters. The total amount of water in the ocean in non-glacial time, e.g. during the Mesozoic, was 5 % greater than that in glacial periods according to Fairbridge (1964). If we accept that all excess water was provided by ice melting, then the amount of deuterium in the ocean in non-glacial time would be 10 ‰ less than during glaciation. Fairbridge estimated the amount of continental ice during glaciation to be 80 million km^3, which exceeds by a factor of 2 the data given by other authors. According to Bowen (1966) the excess of ice during the Pleistocene was $40.2 \cdot 10^6$ km^3, which corresponds to an ocean volume increase of 4 % in non-glacial times. According to Emilliani (1970), the increase in the oxygen-18 content of the ocean in glacial time was equal to 0.5‰. In order to carry out these calculations the increase in the volume of ice during Pleistocene was assumed to be equal to $40 \cdot 10^6$ km^3 and the average isotopic composition of ice to be $\delta^{18}O = -15$‰.

Emilliani assumed that during Pleistocene glaciation the average oxygen-18 content in the North American ice sheet was −9 ‰. At the same time Yapp and Epstein (1977) found from the analysis of the hydrogen isotopic composition in the cellulose of plants of glacial age that, for the Pleistocene glaciers of North America, the characteristic values are $\delta^{18}O$ from −12 to −15 ‰. Assuming during the Wisconsin maximum glaciation the value of $\delta^{18}O = -15$ ‰ for North American glaciers, $\delta^{18}O = -30$ ‰ and $\delta^{18}O = -40$ ‰ for all the others, Yapp and Epstein calculated the average enrichment of ocean waters with ^{18}O as +0.8 ‰. This value is in agreement with the results obtained by Craig and Gordon (1965). A greater value, $\delta^{18}O = +1.1$ ‰, of the average enrichment of the ocean in that period was given by Shackleton and Opdyke (1973). This result was obtained on the basis of an oxygen isotope analysis of the shells of foraminifers of Pleistocene and recent ages in equatorial parts of the Pacific Ocean. But as Yapp and Epstein (1977) pointed out, the deep ocean waters could have been depleted in ^{18}O by 0.2 ‰.

Therefore, the variations in concentrations of deuterium and oxygen-18 in the oceans during glacial and non-glacial times do not exceed ± 10‰ in deuterium and ± 1 ‰ in oxygen-18. The isotopic composition of the oceans has remained practically unchanged at least during the last 250 million years.

9.3 Isotopic Composition of Atmospheric Moisture

The main factor controlling the fractionation of isotopic species in surface waters is the difference in the saturated vapour between various water molecules: $p_{H_2{}^{16}O} > p_{H_2{}^{18}O} > p_{HDO}$. The fractionation factor in isotopic species of water molecules under equilibrium conditions is determined by the ratio of the saturated vapour pressure of light (p) and heavy (p') components (1.49): $\alpha = p/p'$. The fractionation factor α at 20 °C is 1.08 for HDO and и 1.009 for $H_2{}^{18}O$. In this case, the vapour in equilibrium with water will be depleted in deuterium by 80‰ and in oxygen-18 by 9‰. The isotopic composition of the vapor is $R_v = R_w/\alpha$. If water is taken as a standard, then:

$$\delta_v = (R_v - R_w)/R_w = 1/\alpha - 1, \tag{9.8}$$

and the isotopic composition of the water relative to the vapour in equilibrium (vapour is taken as a standard) will be, by analogy, equal to:

$$\delta_v = \alpha - 1. \tag{9.9}$$

The deuterium and oxygen-18 fractionation factors increase when the temperature decreases. The dependence of the factor α on temperature in general is expressed by the equation:

$$\alpha = a \exp(b/RT). \tag{9.10}$$

For oxygen, this dependence in the temperature range from −20 to +100 °C may be described by Zhavoronkov's equation:

$$\alpha_{^{18}O} = 0.982 \exp(15.788/RT).$$

The most important process involved in the formation of the isotopic composition of atmospheric precipitation is the condensation of water vapour. In a closed system a small amount of initial condensate obviously has the same isotopic composition as that of liquid in equilibrium with vapour. With the further condensation of moisture the isotopic composition of the condensate changes in accordance with the following equation (Dansgaard 1964):

$$\delta_k = (1/\alpha_0)(1/\varepsilon F_v + 1) - 1, \tag{9.11}$$

where

$\varepsilon = (1/\alpha) - 1$; α_0 is the fractionation factor at the beginning of fractionation; F_v is the remaining part of the vapour.

If the temperature remains constant, then $\alpha_0 = \alpha$. The isotopic composition of the remaining vapor phase δ_v may be described by the equation:

$$\delta_v = (1/\alpha_o \alpha)[1/(\varepsilon F_v + 1) - 1] \tag{9.12}$$

It follows from Eqs. (9.11) and (9.12) that with a decrease in temperature of the system the vapour and therefore the liquid phase being formed by it, becomes depleted in heavy isotopes of hydrogen and oxygen to a greater extent.

Equations (9.10) and (9.12) describe the closed system in a state of equilibrium. If the condensate precipitates from the vapour phase directly into the solid one, the isotope exchange does not take place between the sublimate and the vapour. In this case the isotopic composition of the condensate and the vapour phase may be given by the Rayleigh equation (Dansgaard 1964):

$$\delta_c = (\alpha/\alpha_0) F_v^{\alpha-1} - 1, \tag{9.13}$$

$$\delta_v = (1/\alpha_0) F_v^{\alpha-1} - 1. \tag{9.14}$$

It follows from Eqs. (9.13) and (9.14) that, in this case, the vapour phase becomes depleted in heavy isotopes to a greater extent during the cooling of the system. Figure 9.4 shows the dependence of isotopic fractionation of vapour and condensate as a function of the remaining amount of vapour F_v.

If Eq. (9.14) is expressed in logarithms and differentiated with respect to temperature t, the equation describing the dependence of the rate of isotopic fractionation $d\delta/dt$ on temperature changes is obtained. In other words one obtains a relationship describing the dependence of the isotopic composition of precipitation on the temperature of condensation. The dependence is experimentally determined between the concentrations of oxygen-18 and deuterium isotopes in precipitation and the surface average temperature t_a of air. It is expressed for a broad range of temperatures by the equations:

$$\delta^{18}O = 0.695 t_a - 13.6‰, \tag{9.15}$$

$$\delta D = 5.6 t_a - 100‰. \tag{9.16}$$

The dependence of isotopic composition on annual average temperature expressed by Eqs. (9.15) and (9.16) is only true for coastal regions with mild and cold climate. With distance from the sea the continental effect appears. In this case the calculated and experimental data do not often coincide. For example, for Vienna the annual average is $\delta^{18}O = -9$ ‰ whereas Eq. (9.16), using the annual average temperature (+9.5 °C), produces $\delta^{18}O = -7$‰ (Drost et al. 1972). Mook (1970) observed the continental effect even in the Netherlands.

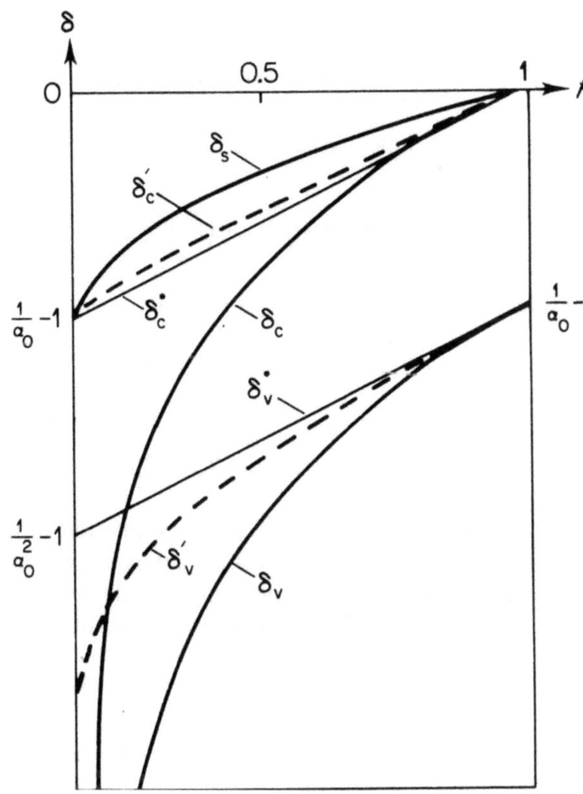

Fig. 9.4 Isotope separation in a vapour–condensate system as a function of the vapour portion F_v remaining; $\delta_v^*(\delta_c^*)$ is the equilibrium curve between the liquid and vapour phases for isothermal condensation; $\delta_v'(\delta)_c'$ is the same for condensation at cooling; $\delta_v^*(\delta_s)$ is the equilibrium curve for the sublimation process; δ_s is the averaged isotopic composition of the solid phase; δ_v (δ_c) is Rayleigh's curve for sublimation or condensation at cooling on the newly formed condensate. (Dansgaard 1964; Ferronsky and Polyakov 2012)

The dependence of oxygen and hydrogen isotopic composition upon the monthly average temperature t_m has been expressed in various ways. For example, for the Moscow region, the dependence is expressed by the empirical equations (Polyakov and Kolesnikova 1978):

$$\delta^{18}O = (0.34 \pm 0.03)t_m - (12.6 \pm 0.3)‰, \ r = 0.82;$$

$$\delta D = (2.4 \pm 0.2)t_m - (101 \pm 2)‰, \ r = 0.89.$$

For Vienna the corresponding equations are (Polyakov and Kolesnikova 1978; Hübner et al. 1979b):

$$\delta^{18}O = (0.40 \pm 0.04)t_m - (13.2 \pm 0.5)‰, \ r = 0.74;$$

$$\delta D = (2.8 \pm 0.3)t_m - (96 \pm 4)‰, \ r = 0.78,$$

$$\delta D = (2.94 \pm 0.19)t_m - (99.3 \pm 16.6)‰, \ r = 0.805.$$

For England we have (Evans et al. 1979):

$$\delta^{18}O = 0.23t_m - 8.62‰,\ r = 0.77.$$

For the stations Thule, Groenendal, Nord and Vienna one has (Gat and Gonfiantini 1981):

$$\delta^{18}O = (0.521 \pm 0.014)t_m - (14.959 \pm 0.208)‰,\ r = 0.893,\ n = 363.$$

It follows from Eqs. (9.15) and (9.16) that:

$$d\delta^{18}O/dt \approx 0.7‰/1°C, \tag{9.17}$$

$$d\delta D/dt \approx 5.6‰/1°C. \tag{9.18}$$

The annual average value of $\delta^{18}O$ in precipitation from different regions is shown in Fig. 9.5 as a function of surface annual average temperature (Dansgaard 1964):

The relationship between the content of deuterium and oxygen-18 in precipitation based on Eqs. (9.15) and (9.16) may be written in a general form as:

$$\delta D = a\delta^{18}O + b. \tag{9.19}$$

The value of the coefficient *a* may be determined from the relation:

$$\alpha = \frac{d\delta D}{d\delta^{18}O} = \frac{d\delta D}{dt} / \frac{d\delta^{18}O}{dt} = \frac{5.6}{0.69} \approx 8.0.$$

Dansgaard's theoretical calculations and the experimental evidence of several researchers have shown that for precipitation such as rain and snow in most regions it ranges from 8.0 ± 0.2 to 8.1 ± 0.4. Some exceptions are the data obtained at island and ship stations located mainly in tropical and subtropical zones. In this case the slope of the line obtained for 15 stations is equal to 4.6 ± 0.4. This fact has not yet been explained satisfactorily on theoretical grounds since the minimal theoretical value of the coefficient *a* in equilibrium during condensation (following Rayleigh's law in the range of temperatures ± 20 °C) is equal to 7.5.

Constant *b* in Eq. (9.19) characterises the degree of disequilibrium in the system caused by the evaporation of initial masses of the ocean water, i.e., the rate of its evaporation. If the evaporation of the ocean water occurs under equilibrium conditions, *b* will be equal to zero. But due to kinetic effects provided by the evaporation of water under non-equilibrium conditions the vapour formed becomes depleted in oxygen-18 to a slightly greater extent. Therefore $b \neq 0$; for most stations on the Earth where experimental determination of D and ^{18}O in precipitation were carried out $b = 10$‰. The values of *b* obtained on some island and ship stations are lower and

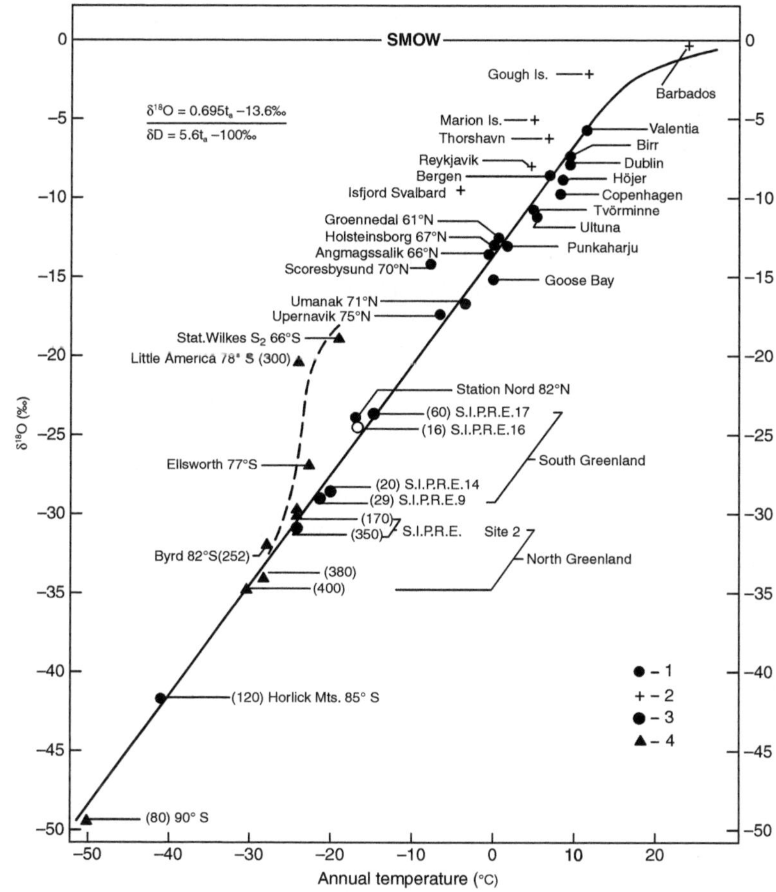

Fig. 9.5 Annual mean values of $\delta^{18}O$ in precipitation as a function of air temperature of the Earth's surface. (*1*) Continental stations of the North Atlantic; (*2*) Island stations; (*3* and *4*) Greenland and Antarctic stations. Figures in the brackets mean the snow layer thicknesses investigated. (Dansgaard 1964; Ferronsky and Polyakov 2012)

sometimes below zero. Parameter b in Eq. (9.19) was defined by Dansgaard (1964) as $d = \delta D - \delta^{18}O$ (where d is the 'excess parameter' determining the deuterium excess in the atmospheric precipitation relative to its amount in the equilibrium process, when d = 0).

For some regions the value d is > 10 ‰. For example, on the East Mediterranean coast d = 22 ‰ (Gat and Carmi 1970), for Japanese islands $d = 17.5$ ‰ (Sakai and Matsubaya 1977), for Alexandria $d = 15.9$, for Karizimir $d = 23.35$ ‰ and for Invercargill $d = -0.3$ ‰ (Yurtsever and Gat 1981). The parameter d is affected by the process of condensation of the precipitation and its value is decreased with distance from the region of vapour formation and the place of precipitation (continental effect). For Valentia $d \approx 10$, for Vienna $d \approx 5$ ‰ and for Moscow $d \approx 0$.

As per Yurtsever and Gat (1981), in the general case the mean weighted value of $\overline{\delta D}$ and $\overline{\delta^{18}O}$ is described by many parametric dependences in the form:

$$\delta^{18}O = a_0 + a_1 T + a_2 P + a_3 L + a_4 A,$$

where

T	is the mean monthly temperature in °C
P	is the mean monthly amount of precipitation in mm
L	is the geographic latitude in grad
A	is the altitude over sea level in m
a_0, a_1, a_2, a_3, a_4	are the regressive coefficients.

Calculation by the least square method for the 91 IAEA/WMO network stations gives the following values of the coefficients: a_1=0.815; a_2=0.303; a_3=0.722; a_4=0.007.

The relationship between δD and $\delta^{18}O$ for precipitation, plotted from experimental data, is given in Fig. 9.6. This dependence was first reported by Craig (1961a) based on a great number of experiments. Therefore the regression line:

$$\delta D = 8\,\delta^{18}O + 10‰ \tag{9.20}$$

is often called Craig's line.

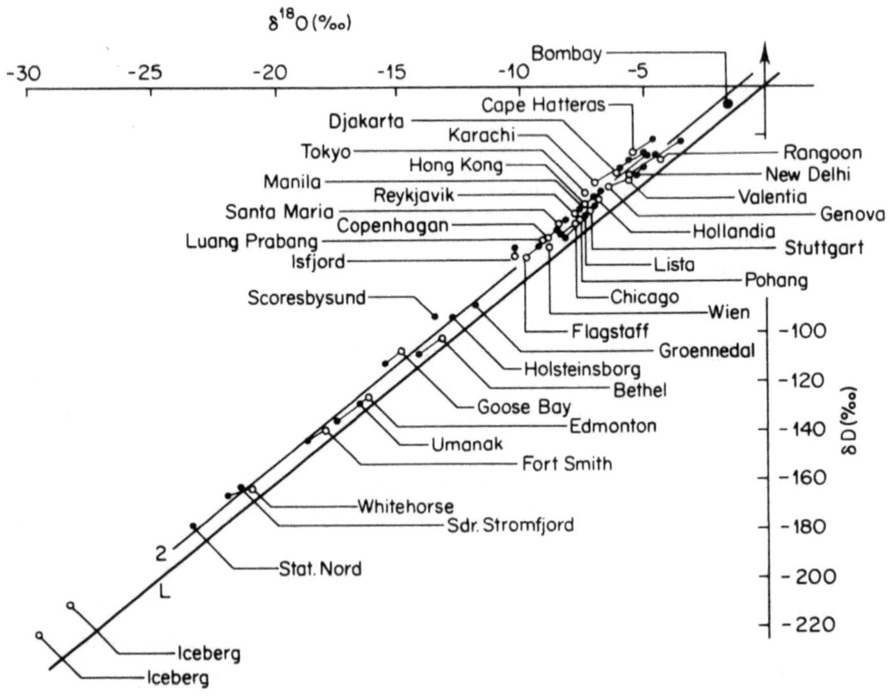

Fig. 9.6 δD and $\delta^{18}O$ relationship in precipitation and continental meteoric water based on the global network data of the Northern hemisphere and expressed by $\delta D = 8\delta^{18}O + 10‰$. (Dansgaard 1964; Ferronsky and Polyakov 2012)

Correlation dependence between the mean weighted (annual) values of $\overline{\delta D}$ and $\delta^{18}O$ for many stations (n=74) calculated by Yurtsever and Gat (1981) have the form:

$$\overline{\delta D} = (8.17 \pm 0.08)\ \overline{\delta^{18}O} + (6.03 \pm 3.08), ‰;\ r = 0.997.$$

This equation is close to Craig's dependence. For stations of the North American continent the dependence obtained by Yurtsever and Gat is:

$$\overline{\delta D} = (7.95 \pm 0.22)\ \overline{\delta^{18}O} + (10.56 \pm 0.64), ‰;\ r = 0.997.$$

It should be noted that Eq. (9.20) is only true for the annual averaged hydrogen and oxygen isotopic data with respect to different regions of the Earth and needs to be corrected for analysis in other time ranges (Gat and Dansgaard 1972). Regression analysis of many years of observation of D and ^{18}O concentrations in precipitation at individual stations on a monthly average basis have shown that for the equationв $\delta D = a\delta^{18}O + b$ the regional coefficients given in Table 9.1 should be taken into account.

For the United Kingdom the following relationship was obtained on a monthly average basis (Bath et al. 1979; Evans et al. 1979):

$$\delta D = 6.6\ \delta^{18}O + 1.3‰;\ r = 0.83;$$

$$\delta D = 6.6\ \delta^{18}O + 1.4‰;\ r = 0.85;$$

For Europe as a whole (Evans et al. 1979):

$$\delta D = 6.9\ \delta^{18}O - 0.3‰;\ r = 0.95.$$

For north-east Brazil this dependence has the form (Salati et al. 1980):

$$\delta D = 6.4\ \delta^{18}O + 5.5.$$

As Table 9.3 shows, the coefficient a for all stations appears to be less than 8. There is a tendency for the coefficient b to decrease with a decrease in the annual decreases to approximately −8.3 and remains constant within the wide range of temperature from 5 to 18 °C. The dependence of the coefficient b upon the annual average temperature t(°C) may be approximated by:

$$b = -(24 \pm 5) + 1.2\bar{t}.$$

As the atmospheric moisture travels further from the Atlantic across the Eurasian continent b decreases by approximately 0.7‰ per 100 km and the decrease of the relative values of D and ^{18}O concentrations is 2.5±0.5‰ and 0.35±0.05‰ per 100 km respectively. A similar relationship for Europe and Africa was reported by Sonntag et al. (1979).

Table 9.3 Calculated numerical factors (δD–$\delta^{18}O$ relationship) for monthly averaged precipitation of some observational stations. (Polyakov and Kolesnikova 1978)

Type of station	Station	Annual averaged data			Number of months of observation	Monthly averaged factors		Correaltion factor r
		Precipitation (mm)	Vapour pressure (mb)	Temperature (°C)		*a*	*b*	
Island	Havaii	3470	21.3	23	92	4.0±0.5	3.0±1.1	0.66
	Azores	1200	15.4	17	48	5.5±0.6	4.0±1.0	0.92
	Valentia	1400	10.7	11	72	6.1±0.6	−3.7±0.5	0.80
	Adak	1400	6.0	5	66	6.1±0.6	−8.9±5.4	0.80
	Isfjord	350	3.7	−4	42	6.1±0.4	−8.5 ± 3.7	0.93
	Nord	150	1.1	−18	78	7.2±0.2	−8.2 ± 3.4	0.99
Coastal	Gibraltar	815	15.4	18	48	5.6±0.4	−0.6±1.4	0.91
	Bethel	490	4.3	−2	48	5.9±0.2	−39.7±2.8	0.97
	Barrow	110	1.7	−12	66	5.5±0.5	−39.7±9.5	0.84
Continental	Vienna	660	8.5	10	108	6.1±0.3	−11.5±3.5	0.92
	Moscow	575	6.5	4	60	6.0±0.5	−26.2±6.7	0.89

The dependence of b upon temperature is probably the result of the processes involved during the non-equilibrium condensation of the precipitation and during the partial evaporation of the liquid-droplet part of the moisture cloud whilst the primary source of moisture vapour is continental.

As pointed out by Polyakov and Kolesnikova (1978), for all regions with an annual average temperature $\bar{t} < 20°C$ the relationship between δD and $\delta^{18}O$ in the total annual precipitation follows the empirical relationship:

$$\delta D = 8\delta^{18}O + 10 - 0.7l^2 ‰,$$

where

l is the distance from the ocean (the source of water vapour) in thousands of kilometres. For the coastal and island stations $l = 0$.

Therefore, Craig's equation (9.20) branches out on sets of parallel lines with varying slopes, which are dependent on the distance from the ocean.

The relationship between the isotopic composition of the monthly average precipitation for the stations with annual temperatures in the range $-15\,°C < t < 20°C$ can be approximated by the empirical equation:

$$\delta D = 6\,\delta^{18}O - 0.7l^2 + 0.7\bar{t} - 7‰.$$

The relationship between δD and $\delta^{18}O$ for different regions is often used for the interpretation of data for hydrogeological reconstructions. Craig's equation, which determines the global relationship between the deuterium content and oxygen-18 on an annual average basis, is often employed. As seen from Table 9.3, the regional effects are not taken into account, which might lead to appreciable errors.

The description of the formation processes of isotopic composition of precipitation based on equilibrium Rayleigh fractionation is the approximate natural model of moisture condensation and precipitation. Nevertheless, in a series of cases, the use of Rayleigh's formula for natural processes coincides satisfactory with the experimental data. Thus Eriksson (1965) suggested that atmospheric moisture be considered as the uniform vapour phase in which the liquid phase forms under isothermal conditions in accordance with the Rayleigh mechanism. With such an assumption the average isotopic composition of atmospheric precipitation R_p is a function of the atmospheric moisture W and the water content in the vertical atmospheric column, i.e.,

$$dR_p / R_p = (\alpha - 1)(dW / W). \tag{9.21}$$

Under isothermal conditions $\alpha = \text{const}$. Integrating Eq. (9.21) and substituting $lnR_p = ln(1+\delta_p) \approx \delta_p$, one obtains:

$$\delta_p = 2.3(\alpha - 1)\lg W + B. \tag{9.22}$$

Equation (9.22) describes the process of advective transition of moisture in the atmosphere. For the turbulent process Eriksson (1965) suggests the substitution of $\sqrt{\alpha-1}$ for $\alpha-1$; then:

$$\delta_p = 2.3(\sqrt{\alpha-1})\lg W + R. \tag{9.23}$$

The experimental justification of Eqs. (9.22) and (9.23) carried out by Eriksson by comparing the annual average isotopic composition of atmospheric precipitation for a series of island stations showed that, despite the considerable scatter of the experimental evidence, the points in the diagram $\delta^{18}O-\lg W$ are predicted by the above equations for $\delta_{^{18}O} = 1.009$.

Brezgunov (1978), on the basis of analysis of a large amount of experimental data, obtained an empirical relationship between δD and lgW for regions of meridional water vapour transition above the ocean and inland transition above Europe and North America. For winter, this relationship, in the case of the sea profile, is:

$$\delta D = 106 \lg p - 138‰, \tag{9.24}$$

and for the inland profile it is:

$$\delta D = 193 \lg p - 215‰. \tag{9.25}$$

For summer, in the case of the sea profile the relationship is:

$$\delta D = 174 \lg p - 219‰, \tag{9.26}$$

and for the North American continent it is:

$$\delta D = 313 \lg p - 456‰. \tag{9.27}$$

Brezgunov noted that the greater rate of δD decrease in summer precipitation with the decrease of lg p (dδD/dlg p), for the stations of the ocean and North American inland profiles, in comparison with winter precipitation, may be explained by the additional isotopic fractionation provided by raindrop evaporation below the level of the clouds.

The dependence of the content of heavy hydrogen and oxygen isotopes on vapour condensation temperature leads to seasonal deviations of deuterium and oxygen-18 in precipitation reaching a maximum in summer and minimum in winter. For the same reason the oxygen and deuterium concentrations in meteoric waters decrease in higher latitudes and altitudes.

Because of the dependence of isotopic composition of precipitation on temperature, deuterium and oxygen-18 concentrations in meteoric waters decrease with an increase of latitude and absolute mark of the land. The highness effect develops differently in regions with different climatic conditions. The gradient values ΔδD/Δh and $\Delta\delta^{18}O/\Delta h$, as a rule, are varying in ranges of 1.5–4‰ (per 100 m for δD) and 0.15–0.5‰ (per 100 m for $\delta^{18}O$) (Yurtsever and Gat 1981).

The isotopic balance in the global hydrologic cycle at evaporation and condensation of water can be considered on the basis of available experimental data.

The unity of natural waters on the Earth consists in their genetic relationship, which has continued during the whole history of the hydrosphere. The principal reservoir of the hydrosphere—the ocean—contains about 97% of the total amount of water, taking no account of the water being bonded in rocks. The other 3% is represented by the polar and continental glaciers, which have accumulated about 2.2% of water in the solid phase, the underground waters (0.7%) and the surface continental waters, which make up less than 0.05% of the total amount of water. The water vapour in the atmosphere represents a small part of the water (about 0.0015%) but is the most dynamic element of the hydrosphere, responsible for the genetic relationship between the ocean and the continental surface and groundwaters. While the total mass of the atmosphere is about $5.2 \cdot 10^{21}$ g, the mass of water vapour there equals about 1.3 · 1019 g, which amounts to about 0.25%. The mean synoptic rate of motion of the air in the atmosphere is 10^3 cm/s and the time of the average water cycle is equal to 11 days. Thus the process of hydrosphere mixing proceeds rather intensively. In the previous section it was pointed out that out of $\sim 4 \cdot 10^{20}$ g of annual average precipitation a quarter falls over the continents and three-quarters over the oceans. At the same time, evaporation from the ocean surface exceeds that from the continents, being 85 and 15%, respectively. The difference between precipitation and evaporation over the continents is equal to 10% or $\sim 3.7 \cdot 10^{19}$ g, which corresponds to the mass of water that is being annually exchanged between the ocean and continents through the atmosphere. The continents return to the ocean the same amount of water in the form of surface and groundwater runoff.

The most important problem in hydrology is the determination of natural principles governing the process of the water cycle on the Earth and the variation of the ratio of its elements with time. The existence of the evaporation—condensation—atmospheric precipitation process is absolutely indisputable from the moment when water appeared on the Earth. But now the volume of the hydrosphere and the water balance within it have changed with time, if its origin is a result of the continuous degassing of water from the Earth's interior. If the bulk of the Earth's ocean has not changed during the greater part of its history then the question arises as to the timing and mechanism of its appearance. The temperature history of the Earth has played a decisive role in the origin and evolution of the hydrosphere and also in the evolution of water cycle elements. The geological and paleontological facts give evidence on secular variations of temperature in the range of 6–8 °C or more on the Earth's surface. The causes of these variations remain unknown. In the meantime they play an important role in the evolution of the surface 'shells' of the Earth since the atmosphere, the hydrosphere and the biosphere are very sensitive to these small temperature variations. In extreme times of temperature variation catastrophic phenomena have taken place. Temperature variations in the surface shells of the Earth were observed up to Archean times, i.e., approximately $2.7 \cdot 10^9$ year ago. The age of the most ancient rocks discovered on Earth is about $4 \cdot 10^9$ year.

Any other evidence of temperature processes that took place earlier have not been observed as they happened in the times, of high-temperature evolution of the Earth.

The study of oxygen and hydrogen isotopic ratios in water and other objects (rocks, gases, organic and mineral remains of buried organisms, etc.), which have interacted with water in the past, throws light upon the origin and evolution of the hydrosphere. In the present section we consider some principal facts concerning the formation of the isotopic composition of water, which is constantly being removed from the ocean into the atmosphere, transferred in the form of water vapour and precipitation and finally rained out over the ocean and continents. This will help us to understand better the mechanism of water movement in nature and, on the basis of data of isotopic composition of water molecules observed in natural conditions, to produce more reliable accounts of the global and regional water balances. Besides, the understanding of the principles of the distribution of stable isotopes of water during its circulation broaches the question of the hydrosphere's evolution and origin.

The problem is to find out what values of isotopic ratios are characteristic for water vapour coming from the ocean surface into the atmosphere and how the isotopic composition of various forms of vapour, from which precipitation is formed, is distributed when falling over continents and oceans. Finally, the problem is reduced to the composition of water balance on a genetic basis.

For the last 50 years, isotopic composition of hydrogen and oxygen in precipitation in a global scale has been studied on the stations of the IAEA/MWO isotopes-in-precipitation network, which are located in different climatic zones. This international project started in 1961 and includes more than 100 stations (see Fig. 9.7). The network covers continental, coastal, island and ship stations. About half of the samples and the meteorological data from all network stations are sent to the IAEA Laboratory in Vienna for analysis. The remaining samples are forwarded to co-operating laboratories in member states. The IAEA has acted as the collection agency for data of stable isotopes and tritium in the hydrological cycle. All the information is available in special report series publications and on the IAEA Internet site.

The picture of the ^{18}O global distribution in atmospheric precipitation prepared by Yurtsever and Gat (1981) and complemented by us with data of the former USSR area is presented in Fig. 9.8. It gives a general view about distribution of ^{18}O concentration in precipitation before 1981 and does not take into account possible regional changes because of climatic variations.

The global estimation of the average isotopic ratios of hydrogen and oxygen for ocean vapour may be given from the condition of mass balance of the evaporating and precipitating water. From the material balance of evaporation-precipitation (E=P) it follows that $\delta_E = \delta_P$, i.e., when the water comes back into the ocean its isotopic composition should be restored. When considering the isotopic balance of the ocean during the evaporation-precipitation cycle, continents play an insignificant role since they obtain, in the form of precipitation, only 10 % of the water evaporated from the oceanic surface. Thus, Craig and Gordon (1965) proposed to consider the ocean as a closed system. It is logical that for separate regional oceanic

Fig. 9.7 Schematic map of the IAEA/MWO isotopes-in-precipitation network stations in 1975. (Ferronsky and Polyakov 2012)

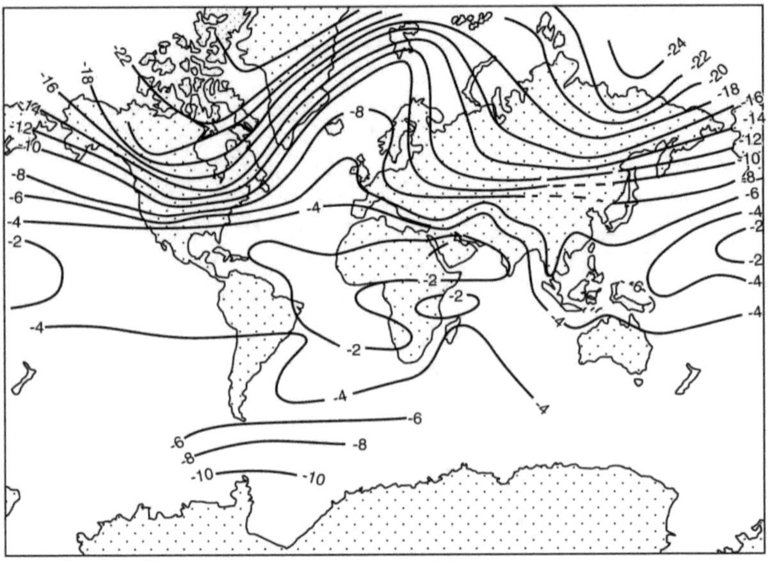

Fig. 9.8 Global distribution of $\delta^{18}O$ in atmospheric precipitation based on data having at least two series of observation. (after Yurtsever and Gat 1981; Ferronsky and Polyakov 1983, 2012)

conditions the influence of continental runoff and contribution of melt-water from polar glaciers should be considered.

On the basis of experimental data on precipitation distribution, the amounts of which are in accordance with latitudinal belts, we can accept the following approximate values of latitudinal ^{18}O abundances:

Latitudes	Ratio of precipitation	$\delta^{18}O$ (‰) (mean value)
0–20°	0.5	−2
20–40°	0.4	−5
40–90°	0.1	−15

Using these data, Craig derived the following theoretical values of isotopic ratios for oceanic vapour and precipitation, which are equal to $\delta^{18}O = -4$‰ and $\delta D = -22$‰. These magnitudes are close to the average ones in continental precipitation, in contrast to high latitudes where the above-mentioned values are significantly lower. But the amount of precipitation for these regions of the isotopic balance for the North American continent (Craig and Gordon 1965) have mean values of $\delta^{18}O = -5$‰ and $\delta D = -30$‰, which agree with the accepted data above.

In the case of equilibrium evaporation from the ocean surface without the influence of atmospheric moisture and kinetic effects, the mean values of isotopic ratios are $\delta^{18}O = -9$‰, $\delta D = -70$‰, as reported by Epstein and Mayeda (1953) and Dansgaard (1964). The experimentally observed data on the isotopic composition of vapour over the ocean (Redfield and Friedman 1964) is evidence in favour of the isotopic ratios being considerably lower than those suggested by the material balance and even lower than the values corresponding to equilibrium conditions ($\delta^{18}O$ is varying from −11 to −14‰). According to Craig and Gordon (1965) the given facts may be explained by the character of steady global circulations of atmospheric air and also those of condensation of atmospheric precipitation. This process may be outlined as follows. Dry atmospheric air descends to the ocean surface in trade wind regions where it picks up the water vapour and moves it to the equatorial belt, where the majority of the vapour rises to high atmospheric layers and precipitates its moisture by cooling and condensation. The remaining vapour moves away from the trade winds to high latitudes where it undergoes an analogous process. This general study system of atmospheric circulation, in which the isotopic variations and the humidity of air are reflected in transport and mixing conditions, is restricted by the local vertical convective flows, rich in water vapour, which form areas of cumulus clouds saturated with moisture. Therefore, we never observe in nature the full saturation of any atmospheric layer with moisture but only local domains of condensation in a general flow of circulating masses of air and water vapour. In these conditions, vapour above the sea will never be in equilibrium with water even if there is no kinetic effect.

The isotopic variations in vapour and precipitation are explained by Craig and Gordon with the help of a simplified single-stage precipitation model. They assumed that there is a homogeneous layer of atmospheric moisture and in the precipitation region there is a large ascending flux of moisture being evaporated from the ocean surface. Thus, compared with this flux, an insignificant amount of moisture precipitates. In such a case, the isotopic composition of precipitation will be formed by a single-stage equilibrium fractionation between the condensing moisture and atmospheric vapour. From the isotopic balance equation one obtains the following values of δD and $\delta^{18}O$, characteristic for elements of the considered system. The slope of the straight line precipitation ($\delta^{18}O = -4$‰, $\delta D = -22$‰)—atmospheric vapour ($\delta^{18}O = -13$‰, $\delta D = -94$‰) is equal to 6 and for the line atmospheric vapor-equilibrium vapour ($\delta^{18}O = -8$‰, $\delta D = -67$‰) the slope is equal to 5.5. The slope of the line of atmospheric vapour-surface seawater ($\delta^{18}O = +1$‰, $\delta D = +5$‰) is about 6.5. For the isotropic ratios in the equilibrium vapour-evaporating vapour system, ascending from the ocean surface, both kinetic effects and atmospheric humidity are important. If the kinetic effect is low enough and humidity is large enough the relationship may be reversed. At normal conditions for the observed parameters in the ocean-atmosphere system the equilibrium vapour will be lighter than the ascending flux of evaporating moisture. In nature the reverse may occur only in high latitudes.

Difficulties have arisen from the instability of evaporation and moisture exchange for different oceanic regions as one tries to develop the model, describing formation of isotopic composition of moisture transferred in the atmosphere. The most important oceanic region from the viewpoint of moisture exchange is the trade wind region between 30°N and 30°S, which incorporates about 50% of the total area of the Earth and receives three-quarters of the annual amount of precipitation and evaporation. The structure of the atmosphere in this region, derived from numerous experimental investigations and the characteristic average values of isotopic ratios of individual atmospheric elements, according to Craig and Gordon (1965), are shown in Fig. 9.9.

The model consists in two atmospheric layers. Up to 600 m above sea level a lower homogeneous layer can be distinguished, which is well mixed by turbulent stirring. The specific humidity of this layer changes sharply in the first 10–20 m above the sea and then remains almost constant up to 600 m with humidity equal to 75 %. The second cloud layer is characterised by the active process of convective mixing with a large-scale descent of dry air and local jets of rising air rich in water vapour around which cumulus clouds are formed. The thickness of the second layer is restricted by the extent of the moist, preferentially convective mixed layer of the outer trades. About this layer is the trade-wind inversion section, separating the second layer from the dry upper troposphere above. The location of the inversion section varies from an altitude of 2 km in the outer trade wind belt to 4 km near the equator. The two principal layers are separated by an intermediate layer with a thickness of about 100–300 m, which is characterised by a specific humidity that, approximately decreases linearly towards the cloud layer. The cloud layer displays a mostly uniform humidity of about 50 %. If the problem of the isotope balance of the exchanging moisture is considered from the principal viewpoint of a system in a steady state, then this model may, to some extent, represent the relationship

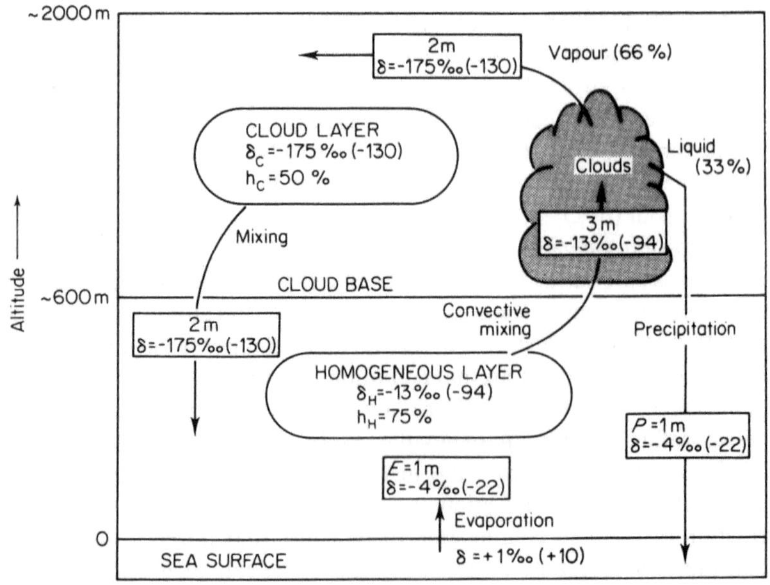

Fig. 9.9 Two-layer model for vapour and precipitation isotopic composition formation and their motion in the ocean-atmosphere-ocean system for trade winds regions. (Craig and Gordon 1965; Ferronsky and Polyakov 2012)

between the ocean and the atmosphere as a whole. Thus both the material and isotopic balances in the system during the evaporation-precipitation process should be taken into account. The loss of moisture evaporated from the ocean surface (10 %) and precipitating over the continents is not considered. Assuming the lower homogeneous and upper cloud layers have common values of humidity equal to 75 and 50 % respectively, evaporation-precipitation amounts to about 1 m per year and the isotopic ratios, according to the observed data, equal $\delta^{18}O = -13$ ‰ and $\delta D = -94$ ‰ for the vapour in the homogeneous layer and $\delta^{18}O = +1$ ‰ and $\delta D = +10$ ‰ for the surface ocean water. The model may be described by the following equations.

The first equation is just the mass balance equation of moisture in the system:

$$P = I(h_1 - h_2), \tag{9.28}$$

where

P	is the amount of precipitation
h_1 and h_2	are humidity of the upper and lower cloud layers
I	is the exchange constant between the upper and lower layers, equal to the flux of moisture between them when the air is saturated.

As shown in Fig. 9.9, the flux of moisture characterised by Eq. (9.28) is equal to a water layer 3 m thick being transported from the lower layer to the upper one and a layer 2 m thick from the upper layer to the lower one.

The second equation relates the isotopic composition and humidity of these layers:

$$h_1(\delta_p - \delta_1) = (\delta_p - \delta_2), \tag{9.29}$$

where

δ_p	is the isotopic ratio for precipitation
δ_1 and δ_2	are isotopic ratios of water vapour in the upper and lower layers.

From the last equation for the upper cloud layer, other parameters being known, we obtain $\delta_2{}^{18}O = -17.5$ ‰ and $\delta_2 D = 130$ ‰.

The third equation, relating the isotopic composition in precipitation and water vapour, may be written on the basis of the model chosen for the formation of precipitation. One may assume that condensation of precipitation takes place in the upper layer from the water vapour ascending from the lower layer for certain values of the parameters describing the cloud layer. But in this case the isotopic composition of the vapour of the cloud layer may be described by Eq. (9.29). To obtain the fixed isotopic ratios for precipitation δ_p and vapour of the cloud layer δ_2, at known vapour pressures, the fractionation factors of ^{18}O and D (1.0135 and 1.124) are required. The initial precipitation temperature corresponding to these values must range between the limits from −10 to −20 °C, which is significantly lower than the observed temperature in nature. The isotopic data show independently that precipitation is formed from the ascending flux of moisture flowing from the lower layer, being a component of the exchange process between the two layers. At the same time the observed temperature criteria lead to a conclusion that condensation of vapour rising from the lower layer is not just a process of equilibrium formation of precipitation but is governed by the more complicated mechanism of inversion distillation similar to that observed in a rectification column.

The atmospheric model, made up of two layers, which was proposed by Craig and Gordon, principally for estimating the isotopic balance of moisture resulting from the interaction between ocean and atmosphere, may be considered as a basis for the development of more detailed models of water movement through the atmosphere based on isotopic investigations both in oceanic regions and in the ocean-continent system. But in contrast to the oceanic conditions in the evaporation-precipitation cycle, where a 10 % loss of evaporating moisture (being precipitated over the continents) may be neglected, a gain of 10 % of ocean water represents 40 % of the continental moisture as a whole. The other 60 % is secondary evaporating moisture from the continental surface, which was previously precipitated. Thus, assuming the previous considerations (see Fig. 9.9), about 40 % of water vapour in a continental region would be contributed by lower and upper atmospheric layers with fixed isotopic composition, depending upon proportions of the vapour mixing. The isotopic ratios of moisture evaporating from the continental surface will have lower values compared with the moisture evaporating from the ocean surface. At the same time the net water runoff from continents into the ocean will be enriched in D and ^{18}O compared with their content in precipitation and will correspond (in accordance with the isotopic balance) to the isotopic ratios of water vapour moving from the ocean surface to the continents.

While considering conditions of formation of isotopic composition of precipitation it is interesting to give data on the evolution of isotopic ratios with time, using as an example the Antarctic ice sheet. Epstein et al. (1970) studied ice samples from a borehole drilled in 1968 at the Antarctic Byrd Station (80° 01′S, 119° 31′W), located 1530 m above sea level. Isotopic studies of the ice were carried out for depths ranging from 99 to 2162 m. The total depth of the borehole was 2164 m. The geomorphologic surface conditions in this region are such that the danger of any catastrophic contribution of ice from other regions is improbable. In addition, there is no evidence that it has been found more than 600 m below sea level.

In Fig. 9.10 the results mentioned above are given, each point representing the mean value of a water sample obtained from the ice core ranging in size from 30 to 151 cm. The exceptions are the last points representing samples taken from the borehole bottom. The relationship between δD and $\delta^{18}O$ for the investigated profile follows the equation $\delta D = 7.9\ \delta^{18}O$. The variation in the age of the ice with depth was calculated from the rate of snow deposition using recent values of accumulation of snow equal to 12 g/cm^2 of water per year. On the left side of the figure the vertical scales of the borehole depth and age are not linear, since the age scale increases disproportionately with the depth scale. On the right-hand side of the figure pictured for $\delta^{18}O$, the age of the layers and their thickness, starting from 1000 m, are given in a linear scale.

From the analysis of the magnitudes and character of isotopic variations in the ice sheet with time the authors of the work provided the following interpretation. If the observed isotopic variations in the upper 300 m during the last 2500 year could be related to some random factors resulting in temperatures 2–3 °C lower

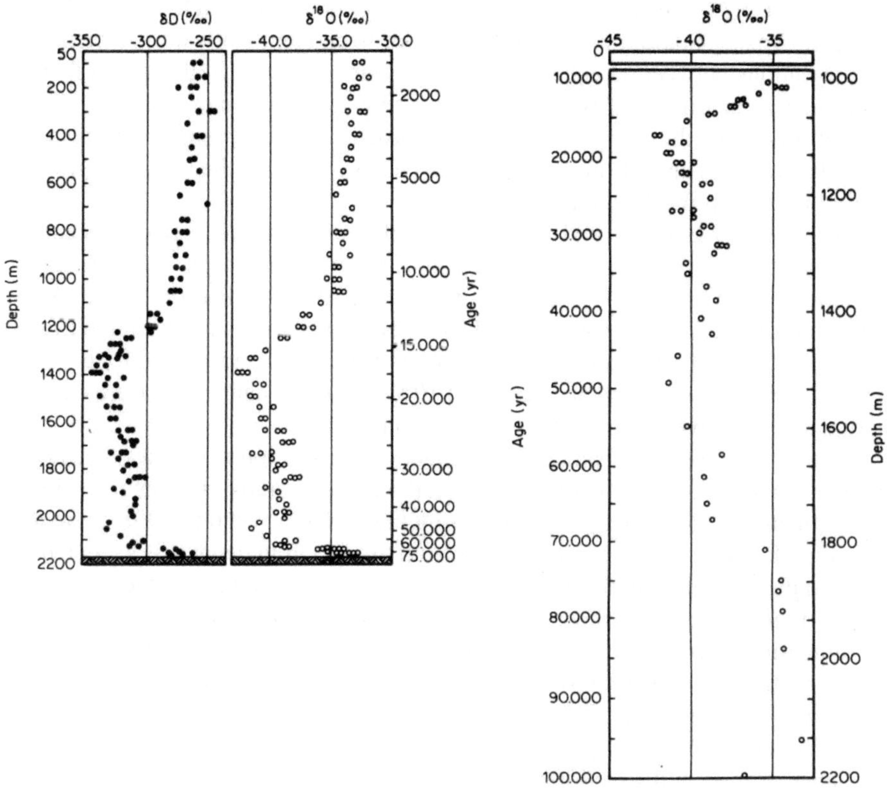

Fig. 9.10 Relationship between deuterium and oxygen-18 content in the ice sheet profile for the Antarctic Byrd Station. (Epstein et al. 1970)

then modern ones, then, in the range of depths from 1050 to 2160 m, they certainly show considerable climatic variations in the past. The corresponding time interval 11,000–75,000 years ago is commonly related to the last period of cooling on the Earth during Pleistocene. Using the variations of isotopic ratios with time it is possible to obtain the colder periods, corresponding to 17,000, 27,000, 34,000 and 40,000 years ago and warmer ones, corresponding to 25,000, 31,000 and 39,000 years ago. It is rather interesting that the last phase of maximum cooling (being 17,000–11,000 years ago) corresponds with the end of the Wisconsin period of cooling for the northern hemisphere. An analogous coincidence in the two hemispheres is also observed for the other cold and warm phases during the whole time period. The data characterise the bottom part of the ice sheet then modern ones for the Byrd station. This warmer period should have ended 75,000 years ago. Emiliani's data (1970, 1978), based on oxygen isotope studies of fossil pelagic foraminifera and the data provided by study of the Greenland ice sheet estimated for the time period of Wisconsin glaciation, are close to those given above. The temperature varies by 7–8°C between glacial and non-glacial periods (Fig. 9.10).

Similar results were obtained by Dansgaard et al. (1969) for a 1300 m ice column from a borehole at Camp Century Station in Greenland and also by Gordienko and

Kotlyakov (1976) while studying an ice column 1000 m in height obtained at the Antarctic Vostok Station. But according to Gordienko and Kotlyakov the temperature variations between Holocene and Pleistocene glaciation in the Antarctic have only been of the order of 5 °C. Thus, it is possible that Epstein et al. (1970) did not take into account the decrease of heavy hydrogen isotopic composition resulting from difference in altitude marks in the ice sheet provided by the ice melting, which could give 500 m for the Byrd station.

According to Emiliani and other authors the temperatures varied in these periods by ~5–6 °C for the equatorial region of the Atlantic ocean, ~7–8 °C for the Caribbean basin and ~3–4 °C for the equatorial part of the Pacific Ocean. As for variations of isotopic composition in glacial-non-glacial times, according to Epstein et al. (1970) and Dansgaard et al. (1971), the Greenland and Antarctic ice sheets were depleted in oxygen by an average of 10 ‰ and the ocean waters were enriched in oxygen by an average of 0.5‰ during glacial periods.

Isotopic composition change in atmospheric precipitation in the past, because of climate change on the basis of a continental ice sheets study, was also considered by Vasilchuk and Kotlyakov (2000). During the period of 1970 to 1998 at the Russian Vostok Station in Antarctic a borehole up to 3623 m in depth was executed and covered the time interval from 0 to 420,000 year. The age of several thousand years for the upper part of ice sheet was studied by annual layer counting. The deeper accumulations were determined by modelling of the ice flowing. The acceleration mass spectrometry techniques based on the cosmogenic radioactive isotopes ^{14}C, ^{10}Be, ^{36}Cl, ^{26}Al were also used.

Hydrogen isotopic composition of the ice core from Vostok Station varies in the range from −420 to −480 ‰. Its higher values (from −420 to −460 ‰) reflect relatively short periods of interglacial warming and more lower values (from −460 to −480 ‰) characterise long periods of glacial cooling. Up to now this is the only borehole covering the last glacial-interglacial cycles: the Würm glacial epoch (10–120 Kyr); Riβ–Würm interglacial (120–140 Kyr); Riβ glacial (140–220 Kyr); Great interglacial (220–320 Kyr); and Mindel glacial (320–420 Kyr). It was found that the last glacial epoch is characterised by three temperature minimums, which happened 20, 60 and 110 Kyr ago. The interglacial peak is at 130 Kyr: 1.5–2.0 °C. The warmer peaks were during the 12th, 16th and 20th centuries and the cooler were during the 12th–15th and 17th–19th centuries. The last interval is called the small glacial period. Paleoclimatic studies of the Earth were carried out on the basis of analysis of ice cores from Camp Century, Dye-3 and Summit (Greenland). A good agreement between the above data of hydrogen and oxygen-18 was found. It means that the paleoclimatic variations characteristic for both hemispheres are very probably a consequence of Milancovich's astronomical effects.

The dependence of heavy hydrogen content and oxygen isotopes on atmospheric precipitation upon temperature makes it possible to carry out paleotemperature studies, not only based on isotopic composition in atmospheric precipitation (for example, accumulated in glaciers) but also using natural chemical compounds of organic and inorganic origin, where water is essential for formation. Such compounds are, for example, clay minerals, formed by weathering of silicate rocks and also the polysaccharides (starch, cellulose) of plants.

The formation of kaolin during the process of weathering of feldspar rocks may be described by the equation:

$$\underset{\text{(albite)}}{2NaAlSi_3O_8} + 2CO_2 + 3H_2O = \underset{\text{(solution)}}{2Na^+} + \underset{\text{(solution)}}{2HCO_3^-} + \underset{\text{(solution)}}{4SiO_2} + \underset{\text{(kaolinite)}}{Al_2Si_2O_5(OH)_4} \downarrow .$$

During clay formation, hydroxides of silicon and aluminium, as described in the above reaction, undergo a stage of precipitation from the water solution, where they attain isotopic equilibrium with water.

Investigations of pure kaolinites taken from the area of weathering, carried out by Savin and Epstein (1970a), have shown that, during their formation, kaolinites and montmorillonites come close to the condition of isotopic equilibrium with meteoric waters participating in the hypergeneous process of geological weathering. This has been substantiated by Lawrence and Taylor (1972) and Taylor (1974). The values of the fractionation factors at temperature ~20 °C indicated that clay minerals were enriched in ^{18}O approximately by 27 ‰ and depleted in deuterium approximately 30 ‰, compared with the water participating in their formation.

The relationship between δD and $\delta^{18}O$ for clay minerals and hydroxides from recent soils in the USA, formed on igneous parent rocks, was studied by Lawrence and Taylor (1971). The $\delta^{18}O$ values have been calculated approximately as they were determined by $\delta^{18}O$ for gross samples of soils, reduced by relative oxygen-18 values obtained by the analysis of parent rock minerals. In addition, the so-called kaolinite line ($\delta D = 7.6\delta^{18}O - 220$ ‰) was used in the interpretation of the results.

It follows from the obtained data that the points corresponding to soil samples are enriched with hydroxides of metals (e.g., gibbsite $Al(OH)_3$). This is probably because the fractionation factors between water and gibbsite are different from those in the water–kaolinite and water–montmorillonite systems.

It has been shown by Taylor (1974), Lawrence and Taylor (1972) and Sheppard et al. (1969) that data on oxygen and hydrogen isotopic content in clay minerals and soil of kaolinite weathering zones of the Tertiary, provide an opportunity to reconstruct the content of these isotopes in tertiary meteoric waters.

In Fig. 9.11 the North American map is shown with isoline δD values for Tertiary meteoric waters calculated by Sheppard et al. (1969) and Lawrence (Taylor 1974) based on the isotopic studies of various Tertiary meteoric–hydrothermal systems and also the upper zone clay and weathering deposits of Tertiary age. Comparing this map with recent δD values for meteoric waters of the North American continent, which are pictured in Fig. 9.12 and based on data collected by Dansgaard (1964), Friedman et al. (1964) and Hitchon and Krouse (1972), one finds that for the Tertiary isolines of δD there is a tendency towards enrichment of about 1–2 ‰. As Taylor (1974) pointed out, these data are in accord with the theory that the climate was warmer in the Tertiary age not only in North America but also in other regions of the Earth and particularly in the former USSR (Petrov 1975).

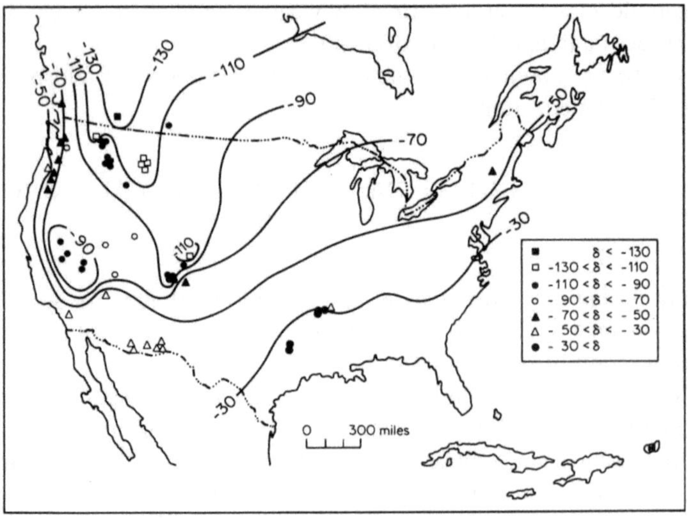

Fig. 9.11 Schematic map of the North American continent showing calculated δD values of Tertiary meteoric waters. (Taylor 1974; Ferronsky and Polyakov 2012)

It should be noted that, although the conclusions of Taylor (1974) on the milder climate in Tertiary (between the Paleogene and Neogene, i.e., about 26 million years ago) have been confirmed by several geological arguments, the differences in the δD values between the recent meteoric waters and atmospheric precipitation during Tertiary, are not likely to be as great as depicted in Figs. 9.11 and 9.12. Yapp and Epstein (1977), citing a private communications with Friedman, considered that all the δD values in Fig. 9.11 should be divided by a factor of 1.04, correcting laboratory error. Besides, the question concerning the period during which the information is covered and enclosed in the hydrogen isotopic composition of hydroxide groups of clay minerals has not yet been solved. There is evidence (James and Baker 1976) to suggest that isotopic exchange between the hydroxide groups and interlayer water takes place at room temperatures. Therefore, caution is needed when making the paleotemperature reconstruction of individual geological periods using hydrogen isotope analysis of hydroxide groups of clay minerals.

Epstein et al. (1976), Epstein and Yapp (1976) and Yapp and Epstein (1977) developed the technique of paleotemperature studies based on hydrogen isotopic content in the cellulose of wood plants. Cellulose is a polysaccharide of the form $(C_6H_{10}O_5)_n$, formed in plants from monosaccharides, e.g., glucose $C_6H_{12}O_6$, which is formed in turn by photosynthesis involving water and carbon dioxide:

$$6_2 + 6_2 \rightarrow_{6126} + 6_2 \uparrow$$

It is obvious that, during photosynthesis the hydrogen of glucose and cellulose being formed is in isotopic equilibrium with meteogenic waters of the region of the plants growth. Experiments show that cellulose extracted from plants by nitration and extraction with alcohol-benzol mixture is depleted by about 20‰ in deute-

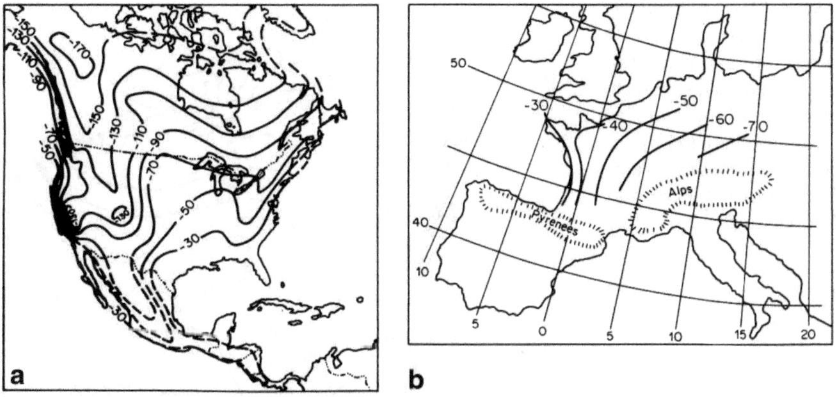

Fig. 9.12 Map of contours of the averaged δD values of recent meteoric surface waters: **a** for North America. (Taylor 1974). **b** for Western Europe. (Sonntag et al. 1979); Ferronsky and Polyakov 2012)

rium compared with water (Epstein et al. 1976). Therefore, the corresponding average isotopic composition of atmospheric precipitation may be reconstructed on the hydrogen isotopes content in plant's cellulose. Yapp and Epstein (1977), taking modern plants as an example, found a good relationship between δD values obtained indirectly from the hydrogen isotopic content in the cellulose of plants and direct measurements of δD in precipitation for various regions of the United States (Fig. 9.12). They showed experimentally that hydrogen isotopic exchange in C–H groups is extremely limited. This circumstance makes it possible to use variations in the hydrogen isotopic content in the fossil cellulose of plants for paleotemperature reconstructions. In particular, the climatic change on the North American continent during the last Pleistocene (Wisconsin) glaciation between 9500–22000 years B.P. has been estimated. The age of fossil plants was determined to be 22000 years ago and the relative deuterium content in atmospheric precipitation was 19‰ higher than the recent one. Yapp and Epstein explained this in terms of a gentle temperature gradient between the ocean and inland surface; cooling of the oceans; higher δD values for the surface oceanic waters; variation of the ratio of summer and winter precipitation; and a positive shift in the δD value in the oceanic vapour provided by the depleted influence of kinetic factors during evaporation of water at lower temperatures (Fig. 9.13).

The reconstruction of the isotopic composition of the glacial lakes Whittlesey and Aquassiz made it possible to estimate the average hydrogen isotopic content (and the average oxygen isotopic content calculated by Craig's equation) of the North American ice sheet. It has been found that δD values for Whittlesey Lake vary from − 89 to − 113 ‰ and $\delta^{18}O$ values vary from − 12 to − 15 ‰. From data obtained while studying the woods that grew on the shore of Lake Aquassiz the average hydrogen and oxygen isotopic content for this lake was found (δD from − 103 to − 129 ‰, $\delta^{18}O$ from − 14 to − 17 ‰). The average $\delta^{18}O$ value of − 15 ‰ for the North American ice sheet during the Wisconsin glaciation was estimated by Yapp and Epstein.

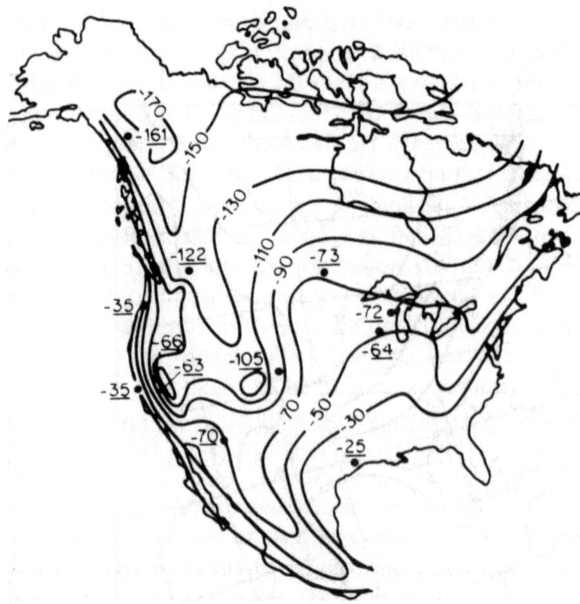

Fig. 9.13 Map of δD values of recent meteoric waters obtained for various locations of the United States by means of hydrogen isotopes in plant cellulose (points) and by direct measurements. (Yapp and Epstein 1977; Ferronsky and Polyakov 2012)

Answering the critical remarks of Wilson and Grinsted (1977), Epstein (1978) explained the marked deviations of the experimental data by the presence of OH-groups that readily exchanged isotopes of hydrogen during Wilson and Grinsted's isotopic analyses of cellulose. As an example of δD determination, Epstein made 19 cuts across the three annual rings of a New Zealand pine Pinus Radiata, which had grown between May 1915 and April 1917 and produced a satisfactory qualitative coincidence of the δD values in plant cellulose with seasonal variation of δD in modern atmospheric precipitation (1963–1967) for the northern region of New Zealand. Epstein made the following conclusions:

1. Variations of δD in the C–H groups of the plant's cellulose reflect hydrogen isotopic variations in atmospheric precipitation.
2. Wilson and Grinsted's statement that a biological thermometer with a temperature coefficient −5 ‰/1 °C permits estimation of variations of limiting temperatures with an error better than 0.1 °C is not supported by the available experimental data.
3. The method of analysis of hydrogen isotopic composition of cellulose, suggested by Wilson and Grinsted (1977), requires further verification.
4. Despite the fact that hydrogen isotopic content in plant cellulose gives agreeable qualitative coincidence with variations in atmospheric precipitation, a number of factors that are related to the kind of plants and the intensity of evapo-transpiration may influence the hydrogen isotopic fractionation, depending on the climatic conditions of the region where the plant is growing.

All these circumstances require further experimental studies, which should account for the influence of the above-mentioned factors upon the isotopic composition of the plant's cellulose.

9.4 Isotopic Composition of Continental Surface Waters

The isotopic composition of natural and artificial surface reservoirs is determined by several factors, the most important of which are (Gat et al. 1968; Merlivat 1970; Fontes 1976):

1. The isotopic composition and the amount of precipitation directly feeding the reservoir.
2. The isotopic composition of the surface and underground sources of water recharge and also the rate of water inflow.
3. The isotopic composition of atmospheric vapour moisture and also the air humidity.
4. The rate of water outflow and also the process of evaporation.

The mass balance of a reservoir during a period of time Δt may be given by the equation:

$$\Delta V = \sum I - \sum Q - E)\Delta t + P, \tag{9.30}$$

where

ΔV is the volume change of the reservoir
I is the component of the surface and subsurface water recharge
Q is the component of the surface and subsurface discharge
E is the mean rate of evaporation
P is the amount of precipitation during the time Δ*t*.

The isotopic balance equation in this case is:

$$R_L \Delta V + V \Delta R_L = \left(\sum R_I I - \sum R_Q Q - R_E E\right)\Delta t + R_P P, \tag{9.31}$$

where

R is the mean isotopic ratio value of the components of recharge and discharge.

The value *R* in Eq. (9.31) can be changed by value δ. Then, the equation acquires the form:

$$\Delta(\delta_L V_L)\Delta t = \sum \delta_I I - \sum \delta_Q Q - \delta_E E + \delta_P P.$$

It is known (Craig and Gordon 1965; Merlivat 1970) that a liquid's evaporation rate in the atmosphere may be described by the equation:

$$E = \kappa \frac{dC}{dz},$$

where

E is the rate of evaporation (vapour flux from the unit of liquid surface)
κ is the coefficient of turbulent diffusion
dC/dz is the vertical gradient of humidity.

For isotopic species of water ($H_2{}^{18}O$ or HDO) one has:

$$E_i = \kappa_i \frac{dC_i}{dz}.$$

The values of the isotopic fractions in vapour R_E in this case may be given by:

$$R_E = \frac{E_i}{E} = \frac{\kappa_i}{\kappa}\frac{dC_i / dz}{dC / dz}.$$

This equation in finite differences, after the coordinate dz is eliminated, obtains the form:

$$R_E = \frac{\kappa_i}{\kappa}\frac{R_L / \alpha - hR_\alpha}{I - h}, \quad (9.32)$$

where

h is the relative air humidity at altitude z above the water surface at a given temperature
R_L and R_a are isotopic ratios for liquid and atmospheric vapour, respectively
α is the equilibrium fractionation factor.

In the expression (9.32) it is accepted that:

$$\Delta C = 1 - h, \Delta C_i = R_i / \alpha - hR_\alpha.$$

Designating $\kappa/\kappa_i = k$, Eq. (9.32) may be written as (Gat et al. 1968):

$$R_E = (R_L / \alpha - hR_\alpha) / k(1 - h). \quad (9.33)$$

Equations (9.32) and (9.33) give an idea about the parameters that determine the isotopic composition of the vapour flux. As Brezgunov (1978) pointed out, these formulae are realistic in the physical sense only when both fluxes have the same sign, i.e., when either evaporation or condensation of water and isotopic species occurs. If the fluxes have different signs, they should be considered separately while estimating the isotopic and mass balance of water in the reservoir.

An additional atmospheric resistance effect of fractionation of isotopes during water evaporation in the atmosphere, being defined by humidity and the parameters of the turbulent diffusion of vapor, Craig and Gordon (1965) evaluated by:

$$\Delta\varepsilon = (\kappa / \kappa_i - 1)(1 - h).$$

The isotopic composition of evaporating water expressed in δ-values after such a transformation as described in Eq. (9.32) will be as follows:

$$\delta_E = \frac{\alpha^* \delta_L - h\delta_\alpha - \varepsilon^* - \Delta\varepsilon}{(1-h) + \Delta\varepsilon}, \tag{9.34}$$

where

$\alpha^* = 1/\alpha$, $\varepsilon^* = 1 - 1/\alpha$; δ_L, δ_a are the isotopic composition of liquid and vapour.

Equation (9.34) can be conveniently used to obtain an equation describing the isotopic balance of reservoirs. To obtain a precise value of δ_E it is necessary to measure the parameters δ_L, δ_α and h directly and know the values of ε^* and $\Delta\varepsilon$. The value ε^* can be calculated easily with the help of an equilibrium factor at given temperatures. The Δε value has been estimated in a number of experimental and theoretical works that were estimated in detail in Brezgunov's review paper (1978). According to experimental data, $\Delta\,\varepsilon_{18_O}$ is dependant on humidity:

$$\Delta\varepsilon_{18_O} = (1-h)16‰.$$

At constant humidity, the ratio $\Delta\varepsilon_D/\Delta\varepsilon_{18_O}$, as shown by the experimental and theoretical estimations of Craig and Gordon (1965) and Merlivat (1970), ranges from 0.5 to 0.85. But in a number of experiments (Craig and Gordon 1965; Gat 1970) $\Delta\varepsilon_D$ exceeds $\Delta\varepsilon_{18_O}$ by 2–4 times. Such deviations between experimental and theoretical values were explained by Craig and Gordon in terms of extra isotopic effects at the boundary of the liquid-vapour system, which might be provided by differences in the condensation (vaporisation) coefficients for isotopic species of the molecules. The values of the parameter $k = \kappa/\kappa_i$ in Eq. (9.33) have been found, experimentally, to range between 1.016 and 1.020 for oxygen-18 and 1.009 and 1.0136 for deuterium (Merlivat 1970). Gat et al. (1968) considered particular values on the basis of work by Craig and Gordon (1965) to be $k_D = 1.009$ and $k_{18_O} = 1.016$. But such an assumption, as was noted above, is based on the fact that the condensation coefficient for isotopic molecules does not undergo visible variations.

It follows from Eq. (9.32) that the isotopic composition of water vapour above the reservoir is determined both by the isotopic composition of evaporated water and by the isotopic content of atmospheric moisture. If under natural conditions the evaporation occurred in accordance with Rayleigh's distillation law, then water and vapour should be in isotopic equilibrium at any moment. In this case the isotopic composition of vapour over the reservoir is equal to $R_E = R_L/a$ (R_L is the isotopic composition of the reservoir water at time t) and the isotopic composition of water during evaporation would be given by the Rayleigh formula (Brodsky 1957):

$$R_L = R_o (V_o / V)^{(\alpha-1)/\alpha} = R_o (V / V_o)^{(\alpha-1)/\alpha}, \tag{9.35}$$

where

R_o and R_L are the ratios of the isotopic species of water initially and at time t
V_o and V are the initial and final volumes of water.

Taking the logarithms of Eq. (9.35), we obtain:

$$\ln R_L = \ln R_o + \varepsilon \ln F,$$

where

ε $1/\alpha - 1$
F V/V_0.

Substituting δ-values for R and taking into account that $\ln(1+\kappa) \approx \kappa$ at $\kappa \ll 1$, one obtains:

$$\delta_L - \delta_0 = \varepsilon \ln F.$$

In other words, during Reyleigh's evaporation the expression for $(\delta_L - \delta_0)$ becomes a linear function of ln*F*. This is convenient for analysing experimental data while studying the change in a liquid's isotopic composition during evaporation.

If under conditions of Reyleigh's evaporation the volume of the reservoir V_0 remains unchanged due to recharge, then the isotopic composition of the water at any time may be found from the equation (Brodsky 1957):

$$\alpha \ln \left[(\alpha - 1) / \alpha - R_L / R_O \right] = V / V_o. \tag{9.36}$$

The maximum enrichment, which might be attained in this case at $V \rightarrow \infty$, is $R_0 \alpha$. Under real conditions changes in the isotopic composition of water in the drying reservoir do not follow Reyleigh's law, because the thermodynamic and isotopic equilibrium between water and vapour here does not occur.

As shown in the previous paragraph, in the case of thermodynamic disequilibrium in a system the rate of transition of molecules HDO and $H_2{}^{18}O$ from the liquid into the gaseous phase will be determined by the difference in the diffusion coefficient of the molecules H_2O, HDO and $H_2{}^{18}O$. In contrast to the processes of equilibrium fractionation when the fractionation factors in the water-vapor system are 1.08 for H_2O–HDO and 1.009 for H_2O–$H_2{}^{18}O$, the difference in the coefficients in the case of diffuse separation is not so great. For the above-mentioned pairs of molecules the coefficients are 1.009 and 1.016, respectively.

The observed relationship between deuterium and oxygen-18 content in precipitation, expressed by the equation $\delta D = 8\delta^{18}O + 10$ ‰, is evidence that moisture condensation occurs under conditions that are sufficiently close to those at equilibrium. Under these conditions the deuterium and oxygen-18 fractionation factors might be assumed to be equal to those for systems in equilibrium: 1.08 for deuterium and 1.009 for oxygen-18 at 20 °C.

During evaporation under non-equilibrium conditions the difference between the deuterium and oxygen-18 content is limited not only by the value of the fractionation factors but also by the ratio of the diffusion rates of the molecules: the coefficient $k = \kappa_i / \kappa$ in Eq. (9.32). This leads to a gentler sloping curve of δD against $\delta^{18}O$ for water in open reservoirs.

As Dansgaarg (1964) pointed out, in this case, the fractionation factor may be given by the expression $\alpha^k = \alpha k$. Hence, $\alpha_D^k = 1.08 \cdot 1.009 = 1.09$; $\alpha_{^{18}O} k = 1.009 \cdot 1.016 = 1.025$.

The value of the ratio $d\delta D/d\delta^{18}O$ may be obtained by differentiating Eqs. (9.35) and (9.36), which are similar to the relationship described by (9.35), since $V/V_o = F$ and $\delta = R_L - 1$ since $R_o = 1$:

$$d\delta D / d\,\delta^{18}O = (\varepsilon_D / \varepsilon_{^{18}O}) F^{\varepsilon_D - \varepsilon_{^{18}O}}, \tag{9.37}$$

where

ε (!−α)/α.

Equation (9.37) can be approximated by:

$$d\delta D / d\delta^{18}O \approx \varepsilon_D / \varepsilon_{^{18}O}, \tag{9.38}$$

since the index $\varepsilon_D - \varepsilon_{^{18}O}$ differs only slightly from zero and at a given temperature is a constant.

The slope of the line describing changes in deuterium and oxygen-18 concentrations in a solution during Rayleigh's evaporation (at t=20 °C $\alpha_D = 1.08$ and $\alpha^k_{^{18}O} = 1.009$) will be:

$$d\delta D / d\delta\,^{18}O \approx (\alpha_D - 1)\,\alpha_{^{18}O} / (\alpha_{^{18}O} - 1)\,\alpha_D \approx 8. \tag{9.39}$$

Under non-equilibrium conditions of evaporation, when kinetic factors are important (at 20 °C $\alpha_D^k = 1.09$; $\alpha^k_{^{18}O} = 1.025$), one has:

$$d\delta D / d\delta^{18}O \approx (\alpha_D^k - 1)\,\alpha^k_{^{18}O} / (\alpha^k_{^{18}O} - 1\,\alpha_D^k) < 8. \tag{9.40}$$

It follows from formulae (9.39) and (9.40) that, for non-equilibrium water vaporisation from a closed reservoir, the values of δD and $\delta^{18}O$ lie below Craig's line plotted for atmospheric precipitation, on a line with a smaller gradient (Fig. 9.14).

At a steady state, when no significant changes in the volume of the reservoir occur, the isotopic composition of well-mixed water tends to a constant value:

$$R_L = \frac{(1-h)R_I s + hR_a/k}{1/\alpha k + (1-h)(1-s)}, \tag{9.41}$$

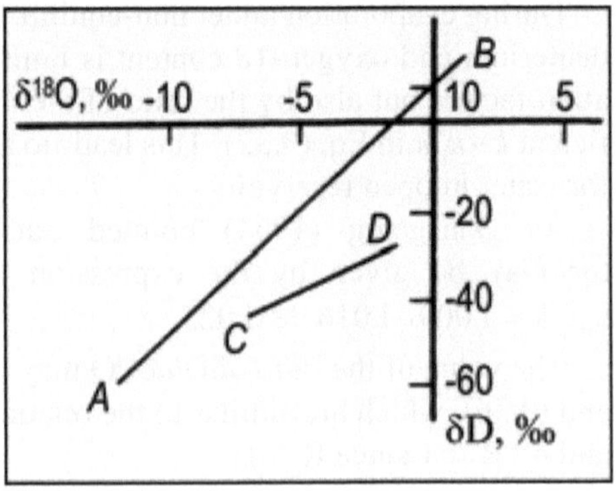

Fig. 9.14 Schematic relationship between δD and $\delta^{18}O$ values for precipitation (AB) and for closed continental basins (CD). Ferronsky and Polyakov 2012

where

s is the ratio of the mean rates of recharge and evaporation.

Expression (9.41) is only true if short time variations of the parameters α, R_I, R_a, h permit one to consider their mean values to be constant in time.

Assuming that water loss from the reservoir takes place only due to evaporation, i.e., $s = I/E = 1$, then Eq. (9.41) becomes:

$$R_L = \alpha\, k[(1-h)\, R_I + h(R_a / k)]. \tag{9.42}$$

Eq. (9.42) gives the value of maximum isotopic enrichment of reservoirs with constant volume. It is obvious that at $h = 0$ (for evaporation when air humidity equals zero) $R_L = \alpha_k R_I$. This expression corresponds to the maximum isotopic enrichment for evaporation of a liquid at a constant level during Rayleigh's distillation where α is substituted by $\alpha k = \alpha^k$.

If the level of the reservoir is reduced only by evaporation and recharge does not occur ($I = 0$), then the integration of Eqs. (9.30) and (9.31) under these conditions yields the following expressions for the isotopic composition of the reservoir as a function of V/V_o:

$$R_L = \frac{[R_0(1/\alpha k - 1 + h) - hR_\alpha/k](V/V_0)^{(1/\alpha k - 1 + h)/1-h} + hR_\alpha/k}{1/\alpha k - 1 + h}. \tag{9.43}$$

The form of Eq. (9.43) is similar to the formula for Reyleigh's distillation (9.35). But, in this case, the kinetic factors, the isotopic composition of the atmospheric vapour and the air humidity should be taken into account.

In drying reservoirs, that is when $(V/V_o) \to 0$, the final isotopic composition of water can be found from the expression:

$$R_L = \frac{hR_\alpha / k}{1/\alpha k - 1 + h}. \tag{9.44}$$

It follows from the analysis of Eqs. (9.43) and (9.44) that under natural conditions of an existing reservoir the isotopic composition of its water does not increase to infinity, as Rayleigh's Eq. (9.35) suggests but tends to some stationary condition determined by the parameters α, k, h and R_α (see Figs. 9.18 and 9.20).

For $h=1$, when the reservoir is in thermodynamic equilibrium with the vapour, the isotopic composition R_α, which is independent of the isotopic composition of the reservoir, has the form:

$$R_L = R_\alpha \alpha. \tag{9.45}$$

Under such conditions, the water in the reservoir attains isotopic equilibrium with the atmospheric moisture. It follows from formula (9.41) that the isotopic composition of water is governed both by evaporation and by recharge of water having an isotopic composition R_I. Therefore, for seas where rivers play an important role in the formation of water bodies, deuterium and oxygen-18 content is lower than that in the ocean. So for the Black Sea $\delta D_{SMOW} = -18.8$ ‰ and $\delta^{18}O = -3.15$ ‰.

As earlier noted, by experimental results and natural observations it was found that the isotopic composition δ_L of an evaporated reservoir in natural conditions tends to some stationary state δ_L^S (in the δ-values). It was shown in the works of Craig and Gordon (1965), Gonfiantini (1965) and Fontes et al. (1979) that the value δ_L^S for a basin with a stationary level is determined by the equation:

$$\delta_L^S = \frac{[(E/I)h\delta_\alpha + (\alpha = h)\delta_I(\varepsilon_v + 1)] + (E/I)\alpha\,\varepsilon_V}{(\alpha - h)(I - E/I)(1 + \varepsilon_V) + \alpha(E/I)}.$$

For drying lakes one has:

$$\delta_L^S = \frac{\alpha\varepsilon_V + h\delta\alpha(1+\varepsilon_V)}{h(1+\varepsilon_V) - \alpha\,\varepsilon_V},$$

where

$\varepsilon_V = \varepsilon^* + \Delta\varepsilon$ is the total effect of enrichment of the evaporated water
a is activity of water in the solution.

The kinetic term $\Delta\varepsilon$ depends on the air relative humidity and the water activity and for $\delta^{18}O$ it is described by the following dependency (Fontes et al. 1979): $\Delta\varepsilon = [14.4(a-h)/a] \cdot 10^{-3}$. The δ-value, substituted into above formula, is expressed in parts of the unit ($\delta = \delta_{measured} \cdot 10^{-3}$).

If the lake discharged water Q then in stationary state for full mixing lakes $\delta_Q = \delta_L^S$. In this case the total water recharge and the evaporating flux have the following relationship: $I = E(\delta_L^S - \delta_E)/(\delta_L^S - \delta_I)$ (Zimmermann 1979).

If the lake volume and also the hydrological and climatic parameters can be accepted constant then the isotopic composition of the lake water in time comes to the stationary state in accordance with the equation (Zimmermann 1979):

$$\ln(\delta_L^S - \delta_L) = (1/\tau)t + \ln(\delta_L^S - \delta_Q),$$

where

- τ is the mean value of the residence time of water in the lake, which is determined by relation V/I
- V is the reservoir volume
- I is the total water recharge that includes the surface and ground inflow and atmospheric precipitation over the lake surface.

For the reservoir that does not reach steady state in isotopic content the value τ can be determined by the time dependence relation ln $(\delta_L^S - \delta_L) = -(1/\tau)t + \ln(-\delta_L^S - \delta_Q)$.

If one plots dependence ln $(\delta_L^S - \delta_L)$ with reference to time then the point should be put on a straight line with the slope $1/\tau$.

The combining of water (9.30) and isotope (9.31) balance, at the assumption that $\delta_I = \delta_P$, gives the following relationship between the rate of the water evaporation E and the total discharge from the lake (Hübner et al. 1979a):

$$E = Q(\delta_L - \delta_Q)/(\delta_E - \delta_I).$$

This equation makes it possible to calculate the values E and Q by one known parameter and experimental values δ_L and δ_Q. In this case the value δ_E can be calculated by formula (9.34). The isotope methods have wide application in the study of lake water balance. The methods are used for indirect assessment of the lake underground recharge and discharge. As a rule, determination of these parameters by traditional hydrological means has difficulties. The examples of such solutions can be found in the IAEA proceedings on Isotopes in Lake Studies (IAEA 1979b).

The most complete regional survey of characteristic deuterium contents in surface continental waters is given by Friedman et al. (1964) and for Canada by Brown (1970). The isotopic composition of water in rivers, lakes and other reservoirs is dependent on the isotopic composition of atmospheric precipitation feeding these reservoirs. The surface continental waters are less subject to isotopic variations with time compared with atmospheric precipitation and springs and wells that do not dry up are the most reliable sources of water for the determination of isotopic ratios characteristic of a given region.

Figure 9.15 shows the regional characteristic of deuterium concentrations in Mississippi river water and the East Atlantic coast of the United States (Friedman et al. 1964). The basin of the Mississippi river covers a considerable area of the eastern region of the USA, characterised by a comparatively low height above sea level and wide ranges of temperature, which determine the initial isotopic composition of atmospheric precipitation. Figure 9.15 shows the isolines of annual snow cover well correlated with relative deuterium content in the surface water. As expected, an even closer correlation is observed between the relative deuterium content and the ratio of the amounts of snow and rain falling down as atmospheric precipitation. In

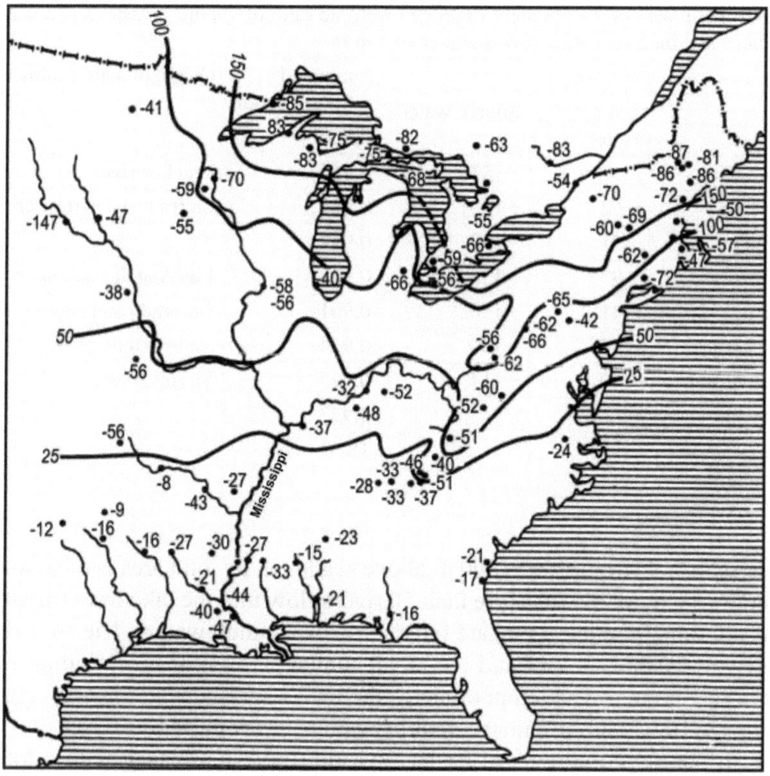

Fig. 9.15 Map of δD values for the Missouri–Mississippi river basin and East Atlantic coast of the United States plotted from the data of 1956–1959. Contours show thickness of snow cover in cm. (Friedman et al. 1964; Ferronsky and Polyakov 2012)

this case, increase in the proportion of snowfall results in a decrease of the deuterium content and vice versa.

Figure 9.15 may serve as an example of regional deuterium distribution. Equation (9.42) may be used as a first approximation in obtaining a picture of oxygen isotope distribution.

The influence of evaporation on the isotopic composition of basins may be demonstrated by investigating closed and particularly saline lakes (Table 9.4). According to theoretical data obtained for this group of lakes, fractionation should occur at 18 °C. But as the authors themselves pointed out, their estimations are imprecise since the initial data on the isotopic composition of the inflow water from tributaries are not accurate enough, the water losses due to underground discharge have not been taken into account and the water in the lakes does not attain complete mixing.

Another example of elucidating the relationship between lake water, sources feeding them and atmospheric vapour is the investigation of hydrogen and oxygen isotope distribution in the basin of the Issyk-Kul Lake (Brezgunov et al. 1979; Brezgunov and Nechaev 1981). The Issyk-Kul Lake is located in a high mountain valley

Table 9.4 Estimates of the apparent hydrogen isotopic ratio of vapour pressures and water from saline lakes and their recharge. (Friedman et al. 1964)

Lakes	δD, ‰		Estimated ratio of vapour pressure	Origin of source water
	Lake water	Source water		
Piranid lake	−30	−87	0.942	Truckee river
Mono lake	−62	−130	0.927	Sierra Nevada near crest
Soda lake	−51	−130	0.918	
Owens lake	−97	−135	0.958	Los Angeles aqueduct
Salton sea	−41	−136	0.901	Colorado and Alamo rivers
Great Salt lake	−89	−130	0.934	Yellowstone plateau
Harney lake	−40	−114	0.923	Silvies river
Mean value			0.923	
Corresponding temperature			18 °C	

in Tien Shan at an altitude of 1600 m above sea level. Its total area is 6200 km^2 with an average depth of 280 m. More than 50 rivers flow into the lake from the slopes of the surrounding mountains and are fed mainly by ice melt waters. The annual runoff of the rivers is about 4 km^3 and the mean annual amount of precipitation is about 250 mm (~ 1.5 km^3). The temperature of the lake's surface water in July–August is equal to 18–19 °C and in January–February it never drops below +2°C.

Due to natural conditions of its location the Issyk-Kul basin, with a total area of about 22,000 km^2, a surface that never freezes and being a source of intensive evaporation (condensation of atmospheric moisture is localised on the slopes of the mountain ridges surrounding the lake), can be considered as a unique natural model of the global circulation of waters.

The authors carried out in the valley from 1974 systematic investigations of the isotopic content in the lake and river waters, sources of groundwaters, atmospheric precipitation, glaciers and water vapour.

The relationship of natural waters of the Issyk-Kul basin is given in Fig. 9.16, plotted in δD–δ^{18}O coordinates, drawn from the data of June 1975. Here the experimental points may be subdivided into three groups. The first group (points 1, 2, 3) corresponds to lake water, characterised by the greatest values of δD and δ^{18}O and the minimum value of parameter b in Eq. (9.41). Close to the line for meteoric waters lie points 9, 10, 11, which correspond to the isotopic composition of river water. Points 4, 5, 6, 7 belong to the third group and lie between those of the first two groups; they correspond to the isotopic composition in precipitation, measured at the coastal stations and forming a sufficiently independent set. Far from this set of points lies point 8, which corresponds to the isotopic content in precipitation at the high mountain station of Bolshaya Kyzyl-Su. Point 12, corresponding to the isotopic composition in atmospheric precipitation obtained in December 1975 for the same station, makes it possible to compare the range of seasonal variations of hydrogen and oxygen isotopic content for the same region. Note that the mean

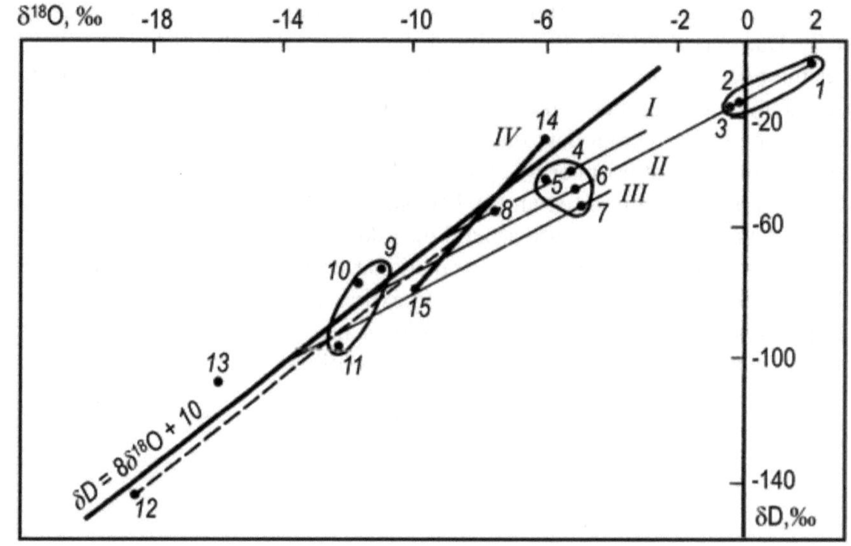

Fig. 9.16 Plot of δD versus $\delta^{18}O$ for natural waters sampled in June 1975 in the Issyk-Kul Valley. (*1*) Kara Kel Lake water; (*2–3*) Issyk-Kul Lake water; (*4*) precipitation from Cholpon-Ata Station; (*5*) precipitation from Pokrovka Station; (*6*) precipitation from Przhevalsk Station; (*7*) precipitation from Rybachye Station; (*8*) precipitation from Bolshaya Kyzyl-Su Station; (*9, 10*) Chon Kyzyl-Su River water; (*11*) Dzhergalan River water; (*12*) precipitation from Bolshaya Kyzyl-Su Station in December 1975; (*13*) calculated values of 'lake' vapour isotopic composition; (*I, II, III*) evaporation lines with a slope of 5; (*IV*) condensate 'mixture' line reflecting the 'lake' and 'regional' vapour mixture degree. (Brezgunov et al. 1979; Ferronsky and Polyakov 2012)

δD and $\delta^{18}O$ values calculated for samples obtained in June and December for the Bolshaya Kyzyl-Su station lie in the domain corresponding to the river water in the δD–$\delta^{18}O$ diagram.

The data given in Fig. 9.16 visibly reflect the process of hydrogen and oxygen isotopic fractionation. Here, the relative values of D and ^{18}O content in waters of the Issyk Kul Lake are considerably greater than in the rivers that are the major sources of the lake water. This is due to evaporation. The straight line in the diagram, connecting data for the river and lake waters, has a slope of about 5, which is characteristic for the line of evaporating water. Maximum values of δD and $\delta^{18}O$ (point 1) are found for the Kara Kel Lake water. The closed lake is small (area is equal to 0.5 km^2) and shallow (depth is ~10 m) and is situated 500 m from the southern shore of the Issyk-Kul Lake. The excess of deuterium and oxygen-18 in the water of this small lake reflects its high rate of water exchange.

On the right of the meteoric water line in the diagram there are points that characterise the isotopic composition of precipitation in the coastal region and for which the parameter b (Eq. 9.41) is negative. Brezgunov et al. (1979) explained this phenomenon in terms of the evaporation of raindrops in the atmosphere below the cloud layer, resulting in enrichment in heavy isotopes of atmospheric precipitation taken near the surface. The reduction of the parameter b to negative values in the summer precipitation is a feature of most high latitude stations, which was noted first by Dansgaard (1964). It was found that, with other things being equal, the lower the relative humidity of air below the cloud layer, the more the rainwater is enriched in heavy isotopes and lower the values of parameter *b* become.

Using the experimental data on the isotopic composition of water in lakes and rivers and those of atmospheric vapour from the five stations situated around the Issyk-Kul Lake, Brezgunov et al. (1979) evaluated hydrogen and oxygen isotopic content in atmospheric moisture around the lake. In order to carry out the calculations they used the model of isotopic fractionation by interaction of the reservoir with the atmosphere and the corresponding expressions (9.34).

Expression (9.34), representing the ratio of heavy isotope flux to the flux of moisture, may be written in terms of δ values as follows:

$$\delta_E = \frac{(\delta_L/\alpha) - \delta_A h - \varepsilon - \Delta\varepsilon}{1 - h + \Delta\varepsilon},$$

where

δ_E, δ_L and δ_A correspond to the isotopic composition in the evaporating flux of moisture, reservoir water and atmospheric moisture

ε and $\Delta\varepsilon$ are values, expressed in δ units, related to equilibrium and kinetic fractionation factors.

For the stationary evaporation process and the model of the reservoir with a constant level, $\delta_E = \delta_I$ (where δ_I is the heavy isotopic content in the inflow sources).

For a reservoir with a constant level, such as the lakes in question, for known δ_L, δ_I values and air humidity above the reservoir and known values of the equilibrium kinetic fractionation factors from the previous expression, it was found that:

$$\delta_A = \frac{\delta_L - \delta_I(1 - h) - \varepsilon - \Delta\varepsilon}{h}.$$

According to this expression, in which the values of ε and $\Delta\varepsilon$ were derived from Dansgaard (1964) and Merlivat (1970), the values $\delta D = -108\,‰$ and $\delta^{18}O = -16\,‰$ were found to be close to the corresponding values of the water vapour determined experimentally.

In addition, by analysis of the data on the isotopic composition of precipitation in the investigated valley and water vapour above the lakes, the atmospheric vapour mixing proportions were estimated. The position of the intersection points of the evaporating lines I, II and III in Fig. 9.16, which have a slope equal close to 5 and which pass through the points with isotopic content of precipitation with mixing line IV, made it possible to estimate the proportion of these two components in the atmospheric vapour (the regional component and that component removed from the lake) in the precipitation that fell in June 1975 in different locations of the Issyk-Kul valley. It was found that in the western part of the valley (Rybachye station) the precipitation was almost completely formed of water vapour from outside the region.

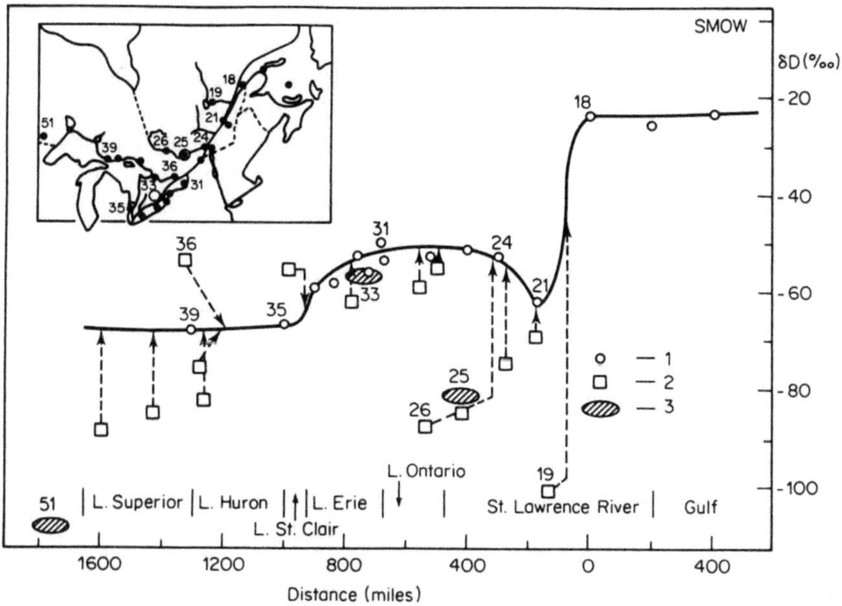

Fig. 9.17 Deuterium concentrations along the Great Lakes–St. Lawrence system in October 1967. Numbers refer to the points of investigation. (*1*) main stream; (*2*) tributaries; (*3*) precipitation. (Brown 1970; Ferronsky and Polyakov 2012)

Precipitation in the eastern part of the valley (Przhevalsky station) was composed of 20 % from lake vapour and 80 % from regional vapour. The maximum contribution of lake vapour (up to 40 %) was found in the central part of the valley at stations Cholpon-Ata, Pokrovka and Bolshaya Kyzyl-Su.

Variation of isotopic composition of the river water due to evaporation may be observed when examining, as an example, the Nile River (Friedman et al. 1964). During flood periods the relative deuterium content in river water near Khartoum is −23 ‰ and near Cairo is −21 ‰. By the calculations of Friedman, the data correspond to an R/R_0 value of 1.002 and a value of 0.97 for the fraction of remaining water. Thus, the Nile appears to have lost about 3 % of its water by evaporation between Khartoum and Cairo (~3000 km). At low water these values are +15 ‰ and +16 ‰ correspondingly. Again, evaporation loss is about 3 %. In the opinion of Friedman the difference in δ values reflects the fact that at high water the Blue Nile is the dominant water source and at low water the Wight Nile predominates.

Another example, illustrating conditions of isotopic composition formation and the influence of evaporation in a large continental water system, is the Great Lakes basin and that of the St. Lawrence River in North America.

Figure 9.17 shows the data on deuterium content in the St. Lawrence River and in the main tributaries of the system and also the data on deuterium content in precipitation, obtained by Brown (1970). As the author pointed out, the deuterium concentration in this water system is the highest compared with all the other reservoirs of water in Canada due to intensive evaporation of water and its enrichment with heavy isotopes. Within the range of these lakes there is a slight (up to 10‰) enrichment in deuterium when water flows from one lake to another. Down the river course the deuterium content drops due to the dilution of its water with higher waters from the tributaries and also due to the discharge of groundwaters of the valley. After Quebec (see Fig. 9.17, after point 21) the deuterium content increases due to contributions of seawaters from the gulf mixing with the river water. Further (after point 18), some variations of isotopic composition are observed, being lower than those in the ocean water.

The following main principles of isotopic composition formation of continental surface waters in a regional scale should be pointed out. Since surface continental waters are represented by a number of separate regional basins in the framework of which accumulation, mixing and runoff of atmospheric precipitation in the ocean occur, the isotopic content of surface inland waters reflects some average regional isotopic ratios in precipitation during a certain period of time. The isotopic composition of water of the basin may reflect both average annual isotopic ratios and seasonal variations in precipitation, which depend upon the climate zone, size of the basin and the supply conditions. A rather important factor in the formation of the isotopic composition of water is the evaporation of water from the basin and the catchment area. This process has particular importance for conditions in lakes where heavy isotope content is always higher than in the source.

The latitudinal and altitudinal effects, which are a feature of isotopic variations in precipitation, are also valid for the surface waters. In high latitude rivers and lakes, waters are depleted in heavy isotopes compared with those in low latitudes. Rivers and lakes of arid zones, being fed by water from high mountain glaciers and snow melting, are distinguished by lower values of heavy isotopes. In contrast, the drying rivers and lakes of arid zones have excessive heavy isotopic contents.

9.5 Isotopic Composition of Water in Evaporating Basins

The problem of deuterium accumulation in closed evaporating basins connected genetically with ocean waters is of great theoretical and practical interest. The study of deuterium behaviour during the evaporative concentration of seawater is important for answering paleogeological questions, since the formation of thalassogenic sedimentary waters might occur not only from basin waters with normal salinity but also from concentrated lagoon brines, shallow bays, etc. If the subsurface waters of connate marine basins can be adequately identified by their 'marine' deuterium content, the history of sedimentary waters formed from waters of saline basins becomes ambiguous.

As shown above, the accumulation of deuterium in nature does not follow Rayleigh's equation but is complicated by kinetic factors and by isotopic exchange between liquid and atmospheric moisture vapour. Whilst studying the behaviour of heavy isotopes of hydrogen and oxygen during the evaporation of seawater Gonfiantini (1965) found an inversion of the isotopic content in the water–vapor system when the water body has decreased by ten times (Fig. 9.18). He correctly explained this phenomenon in terms of the increasing influence of the condensing vapour of atmospheric moisture when the rate of evaporation of the solution decreases by a reduction in the water activity provided by increases in the concentration of salts.

While studying the process of accumulation of heavy isotopes of oxygen during the evaporation of seawater, Lloyd (1966) came to the conclusion that changes in the isotopic composition of liquids may be explained by Sverdrup's evaporation theory that, unlike the equilibrium evaporation theory, assumed the presence of diffuse and mixed layers (Fig. 9.19a, b). A somewhat modified scheme of Lloyds is pictured in Fig. 9.19c, illustrating the influence of the isotopic composition of vapour of a mixed layer on the formation of isotopic composition of evaporating water. If the content of HDO in the mixed layer is greater than, or equal to, their equilibrium concentration with respect to the evaporating liquid (which, at 20 °C, corresponds to a depletion in the deuterium vapor by approximately 80 ‰ in comparison with the liquid), then the outflow of HDO vapour through the diffuse layer occurs less intensively than the outflow of H_2O vapour. In this case Rayleigh's process would predict.

The change in concentration of heavy isotopes in vapour of mixing layers (the degree of deviation of this concentration from the equilibrated value) provides for deviations in the accumulation of deuterium in evaporating water on both sides of the line described by Rayleigh's equation (Fig. 9.19c).

In natural conditions the deuterium concentration in a mixed layer (in atmospheric moisture), as a rule, is appreciably lower than that in equilibrium, which results in a negative deviation from Rayleigh's law. The influence of the isotopic composition of atmospheric moisture vapour upon the isotopic composition of evaporated water is clearly illustrated by Lloyd's scheme. But according to this scheme it is practically impossible to explain inversions in the changes of isotopic composition being conditioned by the increase of saline concentration in the solution. Undoubtedly the rate of evaporation and the isotopic composition of atmospheric moisture are important factors in the formation of the isotopic composition of evaporating liquids (Craig et al. 1963; Gonfiantini 1965; Sofer and Gat 1975; Polyakov and Seletsky 1972).

The isotopic composition of the atmospheric moisture can be considered in laboratory experiments to be independent of the isotopic composition of the evaporating liquid. This is also true within certain assumptions for natural evaporating basins (e.g., where the lagoon is isolated from the sea). In this case the evaporating rate of the liquid will be proportional to the difference between the equilibrium pressure of vapour above the liquid and the vapour pressure of the atmosphere:

$$G = k(p_l - p_a), \tag{9.46}$$

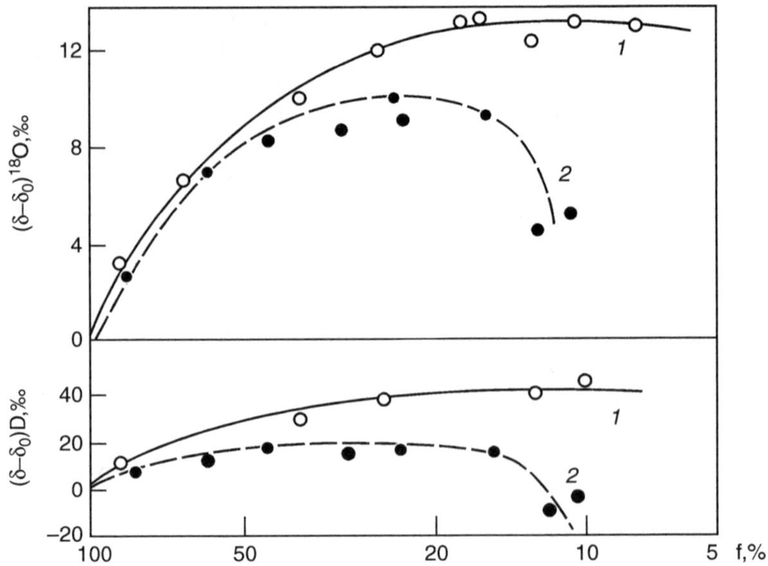

Fig. 9.18 Deuterium and oxygen-18 concentration changes at evaporation of distilled (*1*) and normal sea (*2*) waters. (Gonfiantini 1965; Ferronsky and Polyakov 2012)

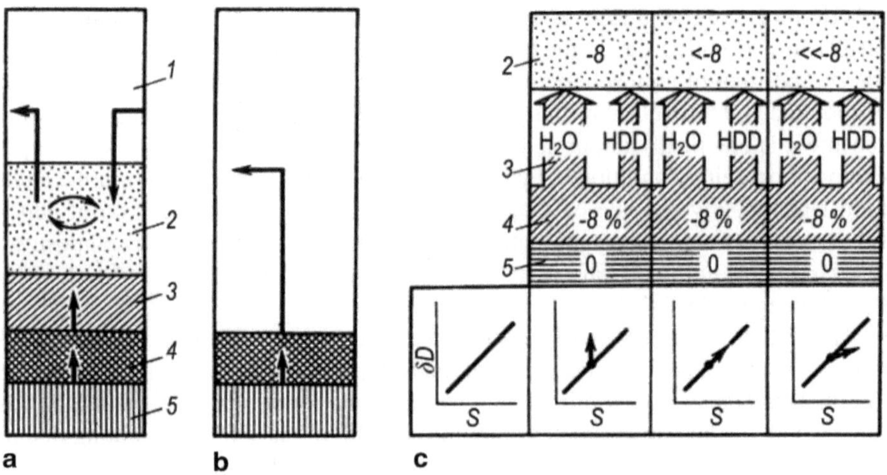

Fig. 9.19 Sverdrups (**a**), equilibrium (**b**) and deviating from equilibrium (**c**) models of water evaporation: (*1*) atmosphere; (*2*) mixed layer; (*3*) diffusion layer; (*4*) equilibrium layer; (*5*) liquid water. (Ferronsky and Polyakov 2012)

where

G is the rate of liquid evaporation
p_1 is the equilibrium pressure of vapour above the liquid
p_a is the vapour pressure of the atmosphere
k is the coefficient of proportionality.

According to Raule's law $p_1 = p_1^0 a$, (where где p_1^0 is the activity of water in the solution). If we assume that $p_a = p_1^0 h$ (where h is relative humidity), then:

$$G = kp_1^0 (a - h)k\ (p_1/p_1^0) - h. \tag{9.47}$$

The activity of water in a solution remains constant during the evaporation of pure water or saturated solutions ($a = 1$ for pure water and $a \approx 0.8$ for saturated solution of NaCl).

If the evaporation has led to an accumulation of salts in the reservoir to such an extent that the vapour pressure above the liquid equals the pressure of the vapour of the atmospheric moisture ($a = h$), then the final isotopic composition of the reservoir water by Eq. (9.44) will be equal to $R_L = R_a a$, i.e., the reservoir will be in equilibrium with atmospheric moisture. If water loss from the reservoir is due only to evaporation and the water body is decreasing then the isotopic composition of the remaining liquid can be calculated by the equation:

$$R_L = \frac{[R_0(a/\alpha k - a + h) - hR_\alpha / k](V/V_0)^{(a/ak-a+h)/(a-h)} + hR_\alpha / k}{a/\alpha k - 1 + h}, \tag{9.48}$$

where

a is the activity of water being changed by evaporation
V/V_o is the remaining body fraction of the solution
V_o is the initial volume
R_o is the initial isotopic ratio of the liquid.

In the limiting case of $V/V_o \to 0$ the final isotopic composition of water will be equal to:

$$R_L = \frac{hR_a / k}{a/\alpha k - a + h}. \tag{9.49}$$

This formula is similar to (9.44) but the former takes into account the activity of water in the solution, where the value of $a \neq 1$.

It follows from Eq. (9.48) that the limiting isotopic composition obtained by evaporating distilled water ($a = 1$) and saturated salt solution ($a < 1$) is not the same as that discovered by experimentation. If water activity in the solution during the experiment (during the seawater evaporation) is not a constant, then the rate of evaporation by Eq. (9.47) will be equal to zero.

Putting the value $a=h$ into Eq. (9.49), one obtains:

$$R_L = aR_a, \tag{9.50}$$

i.e., the limiting isotopic composition of liquid with respect to vapours of atmospheric moisture will be at equilibrium. This result is similar to that obtained from Eq. (9.42) since in both cases it is assumed that the water reservoir is in dynamic equilibrium with atmospheric moisture vapour.

It follows from seawater evaporation experiments (Fig. 9.20) that the isotopic composition of water being evaporated under certain conditions may undergo inversion. This is caused by a decrease in the evaporation rate due to decreasing activity of water in the liquid phase (Polyakov and Seletsky 1972). When $V \rightarrow 0$ the deuterium content tends to a stationary value at which $R_L = aR_a$.

In natural evaporating basins with initially normal seawater salinity, the consequent sedimentation of minerals and the increase in the salt concentration in the residue of the solution occur due to the evaporation of water. The changes in the composition of salts of seawater at different stages of concentration are given in Table 9.5 (Galakhovskaya 1967).

It follows from Table 9.5 that at the stage of halite sedimentation a gradual change in the water salts composition occurs (the sodium chloride water transforms into magnesium chloride water). If evaporation continues the water activity also changes (from 0.8 at the stage of halite sedimentation to ~0.6 at the stage of bischofite sedimentation), resulting in a decrease in the evaporation rate of the brine at the final stage of evaporation. Under conditions of high humidity this process might cause a certain decrease in deuterium concentration in the solution. The magnitude of this decrease depends on the relative humidity, isotopic composition of atmospheric moisture and also on the evaporation rate of the solution. If relative air humidity during evaporation is 70–80 %, then obviously seawater cannot be evaporated to the stage of bischofite sedimentation and the stationary isotopic composition of the solution $R_L = aR_a$ will be attained at earlier stages of evaporation.

In coastal regions (if the saline basin is a marine lagoon) the isotopic composition of vapour R_a will be close to that of equilibrium with seawater, i.e., equal to R_s/a (where R_s is the isotopic composition of seawater).

During the condensation of vapour the liquid phase is enriched with heavy isotopes of hydrogen by a factor α, i.e., the isotopic composition of the coastal evaporitic basin, in any case, does not differ considerably from the isotopic composition of seawater and will be in isotopic equilibrium with the marine water vapour ($\delta D \approx -94$ ‰, $\delta^{18}O \approx -13$ ‰, see previous para). For small inland saline basins (saline lakes), the isotopic composition can differ considerably from the marine one, being depleted in heavy isotopes for two reasons. Firstly, such basins could be fed by groundwaters with an isotopic composition lower than that of seawater. Secondly, the isotopic composition of the vapour of atmospheric moisture is much lower than that in equilibrium with ocean water. Besides, precipitation greatly influences the isotopic composition of some of these basins because of their small storage in comparison with large evaporitic basins of marine origin. Precipitation is decreasing in response to a decrease in the vapour condensation temperature.

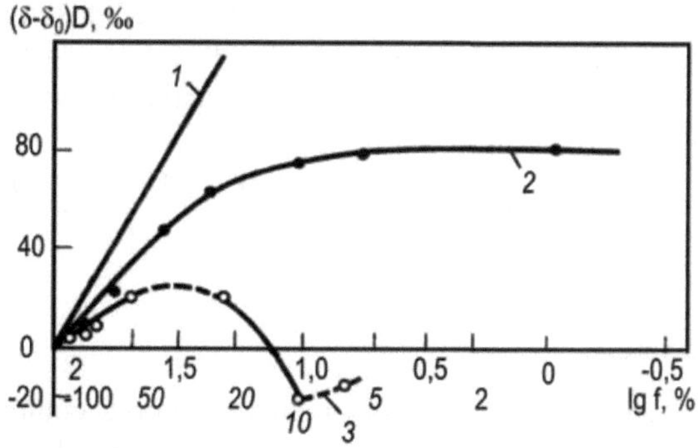

Fig. 9.20 Air moisture influence on the deuterium accumulation in liquid at the evaporation of water with marine salinity (from Atlantic Ocean): (*1*) Rayleigh distillation case; (*2*) the experimental data at air relative moisture equal to 40–50 %; (*3*) the data at relative air moisture equal to 70–80 %. (Ferronsky and Polyakov 2012)

Table 9.5 Seawater composition changes at different stages of concentration

Stage of concentration	Liquid content relative to initial volume (%)	Density (g/cm^3)	Ion composition (mass %)						
			SO_4^{2-}	Cl^-	Ca^{2+}	Mg^{2+}	K^+	Na^+	Σ ions[a]
Calcium carbonate	25	1.10	1.00	7.31	0.12	0.50	0.16	4.05	13.1
Gypsum	15	1.16	1.28	11.36	0.07	0.75	0.25	6.36	19.90
Halite	13	1.21	1.67	15.44	0.03	1.10	0.36	8.52	27.31
Epsomite	1.0	1.30	6.33	14.37	0.11	4.80	1.86	2.16	29.83
Silvine	1.6	1.30	5.53	18.21	0.01	6.39	2.22	1.25	34.44
Bischofite	0.7	1.35	3.03	23.58	–	8.81	0.02	0.28	36.25
Carnollite	0.9	1.32	4.35	19.59	0.03	6.95	1.82	0.77	33.88

[a] Together with the other ions not shown in the table

These conclusions were confirmed by model calculations reported by Vlasova and Btezgunov (1978) for brines with certain isotopic composition formed from standard seawater at different stages of evaporation. Two models were considered by these authors: one that assumes a constant storage (with replenishment) and the exhaustion model (the model of a drying basin). Calculation was carried out using formulae similar to (9.42) and (9.48) taking into account the changes in the water activity in the solution during evaporation. Substituting the isotopic ratios into the replenishing water R_I, in the basin R_L, in the vapour outflow from the basin R_E ($R_E = R_I$) and in the vapour of the atmospheric moisture R_a with the corresponding $1+\delta_i$ values, an expression is obtained that permits the calculation of the liquid's isotopic composition in relative units for given values of air humidity *h*, activity of water in the solution a, isotopic composition of the inflowing water δ_I and the isotopic composition of atmospheric moisture vapour δ_a:

$$\delta_L = \frac{\alpha[k(1+\delta_I)(\alpha - h) + (1+\delta_a)h]}{\alpha} - 1. \tag{9.51}$$

The α-values for a given temperature were taken from reference books and the values of k were taken to be $k_D = 1.009$ and $k_{^{18}O} = 1.016$.

In the case of a drying basin, calculations were made using the formula:

$$\delta_L - \delta_L^0 = \frac{h[\delta_\alpha - k\delta_L^0 - (k-1)] - \Delta\varepsilon_\alpha(1000 + \delta_L^0)}{hk + \Delta\varepsilon_\alpha}\left[1 - f^{hk+\Delta\varepsilon_\alpha/k(\alpha-h)}\right], \quad (9.52)$$

where

$\delta_L^0, \delta_L, \delta_\alpha$ are the relative isotopic contents of water at the beginning of evaporation (f=1), at an arbitrary moment after evaporation has begun and the vapour of atmospheric moisture, respectively; f is the degree of water loss, equal to the ratio of the volume of the remaining water to the initial volume;

$\Delta\varepsilon = 1/\alpha - k$; α is the equilibrium fractionation factor of isotopes at a given temperature; k is the kinetic fractionation factor.

Three constant values of water activity were taken in order to carry out the calculations: $a = 1$, corresponding to evaporation without precipitation of salts when the water body decreases from 100 to 12 %; $a = 0.75$, corresponding to precipitation of halite when the water body decreases from 12 to 3 %; $a = 0.33$, corresponding to precipitation of magnesium salts when the water body decreases from 3 to 0.1 %.

Equations (9.51) and (9.52) were solved for the following parameter values: humidity $h = 25\,\%$, $50\,\%$, $75\,\%$; temperature of evaporating water $t + 10\,°C$, $+20\,°C$, $+30\,°C$; isotopic composition of the atmospheric vapor δ_a for δD −70 ‰, –150 ‰, −200 ‰ and for $\delta^{18}O$ −10 ‰, −20 ‰, −40 ‰; isotopic composition of inflowing water for the model with a constant storage δ_I for δD 0 ‰, − 50‰, − 100 ‰ and for $\delta^{18}O$ 0 ‰, − 5 ‰, − 10 ‰.

The results of the calculations are given in Fig. 9.21 (Vlasova and Brezgunov 1978).

The results show that under natural conditions the isotopic composition of evaporating basins may vary greatly depending on the conditions of evaporation. The climatic conditions of the formation of large saline basins from the sea restrict the range of δD and $\delta^{18}O$ variations. According to Vlasova and Brezgunov (1978), under less favourable conditions, the minimum deuterium and oxygen-18 content in brines of sea origin at the stage of evaporation will not be less than δD = −40 ‰ and $\delta^{18}O = -3$ ‰.

These conclusions enable one to use isotopic data for estimating the genesis of connate brines. Theoretical and experimental data for modern saline basins have been reported to be in agreement with the author's conclusions. They are also in accordance with calculations made by Sofer and Gat (1975) and Gutsalo (1980).

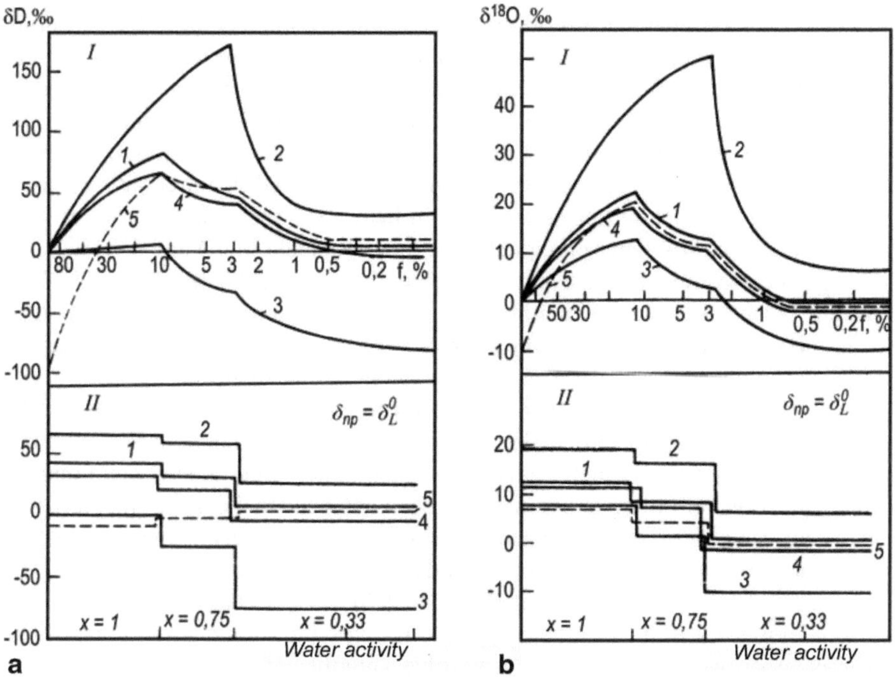

Fig. 9.21 Isotopic composition of hydrogen (**a**) and oxygen (**b**) of evaporated seawater for changed evaporation conditions (calculated curves): (*I*) exhaustion model; (*II*) constant level model. (Ferronsky and Polyakov 2012)

9.6 Isotopic Composition of Water in Unsaturated and Saturated Zones

On the basis of experimental data it has been pointed out that the isotopic composition of shallow groundwaters, which are replenished by infiltration of atmospheric precipitation through the unsaturated zone, are characterised on average by heavy isotopes of typical precipitation of a given region. Under certain conditions there are some differences in isotopic ratios for the two type of waters mentioned above. This is explained by the fact that precipitation in the spring-summer season is partially and often completely re-evaporated from the Earth's surface and from the unsaturated zone both directly and due to transpiration of moisture by plants.

As Zimmermann et al. (1967) have shown, the water that infiltrates from the Earth's surface into the unsaturated zone is already enriched with heavy isotopes compared with precipitation due to the evaporation process. Under identical soil conditions, water enriched in those places, which are covered with plants is, on average, 10‰ greater than in those areas that have plant cover. The above mentioned values, as Zimmermann et al. (1967) noted, are characteristic for Central Europe where the average relative deuterium concentration is about −70‰. The above authors have also shown that variations in the isotopic composition of the soil moisture remain unaffected by the observed ratio between moisture movement velocity in the unsaturated zone, the rate of evaporation of moisture from the surface and the process of diffusion and exchange through transpiration by plants. At the same time, Gonfiantini et al. (1965) reported that the process of water evaporation by leaves results in its isotopic fractionation. It should also be noted that precipitation water in the spring-summer season is enriched in heavy isotopes compared to groundwaters. The autumn-winter precipitation is more depleted in δD and $\delta^{18}O$ and also undergoes enrichment due to evaporation from the surface. Finally, this results in an isotopic balance between the atmospheric precipitation and shallow groundwater.

Analogous investigations on the formation of the isotopic composition in near-surface groundwaters of an arid zone of the south-east Mediterranean coastal region were carried out by Gat and Tzur (1967). They found that groundwaters that are fed by local precipitation are enriched in oxygen-18 by 1–3 ‰ relative to SMOW. Studies carried out with a lysimeter have shown that under experimental conditions an excess of oxygen-18 of up to 2 ‰ has also been observed in the infiltrating moisture due to surface evaporation. In general, surface evaporation results in the visible enrichment of groundwaters in deuterium and oxygen-18 in arid and semi-arid regions (Gat and Tzur 1967; Gonfiantini et al. 1974, 1976; Dinçer et al. 1974).

The knowledge of the formation conditions of groundwater isotopic composition in active water exchange zones helps to solve a number of practical problems. They are as follows:

1. Relationship between surface and groundwater;
2. Groundwater recharge at present time;
3. Groundwater recharge in the past;
4. Identification of an area of groundwater recharge;
5. Relationship between aquifers;
6. Mixing proportions of groundwater of different genesis;
7. Groundwater residence time in an aquifer;
8. Relationship of waters in conjugate hydrologic basins.

More details about solving the above problems can be found in Ferronsky and Polyakov (2012).

9.7 Isotopic Composition of Formation Waters

It follows from the previous chapters that the processes of water evaporation and condensation are of great importance in the fractionation of isotopes of natural waters. At the same time evaporation is primarily an attribute of surface conditions. It might occur in shallow underground waters but it is generally agreed by hydrogeologists that groundwater evaporation does not occur on a regional scale (Zaitsev 1967; Smirnov 1971).

But in local zones underground evaporation is likely to take place. An example of such a phenomenon is the evaporation of groundwater accompanying oil and gaseous deposits (Sultanov 1961). As a rule, these processes of water evaporation occur at elevated temperatures (~80 °C). In this case the isotopic fractionation factors are $\alpha_D = 1.032$ and $\alpha_{^{18}O} = 1.0042$. The vapour phase differs insignificantly in isotopic composition from layer waters to a deposit. The water vapour that has been formed migrates with oil gases. During the migration of the vapour-gaseous mixture through porous layers at lower temperatures, underground fresh water deposits with mineralisation less than 1 g/l and δD and $\delta^{18}O$ values greater than those that are characteristic of meteoric waters might form.

Thus, in the region of the Dnieper-Donets depression at depths exceeding 2000 m, waters with mineralisation of up to 4 g/l and values of δD from −21 to −53‰ and ^{18}O from −2.5 to −4.6‰ were found. In this region, for oil waters with mineralisation ranging from 150 to 330 g/l the deuterium and oxygen-18 content varies within the limits δD from −21 to −54‰ and ^{18}O from +2.0 to −4.6‰ (Yakubovsky et al. (1978).

Analogous waters were also found in the eastern part of the Terek-Sundzha oil and gas region at a depth of 4–5.5 км (Nikanorov et al. 1980). But naturally underground evaporation cannot result in considerable changes of the isotopic composition of the layer waters since the amount of evaporated water is always negligible in comparison with the amount of native water.

The basic process determining the isotopic composition of water undergoing underground circulation is that of isotopic exchange in the water–rock system. The isotopic exchange of water with gases (H_2S, H_2, CH_4, CO_2) and liquid hydrocarbons of oil also takes place. But these processes are considerably less effective than the exchange processes with water-bearing rocks. For example, according to Soyfer et al. (1967), the change in the isotopic composition of hydrogen in groundwaters due to exchange reactions with gaseous hydrogen and hydrogen sulphide is negligible.

The absence of the influence of isotopic exchange in the H_2O–H_2S system upon the isotopic composition of formation waters was demonstrated by Clayton et al. (1966) and Hitchon and Friedman (1969) By indirect means the possible scale of changes of hydrogen isotopic composition in groundwater due to exchange processes with liquid hydrocarbons of the oil series can be estimated. Thus, Mason (1966), with reference to Smith, pointed out that the formation of oil deposits in the USA, in the Gulf of Mexico region, Louisiana and Saint Croix in California, was accompanied by the release of about $3.4 \cdot 10^5$ *t* of liquid hydrocarbons from 1 km^3 of sediments ($\sim 2 \cdot 10^9$ *t*). Assuming that moisture made up 10% of the mass of sediments at the stage of diagenesis, i.e., there were $\sim 2 \cdot 10^8$ *t* of water in 1 km^3 of sediments, then the ratio of hydrogen atomic fractions in water and oil becomes $w/o \approx 300$. Here we assumed that the exchangeable amount of hydrogen in oil per mass unit is twice as large as that of water. It is obvious that for such a low value of w/o the isotopic composition of water remains even when the isotopic equilibrium between hydrocarbons ($\delta D_{п} = -200$‰) and water is maintained such that $\alpha = 1$ ($\delta D_{oil} - \delta D_{water} = 0$). This is confirmed by Clayton's et al. (1966) studies of isotopic composition of groundwater in the oil deposits mentioned above. The results of this research are presented in Figs. 6.3 and 6.5. Some observable changes in the hydrogen isotopic composition are likely to be expected in the region of water-oil contact, when w/o <<300.

The change in oxygen isotopic composition of water due to exchange with carbon dioxide may take place under certain conditions in hydrothermal systems of groundwater gassing CO_2. Thus, Ferrara et al. (1965) observed the negative 'oxygen shift' in the thermal waters of Toscana, which they explained in terms of the exchange of water with carbon dioxide. These results are presented later on.

Isotopic exchange with water-bearing rocks at high temperatures has the greatest influence upon the isotopic composition of water. These processes mainly affect the oxygen isotopic composition of groundwater since, in sedimentary rocks, oxygen content amounts to 40 % whilst hydrogen is only 0.3 %. In granite rocks these values amount to 48 % and 0.12 % and in basalt to 46 % and 0.1 %, respectively (Beus 1972).

The scales of changes of δD and $\delta^{18}O$ values of water provided by isotopic exchange reactions with water-bearing rocks depend on the following factors (Ohmoto and Rey 1974). (1) The initial isotopic composition of water δ_w and rocks δ_r, participating in isotopic exchange reactions. (2) The ratio of the amount of exchangeable oxygen or hydrogen atoms in water to the same in rocks (w/r). (3) The temperature that determines the equilibrium isotopic fractionation factor between water and rocks. As pointed out earlier the effect of isotopic fractionation between two phases, for example between rocks and water, may be approximated by the difference of the values $\Delta = \delta_r - \delta_w \approx 10^3 \ln \alpha$. (4) The degree of exchange in the rock-water system, depending on the average residence time of the water in the aquifer.

Assuming that water and rocks are in isotopic equilibrium at a given temperature, then the final isotopic composition of water δ_w^f may be expressed as (Ohmoto and Rey 1974):

$$\delta_w^f = \frac{\delta_r - \Delta + (w/r)\delta_w}{1 + w/r}.$$

This formula may be used not only for calculating the isotopic composition of water in isotopic equilibrium with rocks but also for estimating the ratio (w/r) when the parameters are known (Sheppard et al. 1969, 1971; Taylor 1974, 1978).

Using the above formula let us estimate the change in the isotopic composition of hydrogen and oxygen of normal seawater ($\delta D_w = 0$, $\delta^{18}O_w = 0$) syngenetic with clay sediments. For example, this process may occur by wringing out the inner-layer of water of clay minerals during the formation of deep waters in sedimentary basins (Karasev and Vagin 1973). Sergeev et al. (1963) showed that the dominant clay minerals in the rocks of Mesocainozoic sediments are those of more ancient ages known as hydromicas. Next in abundance are minerals of the montmorillonite group and then of the kaolinite group. The clay minerals of hydromicas and montmorillonites are very similar in their crystalline structure and hydrogen (water) content of the hydroxide groups coming into the crystalline lattice of the complete dehydrated minerals. For minerals of these groups the values of the hydrogen isotopic fractionation factors are also close to each other in the same way as the oxygen ones. They most readily start the oxygen exchange with water (Savin and Epstein 1970a, b; O'Neil and Kharaka 1976).

These circumstances permit one to consider, in reliable details, isotopic exchange in the montmorillonite-water system as an illustration of the scale of changes in water of δD and $\delta^{18}O$ from initial values. The assumption is made that during clay diagenesis, isotopic exchange occurs under conditions of a closed system, i.e., only the water that was bounded by rocks at the stage of sedimentation is involved in isotopic exchange reactions with clay minerals. Let us take into consideration the following parameters of the process: the humidity of the clay is 20% of the total mass; the hydrogen content in hydroxide groups per mass unit of the dry rock is 0.4% (which is equivalent to about 4 mass % of H_2O). Then w/r ≈ 5 for hydrogen atoms and 0.3 for oxygen atoms. Let us consider further that prior to sedimentation the clay minerals were formed under hypergenous conditions in regions with a temperate climate. The isotopic composition of atmospheric precipitation may then be taken as $\delta D_w = -90$ ‰ and $\delta^{18}O_w = -12$ ‰ and the isotopic composition of the clay minerals of the montmorillonite group, equilibrated with meteoric waters, is $\delta D_w \Delta = -15$ ‰ and $\delta^{18}O_w = +15$ ‰ (Savin and Epstein 1970a). For the purposes of estimation we consider the isotopic composition of pore water, at the initial stage of sedimentation, to be the same as that of normal seawater and isotopic exchange in a system to occur at 100 °C. For these conditions one has $\Delta D_{\text{mineral-water}} \approx -50$ ‰ and $\Delta^{18}O_{\text{mineral-water}} \approx +12$ ‰. The values Δ for oxygen isotopic exchange in the mineral-water system are estimated on the basis of data provided by Taylor (1974). For the isotopic exchange of hydrogen between montmorillonite and water at normal temperatures of hypergenic processes the value Δ will be obtained using Savin and Epstein's (1970a) data. In addition, let us consider that the fractionation factor of hydrogen isotopes between hydroxyl-bearing minerals and water does not undergo any significant changes up to ~400 °C (Taylor 1974; Suzuoki and Epstein 1976).

Now putting δ_w, δ_r, Δ and w/r into the Ohmoto-Rye formula, one obtains $\delta D^f_w = -17$ ‰ and $\delta^{18}O^f_w = +2.3$. Note that the conditions described above are the most favourable for the isotopic exchange (due to filtration of water through the clay). If such filtration takes place the value of w/r calculated per unit of mass of rock should be increased by factor *n* (where *n* is the number of water exchange cycles). For $n=2$, $\delta D^f_w \approx -9$‰ and $\delta^{18}O^f_w \approx +1.2$‰.

From these estimations it follows that the isotopic exchange of water with the clay minerals at layer temperatures below 100°C will not result in significant changes in the content of deuterium and oxygen-18 in groundwater. In natural conditions (w/r = 1) the interaction of water with basic (basalts, $\delta^{18}O \approx +7$ ‰) and acid (granites, $\delta^{18}O \approx +10$ ‰) volcanic rocks at temperatures above 400°C, when the value of $\Delta_{\text{rock-water}} \rightarrow 0$, may result in a maximum oxygen shift from 3 to 5 ‰ when $\delta^{18}O = 0$. The value of the oxygen shift increases as $\delta^{18}O_w$ decreases; the maximum oxygen shift for the above conditions is 10 ‰ when $\delta^{18}O_w = -10$ ‰ and 12 ‰ when $\delta^{18}O \approx -15$ ‰. Other conditions being equal, the maximum oxygen shift will take place at isotopic equilibrium for groundwater in circulation in carbonate rocks composed of limestones and dolomites with $\delta^{18}O_r = +30$ ‰.

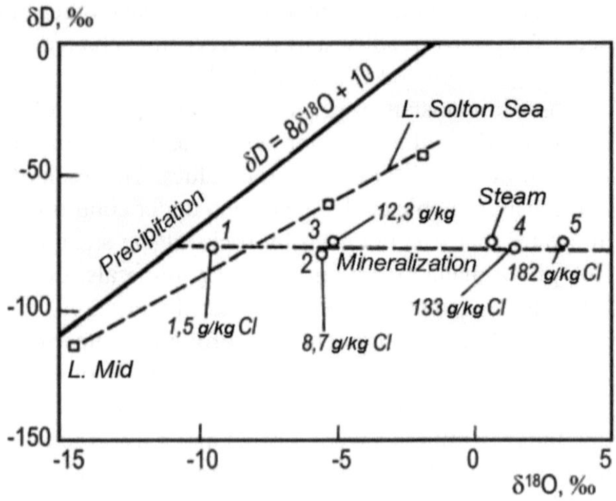

Fig. 9.22 Isotopic composition of waters in the Salton-Sea geothermal area: (*1*) a spring at the base of the Chocolate Mountains (17 °C); (*2*) water from the 'mud volcanoes' on the east shore of the lake (39 °C); (*3*) water from the old CO_2 production wells about 300 m deep (42 °C); (*4*) steam and brine from the two geothermal wells WGS and IID-1; (*5*) brine from a deep well (>300°C). (Craig 1966; Ferronsky and Polyakov 2012)

The hydrogen isotopic exchange with hydroxyl-bearing minerals of siliceous rocks of volcanic genesis does not, in practice, play any significant role in the change of the isotopic composition of underground waters since the hydrogen content in these rocks does not exceed an average of 0.1 % (Beus 1972).

By analysing numerous experimental data in various hydrothermal regions of the world Craig (1963, 1966) showed that deuterium content in hydrothermal waters corresponds to its content in meteoric waters in a region where the former is fed by the latter. According to the data of several researchers (Clayton 1961; Craig 1963, 1966; Hitchon and Friedman 1969; Savin and Epstein 1970a; Kawabe 1978) at higher temperatures water is enriched with heavy oxygen due to isotope exchange with oxygen-bearing rocks (limestones, silicates and so on). This phenomenon of oxygen isotopic exchange has been called the oxygen shift. Figure 9.22 demonstrates the data obtained by investigation of geothermal brines brought out through boreholes in the Salton-Sea region of California, USA (Craig 1966)[1]. It follows from Fig. 9.22 that with increases in temperature of the waters being studied, the deuterium concentration remains constant at the level corresponding to its content in atmospheric moisture in this region, while the ^{18}O content increases sharply for the same conditions. In contrast, on the surface (for the Salton-Sea and Mead Lakes) the content of these isotopes changes in parallel.

Thus, it should be noted that the statement made by several authors (Soyfer et al. 1967) on the parallel behaviour of D and ^{18}O concentration in groundwater cannot be considered as universal. It holds only for waters of the surface cycle as Craig (1961) has shown. For groundwater the behaviour of D and ^{18}O differs due to oxygen isotopic exchange between water and water-bearing rocks, especially at high temperatures and this must be taken into consideration in the course of isotope studies. The variation of ^{18}O concentration in rocks and groundwater is illustrated by another example characteristic of the Larderello region of Italy (Ferrara et al. 1965).

[1] The work was undertaken to refute that the above-mentioned brines are deep ore-bearing fluids.

Sea limestones, for which values of $\delta^{18}O = +30$‰ are typical, are notably depleted in this isotope. At the same time, for water migrating through these limestones the values range from $\delta^{18}O = -7.3$‰ in the region of supply and −4.8 and even −1.9‰ in the areas of emergence at springs on the surface or in the boreholes. Nevertheless, the deuterium concentration is practically a constant equal to that in meteoric waters in the supply region.

Coplen and Hanshow (1973) studied the isotopic composition of oxygen and hydrogen in ultrafilter and sedimentary solution by infiltration of distilled water and NaCL solution through the montmorillonite membrane. The porosity of the disk under pressure reached 35 %. They found that the liquid being infiltrated through the montmorillonite membrane becomes depleted in δ D and $\delta^{18}O$ by 2.5‰ and 0.8‰ respectively. The concentration of NaCL up to 0.01 NH docs not affect the process of isotopic fractionation. Higher saline concentrations are likely to exert influence upon the isotopic separation in the course of infiltration of a liquid through the clayey membranes. It should be noted that the experimental data obtained by Coplen and Hanshow concerning the δD and $\delta^{18}O$ deviations in the infiltrated liquid may be considered as insignificant and the scale of the experiment does not apply itself to natural conditions where, in the range of experimental error, this phenomenon has not been observed.

It worth noting that in the work of Yeh and Epstein (1980), the authors with reference to Coplen and Hanshow, explained that the process of ultra-filtration explains some increase (~20‰) of deuterium content in pore waters in the clays profile opened by boreholes in the Gulf of Mexico. But this phenomenon has another explanation. For example, the observed fact can be related to decrease in depth in the fractionation factor of hydrogen in the water-rock system because of increase in temperature; the effect can also be explained by the w/r ratio decrease due to a reduction of the moisture in the clayey thickness and by the climate changes of North America in the previous epochs.

The formation of deep groundwater of sedimentary basins of the Earth has had a long and complicated history. The deep waters of sedimentary basins were formed within the ancient seas and lagoons located at the edge of continents. A close relationship exists between sedimentary thickness and horizontal tectonic displacements of lithospheric plates (Le Pichon et al. 1973; Bullard 1978) resulting in vertical tectonic displacements of the Earth's crust (Verhoogen et al. 1970).

The main sedimentary basins that originated in various geological times have been related to boundaries of plates, i.e., with those places on the Earth's crust where the maximum vertical tectonic displacements have occurred. It is no coincidence that the majority of giant petroleum fields discovered before 1970 are related to sedimentary basins and located in the shelf areas of modern continents (Moody 1978).

While studying the processes of groundwater formation in a sedimentary basin with the help of commonly used techniques, researchers are normally faced with considerable difficulties. These consist in a marked transformation in the chemical composition of primary marine water both during the stage of surface exposure and as a result of interaction of the groundwater (primary or infiltrational) with water-bearing rocks. Studies of the isotopic composition of sedimentary basin groundwater help to overcome these difficulties to a great extent.

As mentioned earlier (see Para 9.3) during the exposure of a sedimentary basin where the precipitation of evaporites has occurred, the hydrogen and oxygen isotope composition of the water could be a little different from that of the seawater recharging it, due to evaporation. During evaporation there is an increase in the concentration of salts in such a basin resulting in a decrease of water activity and, also, of the evaporation rate. This, in turn, affects the water hydrogen and oxygen isotope composition at the final stages of evaporation. As Sofer and Gat (1975) have shown, the deuterium content in such basins could decrease compared with the primary marine water ($\delta D=0$), depending upon humidity. Thus, at a relative air humidity of 40–50 % and a vapour isotope content $\delta D=-80$ ‰, when the sedimentation of bischofite might occur, the water hydrogen isotope composition is greater than the deuterium content of the ocean water at practically all stages of evaporation. At a humidity of 80 %, when basins in natural conditions may be evaporated to the stage of halite sedimentation, the hydrogen isotope composition of the remaining water might decrease, reaching $\delta D=-20$ ‰.

In the underground conditions at the stage of diagenesis of the sedimentary strata, the isotope composition of the groundwater could also be changing. As a rule, at elevated temperatures compared with the surface, oxygen isotope exchange between groundwater and water-bearing rocks occurs, resulting in an increase of the oxygen-18 content in groundwater. The deuterium content also changes but its variation, provided by the exchange processes with the hydroxide-bearing minerals such as clays, is more diminished than that for oxygen because the oxygen content exceeds the hydrogen content by more than one order of magnitude (Taylor 1974; Ohmoto and Rye 1974).

Concentration of ^{18}O in groundwater is a less conservative parameter. Its high values are explained not only by the marine origin but also by the isotopic shift. This is why a high concentration of ^{18}O cannot be an indicator of marine origin of the studied water. At the same time, the ratio of concentration of D and ^{18}O can be an index of the depth of sedimentary basins.

Many works have been devoted to the principles of stable isotope distribution in deep sedimentary waters and rocks and on the determination of the isotopic content in water using its physical properties (the techniques based on density measurements and others). Later on, due to their accuracy and restricted applicability, these techniques were replaced at first by the photoneutron method and later by modern mass spectrometric methods, which are presently still in use.

Alekseev et al. (1966) and Tyminsky et al. (1966) carried out studies of deuterium distribution in the deep formation waters of the Tashkent artesian basin. With the help of experimental data they found that the infiltrating waters of this basin are depleted in deuterium and the ancient waters are enriched in it.

While studying the absolute ages of groundwater in a chalky water-bearing complex in the Tashkent artesian basin using the helium-argon method, which was suggested by Ovchinnikov, it was found that the most ancient waters (up to 5 million years) are situated in the immersed parts of the depression and the sidewards regions and the northern part of the depression, with rocks of the Cretaceous age at the surface having waters of more recent age (tens or hundreds of thousands of years).

Table 9.6 Distribution of D and ^{18}O in groundwater of artesian basins within the former USSR

Artesian basin	δD, ‰	$\delta^{18}O$, ‰	Reference
Near Baltic	from −119 to −48	from −18.8 to −6.3	Pelmegov et al. 1978
Pripyat Depression	−95 to+2	−12.7 to+4	Tkachuk et al. 1975
Ukrainian Carpathy	−108 to −7	−13.1 to+3	Babinets et al. 1971; Vetshtein et al. 1972; Polyakov et al. 1974
Near Carpathy Mineral Waters type 'Naftusya' and 'Shklo'	−102 to −6	–	Vetshtein et al. 1973
Dnepr-Donetsk	−55 to+11	−7.2 to+0.3	Vetshtein et al. 1973
Near Caspian Depression (north-west)	−107 to −33	−10.1 to −2.4	Alekseev et al. 1975
North Caucasus	−67 to −15	−12 to+5.8	Tarasov 1978
Azov-Kuban	−120 to − 28	−17.7 to+7	Sokolovsky et al. 2007
Caucasus Mineral Waters	−118 to −42	−12.7 to+1.1	Seletsky et al. 1973
Sochi-Adler	−73 to −35	−9.8 to −0.5	Gorbushina et al. 1974
Middle Caspian	−125 to −17	−12.4 to −2.2	Seletsky et al. 1973
West Turkmenian	−79 to −4	−13.4 to+4.4	Seletsky et al. 1973; Alekseev et al. 1975
Yaskhan Lens of Fresh Water	−89.2 to −54.4	−12.4 to −2.5	Seletsky et al. 1973
Mud Volcanoes of West Turkmenia	−57 to−31	–	Seletsky et al. 1973
North Caucasus3	−115 to −28	−14.2 to+5	Alekseev et al. 1975
Moscow	−114 to − 22	− 14.5 to −5	Sokolovsky et al. 2007
Ferganian	−106 to −78	−10.9 to − +1.7	Seletsky et al. 1973
Amu-Darian	−102 to −20	−12.3 to −6.1	Alekseev et al. 1974
Siberian Platform	−180 to −23	−21.5 to 0.0	Pinneker et al. 1973; Pinneker 1974, 1975

It was found that deuterium content grows with increases of groundwater age. In the northern and north-eastern part of the basin relatively young waters occur (up to 60,000 years) having a lower deuterium content (from −160 to −120‰). The maximum deuterium content (from −26 to −7‰) was found in waters at depths of about 2000 m below the land surface, the age of which is about 4–5 million years.

For paleohydrological consideration, the dilution proportions of infiltration and sedimentary waters in chalky sediments of a basin were calculated. The proportions were determined with the help of mixing formulae in common use. The accepted deuterium concentrations were −160‰ for infiltration waters and +26‰ for sedimentary waters. The researchers distinguished zones of abundance of waters with proportions of infiltration waters in the mixture up to 0.25, from 0.25 to 0.5 and more than 0.5.

Using these studies and also investigating the deuterium content in waters of a number of oil and gas field regions the applicability of isotope techniques has been shown for the identification of formation and infiltration waters in artesian basins (Table 9.6).

The isotopic composition of groundwater in the North Caucasus region was studied for two hydrogeological levels (Tarasov 1978). The waters of the neogen-quaternary upper floor are of infiltration origin (δD from −67 to −46‰, $\delta^{18}O$ from −12.0 to −6.8‰). The waters of the lower floor are separated from the upper one by a thick layer of Maykop argillaceous suite. These waters are enriched in heavy isotopes (δD from −55 to −15‰, $\delta^{18}O$ from −4 to+5.8‰) and are similar to thalassogenic ones in terms of origin.

Investigations carried out in the Azov-Kuban artesian basin have shown that the deuterium content in waters of oil and gas field ranges from −68 to 58‰. These values correspond to an intermediate, averaged deuterium content in meteoric (−107.5) and seawaters, which is evidence in favour of the presence of connate sea sedimentary waters up to about 50%. About 50% of Yessentuki's saline-alkali waters (Nagut aquifers) are also of marine origin, which refutes the existing opinion that these types of water originate from meteoric local recharge areas.

A greater portion of ancient seawaters are contained in mud volcanoes in this region, where δD ranges from −56.9 to −54.9‰. At the same time a considerably lower deuterium content has been detected (от −80.8 до −80.0‰) in the iodine waters of the Slaviano-Troitsk aquifers. These data indicate that these waters are diluted with meteoric water to a greater extent, which is in disagreement with the option that there is a genetic relationship between the former waters and those occurring in oil and gas fields.

Within the region of the Caucasus mineral waters it has been reported (Seletsky et al. 1974) that the waters of the Olkhovsk prospecting area have the lowest deuterium content (δD=−118‰). These waters are typically fissured and are completely recharged by precipitation.

The 'Narzan' group of waters in Kislovodsk as a whole, are intermediate, in terms of deuterium content (D=−108 to −76‰), between the waters of the Olkhovsk prospecting area and waters of the deep wells situated in the Yessentuki area.

The 'Yssentuki-20' spring is clearly distinguished by its deuterium content (δD=−117‰) from other waters in the region of the Yssentuki aquifer. The low deuterium content is the result of the spring being recharged by groundwater flowing from the sandy-pebble sediments of the Podkumok River, starting at a height of more than 2 km. The other waters in the Yessentuki area have greater values of δD (up to −42 ‰). This suggests that they also contain ancient waters of marine genesis from the more deeply lying complexes, in particular the Valazhinsky complex that exerts great influence on the formation of all main aquifers of the Caucasus mineral waters.

The relative deuterium content in formation waters of the oil-gas fields in the South Mangyshlal Peninsula ranges from −82.3 to −16.5 ‰. The waters most depleted in deuterium have been contaminated by fresh water pumped into the productive oil-bearing layers. In water of the West Turkmenian artesian basin δD ranges from −76.6 to −37.3 ‰. The elevated deuterium concentrations in waters of the Jurassic complex in the Mangyshlal Peninsula is evidence that waters of marine genesis participate in their formation to a greater extent compared with waters of Tertiary sedimentation in Western Turkmenia.

The investigations of deep waters of Pliocene sediments in the Western Turkmenia artesian basin (Seletsky et al. 1973; Alekseev et al. 1975) have shown, by oxygen-18 and deuterium isotope analyses that brines in red coloured sediments contain a large portion of marine waters.

The hydrothermal ore sediments in the Cheleken area (Gutsalo et al. 1978; Esikov et al. 1979) have been reported to be coloured sediments (δD from −41 to −31 ‰, $\delta^{18}O$ from −3.5 to −2.1 ‰). They are likely to be similar in origin to connate marine waters, being metamorphosed in salt composition. However, in order to elucidate the conditions of high salt saturation and enrichment with heavy metals, further prospecting of this area is required.

While studying the isotope composition of the Yaskhan Lens fresh water in West Turkmenia (Seletsky et al. 1973), considerably different conditions of lens water formation from those of the underlying Kara Kum saline waters have been reported. The lower δD and $\delta^{18}O$ values in waters of the diffusion zone of the lens are evidence that waters of pluvial age exist in the underlying aquifer, whereas the waters of the lens are likely to have been formed in conditions of arid climate.

In the Fergana artesian basin more careful studies of the sulphide mineral water fields of Chimion and Obi-Shiro (Gorbushina and Tyminsky 1974) have been carried out. The deuterium content variations were reported to be insignificant (δD from −98 to −70 ‰), although there are large differences in mineral content (0.7–120 g/l) and the δD values (from −16.1 to −3.7 ‰) show that processes resulting in oxygen isotope shift are involved. The low deuterium content in the Fergana artesian basin waters probably reflect the importance of the mountains surrounding the depression is supplying infiltrating waters to recharge the basin complexes.

In the Amu-Darya artesian basin (Alekseev et al. 1974, 1975) it has been reported that δD and $\delta^{18}O$ variations are respectively −52 to −20‰ and −6.3 to +6.1‰ in waters of the Jurassic aquifer; −102 to −43 ‰ and −11.8 to −5.1 ‰ in the Lower Cretaceous aquifer; −93 to −57‰ and −11.7 to −6.8 ‰ in the Upper cretaceous aquifer; and −97 to −60 ‰ and −12.3 to −8.4 ‰ for the surface waters. On the basis of these data the authors reported that waters of the Jurassic sediments are preferentially of marine genesis and the Cretaceous complex waters originated due to the dilution of marine connate waters with modern meteoric ones, which follows from the fact that the corresponding points fit the line plotted for $\delta D - \delta^{18}O$ coordinates connecting the plots of ocean and meteoric waters. Assuming $\delta D = -25$ ‰ for seawaters and $\delta D = -90$ ‰ for meteoric waters, Alekseev et al. calculated that the proportion of infiltrational waters composing sedimentogenic waters may vary from 9 to 94 %.

The relative deuterium content in highly saline calcium chloride brines in the Angara-Lena artesian basin ranges from −70 to −23 ‰ and in this respect are similar to the waters of petroleum and gas provinces in South Mangyshlak and West Turkmenia. The deuterium contents in sodium chloride brines of this basin are considerably lower. They vary from −168 to −75 ‰ and differ insignificantly from local meteoric waters. The deuterium content of waters here is even lower than that sampled in the artesian basin.

The oxygen-18 concentrations in the considered basins are characterised by the following ranges of values. In the Fergana artesian basin they range between −10.9 to + 1.2 ‰; in the Western Turkmenian basin from −7.6 to + 3.4 ‰. The data obtained from these two basins support the idea that a large portion of infiltrating waters dilutes the Fergana basin waters and the process of oxygen isotope exchange with water-bearing rocks takes place in these two basins. There are no great differences in the range of ^{18}O values compared with the amplitude of deuterium content variations. By comparing the data from both isotope studies in the West Turkmenia artesian basin the relatively high ^{18}O content can be explained both in terms of the inheritance of marine concentrations and isotopic exchange with rocks. In the Fergana basin it can be explained as a sequence of oxygen isotopic exchange.

In the Angara-Lena artesian basin the ^{18}O content ranges between −17.0 to −0.6 ‰, which agrees with the deuterium content variations for this basin. Using the above-mentioned knowledge corresponding to the examined basins, it was concluded that the calcium chloride brines in the Angara-Lena artesian basin contain a considerable proportion of waters of marine genesis, thus being distinguished from all other waters of this region. The brines of sodium chloride composition do not differ from waters that are obviously of meteoric origin. Thus, the experimental data on water isotope composition disagree with the existing opinion that initial sedimentary water resources of this region have been completely renewed.

The isotope studies in the West Turkmenian artesian basin suggest that the origin of the red coloured Pliocene aquifer was not completely sedimentary in origin and that the degassation process from the interior may have occurred.

The isotope data obtained for waters of the Alb-Senoman and Jurassic complexes in the South Mangyshlak Peninsula are in agreement with the hypothesis of the sedimentary origin of the latter and in conflict with the hypothesis suggesting a common hydrodynamic model for all the complexes.

The low deuterium content in waters of the Fergana artesian basin reflects the important role of the mountain surrounding the depression in recharging all complexes in the basin from infiltrating waters.

The waters of mud volcanoes are distinguished from all other deep waters occurring in the Earth's sedimentary shell. The deuterium content in them, according to mud volcano studies in Western Turkmeniya and Northern Caucasus, normally corresponds to the deuterium content of waters of oil and gas-fields of the artesian basins. As to the oxygen-18 concentrations, they attain their highest values for all known groundwaters, not only in the former USSR area but on the Earth as a whole. The $\delta^{18}O$ values detected in them range from +3.4 to +5.0‰. A more detailed discussion on the above results may be found in Seletsky et al. (1973), Alekseev et al. (1975) and Esikov et al. (1979).

While studying the formation conditions of the Matsesta mineral waters in the Sochi-Adler artesian basin (Gorbushina et al. 1972, 1974) it was found that $\delta^{18}O$ values in these waters range from −9.8 to −2.0‰, in the Black Sea waters from −4.0 to −2.3‰ and in local lakes and rivers from −14.5 to −10.9‰. On the basis of these data it was concluded that recharge to mineral springs up to depths of 500 m occurs due to precipitation and melt waters flowing from the snow-packs and deeper mineral waters (at depths of 1200–2000 m) are formed due to the dilution of seawater with surface water.

In the sulphide waters of Jurassic sediments in the same region (Gorbushina et al. 1972) the δD values have been reported to vary from −51 to −35‰ and $\delta^{18}O$ from −5.8 to −0.5‰. In the sulphide waters of chalky sediments these values for deuterium and oxygen-18 range from −73 to −52‰ and from −9.7 to −5.0‰ respectively. Comparing these data with those of the surface waters of the region the authors concluded that the sulphide waters of the Jurassic sediments are connate marine waters and the waters in chalky sediments are marine and meteoric waters mixed in some proportion.

For the conditions of the distant parts of the Baltic artesian basins the hypothesis of the glacial origin of the Strelninsky aquifer and the weathering core water has been confirmed, as has the assumption of the hydraulic isolation of the Strelninsky and Gdovsky aquifers from each other and from the overlying water-bearing complexes (Sobotovich et al. 1977). Moreover, isotopic data on the solutions show that the formation waters of the Strelninsky aquifer are situated in the reduced water exchange zone and their recharge by the surface waters is about 5% per thousand years. Analogous waters with anomalous light isotopic composition (δD up to −170‰ and $\delta^{18}O$ up to −22.2‰) were also found in Cambrian-Vendian sediments near the shore of Tallinn in Estonia Bay (Yezhova et al. 1996).

Within a complicated region of the southern slope of the Ukrainian crystalline shield dipping towards the Black Sea depression, the hydrogen isotope composition was investigated in a profile of the South Belozersk iron ore deposits area (Voytov et al. 1976). It was found that waters of the Buchakovsky and chalky aquifers and weathering core waters are depleted in heavy isotopes (δD = −107 to −90‰) compared with the fissured waters of the iron-siliceous formation (δD = −92 to −70‰ and even −33‰). On the basis of these data the authors assumed that the mine waters are a mixture of two water types. The first being a relict of the pluvial period, since they are considerably lighter than the recent waters of the river runoff (δD = −77 to −61‰).The second type of water has modern meteoric origin.

Within the north-western region of the Caspian depression, deuterium and oxygen-18 abundances have been studied in a profile from surface waters to those of the deep Paleozoic aquifer. In this region high δD values (up to −33 ‰) are observed in waters of Devonian sediments. The reduction of the δD values (−64 to −47 ‰) is a feature of the carboniferous sediments as a whole (for depths from 500 to 3200 m). At a depth of 50–300 m the deuterium content is even lower (δD = −107 to −74 ‰). On the basis of the isotope analysis data (Alekseev et al. 1975) it was reported that within the Don-Medveditsk heights the water of the carboniferous sediments were formed under a certain influence from infiltration processes involving the dilution of the marine sedimentary waters with those of the surface discharge. The influence of the infiltration processes is observed up to depths of 1200–1400 m. The waters of the Devonian sediments are the connate marine waters, which have been insignificantly diluted by meteoric waters probably during the short periods of infiltration regime.

The $\delta^{18}O$ values have been found (Vetshtein et al. 1973) to range from −7.2 to + 0.3 ‰ in waters of deep (1900–4000 m) Devonian, carboniferous and Lower Permian brines (150–300 g/l) of the oil and gas in the Dnepr-Donetsk depression, whereas in the surface waters of this region the $\delta^{18}O$ values range from −11.1 to −8.6 ‰. The relative deuterium content ranges from −55 to +11 ‰. This suggests these waters are 'connate' waters of the sedimentary basin (metamorphosed by the salinity content). Using the isotope evidence while studying the genesis of groundwater in the Dnepr-Donetsk depression, the authors found that the hypothesis of the juvenile infiltration origin of the Paleozoic waters helps to explain the fact that the content of the stable isotope of water (solvent) is close to that in recent seawaters.

The carbonic acid saline groundwater in the Ukrainian Carpathy region (Babinets et al. 1971; Vetshtein et al. 1972) mineralised to 4–13 g/l and taken from the internal flish zone have, in three cases, shown a high oxygen-18 content (−3 to +6‰) and in the fourth case it was −8.6‰. The authors related the first three samples with relicts of the ancient seas and the last one with water of meteoric origin. In aquifers of the same region used for balneological purposes (Vetshtein et al. 1972) δD varies between −108 and −67‰ and $\delta^{18}O$ from −13.1 to −9.4‰. Therefore these waters are recharged from surface waters despite considerable differences in gas and salt composition.

Deuterium and oxygen-18 content was studied in deep waters of the Pripyat depression oil field (Tkachuk et al. 1975). It was found that waters from the aquifer, situated above the salt-bearing sediments, have a δD content between −95 and −75 ‰ and $\delta^{18}O$ content between −12.7 and −10.5 ‰ and are of meteoric origin. The brines situated between the salt-bearing sediments with δD values ranging from −65 to −10 ‰ and $\delta^{18}O$ from −9.7 to + 4‰ and waters below the salt-bearing sediments with δD values ranging from −49 to − 8 ‰ and $\delta^{18}O$ values ranging from −6.4 to + 2 ‰, were attributed to ancient marine origin by the authors.

While studying the oxygen-18 content in saline formation waters sampled in the Siberian Platform, it was found (Pinneker et al. 1973, 1974, 1975) that in the sulphate hydrocarbonate fresh and saline waters $\delta^{18}O$ ranges from −20.3 to −17.8 ‰. In the sodium and calcium chloride saline waters and brines of salt-bearing and above salt-bearing sediments, sulphate hydrocarbonate ranges from −20.1 до

Table 9.7 Deuterium content in Japanese gas-field formation waters. (Kobayakawa and Horibe 1960)

Sampling area	Sampling depth, below surface (m)	δD (‰)
Niigata wells	180	−48
	460	−77
	800	−42
	1500	−47
Shinagawa wells	1100	−9
	1300	−7
Chiba wells	136	+15
	505	−5
	507	−8
	508	−8
Miyazaki mine	–	+2

−15.5 ‰; for the calcium chloride brines from −7 to −1.3 ‰; and for the calcium chloride brines below the salt-bearing sediments from −8.7 to −7.3 ‰. The relative ^{18}O content in the surface waters in this region ranges from −19.9 to −15 ‰ (in the Baikal Lake $\delta^{18}O = -16.5$ ‰ is characteristic of meteoric waters south of Eastern Siberia). On this basis the authors concluded that the sulphate and hydrocarbonate calcium-sodium waters have a completely meteoric origin and the deep brines originate from waters of marine genesis, being markedly diluted with meteoric waters. In spite of the low temperatures (below 44 °C) and almost complete absence of carbonic acid in brines, oxygen isotope exchange with rocks had obviously not markedly affected the formation of their isotopic composition.

The study of the deuterium content in concomitant waters of Japanese gas fields was reported by Kobayaki and Horibe (1960) (Table 9.7): data recalculated relative to the SMOW standard. The deuterium concentrations in the formation waters of Niigata, Shinagava, Chiba and Miyazaki Prefectures are approximately equal to those of seawater. The authors suggest that this results from the penetration of seawater into the gas field waters. Unfortunately, Kobayaki and Horibe did not give the geological structural peculiarities of the profiles studied. It is obvious that such a conclusion, though probably true, cannot be drawn from the deuterium content alone, since similar deuterium concentrations could be exhibited by ancient sedimentary waters.

In the concomitant waters of the Niigata field the deuterium concentrations were reported to be close to the Tokyo standard (0.01489 ± 0.00005 at. %), which is the tap water of Tokyo University. The similarity of the isotope ratios for those two waters should be explained by the percolation or pumping of the local surface waters into the productive gas layers. The particularly low deuterium content has been found in waters at depths of about 400 m. The different chemical composition of these waters and those lying above or below the mentioned level indicates differences in their origin.

Table 9.8 The results of oxygen isotope measurements in the formation of some oil and gas fields in the United States. (Degens et al. 1964)

Geological age of water-bearing rocks	Number of samples	Total dissolved solids (g/kg)		$\delta^{18}O$ (‰)	
		Variation range	Average	Variation range	Average
Marine sediments					
Cambrian	12	77–206	146	−9.98 to (−1.39)	−4.1
Ordovician	4	231–262	256	−0.06 to (−1.77)	±0.4
Devonian	1	132	132	−0.46	−0.5
Carboniferous	6	79–269	153	−2.63 to (±2.05)	±0.7
Cretaceous	9	2.9–5.4	4.0	−11.67 to (±10.29)	−11.08
Tertiary	2	103–104	104	±3.04 to (±3.34)	±3.2
Freshwater sediments					
Tertiary	10	0.2–15.8	5.6	−16.86 to (−3.03)	−12.7
Data for comparison					
Oceans	–	–	35	–	0.0
Surface waters (Utah)	–	–	0.5	–	−16.8
Great Salt Lake	–	–	220	–	−7.4

Degens et al. (1964) studied the oxygen isotopic composition in 44 samples of different ages from water-bearing layers in the oil and gas-bearing fields of Oklahoma, Texas, Colorado and Utah, USA. The data are shown in comparison with those for surface water in Table 9.8 and are recalculated relative to the SMOW standard.

Both in highly and partly mineralised waters there is a tendency for the ^{18}O content to increase with increasing mineralisation. The $\delta^{18}O$ values in highly mineralised waters are similar to those for modern ocean waters. Negative deviations from the ocean mean values of ^{18}O agree well with decreasing mineralisation, resulting in the penetration of the recent meteoric waters or dilution with ancient infiltration waters, during the geological history at positive tectonic dislocations. The positive deviations of ^{18}O content in some samples were explained by the authors to be a result of continuous evaporation before the waters were connected, for example, during the isolation of small parts of the ocean from the whole basin, i.e., during lagoon formation.

From this data the authors (Degens et al. 1964) concluded that the brines under study represent ancient marine water, assuming that the ratio in seawaters has remained more or less unchanged since Cambrian times. This suggestion is confirmed by the fact that calculations of fossils of Paleozoic age, according to Compston, have the same ^{18}O content as analogous modern fossils. Further, Degens pointed out that the similarity of the isotope characteristics of ground brines and less mineralised recent ocean waters, which suggests that concentration of inorganic salts was not completed by evaporation. The completion of such a process might possibly occur during the process of compaction of sediments accompanied by ion infiltration

Table 9.9 Deuterium and oxygen-18 content in some saline waters and brines of California, United States. (White 1965)

Sampling location	Chemical composition	δD(‰)	$\delta^{18}O$ (‰)
Cympric oil-field, 900 m depth from	$M_{17.7}\frac{Cl}{Na}$	−17.1	+2.93
Miocene sandstones			
The same place, 1400 m depth	$M_{17.7}\frac{Cl}{Na}$	−16.1	+3.14
The same place, 1600 m depth from	$M_{25.9}\frac{Cl}{NaCa}$	−11.4	+5.93
Eocene sandstones			
El Dorado oil-field, 870 m depth from	$M_{146.4}\frac{Cl}{NaCa}$	−22.0	–
Mississippian limestones			
Wilbur oil-field, from Cretaceous rocks	–	−17.9	+3.07
Wilbur Springs, mixture of 'relict' and meteoric waters	–	−22.2	+5.58
Tuscany Springs, 'relict' waters	–	−13.8	+5.27
Searls Lake, non-marine evaporitic brines	–	−26.0	+4.20
Sulphur Bank, metamorphic waters	–	−24.1	+5.62
Salton Sea, thermal waters	–	−75.3	+3.27

through the clays. Degens showed that isotopic data might serve as an indicator of the proportions of meteoric and 'connate' water in samples under study.

White (1965) also reported his results of isotopic composition of formation waters of oil-field sedimentary rocks. He carried out rather detailed investigations in the course of which both oxygen-18 and deuterium isotopes were determined (Table 9.9).

White, citing Friedman, pointed out that the majority of highly saline oil-field brines have a lower deuterium content than those of seawater, ranging from 0 to −50‰. As Table 9.9 shows, the deuterium variations for the investigated waters from oil-bearing sediments fall within these limits and are typical. Brines with lower deuterium content are not considered to be of seawater origin.

From Table 9.9 it follows that according to the results obtained heavy oxygen is present in larger amounts than in modern seawaters. The author explained this in terms of equilibrium exchange reactions of ^{18}O between water and water-bearing rocks at high temperatures, which takes place in deep parts of pressurised water systems. Therefore, White stresses that the temperature of waters and the degree of equilibrium between the brines and solid phases are factors that should always be taken into account during more detailed investigations.

Miller et al. (1966) pointed out that oil-field brines are similar in heavy oxygen isotope content to normal seawater and may exhibit mineralisation to a great extent.

The insignificant enrichment in ^{18}O content might be due to low evaporation occurring during sedimentation. They pointed out another process is involved, that of isotopic exchange with water-bearing rocks and also emphasized that ^{18}O decrease in oil brines is always accompanied by the process of dilution of the aquifer with meteoric waters.

The data on ^{18}O isotope ratios were used by these authors to explain the origin of hot acid highly mineralised (310 g/l) brines, localised in the central part of the Red Sea at a depth of about 2000 m. Before the investigations it was supposed that studies would confirm one of three possible hypotheses of their origin: (1) due to inflow of brines, concentrated by local surface evaporation; (2) due to evaporation from the whole surface of the Red Sea during its isolation stage; (3) due to submarine outflow of the brines.

The ^{18}O concentrations in brines and surface waters in the Red Sea and other seawaters insignificantly enriched by evaporation are similar to each other. Therefore, the observed values of oxygen isotope ratios do not support the idea that the investigated brines are evaporates of the normal Red Sea waters. In the last case the ^{18}O content should be considerably higher. In relation to this, the authors Miller et al. (1966) concluded that the hypotheses of submarine discharge of deep groundwaters into the Red Sea are more convenient. Later, Crag (1969) studied in detail the abundances of deuterium, oxygen-18 and argon dissolved in water and the temperatures and salinities of the Red Sea waters. He concluded that Red Sea brines are forming in the near-surface layer of the southern part of the sea near the Bab-el-Mandeb Strait under conditions of high temperature and salinity. An important role of the process that took place in the past, during glacial times when the sea level had markedly dropped and in more distant times when the sea was dried up forming large amounts of evaporates, has been emphasised by the author. Craig's model uses a similar argument. White (1974) pointed out that, in accordance with this model, the existence of the groundwater flow 500–900 km long and passing along the Middle Rea Sea valley should be assumed. Moreover, the brines, in some incomprehensible manner, avoid some other basins that are deeper than those containing hot brines.

There are also some other models, explaining the formation of the chemical and isotope composition of the Red Sea brines, which are free of the hydrogeological contradictions present in Craig's model. For example, White (1974) assumed the hot brines to be recharged from the nearby edges of the Red Sea, which are represented by highly evaporated seawaters often diluted with meteoric waters due to discharge from inland Saudi Arabia. In any case White noted that the hot brines are genetically connected with the Red Sea waters.

The most detailed studies of isotopic composition of the formation waters in the United States were carried out by Clayton and Graf with co-authors (Clayton et al. 1966; Graf et al. 1965, 1966). They studied formation waters in the Illinois, Michigan, Alberta and Gulf Coast artesian basins, which have a relatively simple geological structure, vast geologic documentation and a considerable number of boreholes available for sampling. In the profile of the Michigan basin there are thick deposits of salts and anhydrites and in the Illinois basin only anhydrites of restricted thickness were found. According to the data of isotope studies, it has been discovered that meteoric waters play a significant role in the formation of deep saline waters

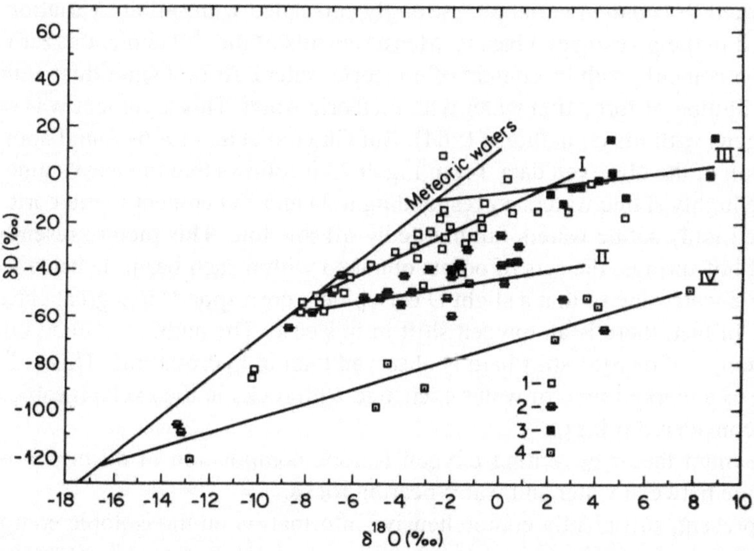

Fig. 9.23 Relationship between $\delta^{18}O$ and δD values for Illinois (*I*), Michgan (*II*), Gulf Coast (*III*) and Alberta (*IV*) basins. (1) Ocean water; (2) meteoric waters throughout the world. (Clayton et al. 1966; Ferronsky and Polyakov 2012)

of the Alberta and Gulf Coast basins. Ninety-five water samples were taken in order to determine D and ^{18}O content. Deuterium content variations of about ±20‰ were indicated in each basin. In the Michigan and Golf Coast basins the total mineralisation is, in general, insignificantly related to the deuterium concentration. In the Illinois and Alberta basins the deuterium content increases proportionally with mineralisation.

In general, while comparing data obtained for individual basins, marked variations in deuterium content were observed. For example, the Gulf Coast basin waters are similar in deuterium content to seawater, while in the Alberta basin the deuterium content of water is about 100% lower than that in the ocean.

More objective conclusions on the water's genesis, as emphasised by the authors, may provide a hypothesis that original marine waters were pressed-off during compaction from syngenetic rocks and were then mixed with meteoric waters formed under climatic conditions close to modern ones.

Clayton et al. (1966) reported that the main process affecting the observed D/H variations from basin to basin is climatic change with time, which has altered the isotopic composition of meteoric waters. The effect of climatic variations can be demonstrated by a few extreme values indicating very low D and ^{18}O contents, which correspond to climatic conditions during the Pleistocene glaciation.

The authors concluded that the processes of isotope exchange between water and other hydrogen-bearing geological objects as a result of hydrogen dispersion to be of no importance since only slight variations in deuterium content were detected in each basin. These slight variations should more likely be due to isotope fractionation by filtration of water through micropores in clay minerals (Clayton et al. 1966).

Marked ^{18}O content variations, strongly correlated with salinity variations, were detected in the investigated basins. Measurements of the ^{18}O content at zero mineralisation coincide with its content of meteoric waters. At first sight this seems to be to the dilution of formation water with meteoric water. This argument was reported by Degens with his co-authors (1964). But Clayton et al. (1966) found another explanation of the observed data. From Fig. 9.23 it follows that the least saline waters and the highly saline waters corresponding to D and ^{18}O content in meteoric waters and the highly saline waters fall markedly off this line. This picture resembles the case of hot springs: the sets of points obtained within each basin lie aside from the meteoric water line so that a slight D enrichment corresponds to a great enrichment in ^{18}O. In fact, there is an oxygen shift in this case. The authors pointed out that a wider range of oxygen shift here is observed than in hydrotherms. This is likely to be due to a marked time of water exchange with rocks in the exchangeable system of the considered brines.

The main factor governing oxygen isotope composition in oil brines is water exchange between water and water-bearing rocks.

At present, sufficiently comprehensive information on the isotopic composition of deep groundwater in sedimentary basins in the USA, former USSR territory and other countries is available, making it possible to further develop isotope techniques and apply them in solving problems of deep groundwater origin. The most successful solution of problems concerning the conditions affecting the formation of sedimentary basin waters may be attained by the interpretation of both isotopic composition data and other hydrological and hydrochemical evidence (IAEA 1976; Bath et al. 1979; Sonntag et al. 1979 and others).

9.8 Isotopic Composition of Groundwater in Volcanic Regions

As well as deep groundwaters of sedimentary basins, groundwaters of modern volcanic regions are of interest to specialists using isotope techniques in their studies, in order to elucidate the problems of their formation and solve a number of applied problems of interest concerning their utilisation. Let us consider the most important results obtained during these investigations.

9.8.1 Isotopes in Studying the Origin of Thermal Waters

Kirshenbaum, Graf and Forstat (Kirshenbaum 1951) determined deuterium content in two specimens of vapour condensate from the Steamboat thermal springs in Nevada (United States) as far back as 1945. The high precision of deuterium measurements, which were carried out using a technique independent of oxygen isotope concentration, allowed them to account for the obtained data as well as subsequent mass-spectrometer measurements. The deuterium content in the samples

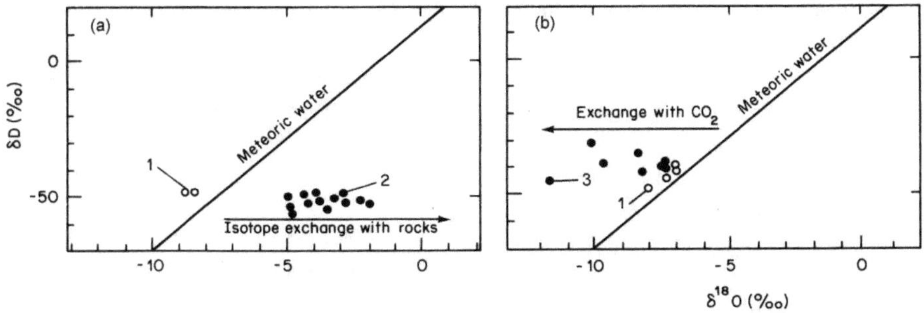

Fig. 9.24 Isotopic composition of selected steam and represented local meteoric water sampled-from the Larderello area (**a**) and Tuscany (**b**): (*1*) atmospheric water; (*2*) water vapour; (*3*) hydrogen-sulphide springs water. (Ferrara et al. 1965; Panichi et al. 1974; Ferronsky and Polyakov 2012)

was −120 and −107‰ respectively (in the SMOW standard). The low deuterium content was explained by the researchers as being a consequence of different HDO and H_2O volatilities during the evaporation of thermal waters and not a reflection of the isotopic content of water before its discharge to the surface where it could be subjected to evaporation.

Boato et al. (1952) studied the hydrogen isotope abundances in the Larderello region hydrotherms both in water vapour molecules and molecular hydrogen, which amounted to 1/1000 of the total vapour from the tube. Collecting vapour from all the wells showed that the deuterium content varies within ±0.1 of the working top water standard (water from the local water-supply system). The data for samples taken from an individual water-vapor borehole varied within the same limits. The deuterium content in hydrogen gas was reported to be lower than in water vapour, equal to 0.79±0.05 of the standard (Rankama 1963). Boato et al. (1952) pointed out that at 600±100 °C (corresponding to above-critical temperatures at considerable depths) the reaction $H_2O + HD \rightarrow HDO + H_2$ takes place. In such isotopic exchange reactions deuterium is concentrated in water whilst gaseous hydrogen is depleted in deuterium. But the expected enrichment of vapour in deuterium has not been found due to the small amount of molecular hydrogen (0.001 for the vapour volume) whereas the low deuterium content in the gaseous hydrogen was detected.

Later on, the isotope composition of the Larderello region hydrotherms was investigated by Ferrara et al. (1965) and Panichi et al. (1974). Because some Italian hydrogeologists believe that Larderello therms are recharged by atmospheric precipitation, typical samples of atmospheric water were taken, in the region of their supply several kilometres to the south of Larderello. The D and ^{18}O values there range from −48‰ to −41‰ and from −7.3 to −8.1‰, respectively. The concentration of the same isotopic composition in vapour carefully sampled from several boreholes in Larderello varied from −48 to −43‰ and from −4.8 to −2.3‰ respectively.

Plotting the meteoric water line for the region $\delta D = 8\ \delta^{18}O + 10$ and the results of isotope analysis of atmospheric water and sampled vapour on the $\delta D - \delta^{18}O$ diagram, one obtains the picture shown in Fig. 9.24. It follows from this figure that

Table 9.10 Isotopic composition of Tuscany springs water (the data recalculated to the SMOW standard). (Ferrara et al. 1965)

Spring location	Type of spring	Sampling data	δD (‰)	$\delta^{18}O$ (‰)
Momialla (Florence)	Hydrogen-sulphide, t=13–18 °C	23.09.1965	−40	−7.3
Petriole (Siene)	Hydrogen-sulphide, t=44 °C	7.04.1965	−41	−7.0
Bollore (Florence)	Hydrogen-sulphide, t=45°C	22.09.1965	−39	−9.3
Le Puzzolr (Siena)	Hydrogen-sulphide, cold	08.06.1965	−32	−9.9
Perdgine (Arezzo)	Mofette	26.03.1965	−46	−11.7
Doccio (Siena)	Hydrogen-sulphide, cold	07.04 1965	−38	−8.1
Torrike (Florence)	Mofette	22.09.1965	−42	−7.8

deuterium content in vapour is analogous to that in atmospheric waters whereas the ^{18}O content is shifted towards deuterium enrichment of the vapour. This fact was accounted for by the authors in terms of the process of oxygen isotope exchange between thermal water and water-bearing carbonates. All the carbonate specimens taken during well drilling in Larderello were depleted in oxygen-18 compared with common marine limestone.

Final evidence connected with oxygen-18 and deuterium abundances in thermal and meteoric water supports the hydrogeologists concept that water vapour from the Larderello springs should be of meteoric origin and that its "juvenile" component (if it exists) should not exceed 5 %. The deep groundwater circulation, assumed for interpretation of the data on ^{18}O concentrations, is also in accord with the geologists' assumption that there is a continuous convective water circulation due to the existence of a large deep heat source. In this case deep groundwater may emerge at a depth of about 5000 m. The authors also discussed a question regarding isotopic composition of hydrogen-sulphide springs and mofettes of Tuscany (Table 9.10).

Large amount of gases, among them CO_2 is predominant, are a feature of waters from all the springs indicated in the table. From Table 9.10 and Fig. 9.24b it follows that deuterium content in spring waters is generally in accord with its content in meteoric waters. The only exception is the Le Puzzole hydrogen-sulphide spring that is insignificantly enriched in deuterium and situated not far from Pienzo (Siena). This may be accounted for by evaporation of water, since the sample was taken from the surface of the spring's catchment area.

The ^{18}O concentration in spring waters, compared with meteoric waters, is shifted towards depletion in this isotope. The authors explained this by isotope exchange between water and carbon dioxide, which follows the reaction $C^{16}O_2+2H_2{}^{18}O \leftrightarrow C^{18}O_2 + 2H_2{}^{16}O$. The fractionation factor of this reaction at a temperature of 25°C is equal to 1.04. This means that $^{18}O/^{16}O$ in the carbon dioxide proceeds easily and quickly. Therefore, if the ratio of $^{18}O/^{16}O$ for carbon dioxide was lower than in water, then the former should tend towards enrichment in heavy oxygen at the expense of the water.

Table 9.11 Deuterium content in Yellowstone National Park gas samples (the data recalculated to the SMOW standard). (Friedman 1953)

Sampling place	Type of sample	δD (‰)
Daisy Geyser	Water	−158
Punch Bowl Spring	Water	−158
Hurricane Vent Spring	Gas	−500
Daisy Geyser	Gas	−278
Punch Bowl Spring	Gas	−240
Iron Creek	Gas	−240
Kaleidoscope Geyser	Gas	−247

Thus, the lower the ratio of the amount of water to carbon dioxide involved in the reaction the more the ^{18}O content of the water decreases. The water from the well in Perdgine (Arezzo), where the maximum deviation of ^{18}O content from the initial value of one has been found, provides an example of such a case. The water from the Perdgine well has the maximum gas-water ratio and is used in obtaining carbon dioxide. The process of oxygen exchange between carbon dioxide and water may be supposedly considered as being predominant over the opposite process of oxygen exchange between water and rocks reduced at low temperatures.

Friedman (1953) investigated thermal waters and gases in the Yellowstone National Park (United States) (Table 9.11).

It follows from Table 8.11 that the deuterium content in gases is markedly lower than in waters, although there are also comparatively low concentrations of this isotope in certain gases. The results of measurements obtained from Hurricane Vent Springs are distinguished among the other gas samples. Hydrogen-bearing gases (0.5 % of molecular hydrogen and 20 % of methane) were also detected in this spring, whereas there were no molecular hydrogen gases in the water from any other studied springs. According to Friedman, the same mechanism of interaction between water and gaseous phases, as was proposed for the Larderello thermal waters, is evident. Assuming that gases from all the springs were in isotopic equilibrium with water (taking no account of methane from the Hurricane Vent Spring), a minimal temperature of about 400°C, corresponding to the isotope exchange reaction $H_2O + HD \leftrightarrow HDO + H_2$, has been calculated using hydrogen.

White et al (1963) investigated the thermal Steamboat Springs (United States). The isotopic composition of the springs exhibits slight seasonal variations. The content of heavy isotopes rises markedly in June, reaches a maximum in August and drops to a normal level in October. The maximum corresponds to the warmest period with the most intensive evaporation from the surface waters including a nearby lake. The non-equilibrium character of the evaporation process from the open reservoir, which is very sensitive to vapour-liquid system isotopic composition, manifested itself in the course of the spring water isotope analyses. The mean relative deuterium content in water was −90‰ and oxygen-18 was −11 ‰. At the same time in the other springs (e.g., the Galena Spring) δD and $\delta^{18}O$ were about −113 and −15.3 ‰ respectively, i.e., the heavy isotope content follows the relationship $\delta D = 8\delta^{18}O + 10$, which is a feature of atmospheric precipitation. In hot springs

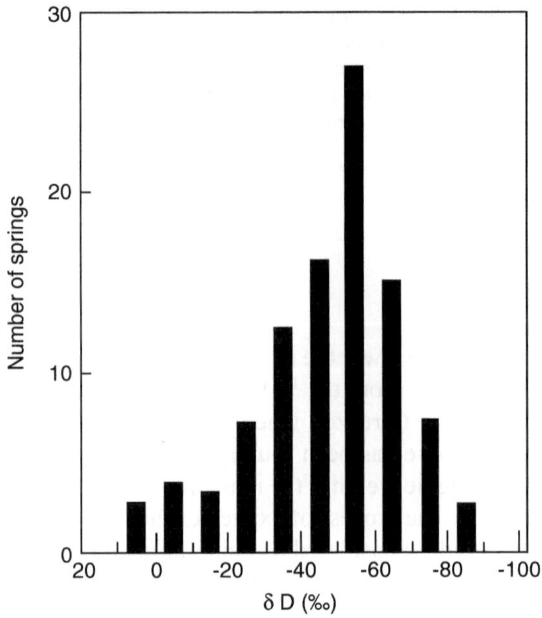

Fig. 9.25 Histogram of deuterium content in groundwater of Japan. (Kobayakava and Horibe 1960; Ferronsky and Polyakov 2012)

the authors detected the effect of oxygen exchange between silicates and water, being about 2.0–3.5‰. The majority of water appears to be absent. If there is any it amounts to less than 5% and is out of the range of the technique's accuracy.

Deuterium content investigations were carried out in the thermal springs of Japan (Kobayakava and Horibe 1960). At first, the aim of this measurement was to discover the springs whose water, rich in deuterium, could be used in heavy water production. As Kobayakawa pointed out a tendency of trying highly productive springs with the highest temperatures (these conditions are most favourable for technological schemes of deuterium extraction from water) was not completely successful from a geochemical viewpoint. About 230 measurements of deuterium concentrations in various samples of water were carried out. The deuterium content varied from −103 to +3‰ relative to the SMOW standard. In 70% of samples deuterium concentrations ranged between −80 to −40‰. The histogram of deuterium content in these samples is shown in Fig. 9.25

On the basis of the results obtained for thermal waters of individual groups of springs some trends are discernible. For example, the deuterium concentration in all the springs of the Ibusika group is proportional to the concentration of chlorine ions and inversely proportional to the temperature. Besides, deuterium content in the water of these springs is markedly higher than in the Tokyo tap water standard (0.0148 at. %). According to the location of these springs it is reasonable to assume that their water is diluted by seawater. But if there has been just dilution of the surface waters with seawater, the deuterium concentration in water of these springs should fit the line, reflecting the different proportions of sea and inland waters. But in fact deuterium concentrations in waters of these springs are higher than the corresponding chlorine ion concentrations. Thus, one may assume either that only the chlorine ions are removed from the seawater in underground conditions or that only that portion of water that originally had a higher deuterium content appears on the surface. Since the process of chlorine ion extraction is unlikely, the authors consider that water discharging at the surface had a higher deuterium concentration before mixing with seawater. In support of this statement it has been pointed out that condensational water from the Hombosudzigoku spring has rather a high deuterium concentration (−40 ‰).

The groundwaters in the region of the Tamagawa group spring, used for potable supply, have a deuterium content −73 ‰ and the thermal springs in this group, Abuki and Hisogawa, have a deuterium content of water of −50 and −53 ‰. The condensational water of fumaroles has approximately the same composition. The water, formed as a result of the mixing of groundwater and fumarole water, exhibits an intermediate deuterium content, equal to −68 ‰.

Finally, Kobayakawa pointed out that on the basis of his results it is impossible to draw definite conclusions. For the complete study of the hot spring water's genesis one should investigate the chemical composition, deuterium content and heavy oxygen content in groundwaters and spring waters.

Friedman et al. (1963) studied various natural waters in Iceland among which were hydrotherms and waters of deep boreholes. On the basis of interpretation of deuterium concentrations in hot and groundwaters and in precipitation it was found that water from the borehole and geyser near Reykjavik is formed from the glacier located a considerable distance northward of them. The same conclusions were drawn by Craig who measured D and ^{18}O in concentrations in hot springs and surface waters around a power station. It was found that the catchment areas of the thermal and cold springs are different. The lowest deuterium content was detected in two boreholes in Langaland, being −115 and −117 ‰. Rather high deuterium concentrations, which were detected in some boreholes (up to −8 ‰), may result from the dilution of hot water with oceanic water.

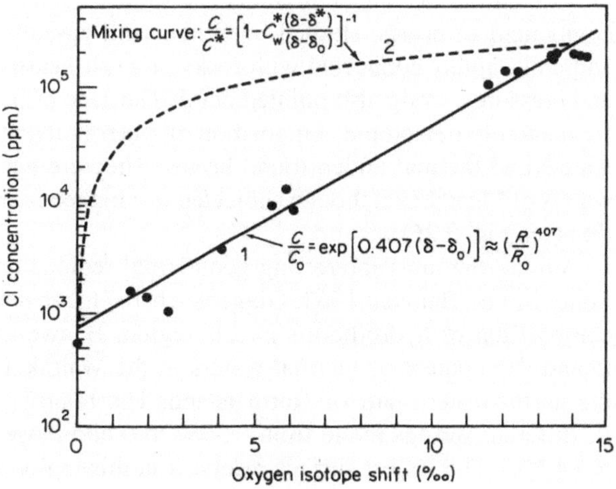

Fig. 9.26 Relationship between chloride concentrations and oxygen isotope shift ($\delta - \delta_0$, $\delta_0 = -11$ ‰) in the Salton Sea geothermal brines: (*1*) salt leaching by surface water curve; (*2*) mixing of surface water and hypothetical deep water curve. (Craig 1966; Ferronsky and Polyakov 2012)

The results of investigations by Arnason and Sigurgeirsson (1967) indicate that during the movement of thermal waters from the catchment area to the area of emergence, deuterium content remains unchanged. It was also found that deuterium content in hot springs differs markedly from that in precipitation in the region of these springs. The researchers concluded that thermal waters in South-Western Iceland are recharged in a region located at a distance of more than 50 km from the place of their emergence at the surface.

The possible applications of the isotope analysis of hydrotherm waters were indicated by Craig (1966) while studying thermal brines in the process of borehole drilling around the Salton Sea in California. The temperature of these highly mineralised (332 g/l) brines, obtained by extrapolation of the bottom of a 1400 m borehole, was 340°C. When the brines were first discovered in the first borehole in this region White et al. (1963b) assumed they were ore-bearing magmatic fluids (i.e., juvenile waters).

Craig determined D and ^{18}O content in these brines and other surfaces and groundwaters in the region. The isotope ratios obtained (see Fig. 9.26) illustrate the process of oxygen shift observed in nature. In this case, at a constant deuterium content equal to that in local meteoric waters, as usually occurs in regions of hydrotherm formation, the ^{18}O concentration increases in parallel not only with temperature but also with salinity. This fact may be explained in two different ways. Firstly, it may result from simultaneous enrichment of water in ^{18}O and chlorine due to the interaction of surface waters with the water-bearing rocks and, secondly, as a result of deep water (i.e., hypothetical ore-bearing fluid) mixing with the downward moving surface waters. Craig pointed out curves corresponding to both of these cases. These curves are shown in Fig. 9.26. But the experimental data shown in the figure indicate conclusively that in fact only the first case holds. Therefore, formation of brines in this case is related to the dilution of salts of water-bearing rocks by local surface waters. Craig (1966), in a note to this work, pointed out that White, after having acquaintance with his results, agreed that his 'ore-bearing fluid' is of meteoric origin.

The results do not contradicts Smirnov's conclusion (1971) that the mineral component of brines, studied by White with co-authors and Craig, is genetically and indissolubly connected with rocks of a salt-bearing basis, at high temperatures and pressures. Craig also pointed out that surface brines of the Salton Sea Lake differ markedly in isotopic composition of water (solvent) from that of local meteoric waters and thermal underground brines. They are genetically of the same type as waters of Mead Lake, though subjected to a higher degree of evaporation concentration (see Fig. 9.26).

Among the most interesting geothermal regions on the Earth is New Zealand. Some works (Banwell 1963; Giggenbach 1971) are devoted to the study of isotopic composition of hydrotherms in this region. Banwell (1963) reported data on the D and ^{18}O content of thermal waters in the Wairakei region, which is situated in the north-western part of North Island. The relative deuterium concentrations in its different springs range from −55‰ to 0 and those of oxygen-18 vary from −5 to +4.5‰. The largest surface reservoir in this region is Taupo Lake, for which δD is −30 ‰ and $\delta^{18}O$ is −5.2 ‰. Banwell, citing earlier works by Craig and his co-authors, inferred that all the hot springs in this region are recharged by local surface waters.

Giggenbach (1971) investigated the isotopic composition of Broadlands, New Zealand, geothermal waters that are of meteoric origin. It was found that the borehole waters, at temperatures from 170 to 235 °C, have a deuterium content ranging between −41.8 to −33 ‰ and that of oxygen-18 from −4.4 to −3.6 ‰. The water vapour has δD and $\delta^{18}O$ values varying from −45.2 to −38.7 ‰ and from −7.0 to −6.0 ‰ respectively. The difference in isotopic composition between water and vapour corresponded to equilibrium between the two phases at these temperatures.

The authors with their colleagues investigated isotopic composition of hydrothrms in the following three regions: in the valley of the Paratunka River, in the Uzon caldera and on Kunashir Island at Kamchatka. The deuterium and oxygen-18 content of waters in all these regions varies in parallel (if these regions are considered as a whole): the lowest deuterium concentrations correspond to lower concentrations of oxygen-18. Each of these regions has its own peculiarities in isotope abundances.

The largest values of D and ^{18}O are typical for Kunashir Island. The deuterium concentrations, ranging from −80 to −32.9 ‰, do not differ from those in waters of a number of sedimentary basins. This may be the result of the following two reasons. The first and the main reason follows from the fact that precipitation, falling on the island, is a condensed vapour, evaporated from the sea surface. The second reason is a latitudinal effect.

The deuterium content for hydrotherms of the Paratunka River valley varies between −123.4 and −105.7 ‰ and for hydrotherms of the Uzon caldera it ranges from −109 to −68.4 ‰. The observed difference in isotopic composition of waters from these two regions may be explained by the recharge of the first system from surface waters of higher references.

Relatively high oxygen-18 content for hydrotherms of Kunashir Island, in which $\delta^{18}O$ varies from −11.9 to −4.2 ‰, may be explained by its marine origin and not by the oxygen shift, the value of which is negligible. The $\delta^{18}O$ content in hydrotherms of the Paratunka valley varies from −16.9 to −12.6 ‰ and in those of the Uzon caldera it ranges from −13.4 to −5.22‰, which results both from the initial isotopic composition of meteoric waters recharging hydrotherms and, partially, from oxygen isotopic exchange between water and rocks. On the basis of the isotope shift one may come to the conclusion that during the formation of hydrotherms a minimal depth of water circulation is observed on Kunashir Island, a greater depth for the Paratunka River valley and a maximum depth for the Uzon caldera.

The thermal waters in the Kurile-Kamchatka volcanic area were investigated by Baskov and Vetshtein (Baskov et al. 1973; Vetshein et al. 1971; Meniaylov et al. 1981). They found that in this region ^{18}O ranges from −12.9 to +3.4 ‰. The maximum value corresponds to a condensate of fumarole gases. The relative deuterium values detected on the Kurile Islands gave −51 to −88 ‰ and those in the Kamchatka region range between −55 and −113 ‰.

Many researchers, while studying individual questions concerning the genesis and conditions of groundwater formations, at the same time tried to estimate general isotope criteria for the 'juvenility' of groundwaters.

Urey (1957) reported that the Earth's gravitational field is insufficiently strong to retain hydrogen and that our planet gradually loses hydrogen, which dissipates into interplanetary space. But due to the difference in atomic weights of hydrogen isotopes, protium dissipates more easily than deuterium.

In Rankama's opinion this process must result in accumulation of deuterium in the atmosphere and surface waters over geological time. Quoting Harteck and Suess, he reported that juvenile water will be depleted in deuterium compared with vadose water, although different mechanisms of isotopic fractionation and dilution of the magmatic waters with vadose water can make the actual picture rather complicated.

From logical considerations it follows that the hydrogen of water involved in the magmatic activity of the Earth should in principle reflect the isotopic composition of hydrogen, which was present during the final stages of the formation of the Earth's crust. It seems the isotopic composition of water should also 'remember' individual stages of the Earth's sedimentary shell formation. Thus it is of interest to study the deuterium content of parental seated brines and in crystallised water of minerals in the Earth's sedimentary shell, which record data throughout hundreds of millions of years.

There is no conclusive opinion on deuterium content in 'juvenile water'. Friedman (1953), Godfrey (1962) and Ferrara et al. (1965) stated that the value of deuterium content in 'juvenile' water should be lower than in surface water. Kobayakawa and Horibe (1960) reported that 'juvenile' waters should be enriched in deuterium. Godfrey also gave the absolute value of deuterium content (obviously an averaged one) obtained on the basis of interpretation of isotopic composition of water in magmatogenic minerals, equal to about −110‰.

A wide range of possible deuterium variation in juvenile waters from −130 to −25‰ has been reported by Ferrara e al. (1965). They quoted Craig in that even though the isotopic composition of juvenile waters is unknown, in some ways it can be determined on the basis of reasonable assumptions. This value should be based on deuterium variations in minerals.

It is worth noting that the discussed problem of juvenile waters has a purely philosophical character because the problem of the origin of the Earth itself has not yet been solved.

Craig (1963) carried out a comparative analysis of the isotopic composition of water and vapour samples from the main geothermal regions of the world. Figure 9.27a shows isotopic data, obtained in the most famous geothermal regions, which correspond to volcanic vapour only, typically neutral or slightly alkaline hot springs with $pH=5$–9. The deuterium content in these waters and vapour correspond to that in the local meteoric waters. The ^{18}O content indicates a typical enrichment that results from oxygen shift, as the author stated. The range of ^{18}O variations relative to meteoric waters is from zero in the New Zealand springs to 14‰ in the Salton Sea region. The last value is the greatest among all detected for oxygen shift

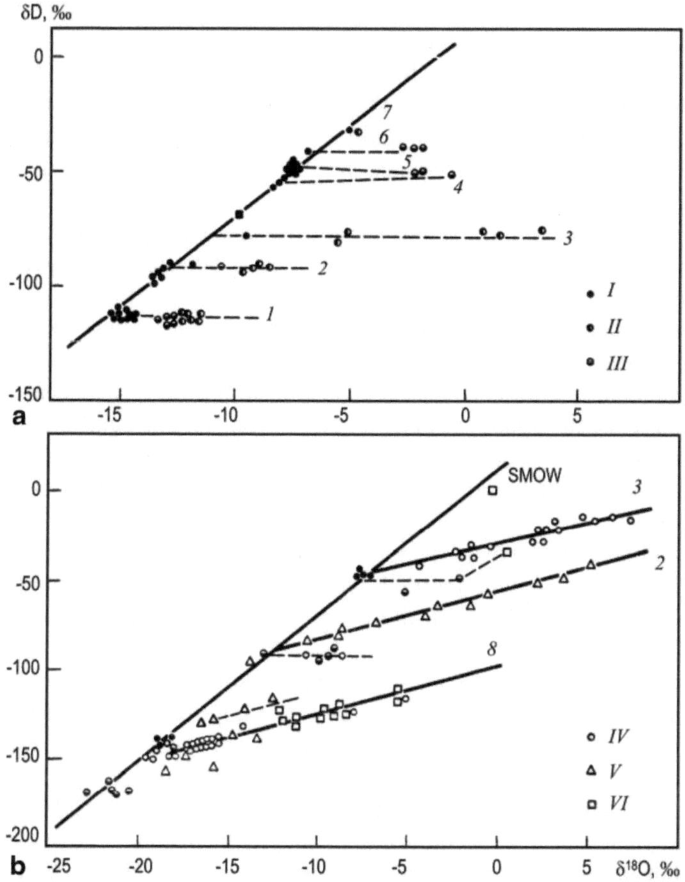

Fig. 9.27 Observed isotope variations in meteoric, near-neutral chloride type geothermal waters and in steam (a) and in acid, alkaline type hot springs, in near-neutral waters and in steam (b) of some geothermal areas: (*1*) Steamboat Springs (USA); (*2*) Lassen Park (USA); (*3*) Salton Sea (USA); (*4*) Hekla (Iceland); (*5*) The Geysers (USA); (*6*) Larderello (Italy); (*7*) Wairakei (New Zealand); (*8*) Yellowstone Park (USA); (*I*) local meteoric waters; (*II*) geothermal waters; (*III*) geothermal steam; (*IV*) acid type geothermal waters; (*V*) Lassen Park; (*VI*) Yellowstone Park. (Craig 1963; Ferronsky and Polyakov 2012)

in hydrotherms. In this case, shown in Fig. 9.27a, for vapour and chloridised waters, the character of the relationship between isotopic ratios for oxygen and hydrogen is similar. This picture is only observed for vapour at high pressures and temperatures. Craig shows that the observed oxygen shift in vapour and water may result from oxygen exchange between meteoric water and rocks during the motion of water. As to juvenile water, it is not detected in the above-mentioned regions according to Craig and its amount lies out of the range of accuracy of isotopic investigations.

The relationship between isotopes of neutral, alkaline and acid waters in the previous regions, complementary to the Yellowstone Park, is shown in Fig. 9.27b. In acid waters the following fact is discernible. The straight lines plotted on the basis of experimental data for individual regions do not intersect at one point corresponding to the isotope content of juvenile waters. These lines pass almost parallel to each other, reflecting the relation between D and ^{18}O, is evidence in favour of the absence of isotope equilibrium during evaporation of meteoric water sampled in regions of springs and in individual experiments, involving the addition of some chloridized water and vapour.

In the studied geothermal regions Craig distinguished, by isotopic composition, two different types of vapour. In all regions vapour indicates an oxygen shift without any marked deuterium shift relative to meteoric waters. Enrichment of vapour in oxygen-18 ranges from 3‰ to 0 in the Wairakei region (New Zealand), where the shift is not detected due to the short residence time of the water. Another type of vapour is found in Yellowstone Park and Steamboat Springs, where deep circulation of water is fixed and vapour has been found to be in isotopic equilibrium with thermal waters. The studied samples from this region show that their isotope content corresponds to vapour-liquid isotope equilibrium in the system at corresponding temperatures.

Sakai and and Matsubaya (1974, 1977) used isotope and hydrochemical techniques while studying hydrothermal systems in Japan. They distinguished four types of isotopically and chemically different thermal water systems, which are described in Table 9.12.

In waters of Arima type, the presence of a prevailing amount of solvent of meteoric origin has been discovered with the help of isotope data (Fig. 9.28). A considerable oxygen shift is also a feature of these waters. In some brines, the $\delta^{18}O$ values are as much as +8‰ but one cannot conclude, as the authors reported, that these brines are of magmatic origin, since thermal waters exhibit higher $\delta^{18}O$ values compared with when they were at equilibrium with granite magma. The oxygen isotopic composition of these waters obviously reflects isotopic exchange with carbonate rocks at about 100 °C. Sakai and Matsubaya pointed out that these waters are similar both chemically and in isotopic composition to the formation waters of sedimentary basins in North America, which were studied earlier by Clayton et al. (1966) and Hitchon and Friedman (1969). (The works of these authors were cited earlier).

The water of Green Tuff formation has been found to be mainly of meteogenic origin. Among other places, similar waters were found in mines. While passing the formation thickness of the green tuff in the Seikan submarine tunnel, a mixture of marine and local meteoric waters was found. From Fig. 9.28 one can see that concentrations of deuterium and chlorine ions increase in parallel, which may be explained by the mixing process of the two types of water. Thermal waters of the Green Tuff type are characterised, in a number of regions, both by oxygen shift and some deuterium enrichment. On the basis of data obtained for the Seikan tunnel the authors concluded that, in a number of hydrotherms of the Green Tuff formation, formation waters are present. These waters are a mixture of marine and local precipitation waters. The increasing deuterium content in waters of this type is in accord with an ordinary process of mixing and the $\delta^{18}O$ increase results from the oxygen isotope shift (Fig. 9.28). It follows from this example that the process of isotopic

Table 9.12 Systems of thermal water in Japan. (Sakai and Matsubaya 1977)

System type	Geology	Chemistry	Example
Arima	Pre-Neogene plutonic and volcanic complex and metamorphic rocks	Na–(Ca)–Cl–HCO_3 High saline, 0 < Cl < 44 g/kg t = 20–97.5 °C	Arima Ishibotoke
Green Tuff	Green Tuff formations Miocene age	Na–Ca–Cl–SO_4–HCO_3 0.5 < Cl < 2 g/kg; t = 31–62 °C	Tottory Owani
Coastal	Quaternary volcanic rocks at ocean coast	Na–Ca–Cl 3 < Cl < 20 g/kg	Ibusuki Shimogama
Volcanic	Quaternary volcanic rocks	H–SO_4 H–Cl–SO_4 NaCl HCO_3 Mixed	Hakone Noboribetsu Tamagawa Satsuma Iwojima

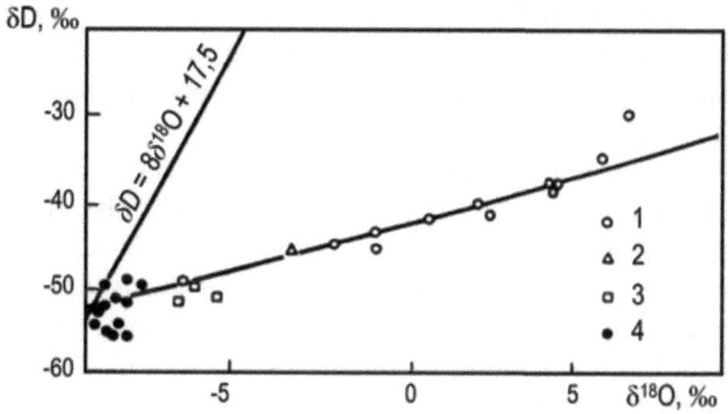

Fig. 9.28 δD and $\delta^{18}O$ plot of the thermal and mineral waters at Arima (*1*), Takarazuka (*2*), Ishibotoke (*3*) and for meteoric waters (*4*) of Japan. (Sakai and Matsubaya 1977; Ferronsky and Polyakov 2012)

exchange has no influence on heavy hydrogen content in thermal waters of this region but result in a considerable enrichment of the thermal waters in oxygen-18.

Coastal thermal waters, as a rule, represent a mixture in certain proportions of marine and meteoric waters. Their chemical composition in a number of gases is markedly metamorphised as a result of the interaction between heated waters and volcanic rocks. In particular, a small amount of magnesium (compared with marine waters), according to Sakai and Matsubaya, may be explained by the following reactions:

Volcanic rocks (glass): $+ 2Mg^{+} + H_2O \rightarrow Mg - \text{chlorite} + Ca^{2+} + K^{+} + H^{+} + H_4SiO_4$.

The sulphates may be released due to the formation of gypsum:

$$Ca^{+2} + SO^{2-} \rightarrow CaSO_4 \downarrow .$$

As a result of this reaction the waters become of the Cl–Na–Ca type.

Table 9.13 Mixing models for acid thermal waters at Higashi (Satsuma-Iwojima volcanic island in the south of the Japanese Archipelago). (Sakai and Mitsubaya 1977)

Locality	δD (‰)	$\delta^{18}O$ (‰)	Cl (mmole/l)	SO_4 (mmole/l)
Meteoric water	−37.0	−6.0	2	1
Seawater	0.0	0.0	535	28.1
Volcanic gas	−25.0	+6.6	370	533
Higashi-1	−32.7	−5.3	52.0	72.0
Higashi-2	−35.0	−2.0	43.7	–
Model 1*	−32.7	−5.3	63.9	8.3
Model 2**	−35.4	−4.3	52.0	7.8

*Model 1 has 88.4% meteoric water +11.6% seawater
**Model 2 has 58% meteoric water +42% volcanic gases

The volcanic type of hydrotherm is likely to be formed from meteogenic waters and volcanic vapour mixed in different proportions. This conclusion is illustrated in Table 9.13.

The data in Table 9.13 include two models of the formation of thermal waters in the Higashi region (see the footnotes to Table 9.13). Sakai and Mitsubaya prefer model 2 but in any case the portion of meteoric waters in thermal volcanic springs is rather large.

The origin of volcanic gases on this island is of specific interest due to their unusual isotopic and chemical composition. The δD values of the fumarole gases are considerably higher than in local groundwaters and precipitation. The oxygen isotopic data are evidence in favour of the isotope equilibrium of the water vapour andesites ($\delta^{18}O = +6.6 \pm 7.0$‰) at magmatic temperatures. This fact, together with the temperature of fumaroles (up to 835°C), supports the conclusion that a marked part of these gases is "magmatogenic" (the author put this word in quotation marks). But the mean values of these "magmatic waters" (δD = −25‰) are considerably higher than the δD values observed in waters of the volcanic gases of Surtsey Island (Iceland), in oceanic basalts not far from the Hawaiian Isles and in hydrogen-bearing minerals originating in the upper mantle, which range from −40 to −60‰ and are typical of the "primary magmatic fluids" (i.e., of juvenile waters). On the other hand, Sakai and Matsubaya pointed out that the condensates of the volcanic vapour from White Island (New Zealand) during high activity are very similar, in a number of parameters, to condensates of the region under study. The observed δD and $\delta^{18}O$ values in the New Zealand volcanoes range from −10 to −20‰ and from 0 to +4‰, respectively. In the condensates of these volcanoes, identical to the Japanese fumaroles, the chlorine content reaches a value of 5 g/kg. The high δD values in the volcanic vapours of the White Island are not related to discharge of 'juvenile' waters but result from the interaction of andesite magmas with marine waters near the surface. This assumption does not contradict the data of Ohmoto and Rye (1974) who studied the isotopic composition of waters of fluid intrusions in pyrite and halkopyrite from a number of deposits in the Kuroko region (Japan). On the basis of variations of δD and $\delta^{18}O$ (from −26 to −18‰ and from

− 1.6 to − 0.3 ‰, respectively) they reported that ore-forming fluids of Kuroko were formed from waters of marine origin, which is in accord with geological evidence. The admixture of magmatic or meteoric water in these fluids does not exceed 25 %.

Therefore, it has been found from the results obtained in the course of groundwater isotope studies in different geothermal regions of the world that meteoric waters are prevalent in the composition of thermal waters such as are found at Larderello (Italy), Wairakei (New Zealand), Hekla (Iceland), Carupano (Venezuela), Yellowstone and Lassen Parks, Steamboat Springs, Salton Sea (USA), The Kurile-Kamchatka region (Russia) plus others. This conclusion can be drawn from the following reasoning. If thermal waters contain a considerable amount of juvenile waters, no difference in their isotopic composition would be observed and their composition would be independent of their geographic location. In this case, the difference in the isotopic composition of the studied thermal waters and, moreover, the quantitative correspondence to local surface waters, shows the close genetic relationship between thermal and meteoric waters. On the assumptions of the above-mentioned authors, the abundance of juvenile water does not exceed 5 %.

As a whole, the isotopic composition of thermal waters, in practically and other case, corresponds to that of surface waters, taking into account the ^{18}O change resulting from the influence of exchange reactions with rocks and may insignificantly change under the influence of shallow, young intrusive bodies due to an input of the juvenile component.

Finally, it must be noted that the supposed range of deuterium variations in juvenile waters should be narrowed to that in basalts. Moreover, apparently, the isotopic composition of thermal waters is a less reliable criterion of the juvenile component than, for example, distinguishing waters of marine and meteoric origin using isotope composition analysis.

9.8.2 Isotopic Geothermometers

Base temperature measurements, using 'isotope' geothermometers with temperature dependence of the fractionation factors of carbon, oxygen and hydrogen stable isotopes between the dissolved gases and water, were carried out in Larderello and some other geothermal system by Panichi and his co-authors (Panichi et al. 1977; 1979). Practically in all hydrothermal fluids, together with water vapour, gaseous components such CO_3, CH_4, H_2, H_2S, N_2, etc. are present. In accordance with the Fischer-Tropsch reaction:

$$4H_2 + CO_2 \leftrightarrow CH_4 + H_2O,$$

and the following isotopic exchange reactions should proceed in hydrothermal systems:

$$^{12}CO_2 + ^{13}CH_4 \leftrightarrow ^{12}CH_4, \Delta^{13}(CO_2 - CH_4);$$

Table 9.14 Equations for base temperature calculation

System	Equation
$\Delta^D(CH_4-H_2)$	$10^3 \ln\alpha.=-90.888+181.269(10^6/T^2)-8.949(10^{12}/T^4)$
$\Delta^D(H_2O-H_2)$	$10^3 \ln\alpha.=-201.6+391.5(10^3/T)-12.9(10^6/T^2)$
$\Delta^{13}(CO_2-CH_4)$	$10^3 \ln\alpha.=-9.01+15.301(10^3/T)-2.361(10^6/T^2)$
$\Delta^{18}(CO_2-H_2O_{vapour})$	$10^3 \ln\alpha.=-10.55+9.289(10^3/T)-2.659(10^6/T^2)$
$\Delta^{18}(CO_2-H_2O_{fluid})$	$10^3 \ln\alpha.=-3.37+4.573(10^3/T)-2.708(10^6/T^2)$

$$1/2C^{16}O_2+H_2{}^{18}O \leftrightarrow 1/2C^{18}O + H_2{}^{16}O, \Delta^{18}(CO_2-H_2O);$$
$$1/4CD_4+1/2H_2O \leftrightarrow 1/4CH_4+1/2D_2O, \Delta^D(H_2O-CH_4);$$
$$1/2D_2+1/4CH_4 \leftrightarrow 1/2H_2+1/4CD_4, \Delta^D(CH_4-H_2);$$
$$1/2D_2+H_2O \leftrightarrow 1/2H_2+D_2O, \Delta^D(H_2O-H_2).$$

Here Δ denotes the difference in isotopic composition of the components A and B, $\Delta=\delta A-\delta B\approx 10^3 \ln\alpha$ (see Chapter 1).

The time taken to attain isotopic equilibrium ranges from 1 year to several years, except for oxygen isotope exchange that proceeds very quickly in the carbon dioxide–water system. Therefore, during the elevation of overheated water up a well or in the course of natural discharge, the isotopic ratios of the components involved in the exchange reactions fix based temperatures in hydrothermal systems even if the water becomes cool at the output of the system.

Panichi et al. (1977, 1979) calculated base temperatures with the help of data on the equations given in Table 9.14, taken from works by the authors.

The geothermometers $\Delta^{13}(CO_2-CH_4)$, $\Delta^D(CO_2-CH_4)$ and $\Delta^D(CO_2-H_2O_{vapour})$ have shown the base temperatures of a corresponding hydrothermal system to be equal to 341±37, 314±30 и 254±25°C, respectively. These values surpass the temperature measured directly at the well head and equal to (216±25°C). The geotermometer $\Delta^{18}(H_2O-CO_2)$, by virtue of its quick attainment of isotopic equilibrium, is indicative of the water temperature at the well head. The geotermometer $\Delta^D(H_2O-CH_4)$ shows barely comparable temperatures resulting from the slow process of isotopic equilibrium in the water-methane system and also from the process of accessory rea ctions accompanying isotopic exchange.

As mentioned above, Friedman (1953) used the hydrogen-water geothermometer for the estimation of base temperatures for the hydrothermal system drained by the Hurricane Vent Spring (Yellowstone National Park, USA). The base temperature of the system of about 400 °C was evaluated using the hydrogen isotopic composition of water and molecular hydrogen.

In the Wairakei geothermal region (New Zealand) Hulston (1977) determined the base temperature to be equal to 260 °C, which coincided with the temperature measured directly in the well using the water-hydrogen isotope thermometer. Similar investigations were carried out by Arnason (1977a) for geotherms of Iceland.

While studying hydrothermal systems, the geothermometer based on oxygen isotope exchange between water and dissolved sulphate ions is in common use. This geothermometer is based on theoretical and experimental studies reported by Lloyd (1968), Mizutani and Rafter (1969), Mizutani (1972) and McKenzie and Truesdell (1977), which are in turn based upon fundamental work by Urey with co-workers (1951). The geothermometer was used for studying geothermal regions in Wairakei, New Zealand (Rafter and Mizutani 1967), Otake, (Mizutani 1972), Larderello, Italy (Cortecci 1974), Tuscany (Longinelli 1968), Shimogana, Japan (Saкai and Matsubaya 1974, 1977) and other geothermal regions of Japan (Mizutani and Hamasuna 1972).

McKenzie and Truesdell (1977) used sulphate–water geotermometers in order to determine base temperatures in the geothermal systems of Yellowstone Park (Wyoming), Long Valley (California) and Ralf River (Idaho). They used the following dependence of the oxygen isotope fractionation factor on temperature (100–350 °C) in the sulphate–water system:

$$10^3 \ln \alpha = 2.88\left(10^6 / T^2\right) - 4.1,$$

obtained earlier by Mizutani and Rafter for temperatures ranging from 100 to 200 °C. The time required for isotopic equilibration of the system depends on the temperature and pH of the system. At pH = 7 and temperature 300 °C equilibrium is reached in 2 years and at 200 °C it is reached in 18 years.

On the basis of the oxygen isotope analyses of sulphates and water, McKenzie and Truesdell determined the base temperatures in a number of springs located in the above-mentioned regions of the United States. In particular, one of the springs in Yellowstone Park with a water temperature of 8 °C is hydraulically connected with a hydrothermal system exhibiting a base temperature ranging from 115 to 135 °C. As a whole, for most hydrothermal springs with temperatures between 70–90 °C at the output, the base temperatures of the system, calculated by the sulphate geotermometer, exceeded 200 °C.

References

Alekseev FA, Gorbushina LV, Ovchinnikov AM, Tyminsky VG (1966) Age of water from the Tashkent artesian basin. In: Alekseev FA (ed Voprosy isotopnoi geologii, vol 3. Nedra, Moskva, pp 40–45

Alekseev FA, Vetshtein VE, Malyuk GA (1974) The isotopic composition of hydrogen and oxygen in groundwater of the Amu-Darya gas and oil-bearing basin as a criterion of its genesis and dynamics. In: Alekseev FA (ed) Yadernaya geologiya. Nedra, Moskva, pp 62–74

Alekseev FA, Gottikh RP, Saakov SA, Sokolovsky EV (1975) Radiochemical and isotopic investigations of groundwater in gas and oil-bearing areas of the USSR. Nedra, Moskva

Arnason B (1977) The hydrogen-water isotope thermometer applied to geothermal areas in Iceland. Geothermics 5:75–80

Arnason B, Sigurgeirsson T (1967) Hydrogen isotopes in hydrological studies in Iceland. In: Isotopes in hydrology: proceedings of a symp. IAEA, Vienna, pp 35–47

Babinets AY, Lugova GP, Markus VI (1971) Oxygen isotopic composition of groundwater of Ukrainian Carpathians region. Dokl AN URSR 7:579–581

Baertschi P (1976) Absolute ^{18}O content of standard mean ocean water. Earth Planet Sci Lett 31:341–344

Banwell CJ (1963) Oxygen and hydrogen isotopes in New Zealand thermal areas. In: Tongeorgi E (ed) Nuclear geology on geothermal areas: Spoletto. Cons Naz delle Ric, Piazzale Aldo Moro, pp 95–138

Baskov EA, Vetstein VE, Surikov SN (1973) Isotopic composition of H, O, C, Ar, and He in thermal waters and gases of the Kurilo-Kamchatka volcanous region as an indicator of their formation. Geokhimiya 2:180–189

Bath AH, Edmunds WM, Andrews JN (1979) Palaeoclimatic trends deduced from the hydrochemistry of a Triassic sandstone aquifer, United Kingdom. In: IAEA (ed) Isotope hydrology 1978: proc symp. IAEA, Vienna, pp 545–566

Beus AA (1972) Geochemistry of the lithosphere. Nedra, Moskva

Boato G, Careri G, Volpi GG (1952) Hydrogen isotopes in steam wells. Nuovo Cim 9:539–540

Bowen RM (1966) Paleotemperature analysis. Elsevier, Amsterdam

Brezgunov VS (1978) Regularity in distribution of hydrogen and oxygen stable isotopes distribution in natural waters during their global circulation. In: Ferronsky VI (ed) Isotopy of natural waters. Nauka, Moskva, pp 10–45

Brezgunov VS, Nechaev VV (1981) Water balance and balance of oxygen stable isotopes in the Issyk-Kul depression. In: Ferronsky VI (ed) Investigation of natural waters by isotope methods. Nauka, Moskva, pp 10–14

Brezgunov VS, Nechaev VV, Erokhin VS (1979) Study of hydrogen and oxygen stable isotope distribution during water exchange in the Issyk-Kul depress. In: Ferronsky VI (ed) Isotope studies of natural waters. Nauka, Moskva, pp 61–69

Brodsky AI (1957) Chemistry of Isotopes Izd. AN SSSR, Moskva

Brown RM (1970) Distribution of hydrogen isotopes in Canada waters. In: IAEA (ed) Isotope hydrology: proceedings of a symp. IAEA, Vienna, pp 3–21

Bullard E (1978) Review of ideas of plate tectonics. In: Fischer AG, Judson S (eds) Petroleum and global tectonics. Nedra, Moskva, pp 9-20 (trans: from English)

Clayton RN (1961) Oxygen isotopic fractionation between calcium carbonate and water. J Chem Phys 34:724–726

Clayton RN, Friedman I, Graf DL et al (1966) The origin of saline formation waters: 1. Isotopic composition. J Geophys Res 71:3869–3882

Coplen TB, Hanshow BB (1973) Ultrafiltration by a compacted clay membrane, 1. Oxygen and hydrogen isotopic fractionation. Geochim Cosmochim Acta 37:2295–2310

Cortecci G (1974) Oxygen isotope ratios of sulfate ions-water pairs as a possible geothermometer. Geothermics 3:60–64

Craig H (1961) Standard for reporting concentrations of deuterium and oxygen-18 in natural waters. Science 133:1833–1834

Craig H (1963) The isotopic geochemistry of water and carbon in geothermal areas. In: Tongeorgi E (ed) Nuclear geology of geothermal areas: Spoletto. Consiglio Nazionale delle Ricerche, Piazzale Aldo Moro, pp 17–53

Craig H (1966) Isotopic composition and origin of the red sea and salton sea geothermal brines. Science 154:1544–1548

Craig H (1969) Geochemistry and origin of the Red Sea brines. In: Degens ET, Ross DA (eds) Hot brines and recent heavy metal deposits in red Sea. Springer, New York, pp 208–242

Craig H, Gordon L (1965) Deuterium and oxygen-18 variations in the ocean and the marine atmosphere. In: Tongeorgi E (ed) Stable isotopes in oceanographic Studies and paleotemperatures: Spoletto, Consiglio Nazionale delle Ricerche, Piazzale Aldo Moro, pp 9–130

Craig H, Gordon L, Horibe Y (1963) Isotopic exchange effects in the evaporation of water. 1. Low-temperature experimental results. J Geophys Res 68:5079–5087

Dansgaard W (1964) Stable isotopes in precipitation. Tellus 19:435–463

Dansgaard W, Jonhnson SJ, Möller J et al (1971) One thousand centuries of climatic record from Camp Century on the Greenland ice sheet. Science 166:377–380

Degens ET, Epstein S (1964) Oxygen and carbon isotopic ratios in coexisting calcites and dolomites from resent and ancient sediments. Geochim Cosmochim Acta 28:23–44

Degens ET, Hunt JM, Reuter JH et al (1964) Data on the distribution of aminoacides and oxygen isotopes in petroleum brine waters of various geologic ages. Sedimentology 3:199–225

Dinçer T, Noory M, Javed ARK et al (1974) Study of groundwater recharge and movement in shallow and deep aquifers in Saudy Arabia with stable isotopes and salinity data. In: IAEA (ed) Isotope techniques in groundwater hydrology: proceedings of a symp. IAEA, Vienna, vol 1, pp 364–374

Drost W, Mozer H, Neumaier F et al (1972) Isotopenmetoden in der Grundwasserkunde, Inf. 61, Büro Eurisotop, Brussels, p 178

Emiliani C (1970) Pleistocene paleotemperatures. Science 168:822–824

Epstein S (1978) The D/H ratio of cellulose in a New Zealand Pinus Radiata. A reply to the criticism of A.T. Wilson and V.J. Grinsted. Earth Planet Sci Lett 39:303–307

Epstein S, Mayeda T (1953) Variation of ^{18}O content of waters from natural sources. Geochim Cosmochim Acta 4:213–214

Epstein S, Sharp RP, Gow AJ (1970) Antarctic ice sheet: stable isotope analyses of Bird Station cores and interhemispheric climatic implications. Science 168:1570–1572

Eriksson E (1965) Deuterium and oxygen-18 in precipitation and other natural waters. Tellus 17:498–512

Esikov AD, Erokhin VE, Chernikova NS et al (1979) Genesis of mud volcanos, south-west of Turkmenia, by hydrogen isotope content. In: Ferronsky VI (ed) Isotope studies of natural waters. Nauka, Moskva, pp 70–74

Evans GV, Otlet RL, Downing RA et al (1979) Some problems in the interpretation of isotope measurements in United Kingdom aquifers. In: IAEA (ed) Isotope hydrology: proceedings of a symp. IAEA, Vienna, vol 2, pp 679–706

Fairbridge RW (1964) The importance of limestone and its Ca/Mg content to paleoclimanology. Intersci Lett, pp 431–478

Ferrara GC, Gonfiantini R, Panichi G (1965) La composizione isotopica della vapore di alcuni soffioni di Larderello e della'acqua di alcune sorgenti e moffete della Toscana. Atti Soc Tosc Sci Nat 15:113–140

Ferronsky VI, Polyakov VA (1983) Isotopy of the Hydrosphere. Nauka, Moskva

Ferronsky VI, Polyakov VA (2012) Isotopes in the Earth;s hydrosphere. Springer, Dordrecht

Fontes JC, Bortolami GC, Zuppi GM (1979) Hydrologie isotopique Hydrologie isotopique du Massif du Mont-Blanc. In: IAEA (ed) Isotope Hydrology, 1978: proceedings of a symp. IAEA, Vienna, vol 1, pp 411–436

Friedman I (1953) Deuterium content of natural waters and other substances. Geochim Cosmochim Acta 4:89–103

Friedman I, Sigurgeirsson T, Gardarsson O (1963) Deuterium in Island waters. Geochim Cosmochim Acta 27:553–561

Friedman I, Redfield AC, Schoen B et al (1964) The variation of the deuterium content of natural waters in the hydrologic cucle. Rev Geophys 2:177–224

Galakhovskaya TV (1967) Distribution of boron, litium stroncium and bromium at evaporation of marine water. In: Valiashko MG (ed) Physical-chemical study of salts and brines. Nedra, Moskva, pp 84–107

Gat JR (1970) Environmental isotope balance of lake Tiberias. In: IAEA (ed) Isotope hydrology: proceedings of a symp. IAEA, Vienna, pp 109–127

Gat JR, Carmi I (1970) Evolution of the isotopic composition of atmospheric waters in the Mediterranean Sea area. J Geophhys Res 75:3039–3078

Gat JR, Dansgaard W (1972) Stable isotope survey of the fresh water occurrence in Israel and Northern Jordan rift valley'. J Hydrology 16:177–212

Gat JR, Gonfiantini R (eds) (1981) Stable Isotope hydrology; deuterium and oxygen-18 in the water cycle. IAEA, Vienna

Gat JR, Tzur Y (1967) Modification of the isotopic composition of rainwater by processes which occur before grounwater recharge. In: IAEA (ed) Isotope hydrology: proceedings of a symp. IAEA, Vienna, pp 49–60

Gat JR, Gonfiantini R, Tongiorgi E (1968) Atmosphere-surface water interaction. In: IAEA (ed) Guidebook on nuclear techniques in hydrology. IAEA, Vienna, pp 175–184

Giggenbach W (1971) Isotopic composition of waters of the Broadlands geothermal field. N Z J. Science 14:959–970

Godfrey J (1962) The deuterium content of hydrous minerals from the East-Central Sierra Nevada and Yosemite National Park. Geochim.Cosmochim Acta 26:1215–1245

Gonfiantini R (1965) Effetti isotopici nell'evapjrazione di acque salate. Atti Soc Tosc Sci Natur Ser A 72:550–588

Gonfiantini R, Gratzini S, Tongiorgi E (1965) Oxygen isotopic composition of water in leaves. In: Isotopes and radiation in soil-plant nutrition studies: proceedings of a symp, IAEA, Vienna, pp 405–410

Gonfiantini R, Dinçer T, Derekoy AM (1974) Environmental isotope hydrology in the Bodna region, Algeria. In: Isotope techniques in groundwater hydrology; proceedings of a symp, IAEA, Vienna, vol. 1, pp 293–316

Gonfiantini R, Conrad C, Fontes JC et al (1976) Etude isotopique de la nappe du Continental intercalaire et de ses relations avec les autres nappes du Sahara septentional. In: Isotope techniques in groundwater hydrology: proc symp, IAEA, Vienna, vol 1, pp 227–240

Gorbushina LV, Tyminsky VG (1974) Radioactive and stable isotopes in geology and hydrology. Atomizdat, Moskva.

Gorbushina LV, Tyminsky VG, Spiridonov AI (1972) On the mechanism of radiohydrogeological anomalies appearance in seismic regions and their significance in earthquake prediction. Sovetskaya Geologiya 1:153–156

Gorbushina LV, Vetshtein VE, Malyuk et al (1974) Hydrogen and oxygen isotopic content in sulphide waters of the Sochi-Adler artesian basin. Geochimiya 9:1102–1106

Graf DL, Friedman J, Meents WF (1965) The origin of saline formation waters. II. Isotopic fractionation by shale micropore systems. U.S. State Geol Surv Circular No. 92

Graf DL, Friedman J, Meents WF (1966) The origin of saline formation waters. III. Calcium chloride waters', U.S. State Geol Surv Circular No. 397

Gutsalo LK (1980) The rules and factors governing changes in isotopic composition of brines during evaporation (in connection with genesis of underground brines). Geokhemiya 11:1734–1746

Hagemann R, Nief G, Roth T (1970) Absolute D/H ratio for SMOW. Tellus 23:172–175

Harpaz Y, Mandel S, Gat JR, Nir A (1963) The place of isotope methods in groundwater research. In: IAEA (ed) Radioisotopes in hydrology: proceedings of a symposium. IAEA, Vienna, pp 175–191

Hitchon B, Friedman F (1969) Geochemistry and oigin of formation waters in the western Canada sedimentary basin, I. Stable isotopes of hydrogen and oxygen. Geochim Cosmochim Acta 33:1321–1349

Hitchon B, Krouse HB (1972) Hydrogeochemistry of surface waters of the Mackenzie River drainage basin, Canada, III. Stable isotopes of oxygen, carbon, and sulfur. Geochim Cosmochim Acta 36:1337–1358

Hübner H, Richter W, Kowski P (1979a) Studies on relationship between surface water and surrounding groundwater of Lake Schwerin (GDR. In: IAEA (ed) Isotopes in lake studies: proc adv group meet. IAEA, Vienna, pp 95–102

Hübner H, Kowski P, Hermichen WD et al (1979b) Regional and temporal variations of deuterium in precipitation and atmospheric moisture of Central Europ. In: IAEA (ed) Isotope hydrology, 1978: proc. symp. IAEA, Vienna, pp 289–305

Hulston JR (1977) Isotope work applied to geothermal systems at the Institute of Nuclear Sciences, New Zealand. Geothermics 5:89–96

International Atomic Energy Agency (1963) Radioisotopes ih hydrology: proceedings of a symposium. IAEA, Vienna

International Atomic Energy Agency (1976) Interpretation of environmental isotope and hydrochemical data in groundwater hydrology: proceedings of an adv group meet. IAEA, Vienna

International Atomic Energy Agency (1979a) Behaviour of tritium in the environmen: proceedings of a symposium. IAEA, Vienna

International Atomic Energy Agency (1979b) Isotopes in lake studies: proseedings of an adv group meet. IAEA, Vienna

James AT, Baker DR (1976) Oxygen isotope exchange between illite and water at 22 °C. Geochim Cosmochi. Acta 40:235–239

Kartsev AA, Vagin SV (1973) The role of clay minerals interlayer water in a history of groundwater formation. Izv Vissh Uch Zaved 3:64–66

Kawabe I (1978) Calculation of oxygen isotope fractionation in quartz-water system with special reference to the bond temperature fractionation. Geochim Cosmochim Acta 42:613–621

Kirshenbaum I (1951) Physical properties and analyses of heavy water. McGraw-Hill, New York

Kobayakawa HY, Horibe Y (1960) Dtuterium abundance of natural waters. Geochim Cosmochim Acta 20:273–283

Kolodny Y, Epstein S (1976) Stable isotope geochemistry of deep sea cherts. Geochim Cosmochim Acta 40:1195–1209

Lawrence JR, Taylor HP (1971) Deuterium and oxygen-correlation clay minerals and hydroxides in Quaternary soils compared to meteoric waters. Geochim Cosmochim Acta 35:993–1003

Lawrence JR, Taylor HP (1972) Oxygen and hydrogen correlation clay minerals and hydroxides in Quaternary soils compared to meteoric waters. Geochim Cosmochim Acta 35:993–1003

Le Pichon X, Francheteau XJ, Bonnin J (1973) Plate tectonics. Elsevier, Amsterdam

Lloyd RM (1966) Oxygen isotope enrichment of sea water by evaporation. Geochim Cosmochim Acta 30:801–814

Lloyd RM (1968) Oxygen isotope behaviour in the sulphate-water system. J Geophys Res 73:6099–6110

Mason B (1966) Principles of geochemistry, 3rd edn. Wiley, New York

McKenzie WF, Truesdell AH (1977) Geothermal reservoir temperatures estimated from the oxygen isotope compositions of dissolved fate and water from hot springs and shallow drill-holes. Geothermics 5:51–61

Meniaylov IA, Vetshtein VE, Nikitina LP, Artemchuk VG (1981) D/H and $^{18}O/^{16}O$ ratios in magmatic water and gas of the Tolbachik great fracture eruption, Kamchatka. Dokl. AN SSSR, pp 258–472

Merlivat L (1970) D'etude quantitative de bilans de lacs á l'aide des concentrations en deuterium et oxygen-18 dans lean. In: IAEA (ed) Isotope hydrol: proceedingsof a symp, IAEA, Vienna, pp 89–107

Miller AR, Densmore CD, Degens TE et al (1966) Hot brines and recent iron deposits in deeps of the Red Sea. Geochim Cosmochim Acta 30:341–359

Mizutani Y (1972) Isotopic composition and undeground temperature of the Otake geothermal water, Kyushu, Japan. Geochim J 6:67–73

Mizutani Y, Hamasuna T (1972) Origin of the Shimogamo geothermal brine. Izu Volcan Soc Japan Bull 17:123–134

Mizutani Y, Rafter TA (1969) Oxygen isotope composition of sulphates, 3 Oxygen isotopic fractionation in the bisulfate ion-water system. N Z J Sci 12:54–59

Moody DJ (1978) Geography and geology of gigantic petroleum fields. In: IAEA (ed) Petroleum and global tectonics. Nedra, Moskva, 112–160 (trans: from English)

Mook WG (1970) Stable carbon and oxygen isotopes in natural waters in the Netherlands. In: Isotope hydrology: proceedings of a symp. IAEA, Vienna, pp 163–189

Nikanorov AM, Yakubovsky AV, Shalaev LN et al (1980) On isotope and chemical anomaly of fresh water in oil fields. In: IAEA (ed) 8th vses symp on stable isotop geochim, Moskva, pp 224–226

Ohmoto H, Rye RO (1974) Oxygen and hydrogen isotope composition of fluid inclusions in the Kuroko deposits, Japan. Econ Geol 69:947–953

O'Neil JR, Kharaka JK (1976) Hydrogen and oxygen isotope exchange reactions between clay minerals and water. Geochim Cosmochim Acta 40:214–245

Panichi C, Celati R, Noto P, et al (1974) Oxygen and hydrogen isotope studies of the Larderello (Italy) geothermal system. In: IAEA (ed) Isotope techniques in droundwater hydrology: proc symp. IAEA, Vienna, vol 2, pp 3–28

Panichi C, Ferrara GC, Gonfiantini R (1977) Isotope geothermometry in the Larderello geothermal field. Geothermics 5:81–88

Panichi C, Nuti S, Noto P (1979) Use of isotopic geothermometers in the Larderello geothermal field. In: IAEA (ed) Isotope hydrology: proceedings of a symposium, IAEA, Vienna, vol 2, pp 613–629

Pelmegov SV, Munaev Ye, Bondarenko GN (1978) Isotopic and geochemical studies of groundwater from a boundary of an artesian basin. Sov Geol 4:119–125

Petrov VP (1975) Stories about white clay. Nedra, Moscow
Petrov VP (1975) Stories about white clay. Nedra, Moscow

Pinneker EV (ed) (1974) Role of isotope investigation at groundwater resources exploration in Eastern Siberia. Groundwaters of Irkutsk Region. Nedra, Leningrad, pp 14–31

Pinneker EV (ed) (1975) Formation of modern hydrotherms in the dead volcanic regions (in the light of isotopic data). Geothermal process in the regions of active structural magmatism. Nauka, Moskva, pp 38–43

Pinneker EV, Vetshtein VE, Dzyuba AA et al (1973) Oxygen-18 content in Siberian platform brines. In: Pinneker EV (ed) Outlines on hydrogeology of Siberia. Nauka, Novosibirsk, pp 86–92

Polyakov VA, Seletsky YuB, Yakubovsky AV et al (1974) Deuterium in the Naftusya' mineral water. Annals VSEGINGEO 59:80–87

Polyakov VA, Kolesnikova LN (1978) Regional spesifics in formation of isotopic content of precipitation. In: GEOCHI (ed) 7th Vses Symp Stab Isotop Geokhim, Moskva, pp 148–149

Rabinovich IB (1968) Isotopic effects in physical and chemical properties of solutions. Nauka, Moskva

Rafter TA, Mizutani Y (1967) Oxygen isotope composition of sulfates: 2. Preliminary results of oxygen isotope variation in sulphates and relationship to their environment and to their ^{34}S values. N Z J Sci 10:815–840

Rankama K (1954) Isotope geology. Pergamon, London

Rankama K (1963) Progress in isotope geology. Intersci Publ, NewYork

Redfield AC, Friedman I (1964) Factors affecting the distribution of deuterium in the ocean. Proc Symp Mar Geochim, pp 149–168

Sakai H, Matsubaya O (1974) Isotope geochemistry of the thermal waters of Japan and its bearing on the Kuroko ore solutions. Econ Geol 69:674–991

Sakai H, Matsubaya O (1977) Stable isotope studies of Japanese geothermal systems. Geotermics 5:97–123

Salati E, Matsui E, Leal JM et al (1980) Utilization of natural isotopes in the study of salination of the water in the Pejeu River valley, Northeast Brazil. In: IAEA (ed) Arid-zone hydrology: investigations with isotope techniques: proc. adv. group neet. IAEA, Vienna, pp 133–151

Savin SM, Epstein S (1970a) The oxygen and hydrogen isotope geochemistry of clay minerals. Geochim Cosmochim Acta 34:25–42

Savin SM, Epstein S (1970b) The oxygen isotopic composition of coarse grained sedimentary rocks and minerals. Geochim Cosmochim Acta 34:323–329

Seletsky Yu B, Polyakov VA, Yakubovsky AV, Isaev NV (1973) Deuterium and oxygen-18 in groundwaters. Nedra, Moskva

Seletsky Yu B, Polyakov VA, Yakubovsky AV, Isaev NV (1974) Preliminary results of deuterium content in certain types of North Caucasus groundwaters. Annals VSEGINGEO 59:70–79

Sergeev EM, Ilyinskaya GG, Rekshinskaya G (1963) On the distribution of clay minerals for their engineering geological study. Vestnic MGU. Ser Geol 4:3–9

Shackleton MJ, Opdyke ND (1973) Oxygen isotope and paleomagnetic stratigraphy of equatorial Pacific core v.28-v.238: oxygen isotope temperatures and ice volumes on a 10^9 years and 10^6 year scale. Quatern Res 3:39–55

Sheppard SMF, Nielsen RL, Taylor HP (1969) Oxygen and hydrogen isotope ratios in minerals from porphyry copper deposits. Econ Geol 64:755–777

Sheppard SMF, Nielsen RL, Taylor HP (1971) Oxygen and hydrogen isotope ratios of clay minerals from porphyry copper deposits. Econ Geol 66:515–542

Smirnov SI (1971) Origin of groundwater salinity in sedimentary basins. Nedra, Moskva

Sobotovich EV, Bondarenko GN, Vetshtein VE et al (1977) Isotope and geochemical estimates of a degree of surface and gtound water interconnection. Naukova Dumka, Kiev

Sofer Z, Gat JR (1975) Activities and concentration of oxygen-18 in concentrated aqueous salt solutions: analytical and geophysical implications. Earth Planet Sci Lett 26:179–186

Sokolovsky LG, Polyakov VA, Golubkova EV (2007) Light isotopes of waters of the Asdov-Kuban artesian basin: conditions of formation and balneological significance. Prosp Prot Miner Resour 5:44–47

Sonntag C, Klitzsch E, Löhnert EP et al (1979) Paleoclimatic information from deuterium and oxygen-18 in carbon-14-dated North Saharian groundwater. In: IAEA (ed) Isotope Hydrology, 1978: proceedings of a simposium. IAEA, Vienna, vol 2, pp 569–580

Soyfer VN, Brezgunov VS, Vlasova LS (1967) Role of hydrogen stable isotopes in study of geological processes. Geokhimiya 5:599–606

Sultanov BI (1961) Deep condensed waters of gas-condensates and their formation conditions. Dokl AH AzSSR 17:1165–1166

Suzuoki T, Epstein S (1976) Hydrogen isotope fractionation between OH-bearing minerals and water. Geochim Cosmochim Acta 40:1229–1240

Tarasov MG (1978) The origin and formation of groundwaters in Near Caucasus Mesozoic sediments by means of hydrogen and oxygen isotopes. Nat Geol Geophys Izuch Zemn Kory, Minsk, pp 62–68

Taylor HP (1974) The application of oxygen and hydrogen isotope studies to problem of hydrothermal alteration and ore deposition. Econ Geol 69:213–298

Taylor HP (1978) Oxygen and hydrogen isotope studies of plutonic granitic rocks. Earth Planet Sci Lett 38:177–210

Tkachuk VG, Vetshtein VE, Malyuk GA, Altshuler PG. (1975) Hydrogen and oxygen isotopes of the Pripyat depression brines and possibilities of their use in oil and gas exploration. Geokhimiya 7:999–1006

Tyminsky BG, Sultankhodzhaev AN, Rozanov IM (1966) Paleohydrological estimations for the waters of the Tashkent artesian basin Uzbek Geol J 3:64–68

Urey HC (1957) Boundary conditions for theories of the origin of the solar system. Physics and Chemistry of the Earth 2:46–76

Urey HC, Lowenstam HA, Epstein S et al (1951) Measurement of paleotemperatures and temperatures of the Upper Cretaceous England, Denmark, and South-Eastern United States. Bul Geol So Am 62:399–416.

Vasilchuk YuK, Kotlyakov VM (2000) Principles of Isotopic Geocriology and Glaciology. Izd MGU, Moscow

Verhoogen J, Turner EJ, Weiss LS et al (1970) The Earth: an introduction to physical geology, vol 2. Holt, Rinehart and Winston, New York

Vetshein VE, Baskov EA, Klimov GI et al (1971) New data on oxygen-18 content in volcanic thermal and mineral waters from Kurili Islands, Kamchatka, and Baykal region. Sov Geol 9:98–108

Vetshein VE, Malyuk G, Lapshin FV (1972) Oxygen and hydrogen isotopic composition of mineral waters in Ukrainian Carpathy as their genesis criterion. Dop. AN URSR 12:1062–1066

Vetshein VE, Gutsalo LK, Malyuk GA, Miroshnichenko AG (1973) On the origin of formation waters in the Dnepr-Donetsk gas and oil-bearing sedimentary basin by oxygen and hydrogen isotopic composition. Geokhimiya 3:327–338

Vlasova LS, Brezgunov VS (1978) The distribution of hydrogen and oxygen isotopic composition in natural brines by model calculations. In: Ferronsky VI (ed.) Isotope study of natural waters. Nauka, Moskva, pp 119–139

Voytov GI, Gureev EV, Erokhin BK et al (1976) Hydrogen isotopic composition of thermal waters from South Belozersk iron ore deposits. Dokl. AN SSSR 231:1226–1229

White DE (1965) Saline waters of sedimentary rocks. In: Young A, Galley JE (eds) Fluids in subsurface environments. Am Assoc Petrol.Geol 4:343–366

White DE (1974) Diverse origins of hydrothermal ore fluids. Econ Geol 69:954–973

White DE, Craig H, Begemann F (1963) Sammary of the geology and isotope geochemistry of Steamboat springs, Nevada. In: Tongiorgi E (ed) Nuclear geology on geothermal areas, Spoleto. Consiglio Nazionale delle Ricerche, Piazzale Aldo Moro, pp 9–16

Wilson AT, Grinsted MJ (1977) The D/H ratio of cellulose as biochemical thermometer (A comment on "Climatic implication of D/H ratio of hydrogen in C-H groups in tree cellulose" by S Epstein and CJ Yapp). Earth Planet Sci Lett 36:246–248

Yakubovsky AV, Isaev NV, Polyakov VA, Tereshchenko VA (1978) On the formation of low-mineralized groundwater with high level of deuterium and oxygen-18 content. In: GEOCHI (ed) 7th Vses Sym. Stab Isotop Geokhim. GEOHI, Moskva, pp 202–203

Yapp CJ, Epstein S (1977) Climatic implication of D/H ratio of meteoric waters over North America (9500-22,000 B.P.) as inferred from ancient wood cellulose C-H hydrogen. Earth Planet Sci Lett 34:33–350

Yeh HW, Epstein S (1980) D/H ratios and late-stage dehydration of shales during burial. Geochim Cosmochim Acta 44:341–352

Yezhova MP, Polyakov VA, Tkachenko AE et al (1996) Palaeowaters of North Estonia and their influence on changes in the resourses and the quality of fresh groundwaters of large coastal water supplies. Geologiya 19:37–40

Yurtsever Y, Gat JR (1981) Stable isotopes in atmospheric waters. In: Gat JR, Gonfiantini R (eds) Stable isotope hydrology. IAEA, Vienna, pp 103–142

Zaitsev IK (1967) Hydrochemical and hydrothermal zoning of the artesian basins of the USSR in connection with underground evaporation hypothesis criticism. In: VSEGINGEO (ed) 5th Meeting of Siberia and Far East, Irkutsk-Tyumen, pp 39–40

Zimmerman U (1979) Determination by stable isotopes of underground inflow and outflow and evaporation of young artificial groundwater lakes. In: IAEA (ed) Isotopes in Lake Studies: proc adv group meet. IAEA, Vienna, pp 87–94

Zimmerman U, Ehhalt D, Münnich KO (1967) Soil water movement and evapotranspiration: changes in the isotopic composition of the water. In: IAEA (ed) Isotopes in hydrology: proceedings of a symposium. IAEA, Vienna, pp 567–584

Chapter 10
Cosmogenic Radioisotopes for Study of the Genesis and Dynamics of Water

Abstract The Earth's atmosphere is penetrated by a continuous flux of charged particles, consisting of protons and nuclei of various elements of cosmic origin. Consequently, a great variety of radioisotopes, referred to as cosmogenic, are produced due to the interaction of these particles with the atomic nuclei of elements that constitute the atmosphere. Transported by air masses, radioisotopes are abundant over the whole gaseous sphere of the Earth. Being mixed with atmospheric moisture, a proportion falls over the Earth's surface, to enter the hydrological cycle as components of surface waters, soil-ground moisture and groundwaters. Another proportion becomes a component of ocean and inland basin waters through exchange at the surface of water reservoirs. Finally, the Earth's biosphere plays an active role in exchange processes, which are of great importance for some cosmogenic isotopes. Cosmic dust is another source of cosmogenic isotopes, as are meteorites that are continually falling onto the Earth's surface. Being in cosmic space these meteorites have been subjected to a bombardment of cosmic radiation. Nuclear reactions accompanying the process produce many radioisotopes. The origin and distribution of cosmogenic radioisotopes, global circulation of tritium, radiocarbon and other cosmogenic radionuclides are discussed in this chapter.

10.1 Origin and Distribution of Cosmogenic Radioisotopes

Cosmic radiation plays the main role in the origin of cosmogenic radioisotopes. Understanding the nature of cosmic radiation has been important for the development of the Earth sciences. In particular, the solution of a number of hydrological and hydrogeological problems, related to natural water dynamics, their genesis and age, has become possible due to investigations of abundances of cosmogenic radioisotopes in the hydrosphere, i.e., isotopes, produced by cosmic radiation.

Cosmic rays of solar and galactic origin are distinguishable. The nature of the nuclear particle flux of solar origin is related to fusion and decay reactions in the solar interior. The origin of galactic and metagalactic cosmic radiation bombarding the Earth's atmosphere is still uncertain.

V. I. Ferronsky, *Nuclear Geophysics,* Springer Geophysics,
DOI 10.1007/978-3-319-12451-3_10

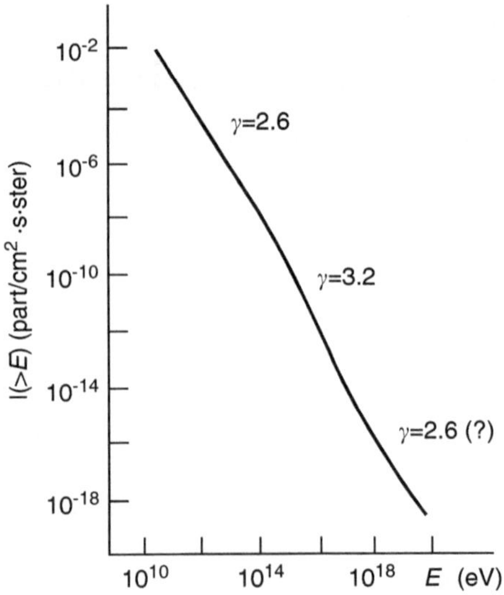

Fig. 10.1 Energy spectrum of primary cosmic radiation. (Ferronsky and Polyakov 2012)

In practical studies of cosmic radiation there have been cases when particles of galactic origin with energies of $10^{19}-10^{20}$ eV have been detected. Some idea of the spectrum of cosmic radiation with energy greater than $E=10^{10}$ eV is given by Fig. 10.1.

Cosmic radiation is made up of about 90% protons, about 9% helium nuclei (α-particles) and about 1% other nuclei. Table 10.1 shows specific abundances of nuclei in the solar system, cosmic space and in cosmic radiation (Webber 1967).

Except for hydrogen and helium, the composition of primary cosmic radiation is poorly understood. The estimations of specific abundances of carbon isotopes show

Table 10.1 Nuclei abundances in the solar system, cosmic space and in primary cosmic rays relative to carbon

Element	Abundances			Element	Abundances		
	Solar system	Cosmic space	Primary cosmic rays		Solar system	Cosmic space	Primary cosmic rays
He	?	400	38	Al	0.004	0.005	0.03
Li	<<0.001	<<0.001	0.27	Si	0.063	0.13	0.11
Be	<<0.001	<<0.001	0.19	P	–	0.002	0.01
B	<<0.001	<<0.001	0.43	$16 \leq Z \leq 19$	0.050	0.02	0.02 – 0.05
C	1.0	1.0	1.0	Ca	–	<<0.001	0.026
N	0.16	0.27	0.46	Ti	–	0.006	0.017
O	1.7	2.3	0.61	Ci	–	0.06	0.030
F	<<0.001	<<0.001	0.09	Fe	–	0.06	0.080
Ne	?	0.80	0.18	Ni	–	0.008	0.015
Na	0.004	0.006	0.08	$Z>20$	–	<<0.001	0.01
Mg	0.005	0.12	0.15	$Z>30$	–	~0.001	< 0.004

that in primary cosmic radiation $^{13}C/^{12}C \approx 1$. Measuring this ratio with the help of photo-emulsion techniques, a value close to unity is obtained.

As for deuterium, no sufficiently reliable measurements of its content in primary cosmic radiation have been carried out so far. Those measurements that were carried out in the upper atmosphere (for atmospheric depths 2–4 g/cm^2) gave values of $^2H/^1H = 0.05$–0.12. The large discrepancy in these data results from the fact that the measurements correspond to different energy intervals and latitudes. But on the whole, the data are in agreement with satellite data ($^2H/^1H \leq 0.06$ or $\varepsilon =$ 25–80 MeV/nucleon), where the effect of the Earth's magnetism and atmosphere are excluded (MeV = 10^6 eV).

The ratio of deuterium abundance to proton abundance in the universe is of the order of $1.4 \cdot 10^{-4}$ (Webber 1967), which may be accounted for by its disintegration. According to spectroscopic measurements, the ratio $^2H/^1H$ is also small in the solar atmosphere, amounting to $4 \cdot 10^{-5}$ (Kinman 1956). Only in the atmosphere of magnetic stars does the ratio increase up to 10^{-2}. The ratio of $^2H/^1H = 10^{-5}\,\rho$, where ρ (particle/cm^3) is the average density of substance through which cosmic radiation has passed (Singer 1958).

While studying the helium isotope composition in cosmic radiation (Appa Rao 1962) the ratio $^3He/(^3He + ^4He) = 0.20$–$0.30$ was obtained in the energy range $\varepsilon = 160$–360 MeV/nucleon. Attempts to measure this ratio for higher energies have failed.

Besides nuclei, α-particles and nuclides of various elements and their isotopes, primary cosmic radiation also contains gamma-ray protons, neutrons, electrons and positrons. The source of gamma-ray protons and neutrons in primary cosmic radiation is supposed to be related to supernova.

Primary cosmic radiation interacts with the Earth's magnetic field, which results in a relationship between the intensity of radiation and geomagnetic latitude. All available data concerning geomagnetic effects of cosmic radiation are in good agreement with the idea that the isotropic flow of charged particles coming from the universe is deviated by the terrestrial magnetic field. Analysing the effect of the Earth's magnetic field on the motion of charged particles, favoured and unfavoured directions of charged particle motion to a given point of the Earth's surface have been found.

The interaction of the Earth's magnetic field with primary cosmic radiation results in latitude variations of its intensity I (Fig. 10.2). This effect, referred to as the latitude effect, is quantitatively expressed by the ratio $[I(90°) - I(0°)]/I(90°)$.

It has been found that the intensity of cosmic radiation coming to the Earth varies with time. Four different types of these time-dependant variations are known (Fireman 1967):

1. Variations related to the 11-year cycle of solar activity. With increasing intensity of the solar particle flux the galactic radiation intensity decreases.
2. Heliocentric variations of galactic radiation, the radial gradient of which is in the range of 1.0–1.5 astronomical units of length (1 a.u. $\approx 1.5 \cdot 10^8$ km), are about +9.6 % per 1 a.u.

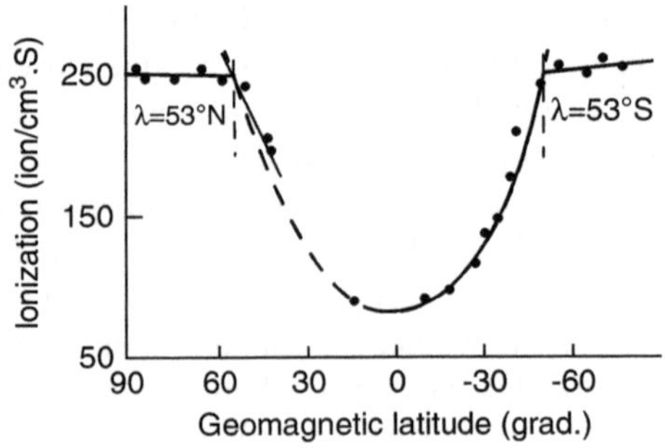

Fig. 10.2 Latitude effect of cosmic radiation for the depth of atmosphere at 50 g/cm^2. (Ferronsky and Polyakov 2012)

3. The secular variations of radiation detected with the help of ^{14}C content vary in the atmosphere in different centuries. They were measured by ^{14}C content variations recorded in the year-rings of trees from various ages.
4. The sporadic flows of nuclear particles of low energies emitted during solar flares.

The intensity of the flux of protons of solar origin, with energies greater than 10 MeV near to the Earth, is equal to 100 protons/cm$^2 \cdot$ sec during a typical solar cycle (relative to the Earth's surface) (Lal et al. 1967). This value was obtained using experimental data at the rate of ^{26}Al production during a time interval of about 10^5 year. The intensity of protons in the solar flux is higher than the galactic intensity by about one order of magnitude, i.e., the latter is characterised by a value of about 10 protons/cm$^2 \cdot$ sec. But while the solar protons have energies of about several dozens of MeV, the protons of galactic origin have energies two orders greater on average. The intensity of α-particle flux, both of solar and galactic origin, is lower than of protons by one order of magnitude. The velocity of nuclear particles of solar origin near the Earth is about 300 km/sec.

In the course of the interaction of high-energy cosmic radiation with the atmosphere the major part of its energy is absorbed and scattered by the Earth's atmosphere. This leads to the production of secondary low-energy radiation composed of mesons, gamma-ray photons, positrons and other particles of various energies. This secondary radiation is mainly composed of less energetic protons and neutrons, which play the main role in nuclear reactions, resulting in the production of cosmogenic radioisotopes in the Earth's atmosphere. The energy threshold of these reactions ranges between 10 and 40 MeV.

The distribution of secondary neutrons and protons in the atmosphere varies both in latitude and altitude. Figure 10.3a shows the experimental data of energy distribution in the neutron flux, characterised by energies lower than 20 MeV, obtained in 1966 during an experiment in Sicily during a quiet Sun period (Boella et al. 1968). A similar thermal neutron distribution was obtained by Korff using a thermal neutron counter in the Princeton region (USA). The data from these measurements are shown in Fig. 10.3b. One can see from the figure that a total flux of neutrons at first increases with altitude and then decreases with atmospheric density due to their escaping from high atmospheric layers. With the transition in latitude from the

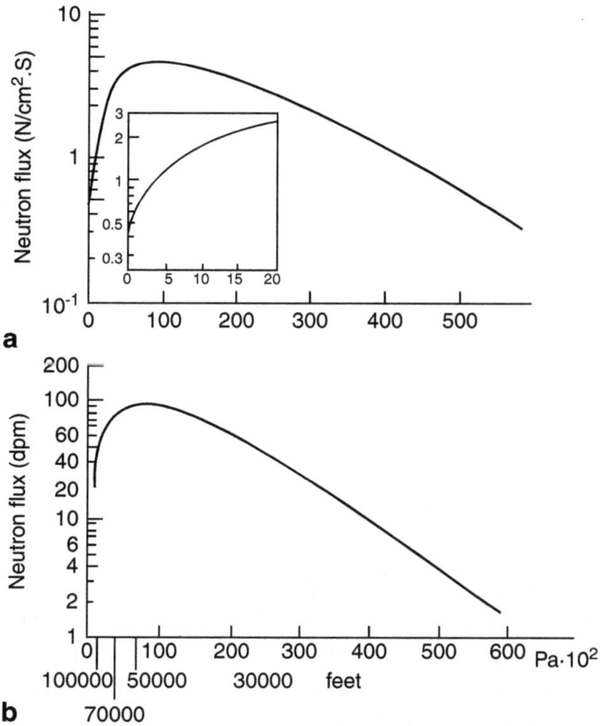

Fig. 10.3 Relation of integrated neutron flux from an atmospheric depth of 0–600 mbar in Sicily for neutron energy lower than 20 MeV (**a**), and for thermal neutrons in the Princeton region (**b**). (Ferronsky and Polyakov 2012)

equator to the poles the density of neutron flux increases. Figure 10.4 shows the experimental data of latitude dependency of neutron flux at an altitude of 1000 m, obtained by Simpson (Libby 1967).

The effect of the production of cosmogenic isotopes in the Earth's crust is negligible even in the upper layers. Nuclear transformations occur here mainly due to the penetration of primary nucleons of light energies, thermal neutrons and $\overline{\mu}$-mesons. The dominating component of this flux varies with depth. At a depth of several metres most nuclear transformations are provided by quick $\overline{\mu}$-mesons and at depths where pressure is greater than 400 kg/cm^2, by the interaction with neutrino flux of both primary and secondary origin (Lal and Peters 1967).

According to Lal and Peters, at depths characterised by pressures greater than 0.7 kg/cm^2, the majority of nuclear transformations occur due to interaction with $\overline{\mu}$ -mesons and 10 % of them are the effect of negative meson capture.

The main proportion of cosmogenic radionuclides is formed in the atmosphere. The main components of atmospheric air are: nitrogen (78.09 %), oxygen (20.95 %), argon (0.93 %), carbon dioxide (0.03 %) and neon (0.0018 %). In the course of interaction between cosmic radiation and atmospheric elements nuclear transformation occurs with the nuclei of neon, oxygen and argon, which plays a major role in the production of cosmogenic radioisotopes. The most typical reactions are: (n, p), (n, 2n), (n, α), (n, γ), (p, n), (p, 2n), (p, pn), (p, 2p).

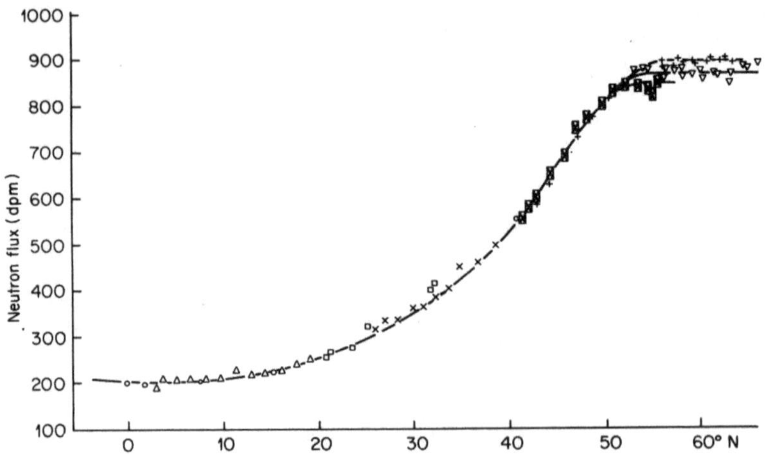

Fig. 10.4 Latitude effect in distribution of neutron flux at an altitude of 1000 m for the North American continent. (Libby 1967; Ferronsky and Polyakov 2012)

Isotopes of cosmogenic origin, which are of interest for the investigations of water circulation patterns in the hydrosphere, are: ^{3}H, ^{3}He (stable), ^{7}Be, ^{10}Be, ^{14}C, ^{22}Na, ^{24}Na, ^{26}Al, ^{32}Si, ^{32}P, ^{33}P, ^{36}Cl, ^{37}Ar, ^{39}Ar, ^{81}Kr. The irradiation of cosmic dust, falling down in amounts of about 10^{9} kg/year on the Earth's surface, by cosmic radiation results in the production of radioisotopes of great importance such as ^{10}B, ^{14}C, ^{22}Na, ^{26}Al, ^{36}Cl, ^{39}Ar, ^{53}Mn, ^{59}Ni.

The rate of production of cosmogenic isotopes varies with altitude and latitude and depends on the intensity of secondary neutron and proton flux in the atmosphere. At the same time the rate of their formation remains constant over time. As a first approximation the ratio of various isotopes produced may be considered to be independent of altitude and time. According to estimations by Lal and Peters (1962) and Young et al. (1970) the rate of isotope production per gram of air, e.g., at 46° N, increases exponentially at an altitude of about 21 km by three orders of magnitude higher than at the surface of the Earth. Further on, it decreases with altitude due to escaping neutrons from the upper atmospheric layers. The isotope production rate increases by about one order of magnitude with transition from the equator to the poles.

Estimations of the production rate of cosmogenic radioisotopes in the atmosphere and in cosmic dust, carried out by a number of authors (Bhaudari et al. 1969; Lal et al. 1967; Lal and Venkatavaradan 1967; Schell 1970), used a value of average density of proton flux with energy $E > 10$ MeV at the upper boundary of the Earth's atmosphere, equal to 100 protons/cm$^{2}\cdot$sec. The latitude and altitude effects of nuclear reactions were taken into account in the course of the calculations.

In order to carry out estimations of steady-state amounts of individual isotopes on the Earth, it was assumed that the flux of cosmic radiation remained constant over a long period of time (being not less than the half-life of a given radioisotope). With a constant cosmic flux and a certain atmospheric composition the production rate of isotopes will also be independent of time.

The steady-state amount of an individual isotope on the Earth follows from the equation:

$$nS = q\lambda,$$

where n is the production rate of a radioisotope relative to the Earth's surface (atom/$cm^2 \cdot sec$); S is the area of the Earth's surface, cm^2; q is the steady-state amount of isotopes or atoms; λ is the decay constant of a given nucleus.

The results of estimations carried out by the above-mentioned authors and clarified by experimental research, related to the production rates and steady-state abundances of various cosmogenic radioisotopes and also their main physical properties, are presented in Table 10.2.

The total steady-state amount of cosmogenic isotopes on the Earth depends on the balance between their production rate in the atmosphere, accumulation over their life and reduction due to radioactive decay.

The main proportion of isotopes (except of the noble gases) is oxidised immediately after production. Among these oxides only carbon dioxide and tritium occur in the atmosphere in a free form. The other isotopes are absorbed by aerosols a short time after their formation. Radioisotopes, contained in aerosols, are removed from the atmosphere fairly quickly by condensation of moisture in low tropospheric layers, whereas those isotopes that remain constant gaseous components of the atmosphere (CO_2, Ar, Kr) are removed from it far more slowly by molecular exchange at the boundary between the atmosphere and oceans.

The ability of cosmogenic isotopes to aid the study of hydrological and hydrogeological processes is restricted by the condition of correspondence between their lifetimes and the duration of a considered process and also by the principles of their displacement in geospheres. The major proportion of cosmogenic isotopes (~70 %) is formed in the upper atmospheric layers, about 30 % of them in the troposphere. There sequential redistribution is caused by large-scale processes involving the motion of air masses in the troposphere and the precipitation of atmospheric moisture on the Earth's surface, together with cosmic dust or in the form of aerosols.

The absolute amount of an individual isotope, or its ratio relative to another radioactive or stable isotope within a natural reservoir, is used for dating or reconstructing events that have taken place in the past and also for elucidating the nature of prevailing physical, chemical and biological processes. Resolution of the investigation of these processes depends largely on the information available concerning the source and rate of production of a radionuclide.

As pointed out earlier, the production rate of a cosmogenic isotope in the atmosphere depends on latitude and altitude. Comparison of the expected production rate with those amounts actually observed in the air for isotopes with adequate lifetimes provides a basis for studying the principles of large-scale circulations and tropospheric fallout. The abundance of an isotope in the hydrosphere depends on its lifetime, its biochemical role and, finally, on the nature of oceanic circulations. Investigations of abundance of isotopes, characterised by different lifetimes and chemical properties, provide an opportunity for understanding the principles of mass-transfer of substance and, in some cases, helps in understanding the geochemistry of an

Table 10.2 Physical parameters and steady-state amounts of cosmogenic radioisotopes on the Earth

Isotope	Production reaction	Half-life	Radiation type and energy (MeV)	Decay product	Production rate (atom/cm^2 · sec	Steady-state amount on the Earth (g)
^3H	$^{14}N(n, ^{14}C)^3H$	12.32 year	β^-; 0.018	^3He	0.25	3500
	$^{16}O(p, ^{14}C)\ ^3H$					
^3He	N, O(^3He)	Stable		–	0.2	$3.2 \cdot 10^9$
^7Be	$^{14}N(n, 3p5n)^7Be$	53 days	γ; 0.48	^7Li	$8.1 \cdot 10^{-3}$	3.2
	$^{14}N(p, 4p4n)^7B$					
	$^{16}O(n, 5p5n)^7Be$					
^{10}Be	$^{14}N(p, 4pn)^{10}Be$	$1.6 \cdot 10^6$ year	β^-; 0.55	^{10}B	$4.5 \cdot 10^{-2}$	$4.8 \cdot 10^8$
	$^{16}O(p, 5p2n)^{10}Be$					
^{14}C	$^{14}N(n, p)^{14}C$	5730 year	β^-; 0.156	^{14}N	2.5	$7.5 \cdot 10^7$
	$^{16}O(p, 3p)^{14}C$					
^{22}Na	$^{40}Ar(split)^{22}N$	2.6 year	β^+; 0.54	^{22}Ne	$8.6 \cdot 10^{-5}$	1.9
			γ; 1.28			
^{24}Na	$^{40}Ar(split.)^{24}N$	15 h	γ; 2.75	^{24}Mg	$1.2 \cdot 10^{-4}$	–
			β^-; 1.4			
^{26}Al	$^{26}Mg(p, n)^{26}Al$	$7.4 \cdot 10^5$ year	β^+; 2.77	^{26}Mg	$1.4 \cdot 10^{-4}$	$1.2 \cdot 10^6$
	$^{26}Si(p, 2p)^{26}Al$					
^{28}Mg	$^{40}Ar\ (split.)^{28}Mg$	21.3 h	β^-; 0.42	^{28}Al	$5.2 \cdot 10^{-5}$	–
^{32}Si	$^{40}Ar\ (split\ ^{32}Si$	~450 year	β^-; 0.1	^{32}P	$1.6 \cdot 10^{-4}$	1400
^{32}P	$^{40}Ar\ (split)^{32}P$	14.3 days	β^-; 1.7	^{32}S	$8.1 \cdot 10^{-4}$	0.4
^{33}P	$^{40}Ar\ (split)^{33}P$	25 days	β^-; 0.25	^{33}S	$6.8 \cdot 10^{-4}$	0.6
^{35}S	$^{40}Ar\ (split)^{35}S$	87.4 days	β^-; 1.67	^{35}Cl	$1.4 \cdot 10^{-3}$	4.5
^{36}Cl	$^{40}Ar\ (split)^{36}Cl$	$3.0 \cdot 10^5$ year	β^-; 0.714	^{36}Ar	$1.1 \cdot 10^{-3}$	$1.5 \cdot 10^6$
^{37}Ar	$^{37}Cl\ (p, 2p)^{37}Ar$	35 days	K–capture	^{37}Cl	$8.3 \cdot 10^{-4}$	–
	$^{40}Ca\ (n, \alpha)^{37}Ar$					
	$^{36}Ar\ (n, \gamma)^{37}Ar$					
^{39}Ar	$^{40}Ar\ (n, 2n)^{39}Ar$	270 year	β^-; 0.565	^{39}K	$5.6 \cdot 10^{-3}$	–
	$^{39}K\ (n, p)^{39}Ar$					
	$^{38}Ar\ (n, \gamma)^{39}Ar$					
^{53}Mn	$^{53}Fe\ (p, 2p)^{53}Mn$	$3.7 \cdot 10^6$ year	K-capture	^{53}Cr	$<10^{-7}$	–

Table 10.2 (continued)

Isotope	Production reaction	Half-life	Radiation type and energy (MeV)	Decay product	Production rate (atom/cm^2 · sec	Steady-state amount on the Earth (g)
	^{56}Fe (p, α) ^{53}Mn					
^{59}Ni	^{59}Co (p, n)^{59}Ni	$7.5 \cdot 10^4$ year	K-capture	^{9}Co	$<10^{-7}$	–
	^{60}Ni (p, pn)^{59}Ni					
^{81}Kr	^{82}Kr (p, n)^{81}Kr	$8.1 \cdot 10^{5\mathrm{r}}$ year	K-capture	^{81}Br	$1.5 \cdot 10^{-7} - 10^{-5}$	–

element's behaviour. Isotopes that have entered into the biosphere, into oceanic floor sediments and into some other objects, thus becoming isolated from the transitional dynamics in the cycle (which is characterised by a continuous process of mixing of substance), can be used to determine the time elapsed since the moment they escaped from the cycle.

In studying the nature of geophysical and geochemical processes it is convenient to divide the atmosphere and upper layers of the Earth into a number of zones or reservoirs characterised by homogeneous pressure, temperature and character of mass-transfer. A schematic description of isotope migration in natural reservoirs is presented in Fig. 10.5 (Lal and Peters 1962).

The rate of isotope production caused by cosmic radiation in ocean waters is negligible. The detectable concentrations of all the isotopes found in these waters

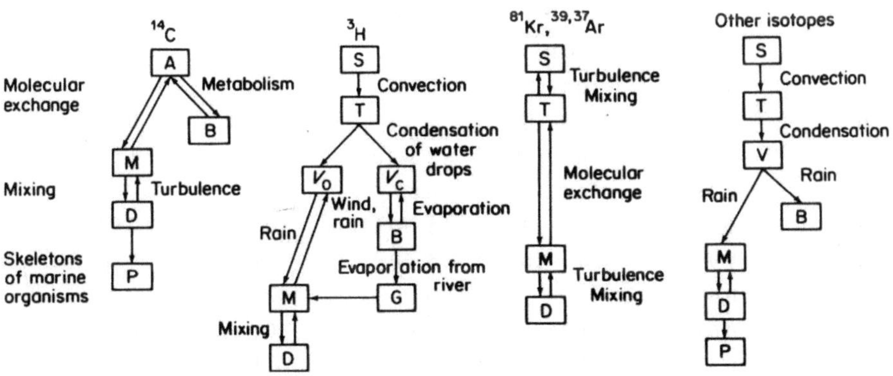

Fig. 10.5 Schematic description and migration of cosmogenic isotopes in the geospheres: (*A*) atmosphere; (*B*) biosphere; (*S*) stratosphere (90–340 g air per 1 cm^2, 1–8 year exchange time); (*T*) troposphere (720–940 g air per 1 cm^2, 30–90 d exchange time); (*V*) water vapour (3–7 g water per 1 cm^2, 4–14 d exchange time); (*M*) mixing layer (75–100 m, 20 year exchange time); (*D*) deep layer (3500 м, 500–3000 year exchange time); (*P*) marine sediments; (*G*) soil water . (Ferronsky and Polyakov 2012)

are the result of exchange between the atmosphere and the oceans. The upper layer of the ocean, receiving cosmogenic isotopes from the atmosphere, is characterised by vigorous movement and quick mixing. It is due to these processes that the geographical inhomogeneity of cosmic radiation is markedly smoothed. Therefore, as a first approximation, the source function of the ocean is equal to zero at depth and constant at the surface. This approximation is accurate for ^{14}C dissolved in water and mixed together with water masses. But there are some other isotopes, such as ^{32}Si and $^{32,33}P$ that are not displaced by water masses.

The use of cosmogenic isotopes for the solution of practical problems is more easily realised if the theoretical data related to their abundances and decay rates in the major reservoirs of the Earth, where their migration occurs, are available. For the calculations that were carried out by Lal (1963), the oversimplified box-model (see Fig. 10.5) was assumed. The model consists in a six reservoir exchange system: stratosphere (170 g/cm^2), troposphere (860 g/cm^2), continental reservoir (biosphere + surface waters), the upper mixed ocean layer (75 m), deep oceanic layer (3500 m) and ocean sediments. The calculations were carried out using the following simplified constants: the average time of air exchange in the stratosphere equals 2 years and the average time of water exchange in the deep oceanic layer equals 1000 years (Table 10.3).

The distribution of an isotope in individual geospheres depends both on its half-life and chemical behaviour. The usefulness of an isotope in hydrological and hydrogeological process studies depends on its half-life, distribution in exchange

Table 10.3 Steady-state inventories and decay rates of cosmogenic radioisotopes in the exchange reservoirs relative to 1 cm^2 of the Earth's surface. (Lal 1963)

Radioisotope	Exchange reservoir				
	Stratosphere	Troposphere	Mixed oceanic layer	Deep oceanic layer	Ocean sediments
^{3}H	$6.8 \cdot 10^{-2}$	$4 \cdot 10^{-3}$	0.50[a]	0.43	0
^{7}Be	0.60	0.11	0.28	$3 \cdot 10^{-3}$	0
^{10}Be	$3.7 \cdot 10^{-7}$	$2.3 \cdot 10^{-8}$	$8 \cdot 10^{-6}$	$1.4 \cdot 10^{-4}$	0.999
^{14}C	$3 \cdot 10^{-3}$	$6 \cdot 10^{-2}$ [b]	$2.3 \cdot 10^{-2}$	0.91	10^{-2}
^{22}Na	0.25	$1.7 \cdot 10^{-2}$	0.62	0.11	0
^{26}Al	$1.3 \cdot 10^{-6}$	$7.7 \cdot 10^{-8}$	$5 \cdot 10^{-5}$	10^{-4}	0.999
^{32}Si	$1.9 \cdot 10^{-3}$	$1.1 \cdot 10^{-4}$	$5 \cdot 10^{-3}$	0.96	$4 \cdot 10^{-2}$
^{32}P	0.60	0.24	0.16	$1.5 \cdot 10^{-4}$	0
^{33}P	0.64	0.16	0.19	10^{-3}	0
^{35}S	0.57	$8 \cdot 10^{-2}$	0.34	$6 \cdot 10^{-3}$	0
^{36}Cl	10^{-6}	$6 \cdot 10^{-8}$	$2 \cdot 10^{-2}$	0.98	0
^{37}Ar	0.63	0.37	0	0	0
^{39}Ar	0.16	0.83	$2 \cdot 10^{-4}$	$3 \cdot 10^{-3}$	0

[a] The value includes amounts present in the biosphere and humus
[b] Includes amount present in the continental waters

Table 10.4 Specific activities of cosmogenic radioisotopes in oceans and the atmosphere. (Lal 1963)

Radioisotope	Average specific activity in oceans		Radioisotope	Average specific activity in atmosphere, disintegrations $min^{-1}\,kg^{-1}$ air	
	Disintegrations $min^{-1}\,t^{-1}$ water	Disintegrations $min^{-1}\,t^{-1}$ element		Stratosphere	Troposphere
^{3}H	36	$3.3 \cdot 10^{-4}$	^{3}H	6	$7 \cdot 10^{-2}$
^{10}Be	10^{-3}	$1.6 \cdot 10^{-3}$	^{7}Be	17	0.63
^{14}C	260	10	^{22}Na	$5 \cdot 10^{-3}$	$6.7 \cdot 10^{-5}$
^{26}Al	$2 \cdot 10^{-5}$	$2 \cdot 10^{-3}$	^{32}P	0.17	$1.4 \cdot 10^{-2}$
^{32}Si	$2.4 \cdot 10^{-7}$	$8 \cdot 10^{-3}$	^{33}P	0.15	$7.6 \cdot 10^{-3}$
^{36}Cl	0.55	$3 \cdot 10^{-5}$	^{35}S	0.28	$7.8 \cdot 10^{-3}$
^{39}Ar	$2.9 \cdot 10^{-3}$	$5 \cdot 10^{-3}$	^{37}Ar	0.19	$2.1 \cdot 10^{-2}$

reservoirs and the technical ability required to measure the expected activity. The specific activities of a number of cosmogenic radioisotopes in the two principal exchange reservoirs, those of the ocean and atmosphere, were evaluated by Lal (1963) and are presented in Table 10.4.

It follows from Tables 10.3 and 10.4 that the major portion of isotopes such as ^{10}B and ^{26}Al is accumulated in the ocean sediments whereas ^{3}H, and ^{7}Be and many other isotopes are absent. The greatest portions of ^{7}Be, ^{32}P, ^{33}P, ^{35}S, ^{37}Ar and ^{39}Ar are concentrated in the stratosphere and troposphere. The surface ocean layer and inland waters contain the majority of cosmogenic isotopes in significant amounts. Their concentration in groundwaters depends on conditions of interrelation between surface waters and groundwaters. Only those isotopes whose lifetimes are longer than that of infiltrated surface water recharge groundwaters. The varying proportions of concentrations of a given cosmogenic isotope in the water-bearing layer and its steady-state amount in precipitation, surface waters, or in the overlying water-bearing layer, provide a basis for studying processes of water circulation and hydrochronology.

An additional source of isotopes in the atmosphere is cosmic dust, the terrestrial accretion rate of which is about 10^{8}–10^{9} kg/year. The use of some cosmogenic isotopes, being components of cosmic dust, provides information on the accretion rate of the Earth in the past. Such isotopes are ^{53}Mn and ^{59}Ni, which are not produced in the Earth's atmosphere and ^{26}Al, the production rate of which in the atmosphere is less than the contribution due to cosmic dust.

Changes in the isotopic composition of cosmic dust and in the outer shells of meteorites result mainly from bombardment by low-energy cosmic particles of solar origin. The outer shell of a meteorite is usually melted and ablated during movement through the atmosphere. Therefore, while studying the total flux of solar cosmic rays and particularly low-energy protons, the most convenient object is cosmic dust accumulated in the ocean floor sediments and polar pack ices.

It should be pointed out that during the last two decades the concentration of the steady-state amounts of cosmogenic radioisotopes in nature has been broken due to the additional production of these isotopes during the course of nuclear and thermonuclear tests in the atmosphere. An intense flux of neutrons is produced at the moment of explosion, which interacts with the atmospheric constituents and results in the production of identical radioisotopes to those produced by the interaction between the atmosphere and cosmic rays.

The International Atomic Energy Agency (IAEA 1973) reported that from 1945 to 1973 936 nuclear tests were carried out, of which 422 took place in the atmosphere. The majority of those tests were before 1963. During the last decade 43 tests have taken place in the atmosphere. The most powerful output of bomb radioisotopes in the atmosphere took place during 1958–1959, i.e., related to the most frequent and most powerful thermonuclear tests. The concentrations of some radioisotopes have increased compared with the pre-thermonuclear steady-state values by one or two orders of magnitude. Thus, the major portion of ^{3}H, ^{14}C and ^{22}Na present in the atmosphere at present is of bomb origin. A steady-state level of concentration of a number of short-lived isotopes, such as ^{35}S and to a lesser extent ^{7}Be and ^{32}P has been distributed but at present their concentration has returned to normal. The distribution of ^{14}C and to a great extent of also ^{36}Cl and ^{81}Kr, has also been distributed.

The effect of these bombs will be manifested as a distinctive 'mark' for a long time in those reservoirs where the processes of water mixing and dilution are slow, such as in groundwater reservoirs. These marks may serve as a good indicator of groundwater motion and also of the individual water-bearing layers between each other and with surface waters.

10.2 Sources of Tritium Discharge into Natural Waters

Among the environmental radioisotopes tritium is the most attractive to those researchers who are studying the principles of water circulation in nature. It is a constituent of water molecules and, therefore, is a perfect water tracer. Interest in the application of tritium for hydrological and meteorological purposes increased greatly during the period of thermonuclear tests during 1953–1962 and also subsequently when a large amount of this artificially produced isotope had been injected into the atmosphere. The bomb-tritium, injected into the atmosphere by instalments after each nuclear test, is a kind of fixed time mark of water involved in water cycling.

Tritium is produced in the atmosphere by the interaction between secondary nuclear particles of cosmogenic origin, mainly neutrons and protons and nitrogen and oxygen nuclei. Neutrons produced by cosmic radiation originally have energies of about several dozens of MeV. Then, due to inelastic scattering on nitrogen and oxygen nuclei, they slow down. At energies greater than 1 MeV the prevailing nuclear reaction is $^{14}N(n, {}^{3}H)\,{}^{12}C$.

Table 10.5 Main reactions with neutrons in the atmosphere

Reaction	Absolute rate, (neutron/$cm^2 \cdot sec$)	Relative rate
Radiocarbon production	4.0	0.56
Tritium production	0.13	0.02
Other reactions	2.2	0.31
Loss	0.8	0.11
Total	7.13	1.0

Table 10.6 Reactions of cosmic-ray production of tritium in the atmosphere

Reaction	Energy of particles (MeV)	Cross-section of reaction (mbarn)	Production rate atom/ $cm^2 \cdot sec$
$^{14}N(n, {}^{3}H)\ {}^{12}C$		11 ± 2	0.1–0.2
$^{16}O(p, {}^{3}H)\ {}^{14}O$	>100	25	0.08
$^{14}N(p, {}^{3}H)\ {}^{12}N$			
$^{16}O(p, {}^{3}H)\ {}^{14}O$	10–100	–	0.01
$^{14}N(p, {}^{3}H)\ {}^{12}N$			
$^{14}N(p, {}^{3}H)\ {}^{12}N$	<10	–	0.05
N, O(γ, ^{3}H)	–	–	10^{-5}

The cross-section of this reaction is about 0.01 barn. Only 3–5 % of all the neutrons generated by cosmic rays in the Earth's atmosphere take part in the production of tritium (Table 10.5).

Beside the above reaction, tritium may be produced by other reactions, the main ones of which are presented in Table 10.6.

It follows from Table 10.6 that the first two reactions of interactions of ^{14}N with medium energy neutrons, fission of ^{14}N and ^{16}O nuclei by protons at energy higher than 100 MeV have the highest cross-section. The production rate of tritium by protons with energies ranging 10–100 MeV has been estimated to be 0.01 atom/ $cm^2 \cdot sec$ because of an absence of reliable experimental data. The contribution of tritium production by the other components of cosmic radiation is 0.1–0.2 atom/ $cm^2 \cdot s$. This value is less than that actually observed, which is 0.3 atom/$cm^2 \cdot s$. An additional tritium input to the atmosphere may take place during intense solar flares. It is most probably formed in the course of the reaction $^{4}He\ (p, 2p)^{3}H$ in the chromosphere of the Sun.

The steady-state amount of tritium on the Earth, formed by cosmic radiation, varies from 3 to 10 kg. The major part of tritium (~93 %) stays in the hydrosphere and only about 7% is in the atmosphere (see Table 10.4). Due to insignificant amounts in natural objects, tritium is commonly expressed in Tritium Units (TU). A tritium unit corresponds to one atom of tritium per 10^{18} atoms of protium, which is equivalent to 7.2 disintegrations per minute per litre of water, or 0.119 Bq/kg. Tritium is a soft β-emitter, characterised by maximum particle energy equal to

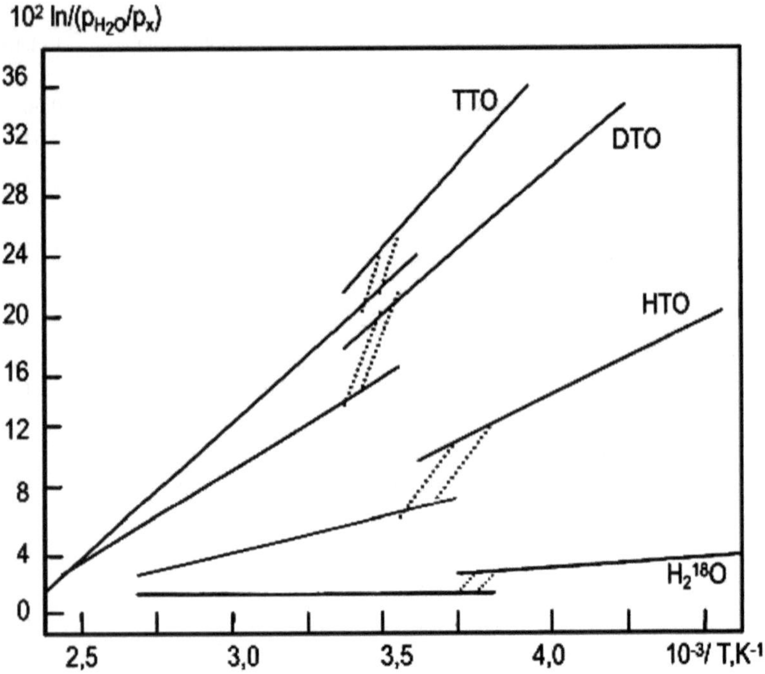

Fig. 10.6 Experimental relationship from vapour pressure of heavy and light water molecules and temperature. (Van Hook 1968; Ferronsky and Polyakov 2012)

18 keV and a half-life of 12.43 year. The final product of tritium decay is the stable isotope of ^3He.

Shortly after production, tritium is oxidised and forms molecules of water HTO. Since the masses of tritium and protium differ, fractionation occurs during phase transitions of water from gases to solid states and vice versa (Fig. 10.6).

It is for this reason that inhomogeneity is observed in tritium distribution between hydrogen-bearing systems in tritium-protium exchange reactions. Experiments involving different types of clays (kaolinite, montmorillonite and silty clays) have shown that in the course of their interaction with water labelled by tritium a marked exchange reaction between tritium and protium is observed. Protium constitutes the clay minerals and hydroxides (Stewart 1965). This effect may be considered as significant in groundwater dating and the large time-scale involved in the investigation of water motion in rocks based on tritium labelling

Before the first thermonuclear tests in the atmosphere (1952), the majority of tritium in nature resulted from cosmic-ray production. At that time only a few measurements of natural tritium on the Earth had been carried out. According to the data of Libby (Kaufmann and Libby 1954), who was the first to study its applicability in hydrology and carried out measurements in Chicago, the average content of environmental tritium in precipitation was about 8 TU. Brown (1961) measured tritium concentrations in the Ottawa Valley (Canada) and found that the mean level

Table 10.7 Distribution of tritium in the geospheres. (Burger 1979)

Geosphere	HTO		HT		CH^3T	
	Ci	TE	Ci	TE	Ci	TE
Ocean, top 100 m	$9 \cdot 10^8$	10–20	$(1–10) \cdot 10^3$	?	–	–
Troposphere, $(3.8 \cdot 10^{18}$ kg air)	$(2–8) \cdot 10^6$	–	$(8–18) \cdot 10^6$	$(3–7) \cdot 10^6$	$(6–20) \cdot 10^6$	$\sim 5 \cdot 104$
Stratosphere, $(1.3 \cdot 10^{18}$ kg air)	$(0.6–5) \cdot 10^8$	(2–8)	$(1–4) \cdot 10^4$	–	< 105	$\sim 5 \cdot 104$

of tritium was 15 TU. According to the calculations of Lal and Peters (1962), this value corresponds to 6 TU. Later on, when some principles of distribution and fallout of corresponding amounts of tritium on the Earth were established, it became clear that tritium content varies within a large range both in space and time. This range may be estimated as being equal to 0.1–10 TU for both hemispheres at a constant rate of tritium production of about 30 atoms/cm^2 min relative to the terrestrial surface (Suess 1969).

Using data obtained by different authors, Burger (1979) reported the data of tritium distribution in individual geospheres given in Table 10.7. Estimations were made of tritium fallout on the Earth with cosmic dust and micrometeorites (Fireman 1967). It was shown that in stone meteorites, the tritium activity equals 200–400 disintegrn/kg min. For iron meteorites this value was 40–90 disintegrn/kg min. Thus, the tritium component contained in meteorites falling down on the terrestrial surface is insignificant and is less than 10^{-5} atoms/cm^2 s.

The results of measurements of tritium concentrations in lunar rocks, carried by the space crafts Apollo-11 and Apollo-12, gave 270–300 disintegrn/kg min (Bochaler et al. 1971), which appeared to be similar values to those in meteorites. The production of tritium in lunar rocks and meteorites occurs due to spallation reactions between cosmic high-energy protons and nuclei or rock-forming elements, such as Fe, Si, Al, etc.

Thermonuclear tests in the atmosphere, carried out since 1952, represent another source of atmospheric tritium. The output of tritium released during a thermonuclear explosion averaged from 0.7 to 5 kg per megaton of thermonuclear fusion and 0.07 kg per megaton of nuclear fission (Miskel 1973). A diagram of tritium injection to the atmosphere from thermonuclear explosions since their beginning is shown in Fig. 10.7 plotted on the basis of data obtained by Eriksson (Schell and Sauzay 1970). Curve 1 accounts for the residence time of tritium in the stratosphere, the main reservoir of accumulation. According to data obtained for precipitation, this residence time is about one year. The residence time in the reservoir is the time required for one half of the tritium present at the beginning to remain in the reservoir. Curve 2 shows the annual variations of the total amount of tritium in the stratosphere.

Energy tests of thermonuclear explosions of megaton energy were carried out by the USA in August 1958 at an altitude of 4–7.5 km. In 1957–1958, at a lower altitude, there were eight megaton tests carried out by the UK. Some portion of tritium produced during these explosions has moved into the stratosphere. The most powerful megaton explosions were made in 1961–1962 by the USA and the former USSR in various places of the globe and at high altitudes. As a result, a large amount of tritium (up to 400 kg) has been stored in the stratosphere (Ostlund and Fine 1979) and its concentration in individual places on the Earth (e.g., White Horse, Canada) in the spring-summer months reached 10,000 TE (Thatcher and Payne1965). It was found later on, after the interdiction of nuclear tests in the three spheres, that the estimation of the residence time of tritium in the stratosphere, which was accepted to be about one year, appeared to be imprecise. Its concentration in precipitation has decreased up to the present time but far more slowly than was assumed theoretically. Thus, the tritium concentrations in precipitation should have reached a natural level by 1970. However, in Western Europe, in the summer months of 1968–1970, the levels were still fixed as they had been at the end of 1960, amounting to about 30 TU.

During the last few years, up to 1970, some increases in tritium content of the atmosphere were observed due to thermonuclear explosions carried out by France and China. But the values of these tritium injections are insignificant compared with the previous ones (see in Fig. 10.7 the maxima corresponding to 1967–1970).

Atomic industries (power and research reactors, plants of nuclear fuel reprocessing, etc.) are also sources of environmental tritium. The output of tritium during uranium fission in different types of reactors depends on the choice of fuel, the energy spectrum of neutron flux and a number of technological factors. Depending on the type of reactor, tritium is produced in the course of the activation of boron, lithium and deuterium atoms by neutrons. In a controlled thermonuclear reactor, which is now under construction, tritium will be the main radionuclide. In this case the major portion of tritium will be ejected from nuclear plants into the environment in a gaseous state (HT, DT, T_2) and partly in liquid phase in the form of HTO.

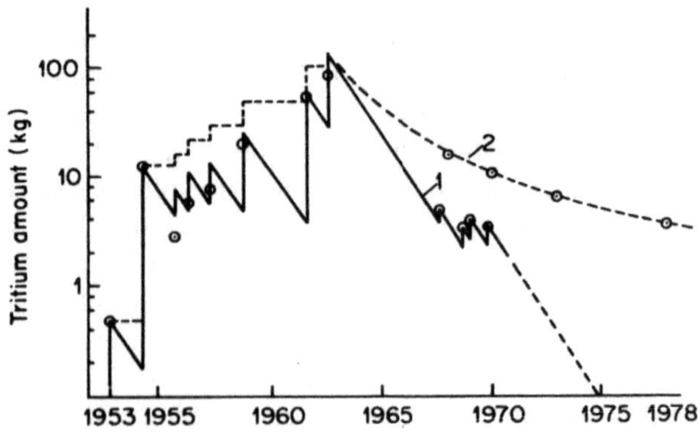

Fig. 10.7 Growth of thermonuclear amounts in the atmosphere taking into account its fallout (*1*), and the stratosphere taking into account its natural decay (*2*). (Ferronsky et al. 1975; Ferronsky and Polyakov 2012)

The gaseous tritium ejected into the atmosphere oxidises quickly and forms water molecules.

According to data obtained by Sehgal and Remport (1971), in the course of uranium and plutonium fission 0.8 atom of tritium is formed per 10^4 acts of ^{235}U fission, 0.9 atom of tritium for that of ^{238}U and 1.8 atoms of tritium for that of ^{239}Pu. Fluss and Dudey (1971) studied the dependency of tritium production on energy of neutrons for ^{235}U. According to them, when the energy of neutrons changes from 175 to 630 keV the yield of tritium increases from 2 to 3.4 atoms per 10^4 acts of uranium nuclei fission.

In the slow-neutron reactors, during ^{235}U fission, the yield of tritium amounts to $8.7 \cdot 10^{-3}\%$ (Taylor and Peters (1972) and in the fast-neutron reactors it corresponds to $2.2 \cdot 10^{-2}\%$ (Dudey et al. 1972). This efficiency of yield corresponds to the VVER and RBMK reactor that provides $1.1 \cdot 10^{-2}$ Ci/day · Mw (t) and for the fast neutron reactors—$2.8 \cdot 10^{-2}$ Ci/day · Mw (t).

During reactions proceeding in the control rods of the reactors, tritium is ejected in accordance with the reactions $^{10}B(\alpha, 2\alpha)^3H$; $^{11}B(n, ^3H)^9B$; $^{10}B(n, \alpha)^7Li$; $^7Li(n, n\alpha)^3H$. The cross-section of these reactions increases with the energy of neutrons. Therefore, the yield of tritium in fast-neutron reactors is considerably higher than in reactors of the other type. Lokante (1971) reported that the tritium yield corresponding to fission reactions amounts to 11,000 Ci and for boron reactions is about 1380 Ci in the 3500 Mw boiling water reactor. In the 300 Mw breeder reactor the tritium yield is 1670 Ci for the fission reaction and 3980 Ci for the boron reaction. The tritium output from the heat-generating elements to the heat carrier depends on the material of the shell. Stainless steel passes up to 60–80 % of the produced tritium and zirconium only 0.1 % (Lokante 1971).

According to Broder et al. (1979), the tritium exhausts from typical VVER-440, VVER-1000 and RBKM-1000 reactors are equal to 0.6, 1.6 and 2.28 Ci/day, respectively. At the Novovoronezhskaya nuclear power station, for example, about 55 % of the total amount of tritium is ejected into the atmosphere, 27 % into surface waters and 13 % into groundwaters. According to data for yearly observations, the tritium concentrations at 1 km downstream of the river are higher by one order of magnitude than in water upstream of the river.

As pointed out above, tritium is produced at nuclear power stations (reactors) both due to the process of uranium fission and due to interactions of neutrons of various energies with constructional materials and coolants. The following substances are some of those used as coolants: light and heavy water, noble gases, melted metallic sodium. The main nuclear reactions, in the course of which tritium is formed, are the reaction of fission of enriched uranium (X) leading to the formation of the fission products $X(n, f)^3H$; $X(n, f)^6He \rightarrow {}^6Li(n, \alpha)\ {}^3H$; $^{10}B(n, 2\alpha)\ {}^3H$; $^2H(n, \gamma)\ {}^3H$; $^9Be(n, 2\alpha)\ {}^3H$ and so on.

In Table 10.8 data on the tritium yield in different types of reactors due to the above-mentioned reactions are given. In Table 10.9, data on tritium input into the atmosphere and surface waters for various nuclear reactors and nuclear fuel processing plants are presented.

Table 10.8 Tritium production in various types of reactors. (Bonka 1979)

Nuclear reaction	Tritium production in Ci/Mw(e) per year					
	BWR[a]	PWR[a]	HWR[a]	AGR[a]	HTR[a]	FBR[a]
Efficiency	0.33	0.33	0.32	0.41	0.41	0.41
Fuel element						
Fission	18	18	20	15	12	30
^{6}Li in fuel, 0.05 ppm	0.3	0.3	0.8	1	0.2	0.1
^{10}B in fuel, 0,05 ppm	$4 \cdot 10^{-5}$	$4 \cdot 10^{-5}$	$3 \cdot 10^{-5}$	$5 \cdot 10^{-5}$	$1 \cdot 10^{-5}$	$2 \cdot 10^{-5}$
^{6}Li in graphite	–	–	–	–	0.5	–
^{9}Be in graphite	–	–	–	–	$1 \cdot 10^{-4}$	–
^{10}B in graphite	–	–	–	–	$3.5 \cdot 10^{-3}$	–
Coolant						
^{1}H in water	$8 \cdot 10^{-3}$	$8 \cdot 10^{-3}$	–	–	–	–
^{2}H in water	$4 \cdot 10^{-3}$	$4 \cdot 10^{-3}$	150	–	–	–
^{10}B in water	–	0.8	–	–	–	–
^{3}He in helium	–	–	–	–	1	–
^{6}Li in sodium	–	–	–	–	–	2
^{9}Be in sodium	–	–	–	–	–	0,01
^{10}B in sodium	–	–	–	–	–	0,01

[a] BWR is boiling-water reactor; PWR is pressurised water reactor; HWR is heavy-water reactor; AGR is advanced gas-cooled reactor; HTR is high-temperature reactor; FBR is sodium-cooled fast breeder reactor

Table 10.9 Tritium emission rates from nuclear power reactors under normal operation and reprocessing plants without tritium retention. (Bonka 1979)

Nuclear facility		Emission rate (Ci/year)	
		Atmosphere	Surface water
Reactor (1000 Mw(e))	BWR	30	150
	PWR	20	900
	HTR	10	900
	FBR	100	200
Reprocessing plant (40,000 Mw(e) full load)	BWR and PWR	$7 \cdot 10^{-5}$	1000
	PWR	$7 \cdot 10^{-5}$	1000
	HTR	$6 \cdot 10^{-5}$	1000
	FBR	$6 \cdot 10^{-5}$	1000

In the report made by the National Council on Radiation Protection and Measurements of the USA (Eisenbund et al. 1979) the following data on tritium yield due to diverse sources are given. The global amount of tritium produced by cosmic radiation equals 70 MCi (1 kg of tritium is equivalent to about 10 MCi), i.e., its production rate is about 4 MCi per year. In 1963 an amount of tritium, estimated at 3100 MCi, was injected into the atmosphere and hydrosphere as a direct result of nuclear and thermonuclear tests. Thus, the natural level of tritium (70 MCi) may be attained as a result of a decay process up to 2030. The production of tritium in nuclear reactors of the PWR type due to fission reactions ranges from 12 to 20 Ci/day per Mw of the thermal power. The activation of light elements gives an additional yield of tritium that averages from 600 to 800 Ci/year per Mw of electrical power. These values are equal to 63 Ci/year for the reactors of the BWR and PWR types with light-water cooling. The average residence time of tritiated water vapour in the troposphere ranges from 21 to 40 days.

The main residence time of HTO molecules in the mixing ocean layer (50–100 m thickness and equal to 75 m on average) is approximately 22 year. The average time of half-removal of tritium from the human body depends upon individual biological features and is equal to hundreds of days. For two arbitrarily chosen and absolutely healthy men of middle age, the time of half-removal was found to be 340 and 630 days.

Some portion of tritium is injected into the environment from research centres, medical institutions and industrial plants dealing with works involving the application of artificial tritium. According to data reported by König (1979), tritium activities of $(1.2–2.1) \cdot 10^3$ Ci/year have been released into the atmosphere from the Nuclear Research Centre in Karlsruhe (Germany) since 1969. Krejči and Zeller (1979) reported that a large amount of tritium is ejected into the atmosphere from the luminous compound industry, producing tritium, gas-filled light sources and tritium luminous compounds. The urine of the operators working on one of the plants producing luminophor contains about 25 Ci/l of tritium (1TU=3.2 pCi/l). In the waste water within the plant area the concentration of tritium is about 0.3 Ci/l and in water at the exit from the cleaning installations the concentration is 0.004 Ci/l. In precipitation at a distance of 50 m from the ventilation system of the tritium department the concentration of tritium is 0.1 Ci/l, at a distance of 200 m it amounts to 0.02 Ci/l and at 2000 m it is equal to 0.001 Ci/l. A large amount of tritium is now used for the production of liquid crystal displays for digital electronic readouts. The annual production of tritium by different industries in the USA is shown in Fig. 10.8 reported by Combs and Doda (1979). According to their estimations the amount of tritium used for the production of backlighted digital watches would reach (and it did reach) 4 MCi in 1979.

A considerable amount of tritium in the environment originates from the nuclear fuel reprocessing industry. Daly et al. (1968) showed that the nuclear fuel reprocessing plant situated in New York State ejects about 200 Ci of tritium per day, 25 % of which is released into the atmosphere, 65 % is contributed in liquid form to the river and 10 % goes into the soil.

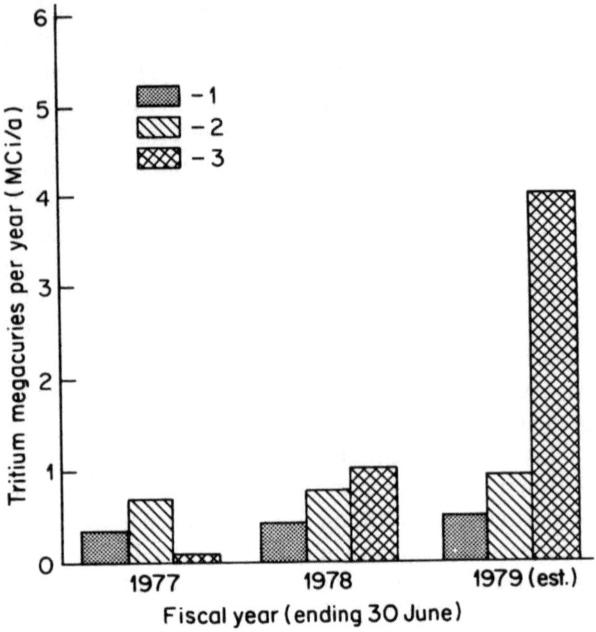

Fig. 10.8 Quantities of tritium produced annually in the United States of America for digital watch lighting (*1*), by power reactors (*2*) and for other commercial products (*3*). (Combs and Doda 1979; Ferronsky and Polyakov 2012)

Taking into account the modern trend of development of nuclear power stations in most countries of the world, it is easy to estimate that by the beginning of the twenty-first century the production of technogenic tritium, which will be continuously ejected into the environment, will overcome the amount of cosmogenic tritium produced in the atmosphere.

In fact, the tritium production rate from all the nuclear plants (power reactors and nuclear fuel reprocessing plants) of the world by 2000 was four times the rate of its natural production by cosmic radiation. But the release of tritium into the environment is negligible since its major portion is collected and buried as radioactive waste. Besides, the tritium produced by nuclear plants cannot reach the stratosphere, where it would be subjected to global redistribution. Therefore, its ejection into the precipitation, surface and groundwaters is of a local character, related to the neighbourhood of the organisation that is studying surface and groundwaters involving tritium measurements.

According to Katrich (1990), some amount of tritium was injected into the environment as the result of the Chernobyl accident. Concentration of tritium over the European part of Russia in May 1986 increased by 2–5 times compared with May 1985. But in June–July the level of contamination dropped to a normal level because the tritium only reached the troposphere.

At least some portion of tritium will be released into the atmosphere due to nuclear explosions used for peaceful purposes (such as the performance of underground oil, gas and water capacities, excavation of rocks in the course of construction works etc.). But the major portion of tritium precipitating in the hydrological cycle in the near future will be, as previously, the bomb-tritium released in the period from 1952 to 1962.

10.3 Global Circulation of Tritium Water

At present the total amount of tritium on the Earth exceeds its pre-bomb level only by 1.5–2 times. This situation is explained by the continuous decay of tritium and isotopic exchange with the ocean's waters.

From the atmosphere, which is the only source of natural and thermonuclear tritium, the tritium water molecules together with air flows enter the troposphere where they form tropospheric moisture. The other effects that determine incoming of tritium into the tropospheric moisture is evaporation from the ocean surface and the molecular exchange between the surface ocean layer and atmospheric moisture. The continental atmospheric precipitation of high tritium content forms river runoff, lake and groundwaters and glaciers. Some part of precipitation evaporates. Concentration of tritium in river water is close to that in precipitation and in lakes it depends on the residence time of water: the longer residence time, then the less content of tritium because of its decay. The same relates to groundwater basins. In glaciers, especially in polar latitudes (Greenland, Antarctic), water looks like it is conserved and in the deep layers tritium is completely decayed. The non-decayed part of the molecules HTO from the rivers, lakes and groundwater arrives to the oceans together with surface and underground runoff.

The oceans' waters are divided into two layers: the upper well mixed layer with a depth of several hundred metres and the lower layer, divided from the upper by the thermocline, with a water exchange time of several hundreds and even thousands of years. The thermocline may be absent in Polar regions and in this case the most favourable conditions for vertical water mixing appear. In the upper layer are observed maximum concentrations of tritium, which lower to the deep and mixed long time and lose tritium at its decay.

Thus, the stratosphere is the source of tritium in the hydrologic cycle and the reservoir for tritium runoff is deep ocean waters and glaciers where it decays. It follows that definite regularities should be expected in distribution of tritium in all chains of the hydrosphere.

10.3.1 Tritium in Atmospheric Hydrogen and Methane

Except water, which is the main object for study of tritium distribution on the Earth, there are two other hydrogen-bearing compounds. They are the molecular hydrogen H_2 and methane CH_4. The study of their behaviour is important for understanding of the geophysical and physical-chemical processes in the atmosphere. Of a special interest are H_2O, H_2 and CH_4 cycles, which have a close relationship in the atmosphere. Their passage from one form to another may be used as a tracer for determining the residence time of hydrogen in its compounds, for estimation of the exchange rate between the hemispheres, for study of the air exchange between the troposphere and stratosphere and for understanding the nature of the compounds origin.

At present the concentration of molecular hydrogen in non-industrial regions amounts to 0.575 ppmv in the northern hemisphere and 0.550 ppmv in the southern hemisphere (Schmidt 1974). On a global scale, 50 % of molecular hydrogen is of anthropogenic origin. The most significant natural source of H_2 in the biochemical process occurring in the ocean is molecule dissociation with hydrogen photosynthesis occurring in the atmosphere (Romanov and Kikichev 1979) also contributing significantly to H_2 concentration.

The majority of tritium in atmospheric hydrogen is of cosmogenic origin. The principal reaction leading to the production of HT molecules, according to estimations made by Harteck (1954) for tritium generated by cosmic rays, is the recurrent photodissociation of TO_2 and the subsequent exchange reaction of the form:

$$T + H_2 \rightarrow HT + H.$$

Only 0.1 % of the tritium produced by cosmic rays exists in the form of HT molecules and 99.9 % is in the form of HTO molecules. The bulk of the HT molecules are formed at altitudes ranging from 10 to 40 km. The total mass of the HT molecules of cosmogenic origin in the atmosphere is about 5 g (Rowland 1959).

Besides the release of tritium from the device itself during thermonuclear explosions, tritium is produced according to the following main reaction:

$$D + D \rightarrow T + H + 4\,\text{MeV}.$$

It is considered that approximately the same relative content of bomb-tritium is contained both in the molecules of H_2O and H_2, which is the result of isotopic exchange between H_2O and H_2 in the expanding and cooling thermonuclear sphere. In this case, the main reactions are:

$$T + O_2 \rightarrow TO_2,$$

$$TO_2 + O \rightarrow TO + O_2 + 61\,\text{kcal},$$

$$TO + H_2 \rightarrow HTO + H + 17\,\text{kcal},$$

$$HTO + H_2 \leftrightarrow H_2O + HT.$$

In the course of underground thermonuclear tests the increase of tritium content in atmospheric hydrogen was not accompanied by an increase of tritium content in the atmospheric moisture. This is likely due to the lack of conditions necessary for oxidising reactions in the medium where the explosion took place.

The main sources of technogenic tritium, as pointed out earlier, are nuclear power plants, which release a considerable amount of tritium in the form of HT molecules.

Tritium was first measured in atmospheric hydrogen in 1948 near Hamburg, where its concentration was found to be equal to $4 \cdot 10^3$ TU (Faltings and Harteck 1950). Later on, due to thermonuclear tests, the concentration of tritium in atmospheric hydrogen sharply increased. It can be seen from figure that the concentration of tritium increased from $4 \cdot 10^3$ to $4 \cdot 10^6$ TU from 1948 to 1973 (Ehhalt 1966; Östlund and Mason 1974). In tropospheric HT its concentration reached maximum values with a delay of about 2–2.5 year. Ehhalt (1966) assumed this to be the effect of removal of HT molecules from the stratosphere, which is the main reservoir of tritium accumulation during thermonuclear explosions, into the troposphere after a certain long time, corresponding to what was observed.

After thermonuclear tests were stopped in the three media, the tritium content in molecular hydrogen remained approximately unchanged from 1963 to 1973 at a level of $(2–4) \cdot 10^6$ TU. The constancy of the HT concentration in the atmosphere can only be explained by ejection of tritium into the atmosphere from some sources in order to maintain the corresponding partial pressures of the HT molecules and compensate the natural losses caused by radioactive decay and other processes of removal.

According to existing estimations (Martin and Hackett 1974) the total anthropogenic release of tritium into the atmosphere should amount to $1.2 \cdot 10^6$ Ci/year in order to maintain the average concentrations of tritium in atmospheric hydrogen at the level of about 80 atoms per mg of air. These sources of anthropogenic tritium release are likely to be atomic industry plants and underground tests.

In contrast to the distribution of HTO, the spatial distribution of tritium in atmospheric molecular hydrogen is characterised by a high homogeneity in the whole atmosphere. According to data of Östlund and Mason (1974), the concentration of tritium was about 50 atom/mg of air in 1971–1972. Only at high latitudes (60° N and higher) does the concentration of tritium increase up to 80 atom per mg of air. According to limited data, the tritium concentration in the stratosphere decreases with altitude following barometric law. Thus, in North Alaska at 70–75 °N the tritium content in the lower stratosphere varies with altitude from 80 to 30 atom/mg of air (Östlund and Mason 1974). The exceptions to these principles are the regions of anomalous release of molecular hydrogen of industrial origin with zero concentration of tritium and also the regions where nuclear industries are located, characterised by raised tritium content. The global content of HT molecules in the atmosphere during the period 1968 to 1978 is shown in Fig. 10.9 (Mason and Östlund 1979).

According to data of Ehhalt (1974), the methane early production is $(5.4–10.6) \cdot 10^{14}$ g, from which 80 % have biogenic origin and the tritium concentration is more than 10^4 TU (Ehhalt 1974). Creation of CH_3T molecules in the atmosphere occurs as a result of nuclear and exchange reactions between HT and CH_4. But as investigations show, these reactions have small efficiency (Begemann and Friedman 1968). It is assumed that the main sources of tritium in methane are

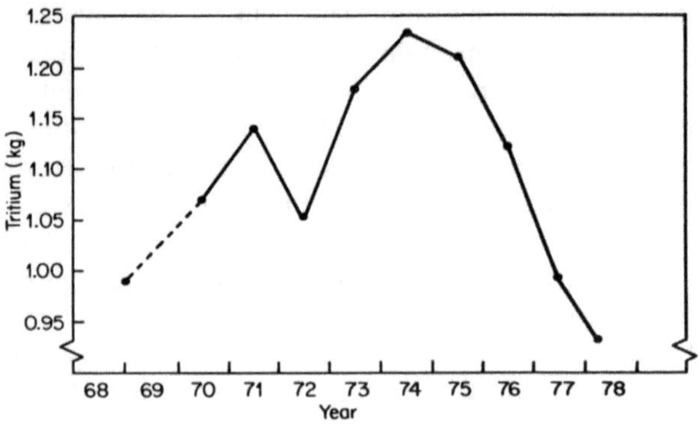

Fig. 10.9 Inventory of global atmospheric HT for 1968–1978. (Mason and Östlund 1979; Ferronsky and Polyakov 2012)

research laboratories and institutions of atomic industry, the technology of which relates to tritium (Burger 1979).

Molecules CH_3T of biochemical reactions have the same T/H ratio as environmental water. It is obvious that atmospheric HT takes part in biogenic methane. As a result of this process, the tritium content in methane correspondingly increases. The exchange time of methane in the atmosphere is about 4–7 year.

10.3.2 Tritium in Atmospheric Water Vapour

As pointed out above, the upper layer of the atmosphere, its stratosphere (15–17 km), is the reservoir where the bulk of natural tritium is accumulated. Despite the small amount of stratospheric moisture it is the main source of tropospheric tritiated water falling to the Earth's surface as precipitation. It will be shown that the stratosphere is also a reservoir accumulating thermonuclear tritium.

Martell (1963), using a supposed production rate of natural tritium of 0.3 atom/ $cm^2 \cdot s$, obtained a value of tritium concentration of about 10^6 TU. The first measurements of stratospheric tritium in the air above Minneapolis in 1955–1958, at an altitude of 14–28 km carried out by Hagemann et al. (1959), showed that the tritium content was equal to $1.1 \cdot 10^6$–$1.52 \cdot 10^7$ atom per gram of air. On the basis of these measurements and with the ratio of T/C^{14} it was found that the amount of tritium in the stratosphere is equal to $6 \cdot 10^{23}$ atoms (6 kg). Later, Scholz et al. (1970) obtained a tritium concentration of $2.2 \cdot 10^6$–$8 \cdot 10^7$ T.U. using their own experimental data.

The most complete studies of tritium distribution in the troposphere were carried out by Ehhalt (1971). Measurements were conducted from November 1965 to January 1967 at continental (Scottsbluff, Nebraska) and oceanic (near California) stations up to an altitude of 9.2 km. The results of Ehhalt's measurements showed that concentrations in water vapour increase with altitude. The lowest concentrations were found at an altitude of 2300 m above sea level in the spring-summer seasons

(1200 TU) and maximum values at an altitude of 9 km (26 000 TU). It was found that the altitude of the seasonal variations at amplitude of 7.5–9 km is greater by a factor of 10 than the variation of tritium concentrations at the Earth's surface.

Detailed data concerning the distribution of HTO molecules with height were reported by Mason and Östlund (1979). Water and hydrogen samples were taken with the help of a molecular trap, placed on a special aircraft, up to a height of 13 km. Differences in the HTO distribution both with altitude and latitude were observed between 1976 and 1977, caused by an atmospheric thermonuclear test conducted by the People's Republic of China on 17 November 1976, which resulted in the release of a large amount of tritium into the stratosphere. According to the estimations of Mason and Östlund, the inventories of atmospheric tritium were about 1 kg of HT molecules and 5.3 kg of HTO at the end of 1977. A major portion of HTO (about 5.1 kg) has been stored in the stratosphere and 0.2 kg has been in transit to the ocean surface through the troposphere.

Bradley and Stout (1970) carried out individual measurements in order to obtain tritium distribution profiles in atmospheric moisture in Illinois State (USA) up to amplitude of 5 km. They obtained three different types of distribution of tritium with altitude. The first type is characterised by an increase of tritium concentration with altitude, the second type by a constancy of tritium concentration and the third type by a decrease of tritium concentration with altitude up to 2.5 km and then by a subsequent increase in concentration. These tritium distributions were explained by Bradley and Stout as the result of different conditions of formation and mixing of atmospheric moisture in the lower troposphere and also in terms of different sources of tritium.

In mountainous regions the vertical distribution of tritium in the atmosphere can be estimated using data of precipitation measurements at different altitudes. These studies were carried out by Romanov (1978) in Caucasus near the Aragats Mountain (Table 10.10). It was found that the average annual concentrations of tritium in precipitation sampled at an altitude of 850–3500 m in 1971 and 1972 increased by factor of three. Assuming the equilibrium conditions of condensation of atmospheric precipitation, it may be assumed that such relationships reflect the actual vertical distribution of tritium in water vapour.

The data obtained by the same author on tritium concentration in the annual layers of the Pamir glacier at an altitude of 4500 m (Table 10.10) were found to be lower. This was explained by a difference in the origin of the atmospheric moisture (Indian Ocean), which forms the sampled precipitation.

Table 10.10 Relationship of tritium concentrations with altitude in Aragats Mountain (Caucasus)

Sampling place	Altitude (m)	Mean annual concentration in precipitation (TE)	
		1971	1972
Oktemberian	850	37	82
Garnovit	1100	114	162
Aragatz	3238	143	201
Pamir (Abramov glacier)	4500	65	72

10.3.3 Tritium in Precipitation

The applicability of environmental tritium as a tracer of air mass circulation in the atmosphere and the formation of precipitation and discharge on the continental surface and in groundwaters is based upon experimental data of tritium content in precipitation on a global scale. This work was initiated in 1961 by the IAEA and the WMO jointly. In order to detect the tritium, deuterium and oxygen-18 content in precipitation, more than 100 meteorological stations linked to the WMO, located in different countries, were involved (see Fig. 9.7). The ocean samples were collected on islands and weather ships. Thus, the network of stations included the most characteristic points of the globe both on continents and oceans in both the northern and southern hemisphere.

In 1965, in connection with the International Hydrological Decade program (1965–1974), the network of stations included additional stations for water sampling from rivers. But the majority of rivers fell out of this network of stations and therefore a representative river network was not established. Up to present, the network includes more than 100 stations plus many national points of observation at which the tritium content is measured. Measurements of tritium content in precipitation continue.

Sampling and analysis were performed according to techniques developed by the IAEA. Samples of atmospheric waters, taken every month, correspond to the monthly average tritium content in precipitation. Tritium content in water samples was measured in low-level counting laboratories in the IAEA and in Canada, Denmark, India, Israel, New Zealand, Germany, Sweden, USA, USSR and other countries.

All these data are being collected by the IAEA and after processing together with the results of analysis of stable isotopes (deuterium and oxygen-18) content, measured in the same samples and also together with meteorological data, are published in special issues of Environmental Isotope Data (IAEA 1969).

The tritium concentrations in precipitation may have a substantial difference in individual fallouts, depending on their origin and trajectory of motion. But in principle, distribution of the fallouts depends on the mechanism of circulation of the atmosphere. Seasonal and annual variations are observed. Seasonal variations are related to the strengthening of the air masses exchange between the spring-summer stratosphere and the troposphere. The effect leads to occurrence of the so-called spring-summer maximum in the annual tritium distribution. Weakening of this process in the winter and autumn leads to occurrence of the autumn-winter minimum.

The long-term variation of tritium concentration of natural origin can be related to phases of solar activity. This is because it occurs at an anti-phase with the intensity of galactic cosmic rays, which are accepted as a source of tritium.

Some researchers have tried to study the correlation between long-term tritium variations, which were observed in Greenland glaciers formed before 1952, with solar activity. But a single-valued result was not found. For example, Begemann (1959) with the Greenland glaciers discovered a negative correlation between tritium concentration and solar maximum activity. Ravoire et al. (1970) with Antarctic snow (1950–1957) found this correlation to be positive. And Aegerter et al. (1967) discovered both types of correlation.

During atmospheric thermonuclear tests, tritium in the form of HTO occurs both in the stratosphere and in the troposphere. A proportion of its amount depends on the height and power of the explosion. The tropospheric component of HTO must have a residence time compared with that of the tropospheric moisture, i.e., equal to several weeks. Experimental data prove this conclusion (Buttlar and Libby 1955). The period of removal of half of the tritium from the troposphere is 45 days. A figure close to this was obtained by Buttlar and Libby by measurements of tritium in precipitation in Chicago.

The stratospheric tritium part is removed substantially lower. The velocity of this process corresponds with the velocity of exchange between the tropospheric and stratospheric air and has seasonal cyclic character. The annual tritium concentration changes in the Ottawa River water are shown in Fig. 10.10 (Brown 1970). Figure 10.11 demonstrates the annual means of tritium concentration changes over Moscow and in the Moscow River (Russia) during the period of 1953–1969, obtained by the authors.

It follows from Figs. 10.10 and 10.11 that the entering of tritium into the atmosphere occurred during the 1954, 1956, 1958 tests. In the period of moratorium from 1959 up to its interruption in September 1961 the tritium concentration within the Moscow region dropped from 760 to 200 TU. The period of removal of the half tritium value for this time interval was equal to about 1 year. In September 1961, the thermonuclear tests were renewed and continued up to December 1962. In that period the main part of bomb-tritium accumulated in the stratosphere, which is observed up to now.

The maximum yearly means of tritium concentrations over the Moscow region reached 3900 TU (Fig. 10.11). After the thermonuclear test ban treaty in the three spheres had come into force the stratospheric tritium reserve started to decrease with a period of 1.2 year up to 1967–1968. After that the decrease slowed down. During 1969–1974, the period became equal to about 3 years. This value was obtained by the authors on the basis of data over vast territories and, therefore, may be considered adequately authentic. One more explanation of the above phenomenon can be redistribution of tritium in the stratosphere between the northern and southern hemispheres. During recent years, this transfer has decreased due to approaching a quasi-equilibrium state.

It was shown in the work of Weiss et al. (1979) that, starting from 1970 over Central and Western Europe, technogenic tritium plays a notable role in formation of its occurrence in precipitation. From here it follows that sampling stations should be placed at an appropriate distance from industrial plants and institutions.

Variations of tritium concentrations in precipitation during spring and summer are determined by specific conditions of the mass air exchange between the stratosphere and the troposphere, resulting from easier connection.

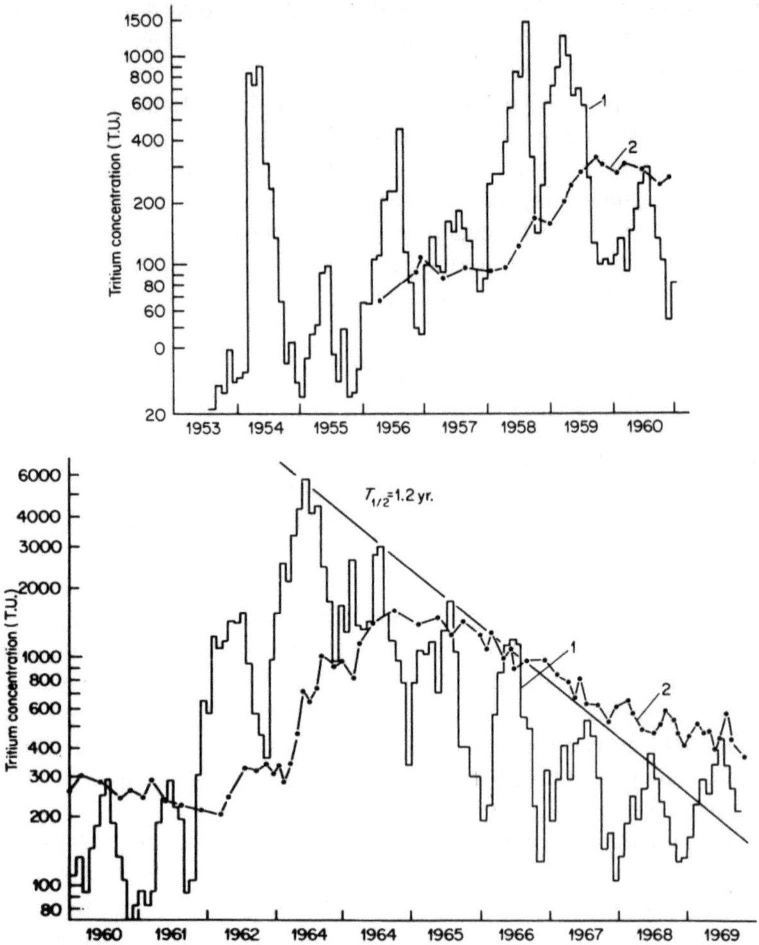

Fig. 10.10 Tritium concentration change during 1953–1969 in precipitation (*1*) and in Ottawa River water (*2*). (Brown 1970; Ferronsky and Polyakov 2012)

The peak of the tritium concentrations in precipitation for the northern hemisphere is observed, as a rule, in June and in the southern hemisphere in September. But very often deviations from this rule occur due to meteorological peculiarities of differed years. These peaks occur very seldom. In this connection, it is better to average data for a number of regions with more or less identical physical-geographical characteristics at the same time interval. Tables 10.11 and 10.12 are an example of such an average. The tables were prepared on the basis of data published by the IAEA (1969) using more than 250 stations of the Earth's globe. For the former USSR area the author's own data were used. Sometimes a substantial deviation in mean monthly tritium values from these seasonal variations is observed.

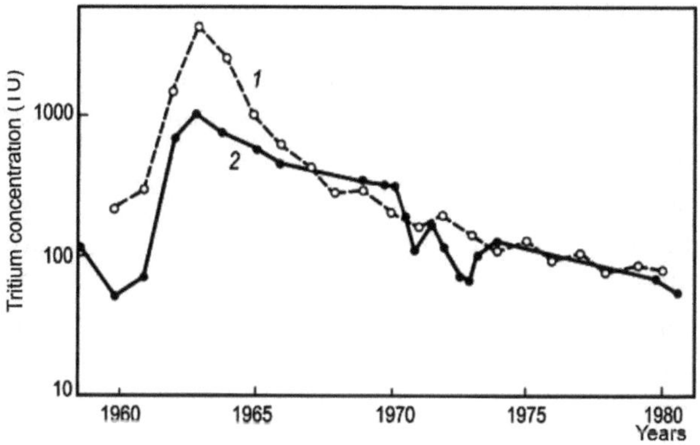

Fig. 10.11 Mean annual distribution of tritium concentration in atmospheric precipitation over the Moscow region (*1*) and in Moscow River water (*2*) 1958–1981. (Ferronsky and Polyakov 2012)

10.3.4 Formation of Tritium Concentrations in the Atmosphere

General principles observed from the picture of tritium concentration, related to latitudinal distribution and seasonal cycles, indicate the existence of a strict mechanism that governs the process of tritium distribution and concentration to precipitation.

In the period of intensive thermonuclear tests, the release of tritium into the stratosphere occurred periodically from individual points where the explosions took place. At the same time the average concentration of tritium fallout does not depend upon the place of injection but is related to latitude and time of the year. This fact resulted in the conclusion that the upper layer of the atmosphere (its stratosphere) is the reservoir where the accumulation and latitudinal redistribution of tritium occurs on a global scale and from which seasonal tritium releases to the lower atmospheric layer (troposphere) take place. Precipitation is formed in the troposphere. Such a reservoir exists for each hemisphere independently and the relationship between them is rather restricted. The mechanism of formation of tritium concentrations is presented as follows. The oceanic water, evaporated from its surface and having low tritium concentration, moves by the rising up air flows with a velocity of about 3–5 м/sec. In the case of unstable thermal atmospheric stratification, the velocity may overreach 10 m/sec. The created water vapour during the day and night is moved several kilometres high. From the other side, the process of mixing of the stratosphere and troposphere leads to interference to the last of enriched tritium water vapour.

The vertical distribution of tritium shows that its interference into the upper layers of the troposphere begins in December and continues up to June-July. In winter time the tritium transfer is made by eddy flows and in spring and summer by high tropic cyclones. In the Antarctic, in winter time, a direct condensation of vapour from the overcooled lower stratosphere is possible (Jouzel et al. 1979). The process of tritium moisture enrichment in clouds is continued due to molecular exchange.

Table 10.11 Seasonal distribution of tritium concentrations in precipitation for stations of the northern hemisphere

Month	Yearly means of concentration in precipitation (TU)					
	1964	1965	1967	1968	1969	mean value
0–20 °N						
January	1.09	1.13	1.26	0.94	0.99	1.07
February	1.17	1.29	1.18	1.06	1.04	1.12
March	1.17	1.28	1.20	1.24	1.23	1.19
April	1.30	1.07	1.27	1.09	1.03	1.13
May	1.31	1.23	1.07	1.16	1.04	1.13
June	1.34	1.40	1.30	1.10	1.09	1.22
July	1.44	1.24	1.13	1.27	1.34	1.35
August	0.96	0.84	0.92	1.09	1.24	0.99
September	0.71	0.66	0.83	0.88	0.92	0.78
October	0.51	0.62	0.66	0.72	0.77	0.64
November	0.51	0.62	0.68	0.75	0.66	0.63
December	0.49	0.62	0.51	0.70	0.64	0.58
20–90 °N						
January	0.89	0.63	0.73	0.46	0.63	0.70
February	1.09	0.98	0.90	0.80	0.76	0.89
March	1.18	1.20	1.09	1.04	1.04	1.08
April	1.46	1.46	1.22	1.19	1.11	1.29
May	1.67	1.54	1.60	1.42	1.49	1.57
June	1.70	1.73	1.58	1.59	1.62	1.64
July	1.41	1.51	1.36	1.44	1.47	1.45
August	0.99	1.10	1.17	1.28	1.29	1.18
September	0.56	0.60	0.88	0.86	0.80	0.74
October	0.43	0.43	0.54	0.60	0.63	0.53
November	0.31	0.35	0.48	0.54	0.50	0.44
December	0.32	0.36	0.46	0.77	0.64	0.50

The idea of the accumulation of radioactive products of thermonuclear tests in the stratosphere, with their subsequent redistribution and injection into the troposphere where precipitation is formed, was first suggested by Libby in 1956 (Libby 1963). Later on, attempts were made to develop this idea and to construct a box model of interacting exchangeable reservoirs: the stratosphere, the troposphere and the ocean. Investigations were mainly aimed at determining the residence time of a radioactive tracer in each of the exchangeable reservoirs. The final goal of these studies was to fix a relationship between the residence time of a tracer in the stratosphere and the motion of air masses in the stratosphere and troposphere.

In estimating the residence time of tritium in the stratosphere, it was assumed that the release of tritium into the troposphere is exponential. Thus, the residence

Table 10.12 Seasonal distribution of tritium concentration in precipitation for stations of the southern hemisphere

	Yearly means of concentration in precipitation (TU)					
Month	1965	1966	1967	1968	1969	Mean value
0–20 °S						
January	1.26	1.01	0.94	1.22	1.15	1.15
March	0.96	1.10	0.89	0.88	0.91	0.95
April	0.70	1.20	0.86	0.82	0.86	0.93
May	0.75	0.89	0.71	0.74	0.76	0.77
June	0.94	0.76	0.73	0.88	0.93	0.85
July	0.93	0.98	1.22	1.04	0.93	1.02
August	1.03	1.06	1.03	0.98	1.11	1.04
September	1.16	0.69	0.94	1.21	1.12	1.02
October	1.03	0.71	1.30	0.99	0.99	1.02
November	0.94	1.28	1.06	0.91	0.87	1.01
December	1.00	1.14	1.17	0.98	1.10	1.08
0–20 °N						
January	1.06	0.94	1.02	1.00	1.05	1.01
February	1.01	0.87	0.79	1.29	0.89	0.97
March	0.67	0.90	0.78	0.67	0.67	0.74
April	0.77	0.64	0.86	0.70	0.70	0.73
May	0.63	0.73	0.79	0.71	0.73	0.86
June	0.72	0.59	0.70	0.71	0.75	0.69
July	0.82	0.88	0.98	0.86	0.84	0.88
August	1.43	1.18	1.27	1.38	1.43	1.34
September	1.56	1.66	1.35	1.54	1.60	1.54
October	1.29	1.43	1.16	1.22	1.27	1.27
November	0.97	1.06	1,02	1.00	1.,06	1.02
December	1.06	1.10	1.28	0.93	0.90	1.06

time of tritium in the reservoir was taken, by analogy with its half-life, to be the time required for one half of the amount to be discharged from the stratosphere.

On the basis of experimental data concerning tritium concentrations in precipitation measured after 1953 at the two stations of Vienna and Valentia, which represent typical continental and coastal regions in terms of fallout, it was found that the residence time of tritium in the stratosphere is equal to one year (Schell and Sauzay 1970). According to data of Brown (1970) for the Ottawa River Valley it is equal to 1.2 year. As a result of the observed residence time of tritium in the stratosphere, its total storage (Fig. 10.2) and half-life removal, it can be estimated that by 1970 the tritium concentration in precipitation should already have been close to natural concentration. At the same time, according to experimental data obtained for Western and North America (see Fig. 10.10) from the beginning of 1967, the tritium content

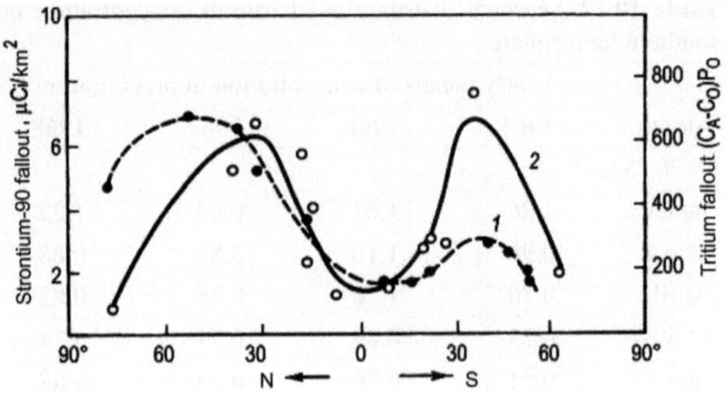

Fig. 10.12 Latitudinal fallout of tritium (*1*) and srtontium-90 (*2*) from the stratosphere over the Atlantic Ocean in 1969. (Ferronsky and Polyakov 2012)

in precipitation in the northern hemisphere remains constant at a level of 150–200 TU and by Romanov's estimations (Romanov 1978), decreases with a half-life removal time of about 3 year. This tritium content is an order of magnitude greater than natural tritium concentrations in precipitation. Different causes of this phenomenon were considered, particularly those related to the injection of tritium from thermonuclear tests conducted in the atmosphere during the preceding years. But in accordance with the measures carried out at a number of stations (e.g., in Tokyo), the increase in tritium concentration in fallout is insignificant because these explosions were not very powerful and provided only small concentrations of tritium to the stratosphere (see Fig. 10.2), where the maximum concentration occurred between 1967 and 1970. Another source of tritium contribution to the atmosphere may be the various installations of the atomic industry. This idea has been well demonstrated by Weiss et al. (1979). But as pointed out earlier, the possible input of tritium into the stratosphere from this source should be negligible and this tritium should have only a local effect upon sampling stations located near to nuclear plants.

Therefore, the residence time of tritium in the stratosphere (1–3 years) does not remain constant and the distribution and mixing of tritium in the atmosphere is a more complicated process than that described by the model of exchangeable reservoirs.

The questions of interpretation of global tritium distribution in precipitation have been studied by a number of researchers (Eriksson 1965a; Libby 1963; Taylor 1968). In one of the works an attempt was made to develop a general model of tracer release from the stratosphere into precipitation in which the injection, exchange, evaporation and condensation of the water vapour would be related to the meteorological parameters of the troposphere.

The model is based on the following physical reasoning. The exchange of air masses between the stratosphere and troposphere occurs mainly due to powerful air streams being thrown down periodically into the troposphere as a kind of trough or trench. This process coincides with the observed spring maximum of radioactive falls in precipitation at mid-latitudes of the northern hemisphere. Therefore, in the general mechanism of air masses exchange between the two reservoirs, the process of stratospheric air transfer to the troposphere due to diffusion is a secondary factor. The periodic rushes of dry stratospheric air into the troposphere, which contain excessive concentrations of radioactive substances and have large potential velocities, determine the mechanism of stratospheric-tropospheric exchange of air masses.

Another process responsible for the ejection of highly radioactive stratospheric air concerns powerful convective storm fluxes rushing into the stratosphere. These air fluxes, containing a large amount of water vapour, are mixing with radioactive substances and return with a high radioactive content as subsequent precipitation.

It has been found that the greatest concentrations of various radioactive nuclei, e.g., ^{90}Sr, ^{14}C and T, in precipitation always occur in the spring-summer period. But the maximum T-values appear with a delay of one month compared with that of ^{90}Sr and one–two months earlier than the maximum ^{14}C-values at the same latitude. The reason for this lies in the differences in their masses and physiochemical properties relative to those of the other atmospheric constituents. In the stratosphere, the behaviour of tritium, a constituent of HTO water molecules, is the same as that of ^{90}Sr and ^{14}C since gaseous T is usually transferred through the air by eddy diffusion.

It has been found that the $T/^{14}C$ ratio in the lower stratosphere exceeds that in humid air layers and is equal to 0.4. While approaching the humid layers of the troposphere, stratospheric HTO molecules may exchange with water at the ocean surface, re-evaporate and transfer into groundwater by infiltration,

The water circulation process in nature is as follows. Water with some tritium evaporates from the ocean surface. The water vapour rises and reaches the temperature of condensation by cooling. Dry air from the low stratospheric layers with a high tritium content exchanges tritium with the rising ocean water vapour, which contains low concentrations of tritium. The time of exchange appears to be an important factor; the longer the vapour stays in the atmosphere, the greater is the probability that it will become enriched in tritium. At some thermodynamic and meteorologic conditions precipitation is formed, which is a mixture of ocean and stratospheric water vapour.

The water balance equation is based on the relationship between radioactive fallout and meteorological parameters for a given atmospheric volume. Near the Earth this equation is:

$$E - P - \Delta F = 0, \tag{10.1}$$

where P is precipitation; E is evaporation–transpiration; ΔF is the derivation of the water flux in the considered air volume.

In order to establish the relationship between the equation and real conditions, it was assumed that water transfer is caused by winds characterised by velocities that vary in space. The balance equation at a concentration of the indicator C and for a vertical column of air relative to the Earth's surface, is as follows (Schell et al. 1970):

$$\iint (EC_E - PC_P - Q_t C_t)\,dxdy = \iiint \left[\frac{\partial (QC_v \cos\theta)}{\partial x} + \frac{\partial (QC_v \sin\theta)}{\partial y} \right] dxdydz, \tag{10.2}$$

where Q_t is an upward loss of moisture; C_E, C_P, C_t and C_v are the tritium concentration in evaporation, precipitation and atmospheric moisture at the border of air volume accordingly; Q is the flux of moisture; θ is the angle of wind direction. The equation may be simplified for computation:

$$<\bar{E}\bar{C}_E> - <\bar{P}C_P> - <\bar{Q}_t\bar{C}_t> = \frac{1}{L}\int_z (Q_2 v_2 C_2 \cos Q_2 - Q_1 v_1 C_1 \cos Q_1)\,dz,$$

where L is the distance between the observation stations and indexes 1 and 2 for the parameters Q, v and C relate to those stations.

The above model was applied for estimation of the relationship between the content of T, ^{90}Sr and ^{14}C in precipitation and meteorological parameters characteristic for a number of sites in Western Europe. Those calculations produced reasonable results.

In order to demonstrate the effects of the global distribution of tritium, Romanov (1978) made a number of assumptions in order to obtain a simpler expression describing the transfer of the atmospheric moisture. The balance equation then becomes:

$$\frac{dW}{dt} = E - P, \tag{10.3}$$

where W is the moisture content in the atmosphere; E and P are the amounts of moisture in evaporation and precipitation per unit of time.

The balance equation of HTO in the atmospheric moisture above the ocean is:

$$d(WC_A) = qdt - C_A Pdt + C_0 Edt - M(C_A - C_0)dt, \tag{10.4}$$

where C_A is the concentration of tritium in the atmospheric moisture; q is the rate of tritium release from the stratosphere; t is the residence time of an air mass over the ocean; C_0 is the tritium concentration in the surface oceanic layer; M is the rate of the eddy transfer of the atmospheric moisture to the surface oceanic layer. The fractionation of tritium during the phase transition is not accounted for.

The left-hand side of the equation expresses the change in the tritium content of atmospheric moisture over the time period dt. In the right-hand side the first term defines the discharge of tritium from the stratosphere, the second term corresponds to the removal of tritium in precipitation, the third term to the release of tritium from the ocean surface layer by evaporation and the fourth term to the injection of tritium into the ocean due to molecular exchange. The last term in Eq. (10.4), according to experimental data obtained by Romanov, is proportional to the gradient of the tritium concentration in the system: atmospheric moisture–ocean surface layer together with the water vapour.

In writing Eq. (10.4) it was assumed that the tritium content in precipitation is equal to its concentration in the whole of the upper atmospheric moisture layer and that the tritium concentrations in the ocean surface layer are little affected compared with concentrations in the atmospheric moisture and the rate of tritium injection from the stratosphere is constant during the whole time of air mass transfer.

Using (10.3), Eq. (10.4) can be rewritten in the form:

$$\frac{dC_A}{dt}+C_A\frac{E+M}{W}=q+C_0\frac{E+M}{W}. \tag{10.5}$$

The solution of Eq. (10.5) is:

$$C_A=C_0+\frac{q}{E+M}\left\{1-\exp\left[-\frac{E+M}{W}t\right]\right\}. \tag{10.6}$$

The integration constant can be defined from the initial conditions: $C_A=C_0$ at $t=0$.

It follows from Eq. (10.6) that the tritium concentration in the atmospheric moisture is greater when the moisture content W is lower and when the interrelationship between the ocean and the atmosphere, expressed by parameters E and M, is weaker. This principle can be observed in nature by analysing experimental data on global tritium distribution in precipitation. The lower values of moisture content, evaporation and molecular exchange in high latitudes result in the observed latitudinal effect of tritium concentration exchange in precipitation.

Taking into account that the rate of water evaporation is proportional to the humidity gradient (p_o–p), the rate of molecular exchange is proportional to atmospheric humidity p and that both processes are identically related to wind velocity, Eq. (10.6) may be rewritten in the form:

$$C_A=C_0+\frac{q}{kp_0}\left[1-\exp\left(-\frac{tkp_0}{W}\right)\right], \tag{10.7}$$

where p_0 is the pressure of the saturated vapour at the ocean surface temperature t; k is the parameter determined from the relationship $M=kp$.

If value t is large enough, then:

$$C_A=C_0+\frac{q}{kp_0}. \tag{10.8}$$

From Eq. (10.8) the velocity q of injection of tritium from the stratosphere can be obtained. The plot of dependence of the velocity from the latitude in Fig. 10.12 is presented. The values of T and ^{90}Sr content in the atmosphere are placed on the plot. It is seen that in both cases the maximum is located in the belt of 40–50 °N, which evidences the common nature of injection to the troposphere of bomb-tritium and strontium-90.

Let us determine the relationship between tritium concentration and the residence time of the air mass above the continent. The balance equation in this case is:

$$d(WC_A)+qdt-C_APdt+C_E\,Edt, \tag{10.9}$$

where C_E is the tritium concentration in the evaporating continental water.

Taking into account the insignificant difference between tritium concentrations in the surface continental water and in precipitation one can assume that $C_E \approx C_A$. In this case, using the balance equation for atmospheric moisture (10.3), a solution of (10.9) is obtained in the form:

$$C_A = \frac{q}{W} t + \text{const.} \tag{10.10}$$

The integration constant can be determined by the initial conditions $t = 0$ *at* which the tritium content in the atmospheric moisture is equal to its content in moisture transported from the ocean. Then:

$$C_A = C_0 + \frac{q}{W} t. \tag{10.11}$$

The last equation accounts for the observed continental effect. In fact the longer the air mass moves above the continent, the greater the concentration C_A becomes and the better the function $C_A(t)$ may be approximated by a linear dependence, which is confirmed by experimental data obtained in many regions of the world.

Further development of the models establishing the relationship between tritium concentrations in precipitation and meteorological parameters requires a deeper understanding of the natural principles governing air mass circulation in the atmosphere.

10.4 Tritium in Ocean Waters

Oceans are the main reservoir of the hydrosphere and the main source of atmospheric moisture on the Earth. From continental runoff, direct falls of precipitation and exchange with the atmosphere, the oceans receive the majority (about 90 %) of natural and bomb-tritium. Therefore, the oceans are the main reservoir of tritium accumulation on the Earth. The distribution of tritium in the surface and deep ocean layers is of interest while studying the principles of water circulation of the oceans together with atmospheric moisture and, in particular, of the ocean itself.

Before the thermonuclear tests the tritium concentrations in the ocean water, measured at different sites, had characteristic values from ~0.5 TU (Kaufmann and Libby 1954). to 1 TU (Begemann and Libby 1957). After the thermonuclear tests in March 1954 tritium concentrations increased at an average to 1.9 TU. It was difficult to measure such a concentration by the time techniques that existed then with appropriate accuracy.

After 1980, systematic measurements of tritium concentrations in the oceans started. The obtained results allowed Östlund and Fine (1979) to calculate the approximate amounts of tritium in different ocean regions (Table 10.13).

Table 10.13 Tritium inventory in the oceans. (Östlund and Fine 1979)

Ocean	Tritium (kg)
North Atlantic	66
South Atlantic	7
Arctic Basin	6
North Pacific	59
South Pacific	14
Antarctic	6
Indian Ocean	6
Total	164

Tritium content in the ocean waters is defined by effect of interaction between the ocean surface and atmospheric moisture, which is developed in precipitation, by evaporation and molecular exchange, by the life time of the surface layer existence, depending mean time on the vertical mixing of water and by interaction of water masses having different origins and tritium concentrations.

The general character of tritium content variations in the surface waters of the ocean reflect the picture of tritium input into precipitation including the rise of tritium content during the period of thermonuclear tests and seasonal cycles in each year. At the same time, it follows from the data that the input to the surface layer is delayed and the amplitude of tritium concentration is decreased because of dilution in a larger water volume. Figure 10.13 shows variation of tritium concentration in a surface water layer along 30 °N in the Atlantic Ocean plotted by the data of a number of researchers (Östlund and Fine 1979; Münnich and Roether 1967).

The corresponding changes in tritium content, because of a series of thermonuclear tests in 1958 and 1961–1962, are obvious. But the peaks are shifted to 2–3 years compared with concentrations in precipitation. The maximum of the amplitudes with higher tritium concentrations is observed in the middle and high

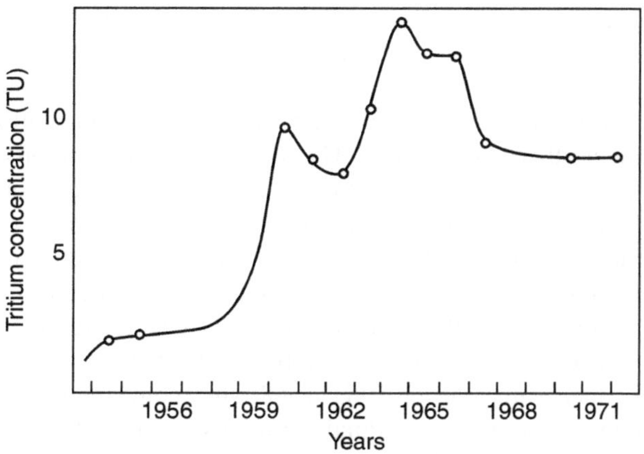

Fig. 10.13 Latitudinal variation of tritium concentration along 30 °N in the Atlantic Ocean during 1954–1972. (Ferronsky and Polyakov 2012)

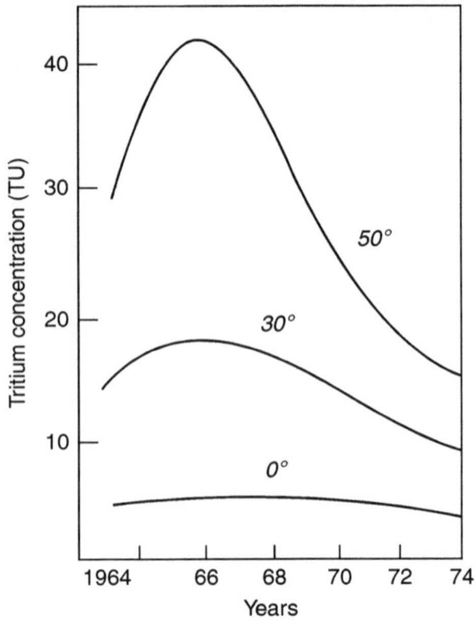

Fig. 10.14 Maximum changes in tritium concentrations in a surface layer of the Pacific Ocean. (Östlund and Fine 1979; Ferronsky and Polyakov 2012)

latitudes (Östlund and Fine 1979) (Fig. 10.14). The effect is strongly smoothed in the equatorial latitudes.

Tritium content variations in the surface layer differ seasonally. In the summer time, surface concentrations are higher because of its high content of atmospheric moisture, higher velocity of molecular exchange and lower dilution of surface layer by deep low-active oceanic water due to the thermocline effect. Dockins et al. (1967) studied tritium concentrations in the Pacific Ocean between 14 °S and 52 °N during 1959–1966. The highest concentrations correspond to the summer maximum in the northern part of the ocean, which are smoothed in the low latitudes. In the southern hemisphere the tritium concentrations in the surface layer decline quickly, which is in agreement with tritium the general picture of tritium falls in precipitation.

Münnich and Roether (1967) studied tritium variations in the surface and depth profiles of the Atlantic Ocean. As for the Pacific Ocean, the latitudinal distribution of tritium in the surface water has maximum concentration in the mid-latitudes of the northern hemisphere. The minimum concentrations of tritium were observed in the equatorial region (not more than 1–3 TU), being obviously close to the natural level observed before the bomb tests. The depth of the tritium mixing layer also increases with latitude from 100 m near the equator to several hundreds of metres in high latitudes. The absolute values of tritium concentrations in the Pacific Ocean are higher than those in the Atlantic Ocean due to more intensive vertical mixing.

Rooth and Östlund (1972) studied the distribution of tritium in the North Atlantic surface waters during 1963–1968. They confirmed the latitudinal increase of tritium content there. Longitudinal variations in concentration are small. They proposed a model applicable for calculation of the vertical turbulent transfer through the ocean thermocline, including advection.

Bainbridge (1963), Michel and Suess (1975) and other researchers, using the ratio between tritium content in the surface ocean layer and in atmospheric moisture, carried out an assessment of residence time in that layer. For this purpose they applied a box model where the ocean was divided into two reservoirs: the first was the surface well mixed layer and the second was the deep layer with lower tritium concentration compared with the surface one. Thus, it was assumed that tritium input is provided only through the dividing border between the atmosphere and the ocean and the tritium loss is a result of the radioactive decay in the well mixing layer. Horizontal advection is refused here because the horizontal tritium gradient in the open ocean is, as a rule, very small.

Applying the box model, Michel and Suess (1975) found that the water exchange time for 12 stations in the Pacific Ocean for the period of 1967–1973 is equal from 7.5 to 26 year. At the same time, they found that the maximum velocities are in the subtropical latitudes and the minimum in the equatorial zone. Values close to the above but calculated by a modified methodology, were obtained by Romanov (1982) for the Black and Caspian Seas (10–15 year) and for the margin seas of the Arctic Ocean.

Vertical tritium distribution in ocean water has its specific features related to its circulation on the globe. A typical picture of vertical tritium distribution is shown in Fig. 10.15. It is specified by the presence of a density jump at the thermocline, which is confined to the surface layer with a relatively high level of tritium content and the deep layer with almost zero tritium content. The two layers are divided by

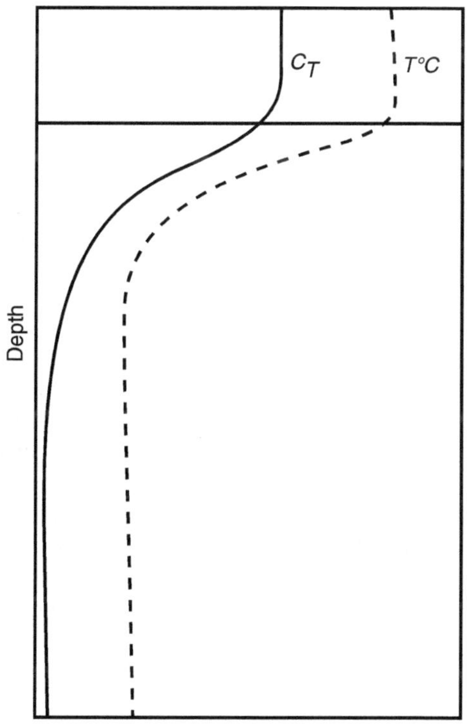

Fig. 10.15 Typical vertical tritium distribution in oceanic waters. (Ferronsky and Polyakov 2012)

an intermediate thin layer where the vertical turbulent mixing of tritium takes place. Such conditions are characteristic for the equatorial and low latitudes.

Applying the tritium gradient of concentration, it is possible to calculate the coefficient of vertical turbulent diffusion K_z. Roether et al. (1970) used this method and for the northern part of the Pacific Ocean obtained $K_z = 0.15\ cm^2/sec$. Rooth and Östlund (1972) for the Sargasso Sea obtained a value of K_z close to the calculated one. In order to obtain this parameter the authors compiled an equation that takes into account the processes of turbulent diffusion and vertical advection. In this case the change of tritium concentration $C(x, y, z, t)$, as a conservative radioactive inform:

$$\frac{dC}{dt} = \frac{K_z \partial^2 C}{\partial z^2} + K_H \left(\frac{\partial^2 C}{\partial x^2} + \frac{\partial^2 C}{\partial y^2} \right) - \bar{V} \nabla C - \lambda C.$$

The coefficient of turbulent diffusion K_H in the horizon plane (at x and y) is taken as the same. The vector of velocity of advective transfer $\bar{V}$ appears to be the vectors sum on all the three space coordinates.

Romanov (1982), in order to determine the coefficient of vertical diffusion for the Black Sea, applied a simplified equation in the form:

$$\frac{dC}{dt} = \frac{K_z \partial^2 C}{\partial z^2} - \lambda C.$$

Its solution for non-stationary conditions is:

$$C = \frac{1}{2} C_n \left[\exp\left(-\sqrt{\frac{\lambda}{K_z}} z \right) \left(1 - \operatorname{erf} \frac{z - 2\sqrt{K_z \lambda t}}{2\sqrt{K_z t}} \right) + \exp\left(\sqrt{\frac{\lambda}{K_z}} z \right) \left(1 - \operatorname{erf} \frac{z - 2\sqrt{K_z \lambda t}}{2\sqrt{K_z t}} \right) \right],$$

where $C_п$ is the tritium concentration in the surface layer of marine water.

Because the tritium concentration $C_п$ in the surface layer is not expressed in analytical form due to occasional character of thermonuclear tests, then a numerical solution with presentation of $C_п$ in a histogram form is used. After that the value of C for each step is found. This was the method used for interpretation of the Black Sea data, where the value of $K_z = 0.1\ cm^2/sec$ was obtained.

In high latitudes, where the process of air masses vertical mixing is more powerful, vertical tritium distribution appears to be more complicated. The same picture of tritium distribution for regional concentrations and divergence of ocean water is observed.

In 1972 tritium content studies at the deep water stations in the western Atlantic Ocean from 3 °N to 74 °N were carried out as part of the GEOSECS program. The results obtained were mainly in agreement with those obtained by Münnich and Roether (1967) and supplemented their profiles to the north and south. Penetration of bomb-tritium, with a concentration more than 0.2 TU, in the north latitudes, was observed up to the depth of 3500 m, whereas in the equatorial region a sharp decrease in content was observed at a depth of only 200 m.

Analogous profiles, obtained by the same authors for the Pacific ocean, show that bomb-tritium here has not fallen so deep as in the Atlantics at the north and most was in the upper layers that circulate in the reverse side in both hemispheres.

The northern and southern currents divide these systems. The deepest tritium penetration (up to 1000 m) is in the northern part of the Pacific Ocean. The most complete mixing, especially up to 500 m, is in the region of 20–40 °N. The maximum tritium concentrations in the near-equatorial current (8–20 °N) are at a depth of 200 m and at the surface. This is because the formed at higher latitudes and moving along constant density oceanic water are dropped. Analogous results and conclusions were obtained in earlier works by Michel and Suess (1975).

Asymmetric distribution of tritium concentrations in both oceans relative to the equator is explained by prevailing fallout in the northern hemisphere.

Note that the marked influence on the distribution of tritium in the surface oceanic layer in near-shore and continental regions effects the river runoff. There is also a good correlation between tritium concentrations and salinity and temperature of sea water.

10.5 Tritium in Continental Surface Waters

The continental surface waters, together with precipitation, are an important transport chain of global water circulation on the Earth. At evaporation they markedly affect the isotopic composition of atmospheric moisture. The surface continental waters are the main source for the groundwaters recharge and to a significant degree determine conditions of formation of isotopic composition and salinity of the marginal seas. A study of isotopic composition of the surface waters helps in obtaining valuable information about the parameters of water dynamics of river and lakes basins. Thus, the study of regularities in tritium distribution of surface waters is an important scientific problem.

10.5.1 Tritium Content in River Water

The factors that determine tritium content in river waters are tritium concentrations in precipitation over an area of the river catchment basin and the residence time of the infiltrating precipitation water in the soil through which it discharged into the groundwater and the river.

Let us express an analytical relationship between the tritium concentration in river water and the water exchange velocity in a river basin. Assume that river water in a basin represents the surface runoff and groundwaters that are well mixed. The age spectral function of river water can be written as $a(t)=\varepsilon \exp(-\varepsilon t)$, where $a(t)$ is the relative portion of precipitation in river water with age t and ε is the exchange velocity (the ratio of total annual recharge to the water volume of the catchment basin). Here value $1/\varepsilon$ has dimension of time (in years). Therefore,

$$\int_0^\infty a(t)\,dt = 1.$$

The tritium concentration in the river water C_r can be expressed by the equation:

$$C_r = \int_0^\infty C_p(t)\varepsilon e^{-\varepsilon t} e^{-\lambda t}\,dt,$$

or (10.12)

$$C_r = \varepsilon \int_0^\infty C_p(t) e^{-(\varepsilon+\lambda)t}\,dt,$$

where λ is the tritium decay constant; $C_p(t)$ is the tritium concentration in precipitation at the time moment t.

At $C_p(t)=\text{const}=C_p$ (e.g., during the pre-thermonuclear period) one obtains:

$$C_r = \frac{\varepsilon C_p}{\varepsilon+\lambda}. \tag{10.13}$$

In order to obtain the solution of Eq. (10.12) when $C_p(t)\neq\text{const.}$ (after the beginning of the tests) the function $C_p(t)$ should be rewritten in the form of a histogram. Then:

$$C_r = \frac{\varepsilon}{\varepsilon+\lambda}\left\{\sum_{i=1}^{n}\left[C_{P_{i=1}} - C_{P_i}\right]\exp\left[(\varepsilon+\lambda)(t_i - t_{i-1})\right]\right\} + \frac{\varepsilon}{\varepsilon+\lambda}C_{Pn}, \tag{10.14}$$

where C_{P_n} is the concentration of tritium in precipitation for the time interval $(t_i - t_{i-1})$; n is number of the time intervals in a period t.

Eriksson (1963) proposed another model for the interpretation of tritium data. It is based on the assumption that groundwaters, which are recharged by precipitation, move in parallel to the watershed. Then the precipitation waters, which fall at different distances from the watershed, do not become mixed. According to this model $a(t)=\varepsilon$ for t ranging from 0 to $1/\varepsilon$, equal to the residence time of infiltration water during its course from the watershed to the bed of the river. Then, the concentration of tritium in the river is as follows:

$$C_r = \varepsilon \int_0^{1/\varepsilon} C_p(t)\, e^{-\lambda t} dt. \tag{10.15}$$

At $C_p(t) = \text{const} = C_p$ one has:

$$C_r = \frac{C_p \varepsilon}{\lambda}(1 - e^{-\lambda/\varepsilon}). \tag{10.16}$$

Note that the ratio of the C_p-values, obtained from expressions (10.13) and (10.16), is equal to 1.3. Putting the expression for $C_p(t)$ written in the form of the histogram into (10.16), one obtains:

$$C_r = \frac{\varepsilon}{\lambda}\left\{C_{p_0}\left(e^{-\lambda t} - e^{-\lambda/\varepsilon}\right) + \sum_{i=1}^{n} C_{p_i} e^{-\lambda(t-t_i)}\left(1 - \exp\left[-\lambda\left(t_i - t_{i-1}\right)\right]\right) + C_{p_n}\left(1 - e^{-\lambda t}\right)\right\}. \tag{10.17}$$

Here the value of t should be less than that of the maximum residence time $1/\varepsilon$.

But a more reliable method for the estimation of water exchange rates in river basins is a balance method, proposed by Eriksson (1965b). In a somewhat simplified form it is as follows (Romanov 1978).

The spectral function of water's age can be represented by a number of fixed values a_0, a_1,..., a_i, where index i corresponds to the portion of infiltration water characterised by the age t_i. Thus the portion of precipitation falling in the year of observation is denoted by a_0, which in the previous year was a_1 and so on.

The balance equation of water is:

$$\sum_{i=0}^{\infty} a_i = 1. \tag{10.18}$$

An expression for tritium concentration in river runoff can be written as:

$$C_r = \sum_{i=0}^{\infty} a_i C_{a_i} e^{-\lambda t}. \tag{10.19}$$

Before the thermonuclear tests, n years ago, $C_a = \text{const} = C$, then:

$$C_r = \sum_{i=0}^{\infty} a_i C_{a_i} e^{-\lambda t} + C \sum_{i=0}^{\infty} a_i e^{-\lambda t}. \tag{10.20}$$

While studying river basins, the second term on the right-hand side of Eq. (10.20) can be neglected, so that:

$$C_r = \sum_{i=0}^{\infty} a_i C_{a_i} e^{-\lambda t}. \tag{10.21}$$

One should write n equations, such as (10.21), for the solution of the problem. In practice, the upper limit of summation can be bounded. For example, the value of $n = 1/\varepsilon$ has been determined beforehand from the conditions $i = 3$, i.e., for the coefficients a_0, a_1, a_2. The other coefficients were made equal to each other. Then:

$$
\begin{aligned}
C_{r_0} &= a_0 C_{a_0} + a_1 C_{a_1} e^{-\lambda} + a_2 C_{a_2} e^{-2\lambda} + \left[1 - \left(a_1 + a_2 + a_3\right)\right] \exp\left\{-\frac{1}{2\varepsilon} + 1{,}5\right\} \frac{1}{1/\varepsilon - 2} \sum_{3}^{1/\varepsilon} C_{a_i} e^{-i\lambda}, \\
C_{r_2} &= a_0 C_{a_2} + a_1 C_{a_3} e^{-\lambda} + a_2 C_{a_4} e^{-2\lambda} + \left[1 - \left(a_1 + a_2 + a_3\right)\right] \exp\left\{-\frac{1}{2\varepsilon} + 1{,}5\right\} \frac{1}{1/\varepsilon - 2} \sum_{3}^{1/\varepsilon} C_{a_{ii+2}} e^{-i\lambda}, \\
C_{r_i} &= a_0 C_{a_i} + a_1 C_{a_2} e^{-\lambda} + a_2 C_{a_3} e^{-2\lambda} + \left[1 - \left(a_1 + a_2 + a_3\right)\right] \exp\left\{-\frac{1}{2\varepsilon} + 1{,}5\right\} \frac{1}{1/\varepsilon - 2} \sum_{3}^{1/\varepsilon} C_{a_{ii+1}} e^{-i\lambda}.
\end{aligned}
\tag{10.22}
$$

Treating the system of Eq. (10.22), one can obtain the values of a_0, a_1, a_2. In a similar way any number of equations can be written and solved.

Applying the balance method, Eriksson (1965b) re-estimated the partition function of the age of runoff in the Ottawa River. Using his solution and the box model, Romanov (1978) reported that the value of ε for the Moscow River in 1964–1965 was 0.195. This value coincides with results obtained on the basis of the laminar flow model, following from Eq. (10.17). In this case the values of the coefficients are $a_0 = 0.25$; $a_1 = 0.18$; $a_2 = 0.14$; $a_i = 0.07$ at $i = 3$–8.

The tritium content in river water, as well as in the atmosphere, is varied in time and space. Long-term distribution of tritium concentrations is governed mainly by that of tritium content in precipitation. Figure 10.11 shows annual means of tritium content in precipitation over the central part of the former USSR territory and in the Moscow River water from 1958 to 1981 (Romanov 1982). The maximum tritium values in river water were observed in 1963 and are accounted by 0.25 part compared with precipitation. The decrease of HTO values in the river water was smoother than in precipitation. After 1971 the tritium level in the river and precipitation were equalising.

Weiss and Roether (1975) carried out analogous observations, starting in 1957, for Rheins River. A study in the Ottawa River, as well as in precipitation, started in 1953. Pre-thermonuclear concentrations were measured only in two tributaries of the Mississippi River near St Louis (Illinois, USA): in January 1952 it was 5.6 ± 0.6 T.U. and in August 1952 it was 1.15 ± 0.08 TU (Stewart 1965).

Figure 10.16 shows variations of tritium content in the Colorado, News, Arkansas, and Potomac rivers during 1963–1964 according to USGS data (Stewart 1965). One can see in the figure that the tritium peak of 1963 is presented in precipitation (1) and in waters of all the rivers (2). The tritium peak in rivers appears with an average delay ranging from several days or even hours for the mountain regions (Fig. 10.16a) to a year or even more for the plain regions (Fig. 10.16b), depending on the geological and geographical conditions of their recharge. The traveltime taken by a tracer to move from the catchment area to the river bed is an important parameter, characterising drainage properties and the capacity of a basin. This parameter and also the general character of tritium variations in the river water

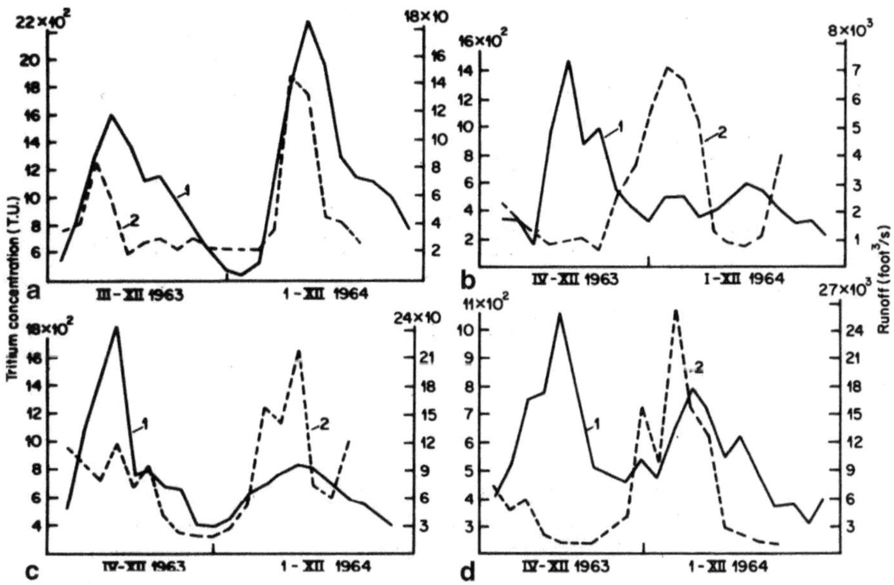

Fig. 10.16 Variation of tritium concentrations in precipitation (*1*) and in river water (*2*) of Colorado (**a**), News (**b**), Arkansas (**c**), and Potomac (**d**). (Stewart 1965; Ferronsky and Polyakov 2012)

and precipitation over time, permits estimations of the residence time of tritium in the drainage to be gained.

The seasonal variations in tritium concentration are markedly lower than in precipitation. The minimum tritium values are observed in the flood period, when the river recharge is mainly provided by melted winter snow, where tritium concentrations are 1.5–2 times lower than the mean annual (Romanov 1982].

After the flood passes, the rivers are recharged by the groundwaters of the catchment, where tritium accumulates by the previous fallouts in precipitation. Such a result was obtained in the Upper Angara River in autumn 1973, where after precipitation with about 250 TU of tritium content the river water activity contained about 350 TU (Romanov 1982].

The seasonal, annual and long-term variation of tritium concentration has been successfully used by many researchers for calculation of hydrographs in different river basins, for determination of residence time of water and water exchange time. For example, Brown (1970), using a box model, found for the Ottawa River that about 75 % of the river volume of water represents the surface runoff with a mean residence time of about one year and the volume rest of about 25 % is the underground runoff that has a residence time of about 6 years.

Weiss and Roether (1975), on the basis of tritium concentrations in the Rheins River water and precipitation during 1961–1973 and by appropriate modelling, discovered three components in the surface runoff separated. It was found that 25 % of the runoff has a value of residence time less than one year, 35 % has 5 years and 40 % do not involve tritium at all. It seems that was old groundwater.

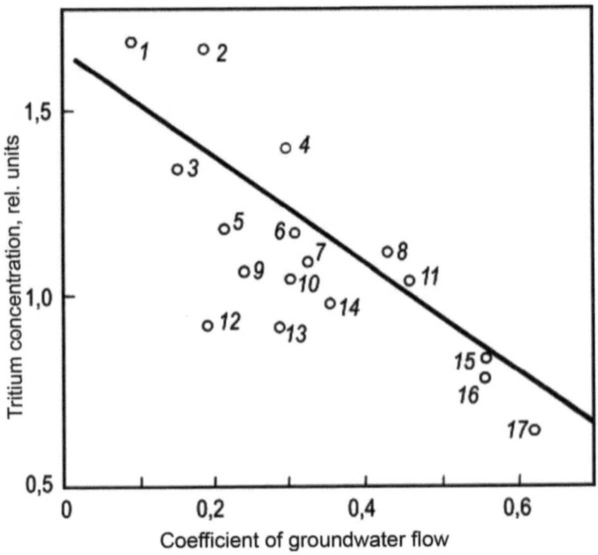

Fig. 10.17 Dependence on tritium concentration in river water from the coefficient of the underground water runoff for the rivers: Indigirka (*1*), Kolima (*2*), Amur (*3*), Lower Tunguzka(*4*), Lena (*5*), Enisey (*6*), Onega (*7*), Dnepr (*8*), Moscow (*9*) Kama (*10*), Upper Angara (*11*), Pechora (*12*), Southern Bug (*13*), Western Dvina (*14*), Ural (*15*), Don (*16*), Oka (*17*). (Ferronsky and Polyakov 2012)

The values of the runoff coefficients of tritiated water are equal to the ratio of the tritium value discharged by the river to the sea and the amount of isotope that fell out with precipitation over the catchment. The runoff coefficient is always equal to the runoff coefficient of the water but sometimes is rather different from it. An attempt was undertaken to compare the ratio of these values and also with the ratio of the annual mean value of tritium concentration and the coefficient of the underground runoff. Figure 10.17 shows the dependence of this value (relative to tritium concentration) on the underground runoff coefficient for the Russian rivers where the values were taken from the work of Domanitsky (1971)

The relationship between these values may be expressed by the regression equation in the form $C_r/C_p = 1.4K_n + 1.65$, where C_r and C_p are the tritium concentration in the river water and precipitation over the catchment area of a corresponding river; K_n is the coefficient of the underground water runoff of the river basin. The correlation coefficient of the regression equation is $r = -0.762$.

The geographical distribution of the relative tritium concentration in river water, for the European part of the former USSR territory, is characterised by a lower or close to unit value. At the same time, for the Siberian and Far East regions this value is within 1.05–1.69 and is increased towards the east. This tendency is well expressed for the northern rivers (Fig. 10.18). The relative tritium concentrations in waters of the north rivers located between Northern Dvina and Kolyma are

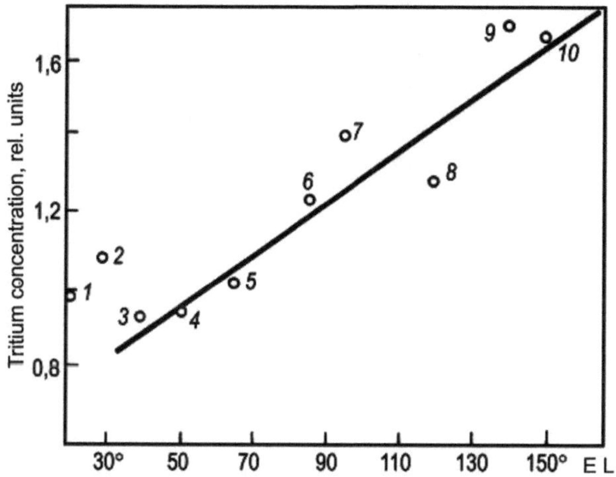

Fig. 10.18 Variation of relative tritium concentration in river waters depending on their longitude: Western Dvina (*1*), Onega (*2*), North Dvina (*3*), Pechora (*4*), Ob (*5*), Evisey (*6*), Lower Tunguzka (*7*), Lena (*8*), Indigirka (*9*), Kolima (*10*). (Ferronsky and Polyakov 2012)

described by the regression equation $C_r/C_p = 7.15 \cdot 10^{-3}\lambda + 0.56$ where the correlation coefficient $r = -0.95$ (λ is the degree of the eastern longitude)

The space distribution of the relative tritium concentration in river waters can be explained by the groundwater discharge where its value is lower. In the case when the relative concentration is higher than the modern precipitation water displaces the groundwater with lower tritium content.

There are experimental data that permit to assess tritium concentrations in groundwaters that take part in the river recharge by use of the equation of isotope balance, such as $C_r = C_p\,(1 - K_{\text{п}}) + C_{\text{п}}K_{\text{п}}$, where $C_{\text{п}}$ is the tritium concentration in groundwaters. Some results of the corresponding calculations are in Table 10.14.

The calculated tritium concentrations in groundwaters are very high and correspond to its content in precipitation fallout 10–15 years ago. In this regards, it is interesting to note that the appropriate groundwater tritium measurements in Yakutiya, carried out earlier, show similar high tritium concentrations (Afanasenko et al. 1973). At the same time, the tritium content in groundwaters of the European part of the former USSR, by multiple studies, very seldom overtopped its value in precipitation (Romanov 1982).

The attempts in finding a relationship between tritium content in river waters and some other hydrological parameters have failed.

From the general picture of space distribution in tritium concentration in river waters the data for Svir, Neva and Angara are of special interest. This is because water of the above rivers represents mainly runoff from large lakes like Onega, Ladoga, and Baykal where formation of isotopic composition is defined by some other conditions.

Table 10.14 Tritium concentrations in groundwaters taking part in river recharge)

River	Tritium concentration in groundwater (TE)	C_r/C_p
Pechora	68	0.65
Onega	101	1.28
Western Dvina	65	0.96
Dniester	93	1.29
Yuzhny Bug	55	0.75
Don	53	0.66
Oka	34	0.44
Moscow	83	1.26
Kama	117	1.17
Ural	78	0.71
Yenisey	193	1.61
Upper Angara	154	1.12
Lower Tunguska	309	2.,38
Lena	364	2.35
Kolyma	325	4.64
Amur	502	3.24

10.5.2 *Tritium in Lakes and Reservoirs*

While studying water exchange in lakes and water reservoirs with the help of a tritium tracer, a number of peculiarities should be taken into account that affect the tritium distribution patterns in them. Among them are the regime of the recharge, the ratio of the catchment area and that of the reservoir itself, the existence or absence of runoff, the temperature regime and many others. All these peculiarities determine whether a reservoir has a steady-state thermocline or a seasonal one, characterised by a complete mixing of water during cold times. Saline and some tropical reservoirs are of the first type. The majority of other lakes correspond to the second type. For the latter the balance equations of water and tritium under a study regime can be written in the form:

$$\frac{dV}{dt} = \sum_{i=1}^{n} R_i + P - E \pm U - A, \qquad (10.23)$$

where *dV* is the lake volume change; R_i is the river runoff; *P* is the precipitation; *E* is the evaporation; *U* is the groundwater discharge; *A* is the surface runoff from the lake.

At a steady-state regime (*dV*/*dt* = *0*), then the equation of tritium balance is written as:

$$\frac{d(CV)}{dt} = \sum_{i=1}^{n} C_i R_i + PC_p - EC_p - M(C - C) - \lambda CV - AC \pm UC_U. \qquad (10.24)$$

Here C, C_i and C_p are the tritium concentrations in the lake water, river runoff and precipitation; $\dot{M}$ is the rate of the turbulent exchange between the atmospheric moisture and the surface water of the lake (Östlund and Berry 1970); C_U is tritium concentration in the groundwater discharge; λ is the tritium decay constant.

Omitting the value $\pm U=0$ due to small groundwater tritium discharge in Eqs. (10.23) and (10.24), one obtains:

$$\frac{dC}{dt}=\bar{C}_i\bar{R}_i+PC_p-\bar{R}C+PC-M\left(C_p-C\right)-\lambda CV, \quad (10.25)$$

where

$$\bar{R}=\sum_{i=1}^{n}R_i;\quad \bar{C}_i=\sum_{i=1}^{n}C_iA_i.$$

Introducing the relative balance components $\bar{r}=\bar{R}/V$; $p=P/V$; $m=M/V$, we obtain the first-order differential equation:

$$\frac{dC}{dt}+C(\bar{r}-p-m-\lambda)=\bar{C}_i\bar{r}+C_pP. \quad (10.26)$$

In integral form, the last equation becomes:

$$C=\exp\left[-(\bar{r}-p-m-\lambda)t\right]\int_0^t\left(\bar{C}_i\bar{r}+C_pp\right)\exp\left[-(\bar{r}-p-m-\lambda)t\right]dt+\text{const.} \quad (10.27)$$

The solution of Eq. (10.27) can be obtained by the summation of the integrand if the variables $\bar{C}_i\bar{r}$ and C_pp are written in the form of a histogram.

The estimation of the water cycle time in Baikal Lake, the largest in the world, carried out on the basis of the analysis of tritium patterns in the lake water, influents and precipitation, showed that this problem can be treated with the help of the box model. Using the balance method and corresponding experimental data on the tritium content in the water sources, it was shown that complete water exchange in this lake lasts 330 year, whereas that in the river basins is 2–3 year (Romanov et al. 1979). These results are in good agreement with hydrological estimations (Afanasyev 1960).

Table 10.15 demonstrates the main hydrological characteristics and the data on tritium concentrations in lake waters of the former USSR territory. In order to exclude the influence of geographic differences on tritium content in precipitation over the lake and river catchment areas, the value of relative tritium concentration is used. This value is equal to the ratio of tritium concentration in water and that over the catchment is of the lake or river. This value is used for comparison with the value of time exchange of the lake or river water (see Table 10.15).

Table10.15 Tritium concentrations and some hydrological characteristics for the lakes of the former USSR territory (1979–1980)

Lake, reservoir	Water volume (km^3)	Annual recharge (km^3/year)	Tritium concentration		Water exchange time (year)	C_i/C_p
			Lake C_i (TU)	Precipitation C_p (TU)		
Ladoga	908	79.8	122	66	11.4	1.85
Onega	295	13.6	138	67	21.6	2.06
Sevan	58.5	1.26	80	64	44	1.26
Issyk-Kul	1730	5.25	35.5	115	330	0.31
Baykal	23000	70.2	40	128	330	0.31
Valday	0.55	0.044	110	62	12.5	1.78
Aral	1023	60.3	217	100	17	2.17
Ribinsk reservoir	25.4	36	86	76	0.71	1.13

The results of a tritium study of lakes on the former USSR territory are shown in Fig. 10.19. Here the maximum relative tritium concentrations in reservoirs were found with the water exchange time of about 17 year (the Aral Sea and Onega Lake). It is interesting to note that 17 year before the lake water was sampled the maximum fallout took place.

The box model can be applied to treatment of experimental data for open water reservoirs. The conditions for such a treatment are as follows: first, equality of the mean value tritium concentration in the system itself and in the discharged water and second, constancy of the volume of the reservoir in time. In fact, the reservoirs with a constant water level and with well mixed water satisfy these conditions.

The box model has been used in calculation of water exchange time for river basins, lakes and reservoirs by many authors (Brown and Grummit 1956; Begemann and Libby 1957; Romanov 1978).

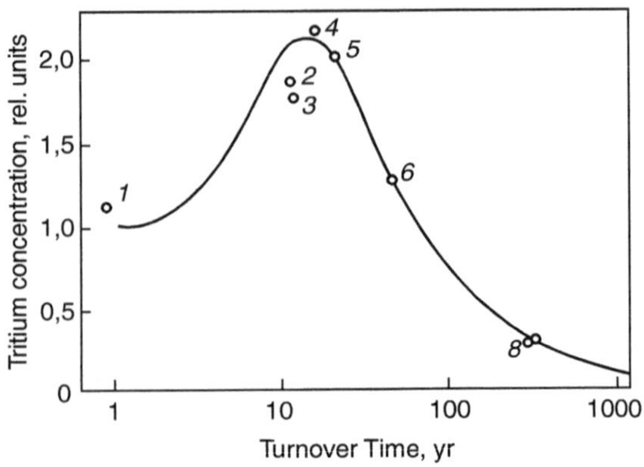

Fig. 10.19 Dependence of relative tritium concentration on the water exchange time for: Ribinsk Reservoir (*1*); lakes Ladoga (*2*); Valday (*3*); Aral (*4*); Onega (*5*); Sevan (*6*); Issyk-Kul (*7*); Baykal (*8*). (Ferronsky and Polyakov 2012)

10.6 Tritium in Groundwaters

The main source of tritium in groundwaters is precipitation water. But, as pointed out by Andrews and Kay (1982), some amount of tritium is created by the nuclear reaction (n, α) on nuclei of ^{6}Li in aquifers, especially those represented by acid rocks. The source of neutrons, in this case, is spontaneous natural uranium and thorium decay. By Andrews and Kay calculations, the above reactions in pore groundwater produce tritium concentrations up to 2.5 TU, which can be reached in a granite massif.

The shortest transition time of precipitation and surface water into groundwaters is observed in the case of a direct hydraulic connection between them, which occurs in the regions of tectonic fractures, fissures and karstic rocks and gravel-pebble sediments. The tritium half-life, equal to 12.43 year, is very often less than the groundwater exchange time. Therefore, the tritium radioactive decay is a significant factor decreasing its concentration. The intensive exploitation of groundwater for economic purposes is accompanied with development of depression funnels and washing of the rocks, intensifying the degree of relationship between the surface and groundwaters and affecting their tritium content.

In recent years, when the natural tritium content was disturbed by injection into the atmosphere of large amounts of bomb-tritium, the possibilities of the absolute dating of groundwater with the help of tritium were lost. At the same time the wide-scale investigations of tritium falls in precipitation on a global scale allow us to use the obtained data for studying the motion of groundwater from a somewhat different viewpoint.

Tritium was released into the atmosphere during the decade of nuclear weapon tests occurred in the form of individual pulses, which correspond to a single powerful or a series of moderate explosions. The tritium falls during the test period and during the subsequent period of time, mirrors its injection into the atmosphere in the form of individual pulses differing in magnitude over the yearly cycle. The knowledge of atmospheric water patterns in the underground part of the hydrosphere together with tritium dating information provides the basis for the solution of different time-dependent problems while studying groundwater dynamics

The study of problems dealing with the transition of tritium 'marks' in groundwater over time is based on systematic measurements of tritium concentrations in the precipitation of a studied region. The occurrence of tritium in groundwaters depends on their recharge conditions.

The most typical case of recharge of tritium from a surface aquifer is percolation of surface water through the unsaturated zone. Here, as a rule, the spring-summer component of the annual precipitation, containing the maximum tritium concentration, does not reach the aquifer. This portion of annual precipitation is lost mainly through evaporation-transpiration and partly by surface reservoir recharge. Some portion of the groundwater storage is similarly lost. During the autumn-winter period and in early spring, when tritium content in precipitation is minimal, the evaporation and transpiration of precipitation water is also reduced. During this period groundwaters are replenished. Thus, in this case tritium concentrations in groundwaters are lower than the annual average concentration in precipitation.

Seasonal variations in tritium concentration have been observed (Andersen and Sevel 1974). The delay time of the peak annual concentrations ranges within about 1–3 months and the peak itself compared with the tritium peak in precipitation is considerably smoothed. The tritium concentrations in water of an unsaturated zone change in depth. Sometimes, the concentration maximum, which corresponds to the infiltrating precipitation of the peak fallout on a definite depth, was observed. This depth was dependent on the filtration properties of the soil of the unsaturated zone (Andersen and Sevel 1974; Atakan et al. 1974; Morkovkina 1978). Such a tritium distribution for determination of the infiltrated water velocity in the unsaturated zone was used.

According to experimental data of Münnich and Roether (1967) and Atakan et al. (1974), who carried out their studies on the alluvial plain of the Rhine River, bomb-tritium dating of shallow groundwaters can be used with a sufficient accuracy if the unsaturated soils and the aquifer itself are homogeneous in composition and properties. The tritium change in depth profiles over time, in an unsaturated zone composed of loess-loam soil and in an unconfined aquifer consisting of fine to medium-size and coarse sand, was studied by the authors. According to the tritium dating techniques, the average recharge rate of groundwaters in the fine to coarse sand material is 200 mm/year, which is in good agreement with that measured on the basis of routine hydrologic techniques.

By tritium distribution in time, the mean residence time of water is determined in the aquifer. Taking into account that aquifer water is drained by rivers then the portion of groundwater discharge to rivers can be calculated (Brown 1970). Or, if this value by some other method is obtained, then by balance relations the mean tritium concentrations discharged to the river are obtained. An attempt was undertaken to obtain these values for the large northern river basins of the former USSR (Romanov 1982). The results of these calculations, presented in the previous paragraph, show that tritium concentration in the groundwaters is increased from west to east of the basin location. And also, for rivers, located to the east behind Ural, their concentration is higher than in modern precipitation. This is because of the continental maximum over the catchment area and possibly due to the permafrost affect of the Siberian surface rocks, where safe tritium concentrations accumulated in the period of maximum fallout during 1962–1966. This idea is proved by the observational data of tritium in groundwater, rivers and precipitation obtained in Yakutia (Table 10.16).

The aquifers of fractured and karstic rocks have intensive recharge from precipitation. In this case there are good conditions for infiltration of water and restrictions for its evaporation. Here irregular distribution of tritium concentrations, due to the existence of a number of hydrologic subsystems or different ways of water transit, is observed.

Devis (1970) studied groundwaters in non-carbonate fractured rocks on the volcanic island CheYu in South Korea. Substantial variations in tritium content, discovered there, allow the mean residence time of the water in the range of 1–8 year

Table 10.16 Tritium content in natural waters of Yakutia. (Afanasenko et al. 1973)

Object	Tritium concentration (TU)
Water from a spring of Upper Nekharan	860
Water from a spring of Yust Nekharan	523
Water from the river mouth Nekharan	382
Snow	252
Precipitation	337

to be found. This is the normal 'age' for groundwaters. A similar study was done in Aragatz Mountain in Armenia by Vlasova and Brezgunov (1978). The transit time for precipitation between the points of recharge and discharge in the summer time was found to be 2–3 months and in the cold period is 7–8 months.

The tritium concentrations in such waters are, as a rule, high and close to precipitation. For example, high tritium concentrations in karstic waters with substantial seasonal variation were found in Southern Turkey (Dinçer and Payne 1971).

A number of researchers have found a good relationship between karstic and surface waters. The velocity of water movement in karstic rocks was found to be from several to hundreds of metres per hour (Fontes 1976).

The anthropogenic activity influence on the groundwater exchange rate is a problem of special interest. For example, concentration of tritium in water of the Middle Carboniferous rocks in the Moscow artesian basin, whose groundwaters have been intensively exploited, in 1978 was equal in average to about 50 TU. At the same time, in marginal parts of the basin, which are closer to the recharge area, the values very seldom exceed 10 TU (Zlobina et al. 1980). This shows that anthropogenic activity may lead to substantial disturbance of the natural circulation of groundwater.

Detailed studies of groundwater motion in saturated and unsaturated zones have been carried out by a number of researchers (Münnich et al. 1967; Andersen and Sevil 1974; Atakan et al. 1974; Allison and Hughes 1974; Verhagen et al. 1979; Morkovkina 1979).

10.7 Dating by Tritium

In view of the interpretation of isotope data and the solution of different problems elucidating groundwater dynamics, different models are widely used. The mathematical ground for the construction of these models is the balance equation of water masses and isotope tracers, together with the water dynamics in the system under investigation. Let us now consider some of the hydrological models.

One of the common problems in groundwater studies is the estimation of the residence (exchange) time of water in a hydrological system, or, as sometimes proposed, the age of the water. Several models applicable for the interpretation of experimental results have been suggested involving tritium tracers for determination of the age of groundwater (Nir 1964; Maloszewski and Zuber 1996).

10.7.1 Piston Flow Model

This model is based on the assumption that portions of water coming into the system follow each other along the flow and do not intermix. The model underestimates the residence time in hydrological systems since water mixing does occur in nature. But it is useful for estimations of the minimal residence time in a system.

According to this model the concentration of radioactive isotope C at a sampling point located a distance x_0 from the recharge zone is defined as:

$$C = C_0\exp(-x_0/\nu T) = C_0\exp(-t/T), \tag{10.28}$$

where C_0 is the concentration of radioactive isotope in recharge water; v is the rate of groundwater motion; T is the lifetime of an isotope ($T = T_{1/2}/\ln 2$).

Introducing a dimensionless parameter $k = t/T$, one obtains:

$$C/C_0 = \exp(-k). \tag{10.29}$$

Thus, the residence time of water from the recharge region of a basin to the sampling point is determined by the ratio C/C_0. Equation (10.28) can be readily transformed into one that corresponds to the common exponential law of radioactive decay:

$$C = C_0 \exp(-\lambda\ t), \tag{10.30}$$

where t is the age of the water, λ is the decay constant of an isotope $\lambda = 1/T = \ln 2/T_{1/2}$.

Therefore, the age of water at a sampling point for the piston flow model is:

$$t = \frac{T_{1/2}}{\ln 2}\ \ln\frac{C_0}{C_t} = \frac{1}{\lambda}\ \ln\frac{C_0}{C}. \tag{10.31}$$

In the framework of this model it has been admitted that C_0 is a constant and that intermixing of waters of different ages does not occur in the system.

10.7.2 Dispersive Model

According to this model the hydrodynamic dispersion and intermixing of waters, entering the system at different times, results in a Gauss' distribution of transition times along the flow. As in the previous case, assuming the spatial distribution of an isotope tracer being dependent on one coordinate x, one has:

$$C\,(x_0,t) = \frac{C_0}{(4\pi Dt)^{1/2}}\exp\left[-\frac{(x_0 - \nu t)^2}{4Dt}\right]dx, \tag{10.32}$$

or

$$C(x_0, x) = \frac{C_0}{(4\pi D_m x)^{1/2}} \exp\left[-\frac{(x_0 - x)^2}{4D_m x}\right] dx, \tag{10.33}$$

where $D_m = D/v$ is the coefficient of hydrodynamic dispersion being characteristic for a given hydrogeological system.

Expression (10.33) represents the concentration of tracer $C(x_0,t)$ at a distance x_0 from the source of recharge. When the tracer passes the average distance $x = vt$, the amount of tracer remains unchanged in time and is equal to C_0dx. The average lifetime of the tracer is T and the concentration at a point x can be estimated from the equation:

$$C(x_0, x) = \frac{C_0}{(4\pi D_m x)^{1/2}} \exp\left[-\frac{(x_0 - x)^2}{4D_m x} - \frac{x}{vT}\right] dx. \tag{10.34}$$

If the concentration of the tracer at the input of the hydrogeological system varies within time, which it does for natural isotopes, the variation of the concentration at the output will depend mostly on the input parameters of the tracer.

Let us consider a hydrological system of volume $V(t)$, having an inflow of water $a(t)$ and an outflow $q(t)$. In this case the water age t is:

$$t = V(t)/q(t).$$

For the steady hydrodynamic state the input concentration of the tracer in discrete form can be given as follows (Martinec et al. 1974):

$$C(t) = \sum_{t=0}^{\infty} p(t) C_a(\theta - t) e^{-\lambda t}, \tag{10.35}$$

where θ is the year of sampling; t is the age of the water; λ is the constant of tritium (0.056 year^{-1}).

The distribution function of the water's age $p(t)$, in a hydrogeological system, is given in fractions of annual replenishment at the output of a system at the moment of sampling.

In the case of the dispersive model the distribution function of the water's age for a semi-infinite aquifer is as follows:

$$p(t) = \frac{2}{\sqrt{\pi D t}} \exp\left[-\frac{(t - t_0)^2}{Dt}\right] - \frac{2\exp(4\pi t_0/D)}{D} \operatorname{erfc}\left(\frac{t + t_0}{\sqrt{Dt}}\right). \tag{10.36}$$

The result of the above expression depends on two parameters, having dimensions of time: $t_0 = x_0/v$ (t_0 is not equal to the average residence time of water in the system) and $D = 4D_m/v^2$, (x_0 is the coordinate of a sampling point), where D_m is the hydrodynamical dispersion. Using a radioactive isotope tracer, one should introduce into Eq. (10.36) a term to account for its decay to the moment of sampling.

Dispersive models have been used by a number of researchers for estimating the residence time of water in hydrological systems (Martinec et al. 1974; Burkhardt and Fhöhlich 1970; Zuber 1994; Maloszewsky and Zuber 1996).

10.7.3 Complete Mixing Model

According to this model, water input at different times mixes quickly. It is not possible to account for water flow lines, as has been done in previous models, only to speak of water residence time in a system obeying exponential distribution. The model, as a rule, gives higher values of the residence time in a system if the distribution function of time is expressed by a continuous function.

In the interpretation of the observed results of tritium concentration changes in the Ottawa River basin, Brown (1961) used a box model that assumes complete mixing of the meteoric and groundwater in the bed flow. In this case, at the output of a hydrological system, the relative fraction of water of a certain age is expressed as an exponential dependency of the form:

$$p\,(t) = (1/\tau)\,\exp\,(-t/\tau), \tag{10.37}$$

where τ is the average value of residence time of water in a system; $1/\tau = \varepsilon$is the water exchange rate, i.e., the ratio of the total annual inflow of water to the volume of the basin.

Due to the function of water distribution (10.37), this model in literature is often called 'the exponential model'.

The distribution of tritium concentration in water at the output of the system, in discrete form and in accordance with Eq. (10.35), is:

$$C(t) = \sum_{1953}^{\theta} \alpha C_a(\theta - t)p(t)e^{-\lambda t}. \tag{10.38}$$

where α is the statistical distribution of the input function, determined as a function of the seasonal precipitation, participating in the recharge of a system compared with the annual amount of precipitation.

During the pre-thermonuclear era when $C_a = \text{const}$ we have expression (10.13) and when $C_a \neq \text{const}$ we have expression (10.17).

In practice it is not always convenient to estimate $C(t)$ by a method of successive approximations. Expression (10.38) can be simplified by letting $e^{-1/\tau} = \kappa$. In this instance, at $1/\tau << 1$, expanding the function $\kappa = e^{-1/\tau}$ in a power series of $1/\tau$, one obtains:

$$K = 1 - \frac{1}{\tau}, \tag{10.39}$$

or

$$\tau = 1 - \frac{1}{K} \tag{10.40}$$

Substituting the last expression into the age function $p(t)$, coming into Eq. (10.38), one obtains:

$$P(t) = (1 - K)\, K^t,\ (K < 1). \tag{10.41}$$

Then, taking $\alpha = 1$, Eq. (10.38) becomes (Dinçer and Payne 1971):

$$C(t) = \sum_{1953}^{t=n} C_t(\theta - t)(1 - K)K^t e^{-\lambda t}. \tag{10.42}$$

Here, as a rule, the summation is carried out over all the years since the beginning of thermonuclear tests, i.e., 1953 and $C_t(\theta - t)$ is accepted as an average tritium concentration in precipitation for a corresponding year. For practical purposes, the values C_t estimated at different values of κ (0.1, 0.2, 0.3, etc.), are indicated in the plot together with the experimentally measured values of tritium in the aquifer. Then, comparing the theoretical and experimental curves, the most probable value of the averaged residence time of water in a basin is determined.

The complete mixing model was used for estimating the residence time of groundwater in the karstic region of the Anatolian coast in Turkey (Dinçer and Payne 1971), for studying aquifers sited near Vladimir (Polyakov and Seletsky 1978), for studying the groundwater discharge characteristics of the Aragatz Mountain region (Vlasova et al. 1978) and for studying the water dynamics in the Moscow artesian basin (Zlobina et al. 1980).

It should also be pointed out that the age distribution of water, described by Eq. (10.37), does not always imply the existence of an underground or surface reservoir with good mixing. This model is also advantageous for studying aquifers drained by the aquifer thickness.

10.7.4 Symmetrical Binominal Age Distribution Model

According to this model the probability of the appearance of water characterised by an age t at the output of the hydrogeological system has the form (Martinec et al. 1974):

$$p(t) = \frac{1}{2N}\binom{N}{t}, \quad t = 0, 1, 2, 3 \ldots \tag{10.43}$$

The parameter N in this case is dependent upon the mean residence time of water in the hydrogeological system and is given in the form: $N = 2\tau - 1$ or $\tau = (N + 1)/2$.

The numerical coefficient *N/t* represents the binominal coefficient defined by the binominal theorem:

$$(a+b)^N = \binom{N}{O}a^N + \binom{N}{1}a^{N-1}b + \binom{N}{2}a^{N-2}b^2 + \ldots + \binom{N}{N-1}ab^{N-1} + \binom{N}{N}b^N.$$

The following properties of this expansion are known: N/0 = N/N = 1 and the sum of all the binomial coefficients is equal to 2^N. Therefore, it is evident from the discrete binomial distribution that in the case of the continuous variation of the parameter t a normal distribution culminates with dispersion and average residence time depending upon the parameter N.

Figure 10.20 indicates some models of age distribution corresponding to the groundwater basin of the Dischma River (Switzerland).

10.7.5 Model of Mixing Waters of Different Ages

This model accounts for the case when discrete differences in the distribution of residence times are observed. For example, young waters in an aquifer may mix with very old waters. If the volume of the whole system is V and the volume of an individual constituent is V_i, then their relative contribution is:

$$p_i = \frac{V_i}{V}, \tag{10.44}$$

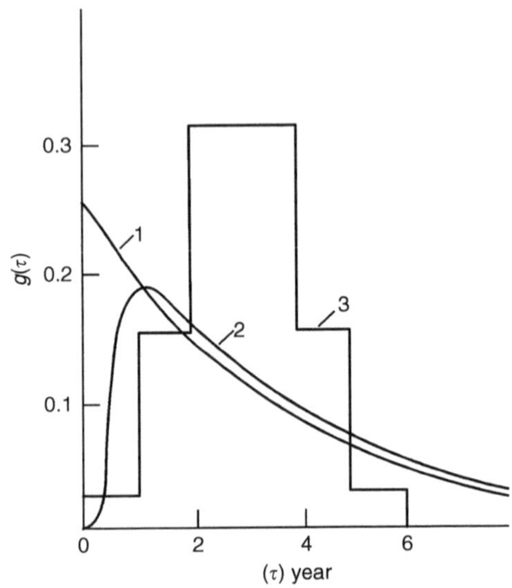

Fig. 10.20 Models of age distribution for the groundwater basin of the Dischma River: (*1*) experimental data, τ = 4 year; (*2*) dispersive, τ = 4.8 year; (*3*) binominal, τ = 3 year. (After Martinec et al. 1974; Ferronsky and Polyakov 2012)

where $V=\sum V_i$ and $\sum p_i=1$.

If the true age of an individual component is t_i, then the age of a mixture is:

$$t_\Sigma = \Sigma p_i t_i = \overline{t}, \tag{10.45}$$

where $\overline{t}$ is the mean weighted age.

If the initial concentration of each constituent is C_{0I}, then at the moment of mixing its concentration is:

$$C_i(t_i) = C_{0i}e^{-\lambda t}. \tag{10.46}$$

After mixing in a system its concentration becomes:

$$C = \sum p_i C_i = \tau = \sum p_i C_{0i} e^{-\lambda t}.$$

While estimating the age of a water mixture, e.g., using the piston flow model, the formal application of expression (10.28) gives the age:

$$t' = \frac{1}{\lambda}\ln\frac{C_0}{C} = \frac{1}{\lambda}\ln\left[\sum p_i \exp(-\lambda t_i)\right]^{-1}. \tag{10.47}$$

Comparing expressions (10.45) and (10.47) we find that actually $t' \neq \overline{t}$. Therefore, it follows that the true age of a mixture of waters of different ages is equal to the average age of its components. Further, the isotope age of the water mixture does not normally equal the true age and its theoretical value has nothing in common with that for a real system in the framework of the piston flow model. In principle, the problem of the mixing of waters of different ages can only be solved while taking into consideration the behaviour of several radioisotopes and counting out the time from the moment the waters become mixed. In this way the problem is reduced to the solution of the system of equations:

$$C_k = \sum^{m} p_i C_{0ik} \exp(-\lambda K t), \ \sum^{m} p_i = 1, \tag{10.48}$$

where C_{0ik} is the concentration of the κth isotope on the ith constituent of a mixture at $t=0$; p_i is the contribution of the ith constituent; λ_k is the decay constant of the κth isotope; t is the time elapsed since the moment of mixing; m is the number of constituents in a water mixture.

Using the stable isotopes only, i.e., those for $\lambda_k=\infty$, system (10.48) is reduced to that describing simple mixing. For more details the reader is referred to Ferronsky et al. (1977); IAEA (1996).

10.7.6 Complicated Model

Such a model is indicative of most natural systems. In this model the output function of a system of one type becomes the input function of a system of another type. For example, the groundwater basin with a normal distribution of transition times is connected to a basin characterised by complete mixing and so on. Studies of the complicated systems, involving tritium and another tracer, require a detailed knowledge of the geology and hydrology of a basin.

Besides the above-mentioned models other combinations and varieties are used in hydrogeological studies.

A number of authors have carried out measurements of tritium concentrations in groundwater up to depths of several hundreds of metres. These studies were carried out in various hydrogeological conditions: in the Vienna basin situated near the Alps, in the hydrothermal regions of New Zealand and Iceland, in the limestone and dolomitic formations of Transvaal in South Africa and elsewhere. It was found that tritium concentrations decrease sharply with depth. Seasonal variations of tritium in the depth of an aquifer were not observed, indicating the continuous replenishment of the aquifer during the year with a constant rate of recharge. Considerable variations of tritium content exist for various boreholes located within a basin at short distances from each other, indicating the various conditions of recharge and rates of inflow. While studying geothermal regions the applicability of the techniques for the estimation of the inflow rate of surface waters to the zone of heating and their subsequent time of circulation, was demonstrated (Theodorsson 1967; Gonfiantini and Panichi 1982; Pinneker et al. 1978).

Tritium techniques are an effective instrument for the investigation of pollution problems, especially release of radio-nuclides into groundwater from nuclear power plants and atomic wastes. The last problem is relevant with active construction of nuclear power plants. In the last decade these studies have intensified (Sokolovsky et al. 2007; Polyakov and Golubkova 2007; Tokarev et al. 2005).

In 1969 Tolstikhin and Kamensky (1969) proposed the helium-tritium method of groundwater age determination. By this method water age is calculated as:

$$t = \frac{1}{\lambda} \ln\left(\frac{{}^{3}He*}{{}^{3}H} + 1\right),$$

where $^{3}He*$ is the heliun-3 concentration in groundwater appearing after tritium decay.

It is not necessary in the determination of the input function $C_a(\theta - t)$ in Eq. (10.38). This method is used, first of all, for the study of 'young' groundwater (Schlosser et al. 1988, 2000; Plummer 2005).

10.8 Radiocarbon in Natural Waters

Carbon has always played an important role in geochemical processes that take place in the upper layers of the Earth and, in the first instance, in the formation of the sedimentary terrestrial layer and the evolution of the biosphere. Radioactive carbon ^{14}C is often used as a tracer of various natural processes such as the circulation of natural waters, their redistribution between natural reservoirs, water dynamics of the hydrosphere and its elements and is applicable for estimating the age of such waters. The age of geological formations and groundwater within the time scale up to 60,000 year is of a great interest for modern geology and hydrogeology. The data of radiocarbon distribution in different carbon-bearing natural objects are used for reconstruction of their paleoclimatic changes and for solving astrophysical problems related to variation in time of cosmic rays. In this chapter, the main focus is the distribution and application of radiocarbon in respect to the dynamics of natural waters.

10.8.1 Origin and Distribution of Radiocarbon in Nature

As was shown in Sect. 10.1, radiocarbon is formed in the atmosphere in the course of the interaction of secondary neutrons generated by cosmic rays, mainly with nitrogen-14, according to the reaction:

$$^{14}N + n = {}^{14}C + p + 0{,}626\ \text{MэВ}. \tag{10.49}$$

The cross-section of the reaction is 1.81 ± 0.5 barn. Table 10.17 shows the main reactions responsible for radiocarbon production in the atmosphere. But their contribution to the total ^{14}C balance comparing with the reaction (10.49) is insignificant.

The radiocarbon produced is usually oxidised to $^{14}CO_2$ after several hours in the atmosphere, which is characterised by approximately the same carbon isotopic composition and takes part in a general global circulation of carbon dioxide. The total equilibrium amount of radiocarbon on the Earth (in the atmosphere, hydrosphere and biosphere) can be theoretically estimated. According to Libby (1955) it is equal to 81 t and the estimations of Lal and other researchers gave 60–75 t, which is equivalent to an activity of about $3 \cdot 10^8$ Ci.

Despite considerable variations in the secondary neutron flux from the equator to the poles, which differs by a factor of 3.5, the ^{14}C isotope is sufficiently homogeneously distributed on the Earth. This effect was well studied by artificial bomb-radiocarbon that evidences the high rate of mixing of the atmosphere. Experimental data show that variations with latitude and altitude do not exceed 3–5 %.

Some amount of ^{14}C can be received on the Earth together with meteoritic matter, where it is produced by interaction with cosmic rays. Lunar soil and rock studies have shown that the carbon content in the sample No 14163 is 109 ± 12 g/t

Table 10.17 Reactions of radiocarbon production in the atmosphere by secondary neutrons action

Reaction	Reaction energy (MeV) +, exothermic; −, endothermic	Abundance relative ^{14}N	Relative rate production in the atmosphere
^{13}C (n, γ) ^{14}C	+8.17	$0.23 \cdot 10^{-5}$	$1.1 \cdot 10^{-9}$
^{14}N (n, p) ^{14}C	+0.626	1.0	1.0
^{15}N (n, d) ^{14}C	−7.98	$0.37 \cdot 10^{-2}$	$3.7 \cdot 10^{-5}$
^{16}O $(n, {}^{3}He)$ ^{14}C	−14.6	0.269	$2.7 \cdot 10^{-3}$
^{17}O (n, γ) ^{14}C	+1.02	$0.99 \cdot 10^{-4}$	$2.3 \cdot 10^{-5}$
^{20}Ne, ^{21}Ne (split) ^{14}C	–	$0.12 \cdot 10^{-4}$	$1.2 \cdot 10^{-7}$

(109 ± 12 ppm together with a correction on the Earth's contamination (Fireman and Stoenner 1982)). The radiocarbon activity in two fractions, extracted by heating in oxygen at T = 1000 °C, was 31.2 ± 2.0 disintegr min^{-1}/kg for the fraction of > 53 μ and 11.2 ± 2.0 disintegr min^{-1}/kg for the fraction of < 53 μ.

The above authors assumed that the values derived in this way for radiocarbon was produced by the action of solar wind. The carbon activity is approximately the same for both fractions (19.2 ± 2 and 21.0 ± 15 disintegr min^{-1}/kg). This is explained by interaction of cosmic rays with lunar rocks and direct partial production of the radiocarbon.

Studies of the dynamics of the carbon exchange between natural reservoirs, based on a ^{14}C tracer and theoretical investigations of radiocarbon redistribution between natural reservoirs (boxes) concluded their number is arbitrarily accepted from one (Gray and Damon 1970; Ralf 1972) to six (Craig 1957; Ekdahl and Keeling 1973). But the majority of works were performed using two or three box-models (Damon, 1970; Yang and Fairhall 1972; Keeling 1972).

Figure 10.21 shows the two-box model used by Sternberg and Damon (1979). In accordance with this model the radiocarbon, being produced in the upper parts of the atmosphere with the rate Q(t) atom. cm^{-2}· sec^{-1}, is immediately moved into reservoir A composed of the atmosphere, biosphere and a mixing layer of the ocean. Between this reservoir and reservoir B of the deep ocean waters the radiocarbon exchange has rates of K_{as} and K_{sa}. Loss of radiocarbon from reservoirs A and B is due to the radioactive decay. The exchange rate K_{as}, as well as N_a and N_s, which are the total amount of carbon per 1 cm^2 of the Earth surface as well as in the reservoirs A and B, are accepted constant in time. The amount of ^{14}C atoms per 1 cm^2 in reservoirs A and B ($n_a(t)$ and $n_s(t)$ accordingly), is the function of $Q(t)$.

The two-box model can be described by two differential equations:

$$\frac{dn_a(t)}{dt} = Q(t) - K_{as}n_s(t) + K_{sa}n_s(t) - \lambda n_a(t), \tag{10.50}$$

$$\frac{dn_s(t)}{dt} = K_{as}n_a(t) - K_{sa}n_s(t) - K_{as}n_s(t) - \lambda n_s(t). \tag{10.51}$$

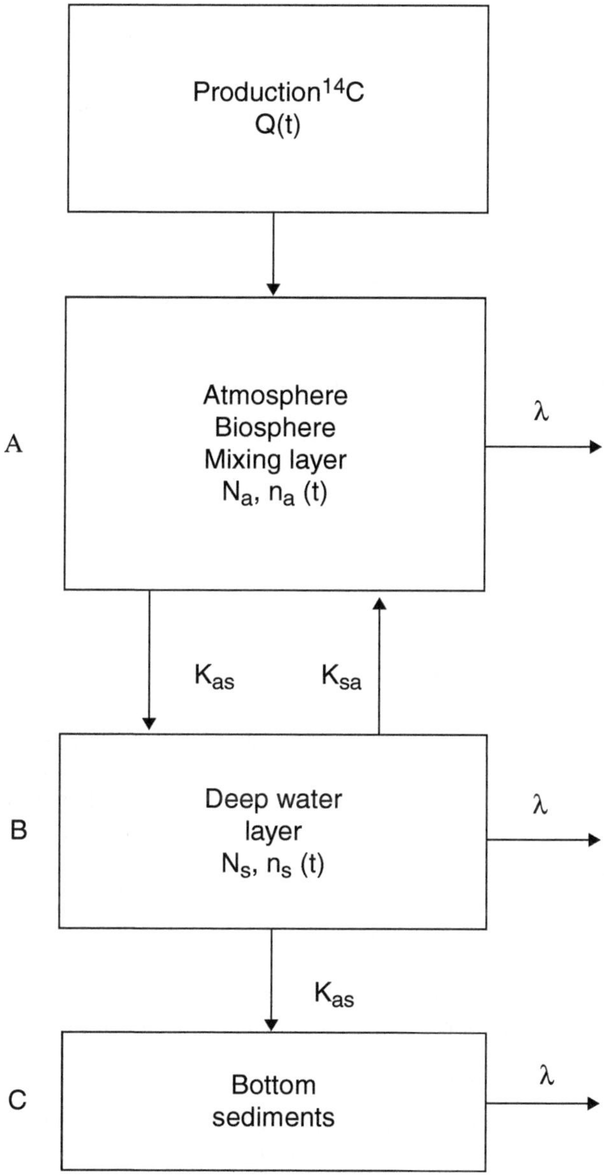

Fig. 10.21 Two-box radiocarbon exchange model. (Sternberg and Damon 1979; Ferronsky and Polyakov 2012)

Note that the reverse value of the exchange rate value is equal to the mean residence time *t* of radiocarbon in the reservoir. Taking into account that the specific activity of radiocarbon A_i in the *i*th reservoir is $\lambda n_i/N_i$, Eq. (10.50) can be written as:

$$\frac{d(\lambda n_a(t)/N_a)}{dt} = \frac{\lambda Q(t)}{N_a} - K_{as}\frac{\lambda n_a(t)}{N_a} + \frac{K_{sa}}{N_a}\lambda n_s(t) - \frac{\lambda^2 n_a(t)}{N_a}, \quad (10.52)$$

$$\frac{dA_a(t)}{dt} = \frac{\lambda Q(t)}{N_a} - K_{as}A_a(t) + \frac{K_{sa}N_s}{N_a\left[\lambda n_s(t)/N_s\right]} - \lambda A_a(t) = = \frac{\lambda Q(t)}{N_a} - K_{as}A_a(t) + \frac{K_{sa}A_s(t)N_s}{N_a} - \lambda A_s(t) \qquad (10.53)$$

Analogously for the reservoir *B* one has:

$$\frac{dA_s(t)}{dt} = \frac{K_{as}A_a(t)N_a}{N_s} - K_{sa}A_s(t) - K_{as}A_s(t) - \lambda A_s(t). \qquad (10.54)$$

In equilibrium conditions the process of carbon exchange between reservoirs *A* and *B* is satisfied to the equality:

$$K_{as}N_a = K_{sa}N_s \qquad (10.55)$$

or

$$K_{as}/K_{sa} = N_s/N_a = \nu, \qquad (10.56)$$

where v is the ratio of the carbon content in reservoirs *A* and *B*.

It follows from Eq. (10.56) that the carbon exchange rate is in reverse proportion to the total carbon reserve and vice versa, the residence time t is proportional to the carbon amount in the exchange reservoirs. The effect of carbon isotope fractionation at its redistribution between reservoirs should be taken into account while studying the parameters K_{as} and K_{sa}. As was shown by Craig (1957), the real value of the fractionation factor $\alpha_{s/a}$ is equal to about 1.012.

On the basis of experimental data, by Eqs. (10.53)–(10.56) one may calculate values of ^{14}C distribution and assess the effect of dynamics of carbon exchange between the reservoirs. For these calculations the values A_q, A_s, Q should be taken as the steady-state condition. The Q parameter, as a rule, is determined by means of the decay rate I of radiocarbon, which relates to 1 cm^2 of the Earth's surface. In the general case, I relates to $Q(t)$ by equation $I = -\lambda\int_0^\infty Q(t)e^{-\lambda t}dt$. It is obvious, if one considers that $Q(t)$ does not depend on time (which is not entirely correct), that $I=Q$. Equating the left-hand side of Eq. (10.53) to zero and replacing $K_{sa}N_s/N_a$ by K_{as}, for the stationary state condition one obtains:

$$K_{as} = \frac{\lambda(Q/N_a - A_a)}{(A_a - A_s)}, t_a = \frac{(A_a - A_s)}{\lambda(Q/N_a - A_a)}. \qquad (10.57)$$

where t_a is the residence time of radiocarbon in reservoir *A*.

On the basis of experimental data Sternberg and Damon (1979) accepted the following mean values: $\overline{A}_a = 14.5$ ppm/g (C), $\overline{A}_s = 12.6$ ppm/g, $Q=I=108$ ppm/cm^{-2} and $N_a=0.361$ g (C)/cm^2. In this case the carbon residence time in reservoir

A is about 50 year. By the same theory, on the basis of more reservoir boxes, the residence time of carbon in the atmosphere exchanging with the fast mixing ocean layer can be estimated. For this case, Sternberg and Damon accepted the following parameters: $\bar{A}_a(1890) = 13.8$ ppm/g (C), $\bar{A}_{\text{п..с.}} = 0.965\bar{A}_s$ and $N_a = 0.125$ g(C)/cm². In this case $t \approx 5$ year. Similar estimates of the carbon residence time in the biosphere, in the fast mixing layer of the ocean and its deep layers, give mean values of 60, 10 and 1500 year correspondingly. These are the approximate results because of low accuracy in the parameters calculated. The values of carbon in a number of exchange reservoirs used for theoretical calculations are given in Table 10.18.

The possibility of ^{14}C production in nitrogen-bearing objects (in wood, for example) *in situ* by reaction with neutrons, generated by cosmic rays or occurring spontaneously in rocks at nuclear reactions, was considered.

Radnell et al. (1979) considered the possibility of radiocarbon accumulation in time in woods by thermal neutrons of cosmic origin irradiation. It was shown that in nuclear reaction with nitrogen-14 (fast neutrons) in bristlecone pine an activity of about $(1.7 \pm 0{,}6) \cdot 10^{-3}$ ppm/g (C) during 8000 year can be reached, i.e., this is only 0.03 % of natural ^{14}C radioactivity in wood of 8000 year age.

Zavelsky (1968) showed that the possibility of ^{14}C generation *in situ* is limiting the upper value of the measured radiocarbon age. According to his calculation, such a limit for a number of objects can be 80,000–100,000 year.

The theoretical specific activity of ^{14}C in modern carbon-bearing samples should be about 17 ppm/g (C) (Libby 1955). Numerous comparable studies in different regions have shown that the specific activity of ^{14}C in the biosphere varies from 14 to 16 ppm/g for inland specimens and from 13 to 17 ppm/g for ocean specimens. The average activity is 15.3 per gram of carbon. For the ocean carbonates this value averages around 16.0 ppm/g, i.e., approximately 5 % higher than for the biogenic ^{14}C. Later on these figures were refined

The major proportion of carbon, which participates in a cycle in the form of dissolved carbon dioxide, carbonates and bicarbonates (H_2CO_3–HCO_3^-– CO_3^{2-}), is in the oceans. If one accepts CO_2 contents in the atmosphere as $N_a = 0.62 \cdot 10^{12}$ t (C), then the ocean contains 65 N_a (C) and the biosphere 2.4 N_a (C), from which 90 %

Table 10.18 The amount of carbon in some exchange reservoirs

Reservoir	Carbon amount (g/cm²)	
	Libby (1955)	Rubey (1964)
Ocean carbonates	7.25	6.95
Ocean dissolved organic substance	0.59	–
Biosphere	0.33	–
Living organisms and non-decomposed organic	–	0.775
substance		
Atmosphere	0.12	0.125
Total	8.29	7.85

is contained in oceanic plankton (Oeschger and Siegenthaler 1979). After death, organic substances are subjected to decomposition. Carbon is cycling through the biosphere about every 300 year. This process can be observed, for example, in biogenic ocean carbonates, the major portion of which is dissolved while precipitating on the ocean floor.

Besides that, the carbon present in the common exchangeable reservoir, which is contained in the sedimentary strata, representing its main terrestrial storage, takes part in the dissolution and admixing of cosmogenic radiocarbon. Sedimentary carbonaceous rocks are, on the one hand, being continuously formed and, on the other hand, are being constantly disintegrated. In this form of solutions and suspended particles the disintegrated rocks are carried out into the ocean. The amount of carbon contained in sedimentary rocks is estimated to be $2 \cdot 10^{22}$ g. If the process of the formation of sedimentary carbonaceous rocks has taken place over the last 3 billion year with a variable rate of formation in various epochs, then, according to different estimations, less than 3 % of the ^{14}C participates in this process at any one time.

An important question in radiocarbon dating concerns the efficiency of ^{14}C mixing in the main reservoirs over time. The homogeneity in ^{14}C distribution in a reservoir can only be attained if the mixing time is short compared with the lifetime of ^{14}C. The mixing time for the biosphere is not more than 300 year and in the atmosphere 10 year (Libby 1955). The mixing time of the Atlantic Ocean does not exceed 2000 year, that of the Mediterranean Sea is about 100 year and for the Black Sea it is about 2500 year. Another fact in agreement with the assumption of complete mixing of the ocean is the magnitude of the heat flux from the oceanic floor, which is equal to that of the terrestrial crust at 30 cal/cm^2 per year (Libby 1955). If this value is correct the absence of a heat inversion near the bottom suggests good mixing of the ocean in the framework of a radiocarbon scale of time.

The constant rate of radiocarbon production and the constancy of the amount of stable carbon in the exchangeable reservoir are of importance in the problem of radiocarbon dating. The constancy depends upon (Stuiver 1965): (a) the variation of intensity of cosmic radiation due to solar activity; (b) the variation of the magnetic dipole and field of the Earth; (c) the climate change of the Earth. Libby (1955, 1967) pointed out that considerable corrections in view of these factors should not be made since the ^{14}C lifetime is relatively small ($T_{1/2}$=5730 year). However, in the cold periods of glacial times the stable carbon content in the ocean may drop and the specific activity of ^{14}C may increase by 5–10 %. The amount of living organic material has no effect on the specific activity of carbon (since its ratio in nature has always been small).

During the last 100 year the content of CO_2 has markedly increased in the Earth's atmosphere as a result of the industrial burning of fossil fuels (coal, oil and gas). This effect is known as 'the industrial effect' or 'Suess effect' consisting in a certain decrease (by about 3 % for the northern hemisphere) of ^{14}C (Houtermans et al. 1967; Oeschger and Siegenthaler 1979). But as a whole the natural equilibrium in ^{14}C content has settled during the last two decades due to thermonuclear explosions conducted in the atmosphere. Due to these, the ^{14}C content in the northern hemispheric atmosphere has more than doubled and become higher in the biosphere and in the surface oceanic layer it has increased by about 20 % (Nydal et al. 1979). In wooden rings of 1963–1965 the ^{14}C increased up to 180–190 % compared with the pre-bomb level (Cain 1979).

10.8.2 Natural Variations of Radiocarbon in the Atmosphere and Biosphere

The specific activity of radiocarbon in the atmosphere and, as a consequence, in the biosphere is mainly governed by variation of the cosmic rays intensity at the Earth's surface. The natural ^{14}C variations can be divided into short-periodic ones, governed by the Sun modulation of galactic cosmic rays and long-periodic ones connected with the geomagnetic Earth's field and climate variation (Dergachev and Kocherov 1977; Sternberg and Damon 1979). Long-periodic variations of the ^{14}C level can also be effected by corpuscular radiation from super-nova flashes and possibly by occurrence of neutron flows during annihilation of meteoric matter. The last problem has a more exotic than practical meaning (Sternberg and Damon 1979). As Dergachev and Kocherov pointed out, the degree of correctness of solar activity (Wolf's numbers W) is high after 1749. Attempts were undertaken to extend the time scale up to 1610 and even to 648 BP (Dergachev and Kocherov 1977). But the correct measurements of the 11 year solar cycle by W numbers have been achieved only since 1749. Attempts have been made to check the last time of the 80 year (century) cycle but difficulty arose in connection with the short period of observations (230 year). In accordance with the calculations, the amplitude of specific ^{14}C activity for the 11 year cycle of solar activity is 5 % and for the century cycle about 1 %. Registration of the 11 year variation in the yearly wooden rings' radiocarbon content is a difficult instrumental task. Stenhouse and Baxter (1979) used two high-stability proportional counters of quartz and metallic for its solution. This allowed obtaining a relative error in the experiments equal to 6.4 ‰. In yearly oak rings over 1840–1890, he found ^{14}C peaks relating to 1851, 1869 and 1880 (at the level of about 1 %). ^{14}C fluctuations at a level of 0.3–0.4 % have not been reached and the 2–3 % variations have definitely not been found. The total ^{14}C activity from 1840 to 1890 decreased on average by 0.03 % per year. After 1890 an increase of ^{14}C concentration in the atmosphere fixing in the yearly wooden rings was observed. The radiocarbon variations in the past and changes in the solar activity during the last 300 year are considered in the work of Dergachev (1975). ^{14}C variation in atmospheric carbon dioxide during the last 1000 year is discussed by Stuiver and Quay (1981). Their results are presented in Fig. 10.22.

It follows from the experimental data that natural reservoirs of carbon do not stay in equilibrium relative to the atmospheric ^{14}C level. Before 1890, variation of the radiocarbon content resulted mainly due to solar activity changes. Since 1890, the ^{14}C activity decrease was connected mainly to dilution of the atmospheric carbon dioxide by the "dead" CO_2 coming from burning fossil fuel. But as was shown by Stuiver and Quay, ^{14}C content in atmospheric CO_2 during the 20th century is not

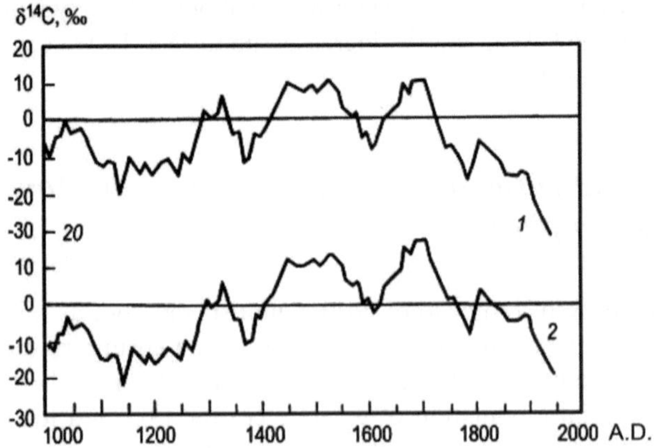

Fig. 10.22 14C variation in the atmosphere in the last 1000 year: non-corrected (*1*) and corrected on the basis of long-periodic geomagnetic field change (*2*). (Stuiver and Quay 1981; Ferronsky and Polyakov 2012)

distinguished by something specific (see Fig. 10.22, (1)). These changes are due to long-periodic variations of the geomagnetic field. If one makes a correction of ^{14}C long-periodic changes by means of a sinusoidal curve of a several thousand years period, then the ^{14}C level in atmospheric CO_2 (and biosphere) in the XX century, at least during the last thousand years, will be minimal (Fig. 10.22, (2)).

These data prove the anthropogenic influence on the increase of total carbon content in the atmospheric reservoir, which has led to a decrease in the specific activity of radiocarbon in the atmosphere, biosphere and a mixing ocean layer. By the data of Stuiver (1980), with reference to Douglas, the ^{14}C level in 1954 differed from the 'standard modern level" (0.95 activity of the NBS standard) by 24 ± 2 ‰.

The study of ^{14}C variation in time by the yearly rings of XVIII and XIX century woods is convenient because at that time the anthropogenic impact on the atmosphere was weaker, the level of solar activity (W-number) change is well known and the climate variations by the synoptic data can be restored. Currently there is a significant amount of data on ^{14}C concentration measurements in the yearly rings of woods from that time. By measurements of Stuiver (1965), over the time period from 1700 to 1870, a decrease in ^{14}C content by 2% from 1700 to 1790 is noticed. The maximum difference in ^{14}C activity within the studied period reaches 2.8%. It is worth noting that due to the small diameter of the rings, Stuiver measured ^{14}C activity in several rings simultaneously. In addition, the studies were performed on the ring slices of four trees from different locations of growth. This may lead to some errors depending on local conditions.

Analogous measurements on ^{14}C changes in yearly wooden rings of pines growing in Lithuania were performed by Dergachev and Kocherov (1977). Figure 10.23 shows their results of a series of measurements on 5 year rings during 1707–1859.

It is seen that the ^{14}C change curve has a sinusoidal form with amplitude of 1.5%. The period of ^{14}C variations for the time interval from 1780 to 1840 is equal

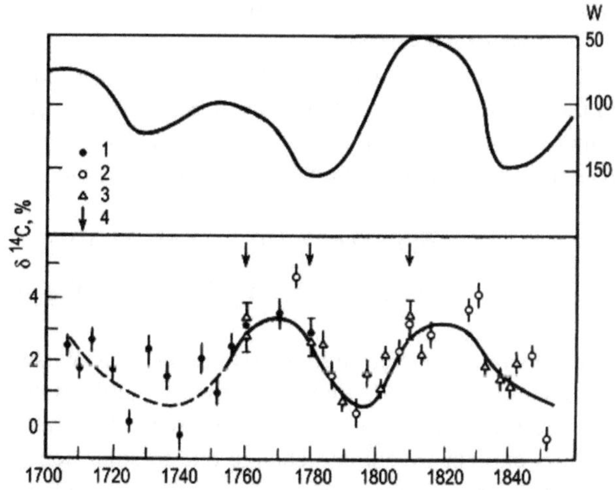

Fig. 10.23 Relationship between δ14C variations in pine rings and solar W-number during 1707–1850: (*1 and 2*) measured by proportional counter; (*3*) measured by scintillation counter; (*4*) calibration points. (Dergachev and Kocherov 1977; Ferronsky and Polyakov 2012)

to about 60 year. The curve itself shows an inverse proportionality in relationship to the solar W-numbers and radiocarbon content. The time shift, equal to about 10 year, is specified by the mean residence time of the carbon in the exchangeable reservoir (see Fig. 10.21, A) and depends on the period. The following expression of the changes rate of ^{14}C generation depending on the W-number for the considered solar activity cycle is proposed by Dergachev and Kocherov:

$$\Delta Q(W)/Q_0 = 0.4 - 0.01\ W,$$

where Q_0 is the equilibrium rate of ^{14}C generation.

Suess (1970) discovered in the yearly bristlecone pine rings a century ^{14}C variation with the period of 181 year within the time period of 7000 year (from the middle of the XX century up to 5000 year BP). The ^{14}C variation amplitude is about 2–3 % and the mean value of the ^{14}C activity is changed in time by sinusoidal law. In connection with solar activity changes and the corresponding climatic variation, it is worth noting that an ice core of 404 m length from the Greenland Camp Century Dansgaard covered the time interval from 1970 to 1200 year and by Fourier treatment $\delta^{18}O$ values equal to 78 and 181 year were derived (cited by Dergachev and Kocherov 1977). Obviously, the climatic changes result from solar activity and the 180 year period corresponds to the Suess data.

It seems the record in observation of the periodic climate changes based on solar activity and registered by isotope data belongs to Libby and Pandolfi (1979). They studied distribution of deuterium and oxygen-18 in the yearly rings of Japanese cedar (*Cryptomeria Japonica*) within the last about 1800 year and derived there eight cycles with periods of 58, 68, 90, 96, 154, 174, 204 and 272 year. The shorter cycles like 11 and 21 year were not discovered due to averaging of the measured wooden rings within 5 years. And also the 174th cycle of the authors corresponds to the 183 year Suess period.

For the last relationship of δD and $\delta^{18}O$ in the Japanese cedar with radiocarbon in the bristlecone pine, measured by Suess for the same time intervals, Libby and Pandolfi discovered an inverse correlation. To the lower values of δD and $\delta^{18}O$ corresponds the higher ^{14}C values. These relations have the form:

$$\delta D = 0.677 \cdot {}^{14}C - 75.5(\mathrm{SMOW}),\ r = 0 - 0.62,$$
$$\delta^{18}O = 0.0613\delta^{14}C - 22.5(\mathrm{PDB}),\ r = 0 - 0.77.$$

The δD value was measured relative to the SMOW standard and $\delta^{18}O$ relative to the PDB standard. This relationship between stable hydrogen and oxygen-18 isotopes and the radiocarbon in wooden rings with ^{14}C amplitude of about 1% the above authors explain by climate change due to solar activity. The relationship between the rate of radiocarbon generation in the atmosphere and solar activity variation seems to have physical bases. But the problem of the influence of solar activity on galactic cosmic rays modulation and on the Earth's climate change has not been yet studied in detail. The search of cyclicity in the long and short-periodic processes has always attracted the attention of researchers. But as Gribbin and Lem (1980) pointed out, the cyclic attraction conceals in itself the risk to discover it, where there is a distribution of random values. To prove this, Gibbin demonstrated by a known graph, drawn by computer, generating random figures from 0 to 9, where deviation of the sliding average value from mathematical expectation appears to be equal to 4.5. On the graph of the random values distribution quasi-sinusoidal short-periodic fluctuations and a long-periodic 'temporal' trend are traced.

The long-periodic variation of ^{14}C content in the atmosphere and correspondingly in the biosphere was reliably fixed by the radiocarbon concentration measurement in the long-living woods. Figure 10.24, taken from work of Sternberg and Damon (1979) and drawn on the basis of results obtained by many laboratories in the world, traces sinusoidal changes in the wood ring's ^{14}C concentration during about 7500 year BP. The averaged data show that the minimal ^{14}C concentration (–5.5‰ from the level of 1890) comes approximately at 1400 year and the maximum (+85‰) at about 6500 BP.

Sternberg and Damon assumed that the observed ^{14}C variation is related to geomagnetic dipole moment changes. The relationship between the rate Q(t) of ^{14}C generation and the strength of the geomagnetic dipole moment has a reverse proportional dependence:

$$\frac{Q(t)}{Q_0} = \left(\frac{M(t)}{M_0}\right)^{-a}, \tag{10.58}$$

where Q_0 is the equilibrium rate of generation of the magnetic moment M_0; α is the coefficient changing within 0.4–0.6.

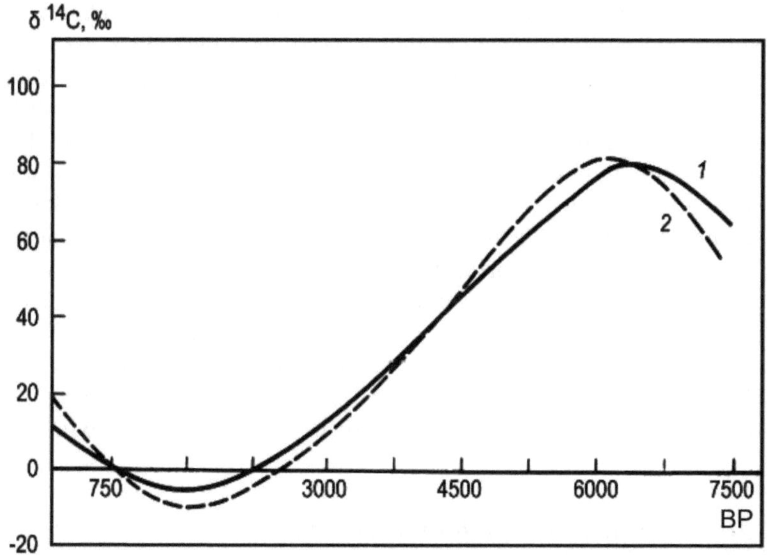

Fig. 10.24 Experimental and calculated data of 14C variation in yearly wooden rings: (*1*)experimental data approximated by a polynomial of the fourth degree; (*2*) the curve (*2*)obtained by calculation. (Sternberg and Damon 1979; Ferronsky and Polyakov 2012)

It was also found by the data of a paleomagnetic study that the magnetic dipole moment is varied in time within (4–12) · 10^{25} Gs · cm^3. It changes by quasi-sinusoidal law and during the last 10,000 years can be described by the equation:

$$M(t) = M_0 + M_1 \sin\omega(T - t + \theta),$$

where M_0 is the mean value of the dipole moment; M_1 is the amplitude variation; T is the period of oscillation; θ is the phase; $\omega = 2\pi/T$; t the time.

Denoting $T - t + \theta$ by t', Eq. (10.58) can be rewritten in the form:

$$Q(t) = Q_0\left[1 + (M_1/M_0)\sin\omega t'\right]. \tag{10.59}$$

Applying the two-box model to the carbon exchanging reservoir (Fig. 10.21), Sternberg and Damon reached good accordance between the experimental and calculating data on the long-periodic variation of the radiocarbon change in the atmosphere at the following conditions: T = 8500 year; $M_{max} = 12.5 \cdot 10^{25}$ Gs · cm^3; $T_{Mmax} = 2500$ year; $\alpha = 0.45$; $\tau_a = 75$ year (see the residence time of carbon in a reservoir in Fig. 10.22); $K_{as} = 5 \cdot 10^{-4}$. The calculations were done by Eqs (10.53), (10.54) and (10.59). The value of ^{14}C (‰) was calculated by the equation $\delta A_a(t) = [(A_a(t) - A_0)/A_0] \cdot 1000$, where A_0 is the activity of the modern carbon standard

Figure 10.24 presents the compared experimental and calculated long-periodic changes of radiocarbon in the Earth's upper atmospheric layers. The authors note that a delay for several hundred years in the phase between the geomagnetic field change and the radiocarbon variation in the exchange reservoirs is observed. This is due to the mean residence time of carbon in reservoir *A*, the ratio of exchanging carbon funds in *A* and *B* reservoirs and the period *T* of the geomagnetic field variation.

In addition to the astronomical and climatic factors effecting the distribution of the natural ^{14}C concentration in the atmosphere and biosphere, the dilution of atmospheric carbon dioxide by endogenic CO_2 may have an influence on local radiocarbon content changes, which is typical to the volcanic and breaking Earth's crust regions. It is suggested that in the modern volcanic regions, ^{14}C content in the biosphere is lower than the present day level. This is because of throwing out of endogenic carbon dioxide in the period of volcanic activity and CO_2 entering from the Earth's crust breaks (Sulerzhitsky and Frolov 1966; Karasev et al. 1981). On this basis Karasev et al. proposed a method of break discovery in active tectonic zones by radiocarbon measuring in plant samples. As was shown by the authors, in a break zone $\delta^{14}C = -50\,\%$ and at a distance of about 10 m $\delta^{14}C$ increases up to $+27\,\%$.

10.8.3 Natural Radiocarbon in Oceans

A number of works have been devoted to the analysis of radiocarbon in the oceanic water's carbonate system and water-bearing organic matter (Fonselius and Östlund 1959; Broecker et al. 1960; Broecker and Olson 1961; Bien et al. 1963; Fairhall and Young 1970; Fairhall 1971).

Starting in 1940–1950s, studies on ^{14}C distribution in the ocean water were difficult due to bomb-radiocarbon entering into the oceans. The artificial mark appearing in culminating tests in 1961–1962 allowed further detailed studying of the dynamics of CO_2 between the atmosphere and oceans exchange but became an obstacle for further study of ^{14}C space distribution in oceans and seas.

Radiocarbon enters into the oceans as a result of carbon exchange between dissolved marine carbonates and atmospheric CO_2. The radiocarbon entering into the surface ocean layer is rather quickly and uniformly distributed in an exchangeable layer ($h \approx 100$ m, $r \approx 10$–15 year). From there, by vortex diffusion and pelagic biogenic carbonate sedimentation, the radiocarbon enters the deep water layers, where its residence time (*t*), by different estimates, is equal to 1500–2000 year. As Craig (1957) noted, CO_2 exchange between the atmosphere and oceans is accompanied by isotopic effects. The mean isotopic content of stable carbon of the dissolver—hydrocarbonates of sea water—has a value of $\delta^{13}C = -7\,‰$ (here and further relative to the PDB standard). Thus, as a result of exchange processes the carbon of the hydrocarbonates becomes heavier by 7 ‰. Enrichment of oceanic ^{13}C can be described by means of the fractionation coefficients $\alpha_{^{13}C} = R_0/R_a$, where R_o and R_a are the ratio of $^{13}C/^{12}C$ in the ocean and atmosphere.

The value R can be presented as $(1 + \delta^{13}C/1000)$. Then the expression for α is rewritten as:

$$\alpha_{^{13}C} = \frac{1 + \delta^{13}C_0/1000}{1 + \delta^{13}C_a/1000}.$$

Analogously, for ^{14}C one has:

$$\alpha_{^{14}C} = \frac{1+\delta^{14}C_0/1000}{1+\delta^{14}C_a/1000} = \frac{\left(^{14}C/^{13}C\right)_0}{\left(^{14}C/^{13}C\right)_a}.$$

It is known (for example, Galimov 1968), that the fractionation constant increases by the square dependency on each additional neutron to the isotopic nucleus. Then, $\alpha_{^{14}C} = \alpha^2_{^{13}C}$. And applying the approximate equality $(1+x)^2 = 1+2x$ at $x << 1$, one obtains:

$$\alpha_{^{14}C} = \frac{1+\delta^{14}C_0/1000}{1+\delta^{14}C_a/1000} = \frac{1+2\delta^{13}C_0/1000}{1+2\delta^{13}C_a/1000}. \quad (10.60)$$

Using the last expression, it is possible to show that if ^{13}C of the ocean bicarbonate is enriched by 7‰, then at such enrichment ^{14}C should be equal to about 14‰. In the other words, $\varepsilon_{^{14}C} = 2\varepsilon_{^{13}C}$, where $\varepsilon_{^{13}C} = \delta^{13}C_0 - \delta^{13}C_a$. Later on this task is discussed in more detail.

Thus, theoretical consideration of the problem of CO_2 exchange between the atmosphere and the ocean shows that the surface oceanic water should be enriched in ^{14}C isotope by about 14‰ compared with atmospheric carbon. Analogously, it is possible to show that the biosphere should have ^{14}C in deficit by about 3% ($\delta^{13}C_a = -25$‰) compared with the atmosphere. As was shown earlier, the normalisation of ^{14}C content is provided for unification of the radiocarbon measurements (−25‰ is the normalising value for the studied samples). The formula for normalisation from Eq. (10.60) was obtained. It was pointed out by Libby (1955) that the oceanic carbonates have ^{14}C activity by 5% higher than biogenic carbon. His statement coincided with theoretical consideration of the natural exchange processes. But later studies have not proved this conclusion. In particular, it was shown that the hydrocarbons of the mixing layer in the Atlantic, Indian and Pacific oceans to the north from 40°S have practically constant ^{14}C content close to that in the biosphere (without the isotope correction) (Fonselius and Östlund, 1959; Broecker et al. 1960; Broecker and Olson, 1961; Bien et al. 1963; Bien and Suess 1967). This fact is explained by an increase of ^{14}C content due to isotope fractionation in the exchanging processes eliminated at mixing with the deep layers.

Stuiver (1980) on the basis of works of Broecker, Östlund, Craig and other researchers, compiled a short report on ^{14}C distribution in ocean waters during the pre-bomb period. It follows from this summary that in nine surface water samples taken in the North Atlantic in 1955 $\Delta^{14}C = -49 \pm 2$‰. The data from 13 stations of the South Atlantic (to the north from 40 °S.), sampled during 1956–1957, have slightly lower values, namely, $\Delta^{14}C = -57 \pm 2$‰. Up to that time of measurements, the decrease of the pre-industrial ^{14}C level in the surface oceanic waters was about 12‰. From this, we can accept that $\Delta^{14}C$ for the surface Atlantic waters before the intensive thermonuclear tests is equal to −40‰ from radiocarbon in the biosphere. It means that this value can be accepted as characteristic for the mixing layer over the thermocline. In the depths from 100 to 600 m the content of radiocarbon decreases exponentially from −40 to −100 and deeper it remains constant. The value −110‰ was characteristic for the Atlantic deep layers in the pre-thermonuclear epoch. The studies of Stuiver (1980) showed that after thermonuclear tests the 'natural' distribution of radiocarbon in the Atlantic Ocean in 1970 was preserved only below 3500 m.

Significant amounts of experimental data were collected by dating of mollusk shells living in normal salinity marine water. Investigations of samples collected in the period between the middle XIX century to 1950 in near-shore areas of different oceans provide the basis for the conclusion that the radiocarbon content (without isotopic correction) in the mixing ocean layer during the pre-thermonuclear epoch was also lower than the biogenic level, determined as 0.95 activity of NBS oxalic acid. By the data of Gillespie and Polach (1979), who studied radiocarbon distribution in mollusk shells from the near-shore of oceans during 1840–1950, the conclusion follows that ^{14}C content in the mixing layer varies in natural conditions from +8 to −11% relative to the modern standard. And without taking into account the effects of fresh river water recharge and the Suess effect, the variation range is narrowing. In average, carbon is depleted by ^{14}C (without correction) by about 15‰. Correcting by 25‰ and by age, the value of $\overline{\nabla}^{14}C = -64.5$‰ (0.95 NBS). Then the "apparent age" of the shells related to the reservoir effect in the mixing layer appears to be 535 year. For the Australian shore area the mean value of $\overline{\nabla}$ $^{14}C = -55 \pm 4$‰ (non-corrected value is ~5‰), which corresponds to the "apparent age" of 450 ± 35 year.

For the shells from the North and South American shore, taken in 1878–1940, the non-corrected value of $\delta^{14}C$ varies from +1 to −8% and the averaged value is about −3% (–30‰). The true ^{14}C value for the mixing layer in the pre-thermonuclear epoch is of great practical interest for radiocarbon dating using mollusk's shells. The ^{14}C differences in mollusk shells are hardly defined by the metabolic processes. They more likely reflect radiocarbon variations in offshore ocean layers, which affect the river runoff and discharge of the continental groundwaters. Sternberg and Damon (1979), after analyses of the great amount of collected experimental data, for theoretical calculation accepted that the mean radiocarbon activity in the mixing layer for the pre-thermonuclear epoch is equal to 0.965 with that in the biosphere in 1890 equal to 13.8 desintegrations/min · g (C).

The estimation of ^{14}C activity in different exchangeable reservoirs before 1950 can be done using the data from the work of Fairhall andYoung (1970) shown in Table 10.19.

On the basis of the above Fairhall and Young data it is possible to conclude that the ^{14}C activity of the mixing ocean layer in the pre-thermonuclear epoch should be 1% lower compared to the biosphere and about 3% than in the atmospheric CO_2. A number of researchers (for example, Oeschger and Siegenthaler 1979) in their theoretical calculations have accepted ^{14}C in the surface ocean layer before thermonuclear tests to be 95% from the atmospheric level in the middle XIX century. The carbon distribution in the deep layers, as the result of thermonuclear tests, inconsiderably changed due to the very long mean carbon residence time in this layer.

Table 10.19 Distribution of natural and thermonuclear ^{14}C at the end of 1962. (Fairhall and Young 1970)

Reservoir	Total amount of carbon (g)	^{14}C before 1950 (atom/g · 10^{10}C)	Natural ^{14}C, (atom · 10^{27})	Thermonuclear ^{14}C
Atmospheric CO_2	$6.8 \cdot 10^{17}$	6.07	41	54
Continental biosphere	$3.1 \cdot 10^{17}$	5.85	18	<1
Humus	$1.1 \cdot 10^{18}$	<5.8	<64	<1
Oceanic living biosphere	$3 \cdot 10^{15}$	6	0.2	<0.1
Oceanic non-decomposed rests	$2 \cdot 10^{16}$	6	1	<0.1
Dissolved organic matter in the oceans	$8 \cdot 10^{17}$	5.3	42	<1
Non-organic matter in the oceans				
upper 100 m	$1 \cdot 10^{18}$	5.8	58	~3
below 100 m	$3.8 \cdot 10^{18}$	5.1	1940	~1
Total	$42 \cdot 10^{18}$		2160	~60

By the data of Fairhall (1971), who summarised his own results and the results obtained during the 1958–1970 expeditions by Broecker, Bien, and Rafter before 1969 in the Atlantic, Indian and Pacific oceans, the non-disturbed picture in radiocarbon distribution below 500 m is observed and also its relative uniform distribution was fixed in the deep ocean layer. Radiocarbon comes to the deep layers from the mixing layer by vortex diffusion (Oeschger and Siegenthaler 1979) and partly due to sedimentation of mollusk shells, which dissolved in depth. By Fairhall's calculations the ^{14}C absolute concentration in ocean water before 1950 amounted to about $1.4 \cdot 10^9$ atoms per litre for the entire depth, for which the ^{14}C concentration depth gradient is equal to zero or correspondingly about 84 % of specific radiocarbon activity in the atmosphere (see Table 10.19). Using Fairfall's data, we may estimate the carbon mean residence time in the ocean deep layers. For this case, the following material balance can be used: $MdA_2/dt = q_1A_1 — q_2A_2 — MA_2\lambda$, where M is the carbon mass in the exchangeable layer; q_1 and q_2 are the carbon mass income; A_1 and A_2 are the specific carbon activities in the exchangeable and deep layers; λ the ^{14}C decay constant. In the stationary state, the left-hand side of the equation is equal to zero and $q_1 = q_2$. Taking into account that $M/q = \tau$, then one has $\tau = (A_1 - A_2)/A_2\lambda$.

Accepting that $A_1 = 0{,}965$, $A_2 = 0{,}835$ and $\lambda = 1.21 \cdot 10^{-4}$, then one obtains $\tau = 1300$ year. If $A_1 = 0.99$, which corresponds to many experimental data for the Atlantic ocean (Fairhall 1971), then $\tau \approx 1500$ year, which corresponds to Libby's (1955) estimation.

10.8.4 Technogenic Radiocarbon in the Atmosphere and Oceans

According to data obtained by different authors (Fairhall andYoung 1970), $6 \cdot 10^{28}$ atoms of ^{14}C have been released into the Earth's atmosphere during thermonuclear tests. Before the tests the ^{14}C content in the atmosphere was estimated to be $4.1 \cdot 10^{28}$ atoms and therefore the total amount of carbon atoms has increased by 2.5 times. Compared with the total equilibrium amount of the isotope on the Earth ($2.2 \cdot 10^{30}$ atoms) the bomb component amounts to 2.5% (see Table 10.19).

On average, at a thermonuclear test $3 \cdot 10^{26}$ atoms of ^{14}C per 1 Mt are produced, which is equivalent to 7 kg of radiocarbon. By radiocarbon thrown into the atmosphere, over 200 Mt of total explosion power was performed up to 1962, which injected into the atmosphere about 1.4 t of artificial radiocarbon.

Numerous measurements (up to a thousand specimens) were carried out in order to determine ^{14}C content variations in the troposphere and stratosphere both with latitude and altitude (Fairhall andYoung 1970; Münnich and Vogel 1963; Fairhall et al. 1969; Hagemann et al. 1959). The most representative results of these studies according to the data of different authors for the troposphere CO_2 generalised from 1963 to 1966 are shown in Fig. 10.25.

There are distinct seasonal variations of ^{14}C concentrations in the northern hemisphere and a considerable latitudinal gradient of concentrations in 1963 indicated in Fig. 10.26 but the gradient quickly decreases to zero by 1967. Both of these effects are a consequence of seasonal variations in the release of bomb ^{14}C from the stratosphere into the troposphere and the longitudinal mixing that occurs in the troposphere. The most intensive release of ^{14}C into the troposphere, as in the case of tritium, is detected in spring and early summer. At the same time the most effective mixing of this isotope occurs in the longitudinal direction.

In contrast to the thermonuclear dust, which is removed as a rule quickly from the troposphere with precipitation, CO_2 has a rather long residence time in the troposphere. Therefore, the levels of bomb ^{14}C in the troposphere air were highest in the middle latitudes of the northern hemisphere and reached maximum concentrations late in the summer.

The longitudinal mixing of ^{14}C extends southwards. Annually, from September to February, the level of ^{14}C in the northern latitudes decreased and in the southern latitudes increased, up to the middle of 1966. Such a process took place until 1966 when the whole troposphere became homogeneous with respect to ^{14}C.

It has been pointed out (Fairhall and Young 1970) that the relative amounts of tropospheric air participating in the circulation at various latitudes should be taken into account when comparing the latitudinal variations of ^{14}C concentrations (Fig. 10.25). The convergence of the meridians and the decent of the tropopause at high latitudes result in a sharp decrease in the volume of tropospheric air with increasing latitudes, compared with the equatorial region. Thus, for the same ^{14}C ejection from the stratosphere its activity in high latitudes will be considerably higher than in the lower ones.

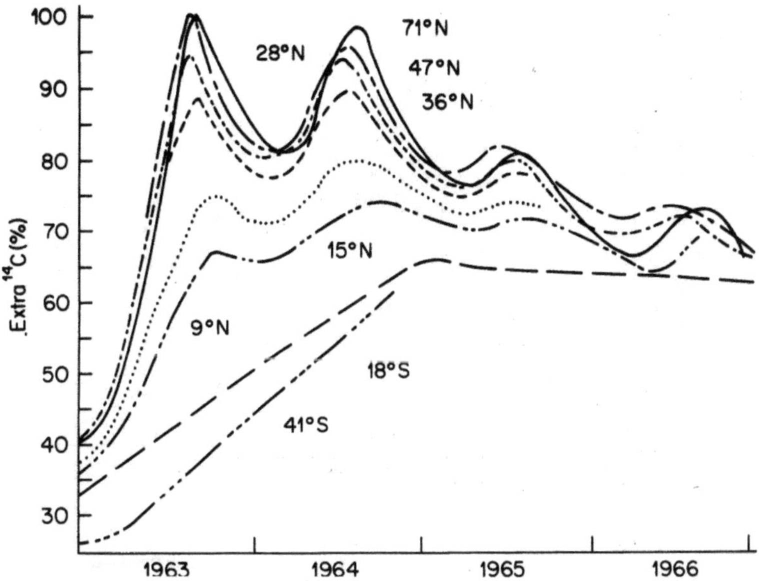

Fig. 10.25 Variation in time of ^{14}C concentrations in tropospheric CO_2. (Fairhall and Young 1970; Ferronsky and Polyakov 2012)

Figure 10.26 shows the variations with time of the absolute amounts of bomb-radiocarbon in the troposphere, stratosphere and atmosphere as a whole for the northern and southern hemispheres.

It follows from Fig. 10.26 that with the decrease in the ^{14}C concentrations in the northern hemisphere the opposite process took place in the southern hemisphere during 1963–1965. The theoretical estimation of the total amount of ^{14}C in the atmosphere based on experimental data shows that the time of half-removal of this isotope is equal to 3.3 year. In the next years the rate of ^{14}C removal from the atmosphere decreased. Its amount in the troposphere up to 1970 compared with 1963 decreased by two times and by 1977 only one third of it was left (Berger 1979). Before 1984 the ^{14}C amount in the troposphere was about 125 % with respect to the pre-thermonuclear level. The residence time of CO_2 in the troposphere calculated by decrease of ^{14}C activity during the period of 1963–1976 was about 10 year (7/0.693) and for 1970–1982 this time was about 15 year, which is close to the earlier estimates made by Arnold and Anderson (14–30 year) (Miyake 1969).

It is assumed that the rate of removal of bomb ^{14}C from the stratosphere is proportional to the difference (gradient) between its concentration in the stratosphere and the surface (mixing) ocean layer. While decreasing the gradient, the velocity of the radiocarbon removal from the troposphere is dropping. On this basis, the conclusion follows that during the 'peak' injections of radiocarbon (and tritium as well) into the atmosphere, the ^{14}C concentration changes, between the atmosphere and hydrosphere, take place in $^{14}CO_2$ but not in general CO_2. The sharp drop of ^{14}C

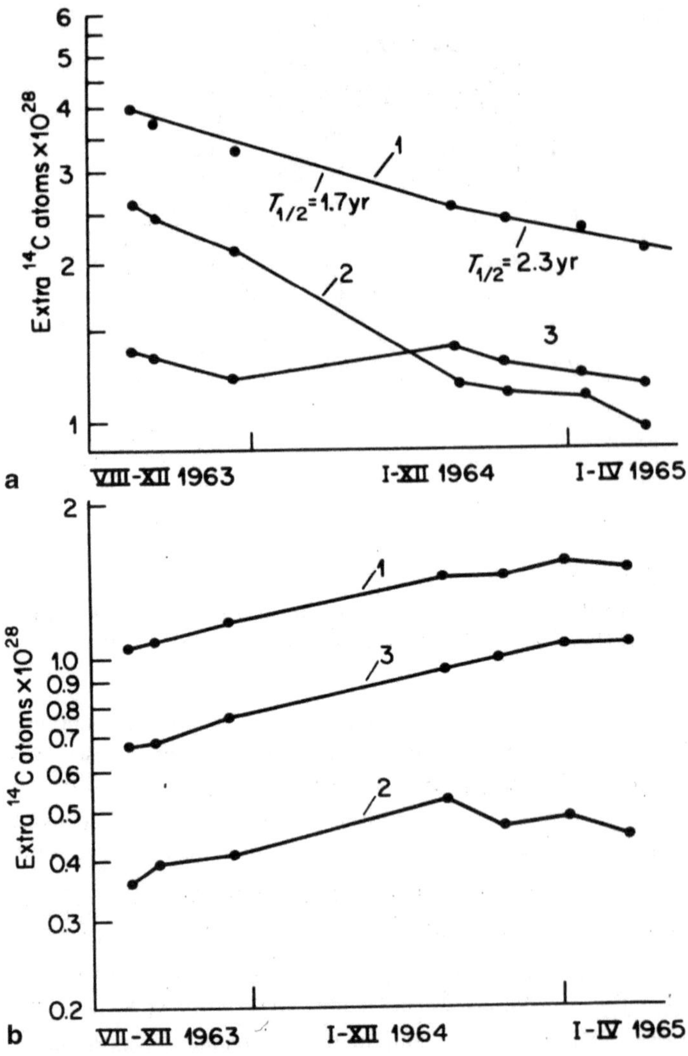

Fig. 10.26 Variation of total amount of bomb 14C in the northern (**a**), and southern (**b**) hemispheres for the atmosphere as a whole (*1*), the stratosphere (*2*) and the troposphere (*3*). (Fairhall and Young 1970; Ferronsky and Polyakov 2012)

half-removal from the atmosphere in the first years after the nuclear tests stopped in the atmosphere is shown in Fig. 10.26a. It is most probable that the ^{14}C half-removal, measured by its activity before 1966, characterised in reality not only the CO_2 exchange process between the atmosphere, ocean and biosphere but also ^{14}C activity decrease in the northern hemisphere as a result of its transfer with air masses into the southern hemisphere.

The altitudinal distribution of ^{14}C specific activity in the atmosphere is not uniform. In 1962, for example, ^{14}C activity in CO_2 of the lower troposphere was about

140% with respect to the natural level. In April 1962, before the USSR thermonuclear tests, at the height of 12.2 km at 36 °N, ^{14}C activity was measured to be three times higher than the natural one (Fergussen 1963). These measurements have shown that considerable amounts of ^{14}C were accumulated in the stratosphere as a result of the nuclear tests. This fact is proven by the summer maximum coming into the troposphere.

By the data of Fairhall and Young (1970) the radiocarbon concentrations in the troposphere and atmosphere are practically equalised. But the observed summer maximums during 1968–1974 evidence against this assumption (Berger 1979). In order to find the correct answer, CO_2 stratospheric sampling at 18 and 21 km heights was carried out during 1975–1977 in the framework of the NASA program. The results of this experiment are shown in Table 10.20.

As is shown in Table 10.20, the spring peak of ^{14}C concentration appears in the lower stratosphere, which looks like the summer maximum in the troposphere. In this regards, one may assume that the ^{14}C enriched air during the summer breaks the troposphere and arrives there from the lower stratosphere. By Berger's opinion, the mean residence time of CO_2 in the stratosphere is about 15–20 year.

In 1963, USSR, USA and the UK signed the treaty on cessation of nuclear weapon tests in the atmosphere, oceans and on the Earth's surface. Since that time, the exchange reservoirs have been insignificantly replenished by artificial ^{14}C during French and Chinese explosions. The peak ejections of radiocarbon into the atmosphere, which created in the atmosphere and stratosphere some kind of marks, allowed a more detailed study of the mass exchange between the atmosphere, oceans and the biosphere and also between different layers of the atmosphere and oceans. A wide program on the study of distribution of ^{14}C caused by the bomb-tests was carried out in the Trondheim Laboratory starting in 1962. In particular, the scientists who took part in the program since 1963 published their results on the study of CO_2 exchange between the atmosphere and oceans. A study on ^{14}C redistribution between the northern and southern hemispheres on a profile of Spitsbergen–Madagascar was also initiated. Before 1967, ^{14}C concentrations in the lower atmosphere of the Earth were practically equalised. This allowed the calculations of carbon exchange cycles to not take into account the time in different parts of the atmosphere.

Table 10.20 ^{14}C content in the stratosphere. (Berger 1979)

Sampling date	Samplingheight	Sampling place	$\delta\ ^{14}C$ (‰)
13.06.75	19.8	Northern Sierra	+127.4
25.09.75	19.8	California	+90.7
03.12.75	18–21	San Francisco—Phoenix	+79.1
02.06.76	18–21	San Francisco—Salt Lake City	+85.6
30.03.76	19.8	San Francisco—Oregon	+107.0
23.09.76	19.8	San Francisco—Denver	+99.2
22.03.76	19.8	Pacific Ocean—Los Angeles	+199.6

The decrease of bomb ^{14}C in the atmosphere occurs due to its absorption by the terrestrial biosphere and CO_2 exchange between the atmosphere and the oceans. The latter reservoir is most important since more than 80% of atmospheric CO_2 is exchanged with the ocean and 20% is replaced through the inland biosphere. In view of the role of the oceans in the absorption of bomb-^{14}C, considerable investigations have been carried out aimed at the measurement of its concentration in surface and deep oceanic waters (Bien and Suess 1967; Fairhall 1971; Fairhall et al. 1969). In contrast to the atmosphere, characterised by quick ^{14}C mixing (less than 10 year), the process of mixing in the ocean is more complicated and requires more time. The radiocarbon technique is advantageous in studying oceanographic processes such as mixing of the oceanic waters and their global circulation.

The spatial ^{14}C distribution in depths of the Atlantic Ocean was studied in the framework of the GEOSECS program (Stuiver 1980). Figure 10.27a shows the location of the stations and also the stations of the research ship 'Meteor' (Germany). The ship's route was adjusted for the GEOSECS program (Roether et al. 1980). The $\Delta^{14}C$ values distribution at the GEOSECS stations in 1972–1973 are presented in Fig. 10.27b, 10.27c. It is seen in Fig. 10.27c that the two main fields of the surface waters immersion into deep layers are observed. The first field is located between 36° and 40 °S and the second between 20° and 30 °S. The upward flow of the deep waters is traced around the equator. The two more localised upward flows are traced between 40–50 °S and 40–50 °N. It follows from Fig. 10.27b, 10.27c that significant amounts of bomb-radiocarbon entered into the surface ocean layer and reached the deep layers in 1972. An analogous picture on the longitudinal profile, obtained during the 23 rd route of the 'Meteor' in 1971, is observed (Fig. 10.28) (Roether et al. 1980).

As pointed out above, the ^{14}C mark that occurred in the atmosphere after the thermonuclear tests during a short period of time, is used to study the atmosphere and oceans dynamical characteristics. The carbon concentration increase in the Earth's atmosphere has a head form of δ-function at the well-known change of carbon concentration in the atmosphere and oceans. This gives the possibility for estimation of the exchange parameters of the reservoir, the main of which are the atmosphere, oceans and biosphere. Each reservoir can be presented by a number of 'boxes' (Dergachev 1977). For example, the oceans can be considered as a two boxes reservoir: the upper one over the thermocline (75 ± 25 m) in which intensive water mixing occurs and a lower, weakly mixing 'box' in which the carbon concentrates due to the diffusion process. The continuous carbon exchange between the deep ocean layer and ocean sediments takes place. The natural border in the atmosphere is the tropopause, which is at a height of 11–12 km. The tropopause divides the troposphere into a layer, well mixing in the vertical direction and a layer where the meso-longitudinal exchange is prevailing. The use of the box models, the possibilities of which were discussed earlier, on the basis of an experimental study of ^{14}C distribution in a reservoir in space and time, allows the rate of exchange processes and the carbon mean residence time in reservoirs to be estimated.

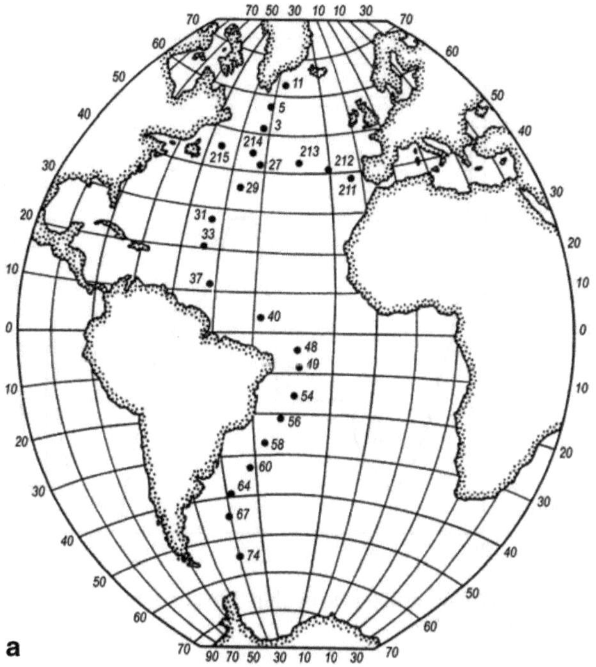

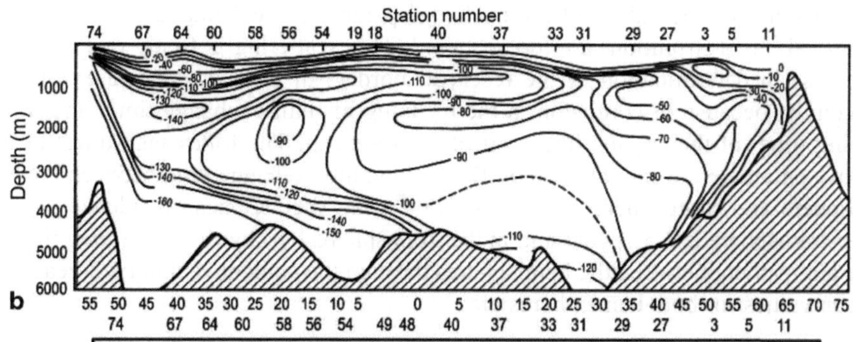

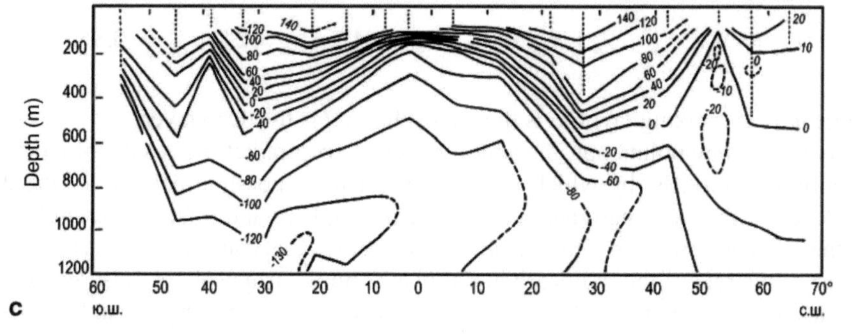

Fig. 10.27 Location of measurement and sampling stations for the GEOSECS program (1972–1973, № №5–74) and the 23rd 'Meteor' route (1971, №№ 211–215) (**a**) (Stuiver 1980; Roether et al. 1980); and $\Delta^{14}C$ vertical distribution in the western part of the Atlantic Ocean in 1972–1973 (**b**) (Stuiver 1980); the same for the depth up to то 1200 m (**c**) (Ferronsky and Polyakov 2012)

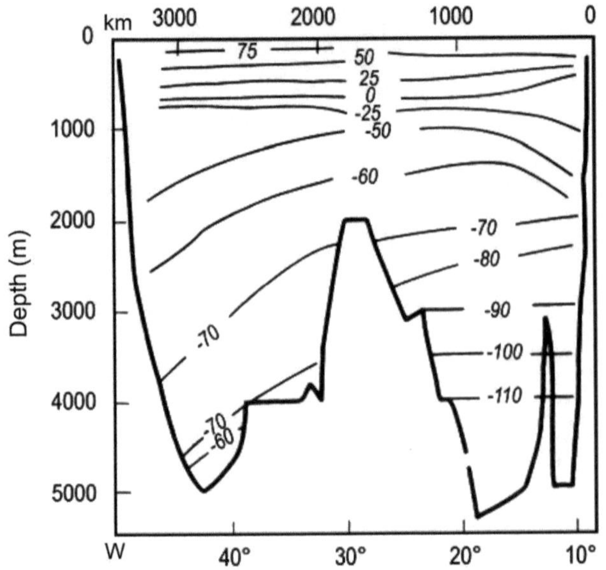

Fig. 10.28 $\Delta^{14}C$ vertical distribution obtained by ship 'Meteor' stations, 23rd route. (Roether et al. 1980); Ferronsky and Polyakov 2012

Dergachev (1977), after analysis of many works on the problem, came to the conclusion that the values of the exchange parameters, calculated on the basis of artificial ^{14}C distribution, are too scattered. This scattering is firstly due to incorrectness in determination of the reservoir borders of the exchange reservoir and secondly, because of confusion in the meaning of 'half-exchange time' and 'half-living time' of the ^{14}C in either part of the reservoir. The following values of the above parameters were obtained by Dergachev: the exchange time between the troposphere and stratosphere is equal to 1.5–2 year; the ^{14}C mean residence time in the stratosphere is 3.5–4 year and in the troposphere 1.5–2 year; the mixing time of the atmosphere between the hemispheres through the equator is about 1.5 year. The mean values for both hemispheres are:

$$\tau_{ma} = 9.1;\ \tau_{ma} = 7.1;\ \tau_{ma} = 8.7;\ \tau_{md} = 4.3;\ \tau_{md} = 2.8;\ \tau_{md} = 3.8\ \text{yr};$$

where *m* is the mixing layer; *a* is the atmosphere; *d* is the deep layer.

As Dergachev points out, the small values τ_{md} contradict the results of calculations by box models, by which $\tau_{md} \approx 10$ year. This is possible because of the dependence of the exchange rate between the atmosphere and different ocean layers on the ^{14}C concentration gradient in exchangeable reservoirs. In this case the decrease of the rate of the radiocarbon removal from the troposphere must lead to an increase in the mean residence time in the atmosphere and mixing ocean layer. The

Table 10.21 ^{14}C (%) content from the standard of modern carbon

Reser-Voir	Time (year)						
	1957	1958	1959	1960	1961	1962	1963
A	104	112	123	122	116	137	193
H_m	98	98	99	99	100	100	102
	1964	1965	1966	1967	1968	1969	
A	188	170	165	160	158	155	
H_m	105	108	110	112	114	118	
	1970	1971	1972	1973	1974	1975	
A	152	150	148	145	142	140	
II_m	116	116	115	115	114	113	

experimental data on ^{14}C distribution in the atmosphere and oceans, obtained in the last years, are evident of such as assumption. The author, applying the Trondheim Laboratory data, on the basis of the one-box model, estimated the residence time of radiocarbon in the mixing and deep ocean layers. Table 10.21 shows the function (A) of the radiocarbon change in the atmosphere and the ^{14}C concentration increase in the mixing ocean layer H_m used in the calculations.

The data for the atmosphere before 1963 were taken on the basis of ^{14}C measurements in alcohol of Georgian and Portuguese wines (Burchuladze et al. 1977; Lopes et al. 1977). For the mixing layer, ^{14}C concentration before 1963 was obtained by extrapolation of the data published in the works of Bien et al. (1963); Broecker et al. (1960). The calculations were carried out by the formula analogous to that used for tritium data interpretation:

$$A_t = \bar{A} + \sum_{1957}^{1975} \delta A_\theta \frac{1}{\tau} e^{-t/\tau-}$$

where A_t is the radiocarbon concentration in the mixing layer at t year after 1957; δA_Θ is the difference (%) between the biogenic radiocarbon level before 1950 (100 %) in the calendar year Θ; $\bar{A}$ is the ^{14}C concentration in the mixing layer (98 %).

The best coincidence of the calculated and experimental data is achieved at the value $\tau \approx 15$ year. The radiocarbon half-life has not been taken into account because of the short time interval compared with the ^{14}C half-life. Certainly, the one-box model used for estimation of the dynamical parameters in the system atmosphere-ocean (Gray and Damon 1970; Ralf 1972) gives an approximate result. But the carbon constant of the reaction transfer rate from the deep to mixing layer by substitution of the value $\bar{A}$ was taken into account. The carbon exchange time of the mixing layer by Eq. (10.55) can be estimated. Taking into account that $N_{п.с.} = 1.3N_a$, then one obtains $\tau = 13$ year, if τ for the atmosphere is equal to 10 year. As in this case the carbon exchange in the system mixing-deep layers was not taken into account, then the more realistic τ value for the mixing layer seems to be 10 year. This does not contradict the three-box model of Dergachev (1977). The ^{14}C distribution in the surface layer of the central ocean parts is of interest. The observations of the Atlantic and Pacific Ocean stations show (Nydal et al. 1979) that ^{14}C concentrations in local areas do not stay constant in time but vary within 10 ‰ (and even up to 20 ‰). A positive correlation between radiocarbon content in the surface layer and the surface ocean temperature is observed.

In the study of Rafter and O'Brien (1972) the ^{14}C distribution in the Pacific surface waters after 1968 is discussed. They found that two belts of increased concentrations ($\delta^{14}C \approx 20\,\%$) near 27 °N and 27 °S and minimal values around the equator ($\delta^{14}C \approx +5\,\%$) are observed.

While considering the ^{14}C distribution in the surface Atlantic waters, Dergachev (1977) derived two maximum values: $\delta^{14}C \approx +7\,\%$ (70 °N) and $\delta^{14}C \approx +19\,\%$ (27 °N). It is interesting to note that the location of the above maximum values is unchanged up to the depth of 500 m. The nature of this phenomenon is not explained. The only thing clear is that the maximum salinity of the oceans coincides with those from Fig. 10.29. The water zones are characterised by higher evaporation of marine water and have a relationship with the global circulation of the air and water masses. It is obvious that before the thermonuclear tests, ^{14}C concentration in the surface ocean waters was not constant and changed under the same effects as after the tests.

The artificial radiocarbon from the mixing layer enters the deep ocean layers. According to Fairhall's (1971) calculations, the process of equilibrium state be-

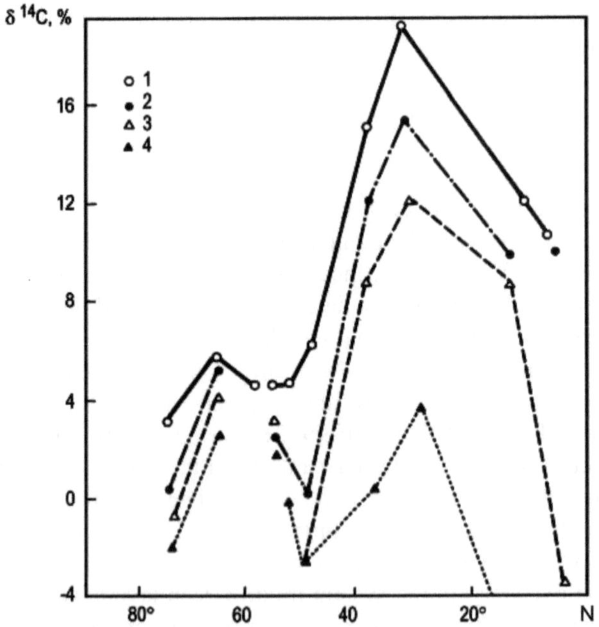

Fig. 10.29 Radiocarbon concentration change in the upper layers in the North Atlantic Ocean: (*1*) surface layer; (*2*) at depth 100 m; (*3*) at 200 m; (*4*) at 500 m (Ferronsky and Polyakov 2012)

tween the atmosphere and oceans will continue for 45 year (starting from 1970). An experimental study of this process will help in understanding the oceanic currents nature and the mixing time of ocean waters.

10.8.5 Forecast of Carbon Dioxide Increase in the Atmosphere

The carbon dioxide increase in the atmosphere, in addition to the ^{14}C specific activity dilution in the exchangeable reservoirs, play independent and important roles. Increase of the carbon dioxide content may lead to a considerable change in the mean yearly temperature on the Earth, which is a consequence of the 'green-house' effect. Carbon dioxide is practically transparent for solar radiation but it absorbs the thermal radiation of the Earth in a number of lines on the infra-red spectrum. If water and carbon dioxide are absent in our planet, then the infra-red radiation leaves it for space and the lower layers of the atmosphere to lose warmth. The carbon dioxide concentration growth leads to the stratosphere cooling and near-earth air growth (Table 10.22).

Oeschger and Siegenthaler (1979) used CO_2 to change in time the parameters for the main reservoirs, obtained on the basis of artificial radiocarbon distribution and to predict the possible increase of CO_2 content in the atmosphere. They proposed models that are based on the following assumptions.

1. The study of ^{14}C content in deep Pacific and Atlantic oceanic waters over 1957–1959 has shown that distribution of radiocarbon in the deep waters was not practically disturbed by the thermonuclear tests.
2. As a result of the thermonuclear tests, ^{14}C activity in surface oceans waters increased by 95 % with respect to the present carbon content (its 'pre-industrial' value for the atmosphere is equal to 100 %) in 1957 and by about 112 % in 1970.
3. About 19 %, with respect to the 'pre-industrial' amount of CO_2, has been injected into the atmosphere before 1970 but its concentration decreased only by 10 %.
4. Because of dilution of carbon dioxide due to combustion of fossil fuel, the ^{14}C specific activity in the atmosphere before 1950 decreased by about 2 %.

Oeschger and Siegenthaler (1979) assumed for their model that the carbon transfer from the mixing layer to the deep waters results from vortex (turbulent) diffusion (Fig. 10.30). Any increase of the partial pressure in CO_2 leads to its redistribution between the atmosphere and oceans. Such a redistribution can be described by a parameter called the 'buffer-factor'. For example, if the CO_2 partial pressure in the at-

Table 10.22 Assuming CO_2 growth in the atmosphere. (Kellog 1980)

CO_2 content change in atmosphere (%)	Assuming year of the change (year)	Growth of near-earth temperature (°C)
+25	2000	0.5–1.0
+100	2050	1.5–3.0

mosphere increases by α%, then the total CO_2 concentration in the ocean waters, to be in equilibrium with the atmosphere, increases in waters only by α/ξ %. The biosphere in the model is represented by the well mixed reservoir, where the amount of carbon is 2.4 times more than in the atmosphere and its residence time is about 60 year. The partial pressure increase of CO_2 in the atmosphere leads to the photosynthetic activity of plants described by the 'growth-factor' ε. If CO_2 pressure in the atmosphere increases by α% then the carbon flow from the atmosphere to the biosphere increases by the value of $\varepsilon\alpha$%. According to Oeschger and Siegenthale (1979), the ε value varies for different conditions from 0 to 0.4. For prediction of the CO_2 increase in the atmosphere the following independent parameters were used (see Fig. 10.30):

N_a	CO_2 in atmosphere	Pre-industrial level	$0.62 \cdot 10^{18}$ g
N_b	CO_2 in biosphere		$2.4\ N_a$
N_m	CO_2 in mixing layer		$1.3\ N_a$
N_d	CO_2 in deep water layer		$64.2\ N_a$
K_{am}	Exchange coefficient in the system atmosphere—mixing layer		1/7.7 year
K_{ab}	Exchange coefficient in the system atmosphere-biosphere		1/25 year
K	Coefficient of vortex diffusion		$3987\ m^2 \cdot sec^{-1}$
ξ	Buffer-factor of CO_2 absorption by the ocean		10
ε	Growth-factor connecting with CO_2 absorption by the biosphere		0.2

The authors give a prediction of the CO_2 content increase in the atmosphere at different input functions. For example, if single CO_2 concentration in the atmosphere increases by α%, then a new equilibrium is reached characterised by α_∞%. The ratio α_0/α_∞ can be obtained by the equation:

$$\alpha_0 N_a = \alpha_\infty N_a + (\alpha_\infty/\xi)(N_m + N_d) + \alpha_\infty \varepsilon N_b,$$

or

$$\alpha_\infty/\alpha_0 = N_a/\left[N_a + (N_m + N_d)\right]/\xi + \varepsilon N_b.$$

At ξ=10, ε=0.2 α_0/α_∞=0.125, i.e., one eighth of the total CO_2 input into the atmosphere. Here, a possible decrease of the α_∞ value as a result of oceanic carbonates dissolution and weathering of the rocks has not been considered. But because of too slow CO_2 transfer into deep ocean layers these processes will continue for a long time. If the CO_2 inflow into the atmosphere stopped in 1970, then $\alpha \approx 10$%, in 2000 the α value would be 7% and after several centuries it would reach an equilibrium value,

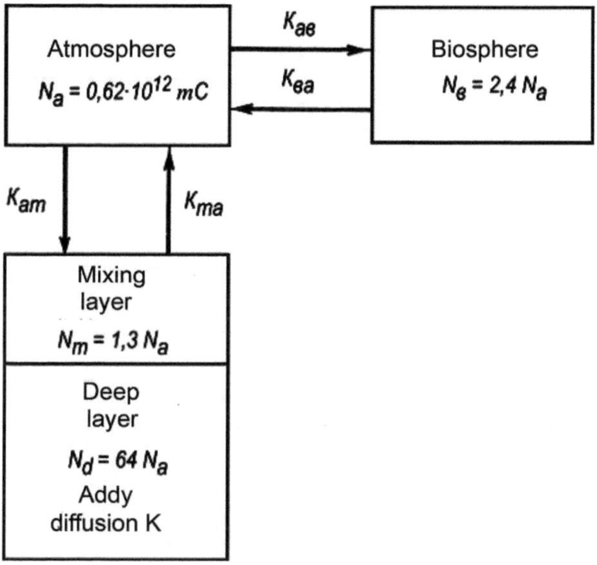

Fig. 10.30 Reservoirs of the carbon exchangeable system that take into account diffusion transfer to the deep ocean layers. (Oeschger and Siegenthaler 1979; Ferronsky and Polyakov 2012)

i.e., $\alpha_\infty \approx 2.3\,\%$. The carbon content in fossil fuel (oil, coal and gas) recalculated by CO_2 is 11.64 times more than CO_2 in the atmosphere during the pre-industrial epoch. According to present-day knowledge, the CO_2 production rate can be described by the equation:

$$P(t) = \frac{d}{dt}\left[\frac{11.65N_a}{1+61\exp(-t/22)}\right],$$

where t is the time in years (t = 0 in 1970).

The production rate P(t) after 1970 comprises 4.5 N_a per year (the mean velocity over 1960–1970 is about 5 % per year). The calculations show that at this rate of fuel combustion more than 2000 %, with respect to the pre-industrial level of CO_2, will be accumulated in the atmosphere up to 2050. This amount of carbon dioxide may lead to an increase of near-ground temperatures from 3° to 4°C (Fig. 10.31). But it is unlikely. In the XXI century, fossil fuel use descended because other sources of energy are now available. In addition, according to recent scientific data on heat and mass exchange between the atmosphere and oceans, the greenhouse effect is by orders of 3–4 overestimated (Sorokhtin 2002).

By Kellogg's (1980) data, in this case the near-ground temperature will increase only by 1–1.5 °C.

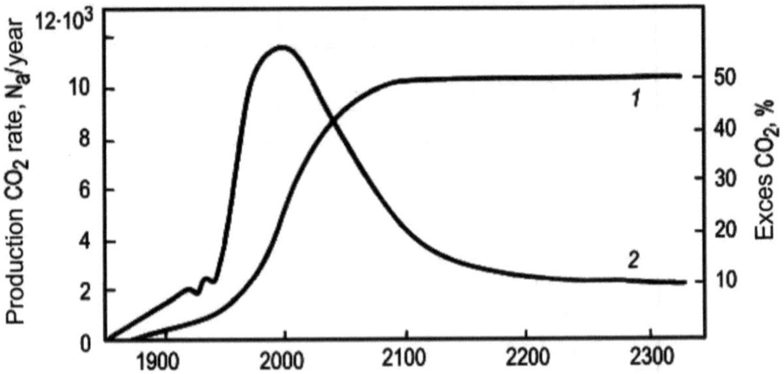

Fig. 10.31 Dependence of CO_2 variation in the atmosphere on present fossil fuel combustion (*1*) and on 30 % reduction of its present combustion (*2*). (Oeschger and Siegenthaler 1979; Ferronsky and Polyakov 2012)

10.8.6 Principles of Radiocarbon Dating

If any system exchange with atmospheric carbon dioxide is finished then the accumulated ^{14}C amount is decreased in time by the radioactive decay law:

$$A_t = A_0 e^{-\lambda t} \tag{10.61}$$

where A_t and A_0 are the radiocarbon activity (or concentration) at time t and $t = 0$; λ is the radiocarbon decay constant; $\lambda = \ln 2/T_{1/2} = 0{,}693/T_{1/2}$; T is the period of the ^{14}C half-life.

If the exchange processes stop, for example due to death, then the formula (10.61) can be used for calculation of the time passed after the process has broken off (the object age):

$$t = (1/\lambda)\ln(A_0/A_t) = 8033\ln(A_0/A_t). \tag{10.62}$$

The half-life period equal to 5568 year was used in Eq. (10.62). In order to pass the 5730 year period, all the calculated age values by Eq. (10.62) should be multiplied by the coefficient 1.03. The method of obtaining radiocarbon dates on organic specimens was developed by Libby (1967).

Radiocarbon dating is based on a number of assumptions, the main of which are: (1) the intensity of cosmic radiation and, as a consequence, the ^{14}C concentration in the atmosphere, remained constant at least during the radiocarbon dating scale (0–80,000 year); (2) the time of exchange of atmospheric carbon with terrestrial carbon is considerably smaller than the half-life of carbon-14 and does not change with time; (3) secondary processes do not affect the isotopic composition of the studied specimens after sampling. This is equivalent to the assumption that the radiocarbon content in a specimen decreases only because of radiocarbon decay.

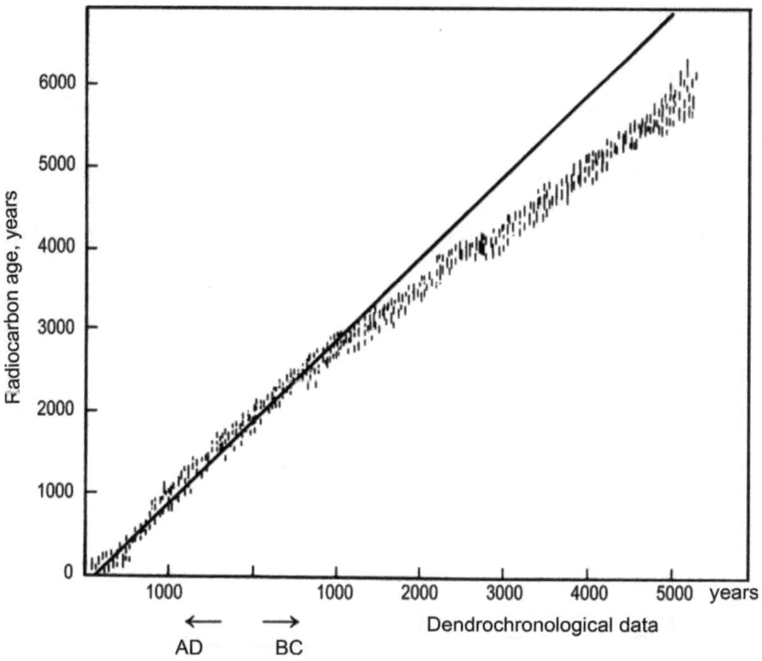

Fig. 10.32 Relationship between dendrochronological and radiocarbon ages of Sequola gigantea and Pinus aristata (Mook 1977; Ferronsky and Polyakov 2012)

In natural cases, the above-mentioned assumptions do not hold strictly, although in most cases the age determined on the basis of radioactive dating are in good agreement with those obtained using other techniques (Libby 1967).

In the case of ^{14}C variations in atmospheric CO_2 and in the biosphere (see Fig. 10.24) over long time periods starting from about 2500 year, the radiocarbon and dendrochronological time scales of the organic specimens differ (Fig. 10.32).

In order to reconcile the radiocarbon results with the dendrochronological ones within 0–8000 year, a number of correlation relations were proposed. According to Wendland and Donley (1971), the age correlation can be done by the third-degree polynomial:

$$T_{corr} = 112 + (0.710 T_{14_C}) + (1.610 \cdot 10^{-4} T^2_{14_C}) - (1.50 \cdot 10^{-8} T^3_{14_C}).$$

For this purpose there are also other correlation equations (Ralph and Klein 1979).

Radiocarbon dating on the basis of organic fragments has a wide application in archaeology, geology, geography, oceanography and other Earth sciences (Suess 1979).

Isotopic composition of the stable carbon in a carbonate system is used for correction of the radiocarbon age of groundwater. The processes of carbon isotope fractionation were considered above.

As follows from Eq. (10.62), the most carbon fractionation is between gaseous CO_2 and hydro-carbonate-ion (HCO_3^--ion is enriched in ^{13}C by about 10% at 0°C compared with carbon dioxide). It is also obvious that the complete fractionation of carbon isotopes in the gas-solution and gas-solid phase system should depend on the medium pH.

Fractionation of the carbon isotopes in the thermodynamically equilibrium processes, stipulated by the kinetic factors (for example, in biochemical reactions), leads to the non-steady state in carbon isotopes of carbon-bearing matter of different natural objects.

Figure 10.33 indicates the isotope variation limits in carbon contained in various objects (Stiel et al. 1979). It is observed from the figure that the most ^{13}C-enriched

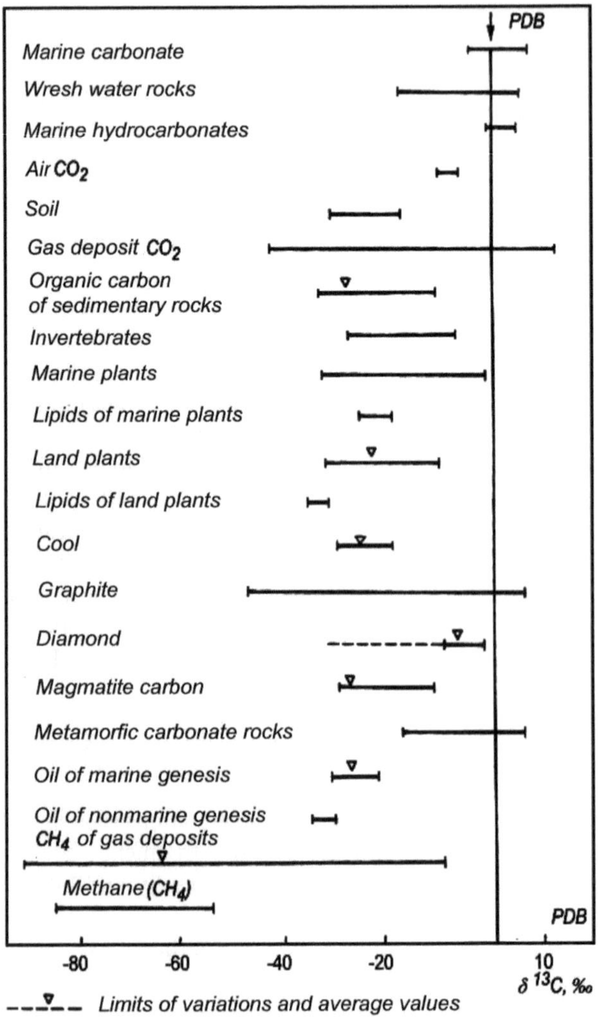

Fig. 10.33 Stable carbon isotope variations relative to the PDB-1 standard in various natural objects. (Stiel et al. 1979; Ferronsky and Polyakov 2012)

and relatively constant by isotopic composition are the carbonates precipitating from oceanic water (mean values of $\delta^{13}C=0$). The PDB-1 isotope standard is composed of these carbonates. Compared with the standard, atmospheric CO_2 is enriched in the light isotope of carbon by −7‰. Plants are even more enriched in the light isotope. The average value for ground plants is $\delta^{13}C=-25$‰. Close to this value is the soil CO_2. A large value of shift ($\delta^{13}C=-30$‰) is detected in oil and maximum enrichment in ^{12}C is observed in natural methane formed in sedimentary rocks ($\delta^{13}C$ up to −90‰).

Since the isotope fractionation factor increases by its square with the addition of each neutron, the corresponding isotope shift in ^{12}C–^{14}C isotope fractionation will be twice as great. Therefore vegetation should be depleted in ^{14}C compared with the atmosphere and have $\delta^{14}C=-50$‰.

10.8.7 Radiocarbon Dating of Groundwater

At present, for groundwater age determination most researchers apply the piston model. It was earlier pointed out that the radioactive isotope concentration in a sampling point *A*, located at a distance x_0 from the recharge, is calculated by the expression:

$$A = A_0 \exp[-(x_0/v)\lambda] = A_0 \exp(-\lambda t), \qquad (10.63)$$

where A_0 is the isotope concentration (activity) in the recharge area; *v* is the velocity of groundwater motion; λ is the ^{14}C decay constant; $\lambda=\ln 2/T_{1/2}$; *t* is the water age, determined as the transit time, during which the isotope reaches the distance from the recharge area to the sampling point.

The time *t*, from Eq. (10.63), can be obtained by the formula:

$$t = (T_{1/2}/\ln 2)\ln(A_0/A) = (1/\lambda)\ln(A_0/A). \qquad (10.64)$$

As was noted earlier, the radiocarbon method of groundwater age determination was applied for the first time by Münnich (1968). The main assumptions of this method are as follows:

1. Radiocarbon concentration in the recharge zone is known and does not change at least over the range of the radiocarbon time-scale.
2. Radiocarbon does not enter the carbonate system of groundwater from outside the recharge zone.
3. If ^{14}C content in an aquifer decreases both due to radioactivity decay and other processes, then the effect of these processes can be taken into account.
4. The leakage of water from one aquifer into another (the mixing of water) is negligible.
5. The rate of movement of dissolved carbonates is equal to the rate of groundwater movement.

The reality of the above conditions is discussed later on.

It is worth mentioning that the term 'age' of groundwater has a great indefiniteness compared with, to-say, 'age' of a solid sample. In the general case, the age of groundwater can be defined by a certain period of time after some hydrogeological event has happened (Dubinchuk 1979). Such an event in radiocarbon dating is the discontinuation of the water exchange with the soil air or the atmosphere. In the framework of this definition, in the radiocarbon scale the natural waters appear to be modern if they are found in continuous exchange with the soil air or atmosphere, where radiocarbon concentration $A_0 = 100\,\%$.

In the general case, the water 'age' is more easily described in the form of the distribution function of residence time in the aquifer, applying corresponding models for a hydrogeological system, which satisfy its real structure (Ferronsky et al. 1977; Dubinchuk 1979). At the present time, for isotope data interpretation and for obtaining the groundwater's age characteristics, in addition to the piston model, there are the following models for mass transfer in hydrological systems: (1) model of complete mixing (exponential or box-model); (2) dispersion model; (3) mixing model of different age waters.

According to the complete mixing (exponential) model (Nir 1964), water input at different times mixes quickly. The relative portion of water of t age in the system is expressed as an exponential function in the form:

$$p(t) = (1/\tau)\exp(-t/\tau), \tag{10.65}$$

where τ is the average residence time of water in a system; $1/\tau = v/q$ is the water exchange rate, i.e., the ratio of the total annual inflow of water to the output of the system. Concentration of the isotope in the system ($A_{\text{system}} - A_{\text{output}}$) in the general case can be written as the following sum in discrete form:

$$A_{\text{system}} = \bar{A}/(1+\lambda\tau) + \sum_{\theta=1952}^{1952+t} \left[A(\Theta - t) - \bar{A}\right]/p(t)e^{-\lambda t}, \tag{10.66}$$

where Θ is the current year; $(\Theta - t)$ the time of input into the water system with a concentration of the isotope (^{14}C and others.) $A(\Theta - t)$, which happened by t year from the sampling moment; $\bar{A}$ is a steady-state concentration before the thermonuclear tests; other definitions are as previous. Applying the formula (10.66), by observed ^{14}C concentrations in the system (A_{system}), the mean residence time of water can be obtained if the function of the isotope distribution $A(\Theta - t)$ at the system entrance is known. The exponential model is used for interpretation of tritium data. But it can also be used for estimation of the dynamical characteristics of groundwater, applying the radiocarbon, because the function $A(\Theta - t)$ (radiocarbon input at the hydrogeological system entrance) is known.

Calculation shows that the effect of ^{14}C concentration change, induced in groundwater by the thermonuclear tests, in so much as the mean residence time in the system, is less. If the function $A(\Theta - t)$ is constant the ^{14}C radiocarbon residence time of water $\tau = v/q$ can be determined by the formula:

$$\tau = (A_{input} - A_{output})/A_{output}\lambda = 8033(A_{input}/A_{output} - 1), \tag{10.67}$$

where A_{input} and A_{output} are the radiocarbon concentrations (in % with reference to the standard of the modern carbon) at the entrance and outlet of the system.

It is not difficult to show that in the framework of the exponential model, if A_{input} is constant in time, then the water age t is related to the residence time τ by the expression $t=(1/\lambda)(1-\lambda\tau)$. Application of the integral or discrete expressions like (10.66) allow to estimate the residence time in the hydrological system at different forms of the distribution function $p(t)$. But, despite the wide application in tritium data treatment, interpretation of the radiocarbon data by such a way is not used.

As a special case, it is necessary to consider the model of different-age water (Evans et al. 1979). If the hydrogeological system consists of n components of water with the ages T_i, then its age can be determined as a mean-weighted value:

$$\bar{T}=\sum_{i=1}^{n} P_i T_i, \tag{10.68}$$

where P_i is the part of water of age T_i.

In such a system the mean-weighted isotope concentration (for example, ^{14}C) is established:

$$\bar{A}=\sum P_i T_i. \tag{10.69}$$

If the source of the radioactive isotope input is the same, then the values A_i will be differed due to partial radioactive isotope decay over the time T_i. Such a condition is satisfied for the radiocarbon concentration where in the recharge area it is constant and equal to A_0. Then the expression (10.69) can be rewritten as:

$$\bar{A} = A_0 \sum P_i \exp(-\lambda T_i). \tag{10.70}$$

Determining the mixing age by concentration $\bar{A}$ and using the exponential formulae, we may obtain the value:

$$T' = 8003 \ln(A_0/\bar{A}) \tag{10.71}$$

Comparing the expressions (10.71) and (10.68), we find that $T' \neq \bar{T}$. It follows from the above that: (1) the true age of mixture of different waters is equal to the mean-weighted age of the water components; (2) in the general case, the isotopic (radiocarbon) mixture of waters is not equal to the true age and its formal determination does not give unique time-information about a system. This problem can be solved correctly only in the case of independent measuring of the date and age for each component. For obtaining such information one may use, for example, the data on distribution of other isotopes (D, ^{18}O, T and so on). Evans et al. (1979) calculated the possible change in radiocarbon age for a two-component mixture depending on a portion of the modern component ($A_0 = 100\,\%$). It follows from their calculations that the groundwater of radiocarbon age equal to 57,000 years may represent a mixture of 99.9 % of water with 100,000 years of age and 0.1 % of modern water. This result only slightly changes if the present-day component has 5,000 years of age. But despite the radiocarbon age of waters mixture not providing true information, it can be used to study a relationship between the aquifers, localisation of recharge areas and so on (Borevsky et al. 1981; Sobotovich et al. 1977; Seletsky et al. 1979; Fröhlich et al. 1977; Deak 1974).

Since under real conditions the initial radiocarbon concentration can always be estimated with some error, corresponding corrections should be done (see Ferronsky and Polyakov 2012).

10.9 The Other Cosmogenic Isotopes in Natural Waters

Besides tritium and radiocarbon, other cosmogenic isotopes are used in studying natural waters. While considering the amounts and production rates of primary cosmic rays and secondary nuclear particles with the three main constituents of the atmosphere, nitrogen, oxygen and argon were accounted for. But it has been pointed out that nuclei of neon, krypton and xenon and also those of volcanic and meteoric origin can make a certain contribution to the production of cosmogenic isotopes.

In the course of neon nuclei fission reactions, ^7Be, ^{10}Be, ^3H and other isotopes are produced. But the yield of isotopes occurring during these reactions is lower than for the argon fission reactions. Besides, in the atmosphere, the ratio of concentrations of $Ne/Ar = 0.02$. The content of krypton and xenon in the atmosphere is even lower and, respectively, the contribution of cosmogenic isotopes occurring from reactions involving these elements is also small. Moreover, only reactions of the (n, γ)–type on nuclei of ^{80}Kr, ^{84}Kr and ^{132}Xe can serve as a source of very small amounts of ^{81}Kr ($T_{1/2} = 2.1 \cdot 10^5$ year), ^{85}Kr ($T_{1/2} = 10.8$ year) and ^{133}Xe ($T_{1/2} = 5.3$ days). An approximate estimation of the cosmogenic ^{81}Kr yield by the (n, γ) reaction produces a value of $3 \cdot 10^{-7}$ atom/cm$^2 \cdot$ s.

Bartels (1972) suggested that the major portion of the cosmogenic isotopes of ^{22}Na, ^{32}Si, ^{32}P, ^{32}P, ^{35}S and ^{36}Cl may be produced by the reactions of cosmic rays on nuclei of the elements released into the atmosphere in the form of volcanic explosions. His suggestion is based on data on the rate of contribution of s number of elements, together with volcanic material, to the terrestrial atmosphere (Table 10.23).

The conditions in which the production of many cosmogenic isotopes takes place were considered by Lujanas (1975, 1979). He pointed out that the atmospheric concentrations of the elements, indicated in Table 10.23, are much less than those of

Table 10.23 Amounts of some elements released into the atmosphere from volcanic explosions

Elements	Released amounts (kg/year)
B	$7.3 \cdot 10^7$
F	$7.3 \cdot 10^9$
Na	$6 \cdot 10^9$
S	$1.7 \cdot 10^{10}$
Cl	$7.6 \cdot 10^9$
K	$2.3 \cdot 10^9$
Ca	$7.2 \cdot 10^9$

the main air constituents. Therefore, the effect of fission reactions on their nuclei will be negligible. But in a number of cases the cross-sections of reactions on slow neutrons are great. In order to estimate the yield of cosmogenic isotopes in these reactions let us write the expression for the production rate P_i of a given isotope relative to the production rate P_c of ^{14}C isotope in the reaction $^{14}N\,(n, p)\,^{14}C$ on slow neutrons:

$$P_i = \frac{P_c \sigma_i N}{\sigma_c N} P_i, \qquad (10.72)$$

where σ_i and σ_c are the effective cross-sections of production of a given isotope and ^{14}C, respectively; N and N_c are numbers of atoms in a nuclei-target and in the atmosphere column with a unit section.

One can find, for example from expression (10.72), that in reaction $^{10}B\,(n, p)\,^{10}Be$, at $\sigma_i = 4010$ barn, $\sigma_i/\sigma_c \approx 2300$ and $N/N_c = 1.9 \cdot 10^{-11}$, the value of $P_i = 8.7 \cdot 10^{-7}$ atom/cm$^2 \cdot$ s is less than the rate of yield of the same isotope in reactions on air nuclei, being equal to $9 \cdot 10^{-2}$ atom./cm$^2 \cdot$ s, by several orders of magnitude. The estimations of the production rates of the other isotopes in reactions on slow neutrons give even smaller values.

The studies of meteorites falling on the Earth's surface show that they also contain radioactive isotopes such as ^3H, ^{10}Be, ^{22}Na, ^{26}Al, ^{36}Cl, ^{37}Ar, ^{39}Ar, ^{40}K, ^{44}Ti, ^{45}Ca, ^{46}Sc, ^{48}V, ^{49}V, ^{51}Cr, ^{53}Mn, ^{54}Mn, ^{57}Co, ^{58}Co, ^{60}Co and others. Table 10.24 shows data on the flux density of some cosmogenic radioisotopes released into the atmosphere on the burning of between 0.5 to 0.7 of the mass of the meteorite material on the assumption that the total amount of meteorite material amounts to 10^9 kg/year. Table 10.24 also shows the cosmic-ray production rates of the same isotopes in the atmosphere.

It follows from Table 10.24 that the flux densities of the isotopes ^3H, ^{10}Be, ^{14}C, ^{22}Na, ^{36}Cl, ^{37}Ar and ^{39}Ar arriving together with meteorites into the atmosphere are lower by several orders of magnitude than the rates of production in the atmosphere due to the irradiation of the atmospheric nuclei of atoms with cosmic rays. At the same time the ^{26}Al meteorite contribution, for example, is comparable in magnitude to its atmospheric production rate.

Table 10.24 Flux density of radioactive isotopes in meteoritic material falling into the atmosphere and the rate of their production by cosmic rays. (Lujanas 1975)

Isotopes	Flux density due to meteorites (atom/cm$^2 \cdot$ sec)	Production rate by cosmic rays (atom/cm$^2 \cdot$ sec)
^{3}H	$1.5 \cdot 10^{-8}$	0.45
^{14}C	$1 \cdot 10^{-5}$	2.5
^{10}Be	$2 \cdot 10^{-4}$	$4.5 \cdot 10^{-2}$
^{22}Na	$8 \cdot 10^{-10}$	$8.6 \cdot 10^{-5}$
^{26}Al	$1.6 \cdot 10^{-4}$	$1.4 \cdot 10^{-4}$
^{36}Cl	$1 \cdot 10^{-5}$	$1.1 \cdot 10^{-3}$
^{37}Ar	$6 \cdot 10^{-12}$	$1.7 \cdot 10^{-3}$
^{39}Ar	$1 \cdot 10^{-8}$	$6.3 \cdot 10^{-3}$
^{53}Mn	$1 \cdot 10^{-3}$	–
^{54}Mn	$6 \cdot 10^{-3}$	–
^{55}Fe	$1 \cdot 10^{-6}$	–
^{59}Ni	$8 \cdot 10^{-6}$	–
^{60}Co	$2.5 \cdot 10^{-10}$	–

References

Aegerter SK, Loosli HH, Oeschger H (1967) Variation in the production of cosmogenic radionuclides. In: Radioactive dating and methods of low-level counting: proceed symp, IAEA, Vienna, pp 49–55

Afanasenko EA, Morkovkina IK, Romanov VV (1973) Age of groundwater in the Tas-Khayat Ridge and Selennyakh depression. Bull Moscow Univ, Ser Geol 5:105–109

Afanasyev AP (1960) Water balance of Baykal Lake. Proc Baykal Limnol Stn 18:85–95

Allison GB, Hughes MW (1974) Environmental tritium in the unsaturated zone: estimation of recharge to an unconfined aquifer. In: Isotope techniques in groundwater hydrology: proceedings of a symposium, IAEA, Vienna, 11–15 March 1974, pp 57–70

Andersen LJ, Sevel T (1974) Six years' environmental tritium profiles in the unsaturated and saturated zones. In: Isotope techniques in groundwater hydrology: proceeding symp, vol 1, IAEA, Vienna, pp 3–18

Andrews JN, Kay RLE (1982) Natural production of tritium in permeable rocks. Nature 298:361–363

Appa Rao MVK (1962) The ^{3}He/(^{3}He + ^{4}He) ratio in primary cosmic radiation. J Geophys Res 67:1289–1392

Atakan YW, Roether W, Münnich KO, Matthess G (1974) The sandhausen shallow groundwater tritium experiment. In: Isotope techniques in groundwater hydrology: proceedings of a symposium, IAEA, Vienna, 11–15 March 1974, vol 1, pp 21–43

Bainbridge AE (1963) Tritium in the Northern Pacific surface water. J Geophys Res 68:3785–3789

Bartels OG (1972) An estimate of volcanic contributions to the atmosphere and volcanic gases and sublimates as the source of the radioisotopes ^{10}B, ^{35}S, ^{32}P and ^{22}Na. Heal Phys 22:387–392

Begemann F (1959) Neubestimmung der natürlichen irdischen Tritiumzerfallstrate und die Frage der Herkunft des natürlichen Tritium. Naturforsch 4a:334–342

Begemann F, Friedman I (1968) Isotopic composition of atmospheric hydrogen. J Geophys Res 73:1139–1147

Begemann F, Libby WF (1957) Continental water balance, groundwater inventory and storage times, surface ocean, mixing rates and world-wide water circulation patterns from cosmic-ray and bomb tritium. Geochim Cosmochim Acta 12:277–296

Berger R (1979) Artificial radiocarbon in the stratosphere. In: Berger R, Suess HE (eds) Radiocarbon dating. University of California, Berkeley, pp 309–321

Bhaudari W, Fruchter J, Evans J (1969) Rates of production of ^{22}Na and ^{28}Mg in the atmosphere by cosmic radiation. Earth Planet Sci Lett 7:89–92

Bien GS, Suess HE (1967) Transfer and exchange of ^{14}C between the atmosphere and the surface water of the Pacific Ocean. In: Radiocarbon Dating and Methods of Low-Level Counting: proc. a symp, IAEA, Vienna, pp 105–115

Bien GS, Rakestrow P, Oldenbourg M, Suess HE (1963) Investigations in marine environments using radioisotopes produced by cosmic rays. In: Radiocarbon dating: proc. symp, IAEA, Vienna, pp 159–174

Bochaler P, Eberhardt P, Geiss J (1971) Tritium in lunar materials. Proc. 2nd Lunar Sci Conf 2:1803–1812

Boella GC, Dilworth M, Panetti M, Scarsi L (1968) The atmospheric and leakage flux of neutrons produced in the atmosphere by cosmic ray interactions. Earth Planet Sci Lett 4:393–398

Bonka H (1979) Production and emission of tritium from nuclear facilities, and the resulting problems. In: Behaviour of tritium in the Environment: proc symp. IAEA, Vienna, pp 105–122

Borevsky BV, Polyakov VA, Subbotina LA (1981) Investigation of regularities in leaching through Neocom-Jurassic strata within the depression funnel of the Bryansk water intake well. In: Isotopes in the Hydrosphere, Abstract Symp, IWP RAN, Moskva, pp 52–53

Bradley W, Stout G (1970) Vertical distribution of tritium in water vapor in the lower troposphere. Tellus 22:699–706

Broder DP, Golubev LI, Ilyasov VM (1979) Tritium distribution in the technology scheme of the New-Voronezh atomic power station. At Energy 47:120–122

Broecker WS, Olson EA (1961) Lamont radiocarbon measurements. Radioscarbon 21:199–216

Broecker WS, Gerard R, Erwing M, Heezen BC (1960) Natural radiocarbon in the Atlantic ocean. J Geophys Res 65:2903–2909

Brown RM (1961) Hydrology of tritium in the Ottawa Valley. Geochim Cosmochim Acta 21:199–216

Brown RM (1970) Distribution of hydrogen isotopes in Canada waters. In: Isotope Hydrology: proc. symp, IAEA, Vienna, pp 3–21

Brown R, Grummit WE (1956) The determination of tritium in natural waters. Canad J Chem 34:220–226

Burchuladze AA, Gedevanishvili DD, Pagava SV, Togonidze GI (1977) Variation of radiocarbon content in the atmosphere for 1950–1975 years measured in Georgian vines. In: Low radioactivity measurements and mpplications. Proceedings of a symposium, High Tatras, 1975, Slovenske Pedagog, Nakland, Bratislava, pp 261–263

Burger LL (1979) Distribution and reactions of tritiated hydrogen and methan. In: Behaviour of Tritium in the Environment, proc symp. IAEA, Vienna, pp 47–63

Burkhard W, Fröhlich K (1970) Grundlagen hydrologischer tritium untersuchungen und Ihre anwendung bei der bestimmung der herkunft in grubenwasser einer eisenerzgrube. Bergakademie, Freiberg vol 22, pp 15–30

Buttlar HV, Libby WF (1955) Natural distribution of cosmic ray produced tritium. J Inorg Nucl Chem 1:75–91

Cain WF (1979) ^{14}C in modern American trees. In: Radiocarbon dating. University of California , Berkeley, pp 495–510

Combs F, Doda RY (1979) Large-scale distribution of tritium in a commercial product. In: Behaviour of Tritium in the Environment: proc symp, IAEA, Vienna, pp 93–99

Craig H (1957) Isotopic standards for carbon and oxygen and correction factors for mass-spectrometric analysis of carbon dioxide. Geochim Cosmochim Acta 12:133–149

Daly IC, Manchester AV, Gabay IJ, Sax NI (1968) Tritiated moisture in the atmosphere, surrounding a nuclear fuel reprocessing plant. Radiol Heal Data and Repts 11:217–229

Damon PE (1970) Climatic versus magnetic perturbation of the atmospheric C-14 reservoir. In: Radiocarbon variations and abcolute chronology, XII nobel symp. Wiley, NY, pp 571–593

Deak J (1974) Use of environmental isotopes to investigate the connection between surface waters in the Nagynusad region, Hungary. In: Isotope techniques in groundwater hydrology: proceedings a symp, vol 1, IAEA, Vienna, pp 157–167
Dergachev VA (1975) Variations in solar activity and radiocarbon content in the atmosphere. Izv AN SSSR, Ser Phys 32:325–333
Dergachev VA (1977) Optimal model for residence time determination of exchange reservoir. In: Low-radioaktivity measurements and applications, proc. symp., IAEA, Vienna, pp 269–277
Dergachev VA, Kocharov GE (1977) The ceqular cycles of radiocarbon variation in the Earth atmosphere. In: Low-Radioaktivity Measurements and Applications. Proceedings a symp, IAEA, Vienna, pp 279–286
Devis DH (1970) Geohydrologic interpretations of a volcanic island from environmental isotopes. Water Resour Res 6:652–671
Dinçer T, Payne BR (1971) An environmental isotope study of the south-western karst region of Turkey. J Hydrol 13:233–258
Dockins KO, Bainbridge AE, Houtermans JC, Suess HE (1967) Tritium in the mixed layer of the North Pacific Ocean.In:Radioactive aating and methods of low-level counting, proceedings a symp, IAEA, Venna, pp 120–160
Domanitsky AP (1971) Rivers and lakes of the Soviet Union. Gidrometizdat, Leningrad
Dubinchuk VT (1979) What is the groundwater age? MOIP, Geol Sect 54:70–79
Dudey ND, Malewski RL, Rymas SL (1972) Tritium yield from fast-neutron fission of ^{235}U. Trans Amer Nucl Soc 15:483
Ehhalt DN (1966) Tritium and deuterium in the atmospheric hydrogen. Tellus 18:249–255
Ehhalt DN (1971) Vertical profiles and transport HTO in the troposphere. J Geophys Res 76:7351–7367
Ehhalt DN (1974) The atmospheric cycle of methane. Tellus 26:58–63
Eisenbund M, Bennett D, Blanco RE et al (1979) Tritium in the environment-NCRP report No 62. In: Behaviour of tritium in the environment: proceedings symp, IAEA, Vienna, pp 585–587
Ekdahl CA, Keeling CD (1973) Atmospheric CO_2 in the natural carbon cycle: 1. Quantitative reduction from records at Mouna Loa Observatory and at the South Pole. In: Carbon and biosphere: 24th Brukhaven symp biol, Springfield, pp 51–85
Eriksson E (1963) Atmospheric tritium as a tool for the study of certain hydrologic aspects of river basins. Tellus 15:303–308
Eriksson E (1965a) Account of the major pulses of tritium and their effects in the atmosphere. Tellus 17:118–130
Eriksson E (1965b) Deuterium and oxygen-18 in precipitation and other natural waters. Tellus 17:498–512
Evans GV, Otlet RL, Downing RA et al (1979) Some problems in the interpretation of isotope measurements in United Kingdom aquifers. In: Isotope Hydrology: proceedings symp, vol 2, IAEA, Vienna, pp 679–706
Fairhall AW (1971) Radiocarbon in the Seas, Rep.No RLO-225-T20-3, US AEC
Fairhall AW, Young YA (1970) Radiocarbon in the environment. In: Radionuclides in the environment. Adv Chem Ser.No 9, Am Chem Soc, pp 401–418
Fairhall AW, Buddemeir RW, Yang IA, Young YA (1969) Radiocarbon from nuclear testing and air-sea exchange of CO_2. Antarctic J 4:14–18
Faltings V, Harteck P (1950) Der Tritium Gehalt der Atmosphere. Ztschr Naturforsch 8:438–439
Fergussen GY (1963) Upper tropospheric carbon-14 levels during spring 1961. J Geophys Res 68:3933–3941
Ferronsky VI, Polyakov VA (2012) Isotopes in the Earth's hydrosphere. Springer, Dordrecht
Ferronsky VI, Danilin AI, Dubinchuk VT et al (1968) Radioisotope methods of investigation in ingeneering geology and hydrogeology. Atomizdat, Moskva
Ferronsky VI, Dubinchuk VT, Polyakov VA et al (1975) Environmental isotopes of the hydrosphere. Nedra, Moskva
Ferronsky VI, Danilin AI, Dubinchuk VT et al (1977) Radioisotope methods of investigation in ingeneering geology and hydrogeology. Atomizdat, Moskva

Fireman EL (1967) Radioactivities in meteorites and cosmic-ray variation. Geochim Cosmochim Acta 31:1197–1206

Fireman EL, Stoenner RW (1982) Carbon and carbon-14 in lunar soil 14163. Proc 12 Lunar and Planet Sci Conf NY 12:559–565

Fluss MJ, Dudey ND (1971) Tritium and helium yields in fast fission of ^{235}U. Trans Amer Nucl Soc 14:809–812

Fonselius S, Östlund HG (1959) Natural radiocarbon measurements on surface water from the North Atlantic and the Arctic sea. Tellus 11:77–82

Fontes JC (1976) Les isotopes du milieu dans les laux naturells. Le Houille Blanche 3/4:205–221

Fröhlich K, Jordan H, Hebert D (1977) Radioactive Umveltisotope in der Hydrologie. Grundstoffindustrie, Leipzig

Galimov EM (1968) Geochemistry of stable carbon isotopes. Nedra, Moskva

Gillespie R, Polach YA (1979) The suitability of marine shells for radiocarbon dating of Australian prehistory. In: Radiocarbon dating. University of Calif. Press, Berkeley, pp 404–421

Gonfiantini R, Panichi C (1982) Geothermal water studies. In: Guidebook on nuclear techniques in Hydrology. IAEA, Vienna, pp 151–162

Gray DC, Damon PE (1970) Sunspots and radiocarbon dating in the middle ages. In: Scientific methods in medieval archeology. University of California Press, Berkley, pp 167–182

Gribbin J, Lem GG (1980) Climate change in hystorical period. In: Climate change. Gidrometeoisdat, Leningrad, pp 122–140 (transl. from Engl.)

Hagemann F, Grey J, Machta L, Turkevich A (1959) Stratospheric carbon-14, carbon dioxide, tritiumю. Science 130:542–552

Harteck P (1954) Relative abundance of HT and HTO in the atmosphere. J Chem Phys 22:1746–1751

Houtermans J, Suess HE, Munk W (1967) Effect of industrial fuel combustion on the carbon-14 level of atmospheric CO_2. In: Radioactive dating and methods of low-level counting: proceedings of symp, IAEA, Vienna, pp 57–68

International Atomic Energy Agency (1973) Bulletin No 4. IAEA, Vienna, pp 10–16

International Atomic Energy Agency (1983) Guidebook on nuclear techniques in hydrology (1983 Edition), IAEA, Vienna

International Atomic Energy Agency (1969, 1970, 1971, 1973, 1975, 1979, 1983, 1986, 1990, 1994) Environmental isotope data: world survey of isotope concentrations in precipitation, No 1–10, IAEA, Vienna

International Atomic Energy Agency (1996) Manual on the mathematical models in isotope hydrology. Tecdoc no 910. IAEA, Vienna

Jouzel J, Pourchet M, Lorius C, Merlivat L (1979) Artificial tritium fall-out at the South Pole. In: Behaviour of tritium in the environment: proceedings of symp, IAEA, Vienna, pp 31–45

Karasev BV, Sokolovsky LG, Kuznetsova LA (1981) Application of carbon isotopes for identification of break zones gassing by carbon dioxide. In: Ferronsky VI (ed) Investigation of natural waters by isotope methods. Nauka, Moskva, pp 155–157

Katrich IYu (1990) Tritium in natural waters after the Chernobyl atomic power accident. Meteorol and Hydrol 5:92–97

Kaufman S, Libby WF (1954) The natural distribution of tritium. Phys Rev 93:1337–1344

Keeling CD (1972) Carbon dioxide cycle: reservoir models to depict the exchange of atmospheric carbon dioxide with oceans and land plants. In: Chemistry of the lower atmosphere. Plenum , NY, pp 251–329

Kellog WW (1980) Global influence on human activity on climate'. In: Climate change. Hydrometeoizdat, Leningrad, pp. 273–302 (transl. from English)

Kinman TD (1956) An attempt to detect deuterium in the solar atmosphere. Month Not Roy Astrophys Soc 116:77

König LA (1979) Impact of the environment of tritium releases from the Karlsruhe Nuclear Research Center. In: Behaviour of tritium in the environment: proc of symp, IAEA, Vienna, pp 591–610

Krejči K, Zeller A (1979) Tritium pollution in the Swiss luminous compound industry. Behaviour of tritium in the environment: proc of symp, IAEA, Vienna, pp 66–77
Lal D (1963) Study of long and short-term geophysical processes using natural radioactivity. In: Radioactive dating: proc of symp, IAEA, Vienna, pp 149–157
Lal D, Peters B (1962) Cosmic-ray-produced isotopes and their application to problems in geophysics. Progr Elem Part Cosm Ray Phys 6:1–74
Lal D, Peters B (1967) Cosmic-ray-produced radioactivity on the Earth. Encycl Phys 46:551–612
Lal D, Venkatavaradan VS (1967) Activation of cosmic dust by cosmic-ray particles. Earth Planet Sci Lett 3:293–310
Lal D, Rajan RS, Venkatavaradan VS (1967) Nuclear effects of solar and "galactic" cosmic-ray particles in near-surface regions of meteorites. Geochim Cosmochim Acta 31:1859–1869
Libby WF (1955) Radiocarbon dating. Chicago University Press, Chicago
Libby WF (1963) Moratorium tritium geophysics. J Geophys Res 68:4485–4494
Libby WF (1967) History of radiocarbon dating. In: Radiocarbon dating and methods of low-level counting: proc symp, IAEA, Vienna, pp 3–25
Libby LM, Pandolfi LJ (1979) Isotopic tree thermometers: anticorrelation with radiocarbon. In: Radiocarbon dating. Univ Calif Press, Berkeley, pp 661–669
Locante J (1971) Tritium in pressurized water reactor. Trans Am Nucl Soc 14:161–162
Lopes JS, Pinte RE, Almendra M, Machado JA (1977) Variation of ^{14}C activity in portuguese wines from 1940 to 1974. In: Low-radioactivity measurements and applications. Proceedings of a symposium in High Tatras. IAEA, Vienna, pp 265–268
Lujanas VYu (1975) On the rate of cosmogenic radionuclides production. Cosmogenic Radioact Isotopes 3:17–25 (Vilnius)
Lujanas V Yu (1979) Cosmogenic radionuclides in the atmosphere. Mokslas, Vilnius
Maloszewski P, Zuber A (1996) Lanped parameter models for the interpretation of environmental tracer data. In: Manual on the Mathematical Models inIsotope Hydrology. Tecdoc No 910, IAEA, Vienna, pp 9–58
Martell EA (1963) On the inventory of artificial tritium and its occurrence in atmospheric methane. J Geophys Res 68:3759–3769
Martin ID, Hackett IP (1974) Tritium in atmospheric hydrogen. Tellus 26:603–607
Martinec J, Siegenthaler U, Oeschger H et al (1974) A new insights into the run-off mechanism by environmental isotopes. In: Isotope techniques in groundwater hydrology: proc symp, vol. 1, IAEA, Vienna, pp 129–142
Mason AS, Öslund HG (1979) Atmospheric HT and HTO: V. Distribution and large-scale circulation. In: Behaviour of tritium in the environment: proc symp, IAEA, Vienna, pp 3–15
Michel RL, Suess HE (1975) Bomb tritium in the Pacific ocean. J Geophys Res 40:4139–4152
Miskel JA (1973) Production of tritium by nuclear weapons'. In: Moghissi F, Carter M (eds) *Tritium*, messenger graphics. Las Vegas, Phoenix, pp 79–85
Miyake Y (1969) Fundamentals of geochemistry. Nedra, Moskva (translation from English)
Mook WG (1977) The radiocarbon time scale. In: Low radioact meas appl: proc symp. High Tatras, Bratislava, pp 193–298
Morkovkina IK (1978) Tritium application in study of groundwater recharge. In: Frronsky VI (ed) Isotopy of natural waters. Nauka, Moskva, pp 165–179
Morkovkina IK (1979) Tritium use in hydrogeological studies. In: Frronsky VI (ed) Isotope investigation of natural waters. Nauka, Moskva, pp 75–84
Münnich KO (1968) Isotopen datierung von Grundwasser. Naturwissenschaften 55:158–163
Münnich KO, Roether W (1967) Transfer of bomb ^{14}C and tritium from the atmosphere to the ocean on the basis of tritium and ^{14}C profiles. In: Radiocarbon dating and methods of low-level counting: proc symp, IAEA, Vienna, pp 93–104
Münnich KO, Vogel JC (1963) Investigation of meridional transport in the troposphere by means of carbon-14 measurement. Radiocarbon dating. Proceedings of a symp, IAEA, Vienna, pp 189–197
Münnich KO, Roether W, Thilo L (1967) Dating of groundwater with tritium and ^{14}C. In: Isotope hydrology: proc. symp., IAEA., Vienna, pp 305–319

Nir A (1964) On the interpretation of tritium age measurements of groundwater. J Geophys Res 69:423–431

Nydal R, Lövseth K, Gulliksen S (1979) A survey of radiocarbon variation in nature since the Test Ban Treaty. In: Radiocarbon dating. University of California, Berkely, pp 313–323

Oeschger H, Siegenthaller U (1979) Prognosis for expected CO_2 increase to fossil fuel combustion. In: Radiocarbon dating. University of California , Berkely, pp 633–642

Östlund HG, Berry E (1970) Modification of atmospheric tritium and water vapour by Lake Tahoe. Tellus 22:463–468

Östlund HG, Fine RA (1979) Oceanic distribution and transport of tritium. In: Behaviour of tritium in the environment. Proceedings of a symposium, IAEA, Vienna, pp 303–312

Östlund HG, Mason AS (1974) Atmospheric HT and HTO: Experimental procedures and tropospheric data 1968–1972. Tellus 26:91–102

Pinneker EV, Romanov VV, Dzyuba AA (1978) The peculiarities of tritium distribution in the near-Baykal Lake natural waters. In: Regional hydrogeology and engineering geology of the Eastern Siberia. Nauka, Novosibirsk, pp 86–92

Plummer LN (2005) Dating of young groundwater. In: Aggarwal P, Gat J, Froehlich K (eds) Isotopes in the water cycle. Springer, Dordrechht, pp 193–218

Polyakov VA, Golubkova EV (2007) Protection study ogf grounwaters by isotopic and hydrochemical data. Prospect Prot of Miner Resour 5:48–52

Polyakov VA, Seletsky YuB (1978) Radiocarbon and tritium study of groundwater dynamics in the Assel-Kliasinan aquifer at Sudogda River region. Geokhimiya 8:1230–1238

Radnell CL, Aitken MJ, Otlet RL (1979) In situ ^{14}C production in wood. In: Berger R, Suess HE (eds) Radiocarbon dating. University of California Press, Berkely, pp 643–650

Rafter TA, O'Brien BJ (1972) C-14 measurements in the atmosphere and in the South Pacific Ocean. In: Proceedings of the 8th international conf. radiocarbon dating, Lower Hutt, Wellington, p 241

Ralf EK (1972) A cyclic solution for the relationship between magnetic and atmospheric C-14 changes. In: Proc. 8th Intern conf. radiocarbon dating. Lower Hutt, Wellington, pp 76–84

Ralf EK, Klein J (1979) Composite computer plots of ^{14}C dates for treering-dated Bristlecone Pine and Sequoies. In: Radiocarbon dating. University of California Press, Berkely, pp 545–553

Ravoire I, Lorius C, Robert J, Roth E (1970) Tritium content in a firn core from Antarctica. J Geophys Res 75:2331–2336

Roether W, Münnich KO, Östlund HE (1970) Tritium profile at the North Pacific (1969) Geosecs intercalibration station. J Geophys Res 75:7672–7675

Roether W, Münnich KO, Ribbat B, Sarmiento JL (1980) A transatlantic ^{14}C section near 40 °N of F/S Meteor. Ergebnisse A 21:57–69

Romanov VV (1978) Regularities in tritium distribution for natural waters. In: Ferronsky VI (ed) Isotopy of natural waters. Nauka, Moskva, pp 46–89

Romanov VV (1982) Tritium use in study of marine and river water mixing. Water Res 5:22–26

Romanov VV, Kikichev HG (1979) Tririum in atmospheric hydrogen. In: Ferronsky VI (ed) Isotope studies of natural waters. Nauka, Moskva, pp 85–92

Romanov VV, Salnova LV, Seryegina LA (1979) Tritium use in studying dynamics of the Baykal Lake waters. In: Ferronsky VI (ed) Isotope sdudies of natural waters. Nauka, Moskva, pp 46–54

Rooth CG, Östlund HE (1972) Penetration of tritium into the Atlantic thermocline. Deep-sea res 19:481–492

Rowland FS (1959) Ratio of HT/HTO in the atmosphere. J Chem Phys 30:1098–1099

Schell WR (1970) Investigation and comparison of radiogenic argon, tritium and C-14 in atmospheric reservoir. In: Radiocarbon variations and absolute chronology, XII nobel symp. Wiley, NY, pp 447–466

Schell WR, Sauzay G (1970) Global sampling and analysis of tritium and stable isotopes. In: Report to panel on procedures for establisching limits for radionuclides in the sea, IAEA, Vienna

Schell WR, Sauzay G, Payne B (1970) Tritium injections and concentration distribution in the atmosphere. J Geophy Res 75:2251–2266

Schmidt U (1974) Molecular hydrogen in the atmosphere. Tellus 26:78–90

Schlosser P, Stute M, Sonntag C, Münnich KO (1988) Tritium/^{3}He dating of shallow groundwater. Earth Planet Sci Latt 89:353–368
Schlosser P, Shapiro SD, Stute M (2000) Tritium/^{3}He measurements in young groundwater: progress in applications to complex hydrological systems. In: Tracers and modelling in hydrogeology: proc intern conf, IASH Liege, pp 481–486
Scholz TG, Ehhalt DH, Heidt LE, Martell EA (1970) Water vapour, molecular hydrogen, methane and tritium concentrations near the stratopause. J Geophys Res 75:3049–3054
Sehgal BR, Remport HH (1971) Tritium production in fast reactors, containing B_4C. Trans Amer Nucl Soc 14:779–780
Seletsky Yu B, Nechaev VI, Polyakov VA (1979) Radiocarbon as an indicator of groundwater recharge and discharge location. In. In: Ferronsky VI (ed) Isotope studies of natural waters. Nauka, Moskva, pp 111–121
Singer SF (1958) The primery cosmic radiation and its time variations. Prog Cosm Ray Phys 4:205–335
Sobotovich EV, Bondarenko GN, Vetshtein VE et al (1977) Isotope and geochemical estimates of a degree of surface and groundwater interconnection. Naukova Dumka, Kiev
Sokolovsky LG, Polyakov VA, Golubkova EV (2007) Light isotopes of waters of the Asdov-Kuban artesian basin: Conditions of formation and balneological significance. Prospect Protect Miner Resour 5:44–47
Sorokhtin OG (2002) Green house effect: myth and reality. Inf Analit Vest, Rus Center, No 1, Moskva, pp 27–28
Stenhause MJ, Baxter MS (1979) The uptake of bomb ^{14}C in humans. In: Radiocarbon dating. University. California Press, Berkely, pp 324–341
Sternberg RS, Damon PE (1979) Sensitivity of radiocarbon fluctuations and inventory to geomagnetic and reservoir parameters. In: Radiocarbon dating. University of California Press, Berkely, p 691
Stewart GL (1965) Experinces using tritium in scientific hydrology. In: Radiocarbon and tritium dating: proceedings of the 6th intern. conf., USAEC, Washington, pp 645–658
Stiel G, Haendel D, Runge A et al (1979) Isotopenverhältnisse und hydrogeologischen Praxis sowie in der Umwelt. Zeits Ang Geol **25**:9–14
Stuiver M (1965) Carbon-14 content of 18th and 19th century wood: variations correlated with sunshot activity. Science 149:533–535
Stuiver M (1980) ^{14}C distribution in the Atlantic ocean. J Geophys Res 85:2711–2718
Stuiver M, Quay PD (1981) Atmospheric ^{14}C change resulting from fossil fuel CO_2 release and cosmic ray flux variability. Earth Planet Sci Lett 53:349–362
Suess HE (1969) Tritium geophysics as an international research project. Science 163:1705–1410
Suess HE (1970) The three causes of the secular C-14 fluctuations, their amplitudes and time constants. In: Radiocarbon variations and absolute chronology, XII nobel symp. Wiley, New York, pp 595–605
Suess HE (1979) A calibration table for conventional radiocarbon dates. In: Radiocarbon dating. University of California, Berkely, pp 777–784
Sulerzhitsky LD, Forova VS (1966) Radiocarbon in woods from the modern volcanous areas. Dokl. AN UzbSSR 6:1421–1423
Taylor CR (1968) A comparison of tritium and strontium-90 in fallout in the Southern hemisphere. Tellus 20:559–576
Taylor JR, Pefers FE (1972) Tritium transport in LMFBR's. Trans. Am Nucl Soc 15:431–432
Thatcher LL, Payne BR (1965) The distribution of tritium in precipitation over continents and its significance to groundwater dating. In: Radiocarbon and t6ritium dating. Proceedings of the 6th intern conf, USAEC, Washington, pp 604–629
Theodorsson P (1967) Natural tritium in groundwater studies. In: Isotope hydrology: proceedings of a symp, IAEA, Vienna, pp 371–380
Tokarev I, Zubkov AA, Rumynin VG et al (2005) Origin of high $^{234}U/^{238}U$ ratio in post-permafrost aquifer. In: Merkel BJ, Hasche-Berger A (eds) Uranium in the environment, mining impact and consequences. Springer, New York, pp 854–863

Tolstikhin IN, Kamensky IL (1969) Determination of of groundwater age by the T-^{3}He method. Geochen Int 6:810–811

Van Hook WA (1968) Condensed Phase Isotope Effects. Isotopenpraxis 5:161–169

Verhagen BTh, Smith PE, McGregore I et al (1979) Tritium profiles in Kalachari sands as a measure of rain-water recharge. In: Isotope hydrology 1978: proceedings of a symp, vol 2, IAEA, Vienna, pp 733–749

Vlasova LS, Brezgunov VS (1978) The distribution of hydrogen and oxygen isotopic composition in natural brines by model calculations. In: Ferronsky VI (ed) Isotope study of natural waters. Nauka, Moskva, pp 119–139

Webber W (1967) The spectrum and charge composition of the primary cosmic radiation. Encycl Phys 46:181

Weiss W, Roether W (1975) Der Tritium abfluss des Rheins 1961–1973. Dt. Gewasser Kd Mitt 19:1–10

Weiss W, Bullacher J, Roether W (1979) Evidence of pulsed discharge of tritium from nuclear energy installations in Central European precipitation. In: Behaviour of tritium in the environment: proceedings of a symp, IAEA, Vienna, pp 17–30

Wendland WM, Donley DL (1971) Radiocarbon calendar age relationship. Earth Plane Sci Lett 11:135–139

Yang A, Fairhall AW (1972) Variations of natural radiocarbon during the last 2000 years and geophysical mechanism for producing them. In: Proceedings of the 8th intern conf on radiocarbon datin. Lower Hutt, Wellington, pp A44–A54

Young JA, Wogman NA, Thomas CW, Perkins R (1970) Short lived cosmic ray produced radionuclides as tracers of atmospheric processes. In: Radionuclides in the environment. Adv Chem Ser No 93, Am Chem Soc, Washington, pp 506–521

Zavelsky FS (1968) One more clarification to radiocarbon method Dokl. AN SSSR 180:1189–1192

Zlobina VL, Kovalevsky VS, Morkovkina IK et al (1980) On the use of helium and tritium mapping for groundwater recharge study. Water Res 1:166–170

Zuber A (1994) On calibration and validation of mathematical models for the interpretation of environmental tracer data in aquifer. In: Mathematical models and their application to isotope studies in groundwater hydrology, IAEA, Vienna, pp 11–41

Chapter 11
Radiogenic Isotopes in Dating of Natural Waters and Sediments

Abstract The production and distribution of radiogenic (natural) radioisotopes and the distribution and separation of uranium, thorium and radium isotopes (^{238}U, ^{235}U, ^{234}U, ^{232}Th, ^{230}Th, ^{238}Th, ^{224}Ra, and ^{226}Ra) in natural waters are discussed in this chapter. Methods of dating surface waters, groundwaters, closed reservoirs and bottom sediments are analysed. The values of the radiogenic isotopes in waters are practically independent of chemical factors. It is mainly determined by the uranium distribution in rocks of a water-bearing complex. Therefore, the uranium isotope ratio serves as some kind of natural indicator of water of a certain water-bearing complex. This fact makes it possible to determine patterns of natural water filtration; to distinguish the rock of the water-bearing complex, to construct models of groundwater circulation, to establish the mixing proportions of waters of different complexes and to examine the interrelations between waters of different complexes.

11.1 Production and Distribution of Radiogenic Isotopes

At present, more than 20 long-lived radioisotopes of heavy elements are known to exist in the Earth's crust. This is evidence of the gigantic processes that resulted in the formation of chemical elements in our galaxy. The main characteristics of these elements are shown in Table 11.1.

Because of the considerable difficulties involved in deriving and measuring very small amounts of radioisotopes, only ^{40}K, ^{87}Rb and isotopes of the uranium-thorium series are of use in practice. But in the future, with the perfection of analytical methods, an increasing amount of natural radioisotopes will be used in the solution of practical problems of isotope geology, hydrology and hydrogeology. Among them the elements of the uranium-thorium series (with atomic numbers from 81 to 92 in the periodic system) are of practical interest. Undergoing numerous sequential nuclear transformations, these elements give rise to three radioactive series (Figs. 11.1, 11.2, and 11.3). The existence in nature of each of these series is determined by the existence of the primary substances, the half-life of which is comparable to the Earth's age. In the uranium-radium series, the parent radioactive isotope is uranium with atomic weight 238 and half-life of $4.51 \cdot 10^{10}$ yr. Uranium-235, characterised by a half-life of $7.13 \cdot 10^{8}$ yr is the originator of own series

V. I. Ferronsky, *Nuclear Geophysics*, Springer Geophysics,
DOI 10.1007/978-3-319-12451-3_11

Table 11.1 Long-lived radioactive isotopes. (Voytkevich 1961)

Parent elements	Daughter elements	Half-life (yr)	Decay type	Relative abundance (%)
^{40}K	^{40}Ar	$1.3 \cdot 10^9$	β (88%)	0.0118
	^{40}Ca	$1.3 \cdot 10^9$	K-capture (12%)	0.0118
^{50}V	^{50}Ti	$6.0 \cdot 10^{15}$	K-capture	0.24
	^{50}Cr	$6.0 \cdot 10^{15}$	β	0.24
^{87}Rb	^{87}Sr	$4.7 \cdot 10^{10}$	β	27.85
^{115}In	^{115}Sn	$5.9 \cdot 10^{14}$	β	95.67
^{123}Te	^{123}Sb	$1.2 \cdot 10^{13}$	K-capture	0.87
^{38}La	^{138}Ba	$1.1 \cdot 10^{11}$	K-capture (70%)	0.089
	^{138}Ce	$1.1 \cdot 10^{11}$	β (30%)	0.089
^{142}Ce	^{138}Ba	$5.0 \cdot 10^{15}$	α	11.7
^{144}Nd	^{140}Ce	$2.4 \cdot 10^{15}$	α	23.8
^{47}Sm	^{143}Nd	$1.0 \cdot 10^{11}$	α	15.1
^{148}Sm	^{144}Nd	$1.2 \cdot 10^{13}$	α	11.35
^{149}Sm	^{145}Nd	$4.0 \cdot 10^{14}$	α	14.0
^{152}Gd	^{148}Sm	$1.1 \cdot 10^{14}$	α	0.205
^{156}Dy	^{152}Gd	$2.0 \cdot 10^{14}$	α	0.057
^{174}Hf	^{170}Yb	$4.3 \cdot 10^{15}$	α	0.163
^{176}Lu	^{176}Hf	$2.2 \cdot 10^{10}$	β	2.588
^{187}Re	^{187}Os	$4.0 \cdot 10^{10}$	β	62.93
^{190}Pt	^{186}Os	$7.0 \cdot 10^{11}$	α	0.0127
^{207}Pb	^{200}Hg	$1.4 \cdot 10^{17}$	α	1.4
^{232}Th	^{208}Pb	$1.39 \cdot 10^{10}$	6α+4β	100
^{235}U	^{207}Pb	$7.1 \cdot 10^8$	7α+4β	0.715
^{238}U	^{206}Pb	$4.51 \cdot 10^{10}$	8α+6β	99.28

and thorium-232 with a half-life of $1.39 \cdot 10^{10}$ yr is the originator of the thorium series. The final decay products in each series are the stable isotopes of lead, ^{208}Pb, ^{207}Pb and ^{206}Pb, respectively.

If there were no separation of elements and isotopes in natural interacting systems, the radioactive series would be at a state of radioactive equilibrium and the content of each element of a series in the subsystems would be determined strictly by the content of the originator of a corresponding series, in accordance with the equation of radioactive equilibrium. But natural systems in a state of radioactive equilibrium are rare. Chemical processes and radioactive decay itself result in the

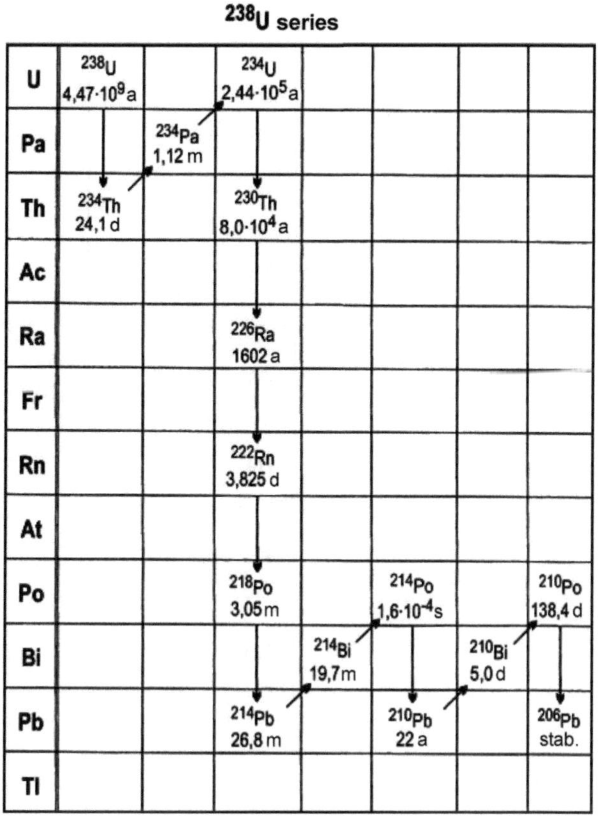

Fig. 11.1 Radioactive uranium-238 series. (Ferronsky and Polyakov 2012)

displacement of equilibrium. A system taken out of the radioactive equilibrium state has a tendency to attain this state again. The observed effect can be employed in studying the temporal characteristics of the processes resulting in this shift. The specific conditions of the formation of isotopic content of heavy radioelements can be of use to solve a number of genetical problems. These problems are related to the origin and characteristics of the shift processes or other geological processes. Let us consider the main chemical and geochemical properties of the radioelements constituting the uranium-thorium series.

Fig. 11.2 Radioactive thorium-232 series. (Ferronsky and Polyakov 2012)

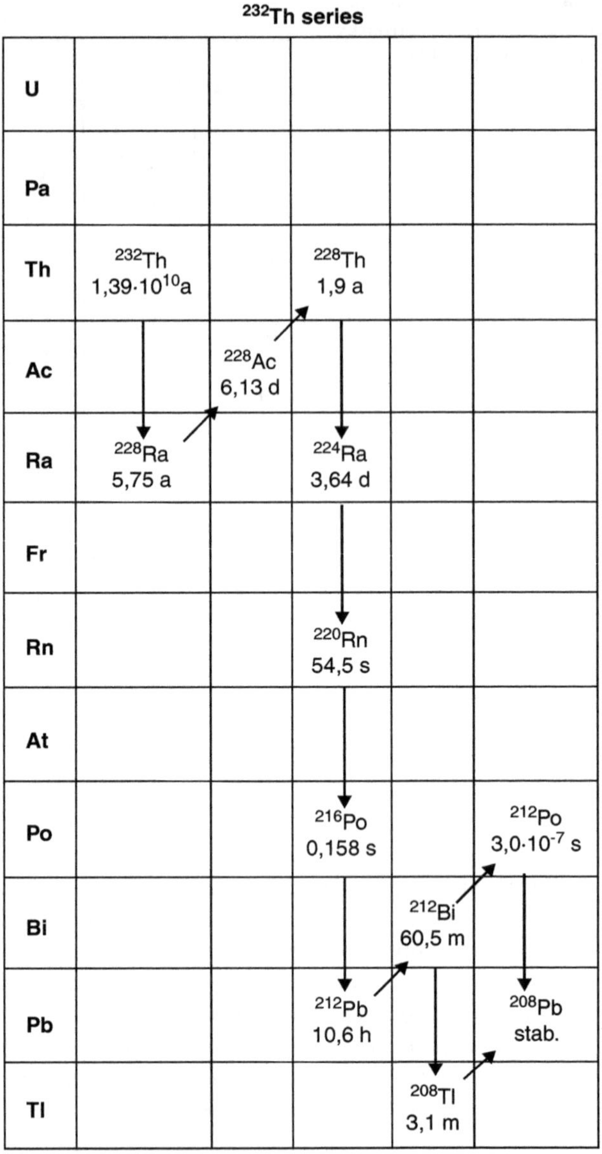

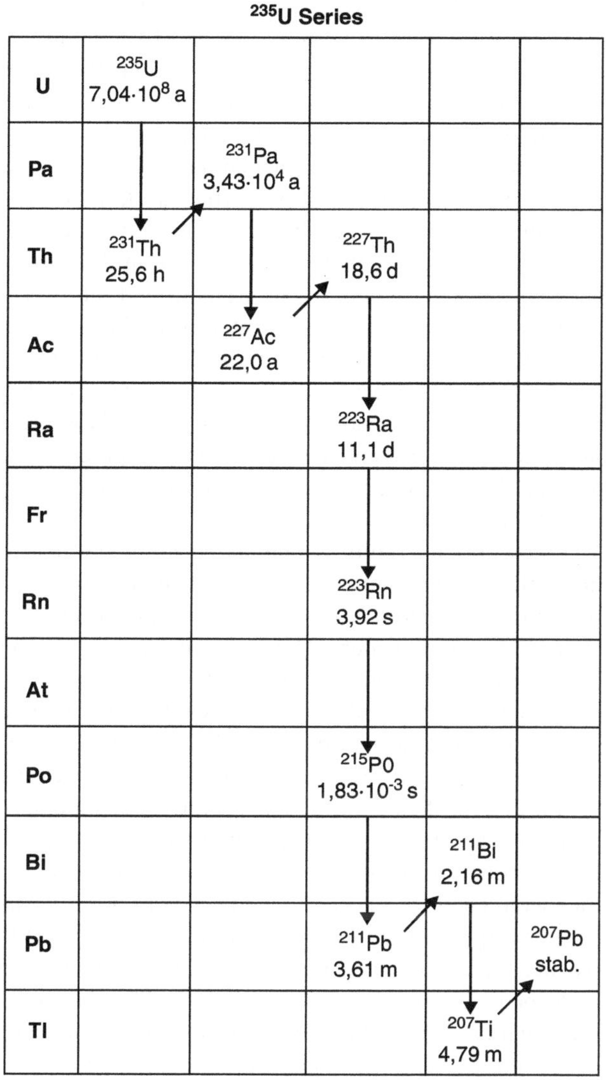

Fig. 11.3 Radioactive uranium-235 series. (Ferronsky and Polyakov 2012)

11.2 Separation of Radiogenic Isotopes

Consider first the main processes resulting in the disequilibrium of radioactive series. According to the idea of Starik (1961), two processes occur at the interaction of solutions with rocks: dissolution, resulting in the extraction of radioisotopes with the destruction of the crystalline lattice; and leaching, resulting in the extraction of radioisotopes without the destruction of the crystalline lattice.

Isotopes extracted by dissolution into natural waters preserve the isotopic composition typical for a given water-bearing layer and the separation of isotopes does not occur in the process. But the separation of elements does occur, since they are released in accordance with the power of formation of soluble compounds. When the radioelements transfer into the solution, the following factors usually affect the solubility: the chemical composition of water, concentration of the pH-ion, the amount of free oxygen, temperature, etc. Uranium and thorium, the parent elements of the radioactive series, are accumulated during magmatic differentiation in acid rocks, where their chemical properties are largely similar. The ratio of thorium to uranium varies over a small range. The four-valent ions of uranium and thorium have approximately equal ionic radii. But in the hypergenesis zone, a sharp difference between the properties of uranium and thorium is observed. Uranium forms readily soluble compounds in water at the last oxidation stages, whereas thorium compounds are practically insoluble and practically no migration of thorium occurs. Therefore, all the secondary formations in the hypergenesis zone are usually enriched in uranium and have little thorium. Therefore, transference of this element only occurs in a suspended form. Radium is practically equilibrated in chloride waters and there is almost no evidence concerning the ability of actinium transference by natural waters.

Extraction of a radioelement by leaching does not result in the destruction of the mineral crystalline lattice. The content and the form of its occurrence in the crystalline lattice are therefore the main factors governing the abundance of an individual element in natural waters. The forms of the parent elements and the decay products present in the mineral crystalline lattice are essentially different. If ^{238}U, ^{235}U and ^{232}Th, the originators of radioactive series, were involved in all the processes (from nuclear fusion to formation of the Solar System and the Earth) and, together with stable elements, had taken part in processes that had resulted in the differentiation of the Earth's substance (formed their individual minerals or are present in them as isomorphic admixtures), then the content of the decay products would have been largely dependent on the content of the parent elements of a corresponding series and they would have been present in the crystalline lattice in another state. On obtaining recoil energy during radioactive decay, a nucleus leaves the crystalline lattice and reaches the micro-capillaries and micro-imperfections of the natural crystals where it can be readily washed out by solutions. They are abundant in rocks in a dispersed form, which is more favourable for extraction by aqueous solutions. On leaching, the solution obtains some portion of the parent isotopes, which make their way into the micro-imperfections and capillaries in the course of the destruction of the crystalline lattice and subsequent diffusion. According to Starik, leaching is the main form of transference of radioelements in natural waters and therefore the decay products should be predominant there. Separation of both the elements and isotopes can occur during the process of leaching. The rate of leaching is largely dependent on the condition of the crystalline lattice (the degree of which it is destroyed). The extraction rate of an element from capillaries and imperfections of the crystalline lattice depends on the chemical properties of an element and on the composition of the leaching solution as well as on the dissolution process. It should be pointed out that the processes of leaching and dissolution do not occur in nature independently. The isotopes of radon, which is a noble gas, are most mobile. In particular cases the radon emanation power reaches 100 %, i.e., almost all the emanated radon is released into natural waters. Numerous experimental data on leaching, obtained both in the laboratory and in the course of studying the content of radioisotopes in natural waters, have shown (Starik and Melikova 1957; Cherdyntsev 1955, 1969) that the separation of radioelements during their transference in natural waters occurs in the following order: Rn > Ra > U > Th > Pa.

11.2.1 Separation of Uranium Isotopes

Among the three natural isotopes of uranium (^{238}U, ^{235}U, ^{234}U) the ratio of abundance of ^{238}U and ^{235}U primordial isotopes is essentially constant. Uranium-234 is a decay product of uranium-238 and, therefore, it has another chemical state in the mineral's crystalline lattice compared with the primordial ^{238}U. In view of this, it should be expected that during leaching uranium-234 will preferentially escape into natural waters. This phenomenon was experimentally found by Cherdyntsev and Chalov in 1953 (Cherdyntsev 1955). Numerous determinations of uranium content in natural waters and minerals deposited from waters have shown that the natural waters exhibit an excess of uranium-234, i.e., the ratio $^{234}U/^{238}U > 1$. During continuous leaching the mineral can become markedly depleted in uranium-234 and then the ratio $^{234}U/^{238}U < 1$. Natural waters, interacting with these minerals, can have uranium isotopic ratios less than one, though these waters are very rare. Uranium leached out of uranium minerals usually displays an equilibrium isotopic composition. Syromyatnikov (1961) accounted for this effect in terms of isotopic exchange. In uranium minerals, the uranium-234 atoms exchange with uranium 238 atoms and substitute for them after radioactive recoil. With increasing dispersion, this effect should be markedly reduced. In effect, the greatest value of the uranium isotope ratio has been detected in those waters leaching uranium out of the minerals in which it occurs in the form of isomorphic admixture. It should be pointed out that shifts in isotope abundances of radiogenic (daughter) elements are characterised by deviation from their equilibrium values and their isotope ratios are described in terms of the activity of the daughter and parent isotopes.

Chalov (1959) showed that uranium isotope separation during transition into the liquid phase can be partly accounted for by uranium-234, which is more favourably oxidized since it is not bounded to the crystalline lattice. But this effect is not widespread. A correlation between the six-valent uranium-238 and uranium-234 isotope ratios in minerals and extraction was found by Chalov only for natural uranium oxides.

A suggestion proposed by some authors that ^{234}Th, an intermediate decay product between uranium-234 and uranium-238, can play a role in the variation of the uranium isotope composition, is unlikely to prove correct. The weak migration power of thorium does not allow for it to be leached from minerals in significant amounts, although, as a decay product, it should be preferentially released into a liquid phase compared with the parent isotope of uranium-238. In fact, the $^{234}Th/^{238}U$

ratio in waters of uranium deposits and ore exposures does not exceed 0.01–0.10. In experiments on comparative ^{234}Th and ^{238}U leaching, Syromyatnikov (1961) always obtained $^{234}Th/^{238}U < 1$. Only in a few cases (Cherdyntsev 1955) have elevated values of the $^{234}Th/^{238}U$ ratio been detected in natural waters and these were in extractions from the thorium minerals reported by Syromyatnikov (1961). But in the case of uranium minerals a great number of other thorium isotopes, being carriers of ^{234}Th, are leached.

11.2.2 Separation of Thorium Isotopes

Let us consider now the separation of isotopes ^{232}Th, ^{230}Th and ^{238}Th, which are of interest from a practical viewpoint. The thorium-uranium fraction of rocks, being on average 3.78 units, varies over limited ranges. At equilibrium conditions, the value of the $^{230}Th/^{232}Th$ ratio in activity units is equal to 0.79. But usually this ratio is much greater. Experiments concerned with the relative leaching of ionium and thorium with ferrithorite have shown (Cherdyntsev 1955) that the relative content of $^{230}Th/^{232}Th$ in extraction is 30 times as much as in minerals. Ionium is a decay product, whereas thorium (a primordial element) is incorporated into the mineral's crystalline lattice. Therefore, an elevated ionium content, compared with thorium, should be observed during leaching, which is actually the case in the natural process.

The ^{232}Th and ^{228}Th isotope separation has some peculiarities. Radiothorium is a daughter product of ^{228}Ra, the lifetime of which is 6.7 year. The radiothorium content in natural waters is usually determined by the content of radium, being extremely mobile and an unstable element. In this case, the isotopic shift is conditioned not by the separation of the isotopes but by the separation of the elements radium and thorium.

11.2.3 Separation of Radium Isotopes

The condition of radium accumulation in natural waters can be written as (Cherdyntsev 1969):

$$n_{\mathrm{Ra}} = q_{\mathrm{U}} a (1 - e^{-\lambda_{\mathrm{Ra}} t}),$$

where q_U is the content of the uranium parent isotope in water-bearing rocks; a is the coefficient accounting for the fraction of the daughter isotope transition in water.

In a first approximation, one can consider that the coefficient a is the same for all radium isotopes. Since all radium isotopes are daughter products then their occurrence in minerals is similar. Then the ratio $^{224}Ra/^{226}Ra$ becomes:

$$\frac{\mathrm{Th}}{\mathrm{Ra}} = \frac{\mathrm{Th}}{\mathrm{U}} \frac{1 - e^{-\lambda_{\mathrm{Th}} t}}{1 - e^{-\lambda_{\mathrm{Ra}} t}}.$$

The lifetime of the ^{224}Ra isotope is 3.64 days and, practically, for all the waters the value $1-e^{-\lambda(Ra\text{-}224)t}=1$. For the young waters (with ages less than the radium half-life, being equal to 1600 years) the value is $1-e^{-\lambda Rat}\approx-\lambda_{Ra}t$. Then, the expression given above becomes:

$$\frac{Th}{Ra}=\frac{Th}{U}\frac{1}{1-\lambda_{Ra}t},$$

i.e., large values of this ratio should be expected for young waters. Such waters ($^{224}Ra/^{226}Ra>1590$) were found by Cherdyntsev (1969) as long ago as 1934. But due to the interphase isotope exchange between the water and minerals this ratio is smoothed down quickly to the normal one (Voytkevich 1961). A similar ratio can be written for the radium isotopes $^{228}Ra/^{226}Ra$ (for the young waters with ages of several years):

$$\frac{^{228}Ra}{Ra}=\frac{Th}{U}\frac{\lambda_{Ra\text{-}228}}{\lambda_{Ra}},$$

i.e., the ratio $^{228}Ra/^{226}Ra$ for these waters can reach the limiting value 240, which really occurs in nature (Cherdyntsev 1973). The interphase isotope exchange also smoothes this ratio down to the normal one, equal to the ratio of the activities of the originators of the radioactive series. In ancient waters radium isotope ratios usually correspond to their parent isotope ratios.

The main factors resulting in the separation of isotopes of radioactive elements in nature were first summarised by Cherdyntsev (1955):

1. The difference in the chemical properties of the elements (the shift of the $^{228}Th/^{232}Th$ ratio from equilibrium).
2. The bond of the daughter isotopes to the crystalline lattice is less than that of the parent isotopes (the preferential escape of ionium compared with thorium and uranium-234 compared with uranium-238).
3. The difference in the half-life times: the recoil atoms of the short-lived isotopes result in a quicker saturation of natural solutions (the anomalously high ratios of $^{234}Ra/^{226}Ra$, $^{223}Ra/^{226}Ra$, $^{228}Ra/^{226}Ra$).
4. The interphase isotope exchange results in a shift of the radioisotope ratio towards radioactive equilibrium ($^{234}U/^{238}U$ ratio shifts towards the equilibrium one in a mineral deposit).

11.3 Distribution of Radiogenic Isotopes in Natural Waters

The content of uranium and thorium, the originators of radioactive series, in natural waters is governed by their content and distribution in rocks. Table 11.2 shows their content in the main types of rocks and stone meteorites according to Cherdyntsev (1969).

Table 11.2 Uranium and thorium content in rocks and stone meteorites

Rock of meteorite	Content (10^{-6} g/g)		
	U	Th	Th/U
Stone meteorites (Chondrites)	0.006–0.03	0.03–0.08	4.0
Basalts	0.59	2.7	4.6
Gabbro	0.96	3.9	4.0
Igneous rocks (averaged values)	1.5	5.4	3.6
Granodiorites	2.0	7.8	3.9
Granites	3.0	13	4.3
Sedimentary rocks	3.0	13.3	4.4
Soils	2.9	9.0	3.2

Among all radioisotopes of the uranium-thorium series, the most comprehensive evidence of their occurrence in natural waters is available for uranium and radium; somewhat less common are data concerned with thorium-232, uranium-234 and radon occurrence; an insignificant amount of data are concerned with the isotopes of thorium ^{230}Th and ^{228}Th. There are only a few data on the content of protactinium, mesothorium, polonium and radioactive plumbum-210. All these are conditioned by considerable methodological difficulties related to the extraction and detection of small amounts of heavy radioelements.

In order to determine the content of an isotope, it should be concentrated out of a large volume of water (of an order of hundreds of litres), and normally in field conditions. For the determination of uranium content, just one litre of water is required. A large amount of the available data on uranium-238 content in the hydrosphere was determined in this way.

11.3.1 Uranium Isotopes in Natural Waters

According to Germanov (Vinogradov 1963) the average value of uranium content in atmospheric precipitation is $(2–3)\cdot 10^{-8}$ g/l. The main carriers of uranium in atmospheric precipitation are dust particles. Therefore, the uranium content in precipitation depends markedly on the regional climatic conditions. Rain waters in arid regions, where a great amount of dust is injected into the atmosphere, has a great content of uranium ($n\cdot 10^{-8}$–$2\cdot 10^{-6}$ g/l). At present the artificial radioisotopes formed during thermonuclear tests are present in the atmosphere. The uranium concentration reached $73.3\cdot 10^{-6}$ g/g in atmospheric fallout near Vilnius (Styro et al. 1970). The correlation of high uranium content with that of ^{239}Pu, which is a typical technogenic isotope, in atmospheric precipitation indicates the technogenic origin of uranium. High uranium concentrations are observed in winter months. The uranium-228 falls in the summer are of local character and are conditioned by dust transport from nearby regions. The uranium content in fallout decreases during this season.

The content of dissolved uranium in rivers (according to Germanov) ranges from $3 \cdot 10^{-8}$ to $n \cdot 10^{-5}$ g/l with an average value of $1 \cdot 10^{-6}$ g/l. The climatic conditions have a great effect upon the concentration of uranium. The general mineralisation and uranium content in rivers in arid regions is greater than that of humid regions. The uranium content increases downstream for some rivers (Dniepr, Don, Volga, Syr-Darya, Amu-Darya).

The occurrence of uranium deposits results in an increase of its content in waters of some small rivers and springs up to $n \cdot 10^{-4}$ g/l. The problem of uranium migration in rivers was studied in detail by Baturin and Kochenow (1969) and Baturin (1968). They carried out a complex study on uranium content in dissolved and suspended forms. According to Baturin's data the uranium content in the deltaic deposits of 12 rivers in the former USSR is in the range $(0.4–3) \cdot 10^{-6}$ g/g. The maximum values are typical for suspensions of the Syr-Darya, Don and Volga rivers and the minimum values for those of Severnaya Dvina and Amu-Darya. The average value, determined in relation to the annual solid runoff, is $1.05 \cdot 10^{-6}$ g/g, which is less than half of the average uranium content in the Earth's crust. Baturin found two general principles for the basins of the Black and Caspian seas: (1) the direct dependency of the relative amount of suspended uranium on sediment discharge (the annual average amount of material from a unit area of the watershed) and also direct dependency on the fraction of mechanical denudation, which is considerably less evident; (2) the inverse relationship between average uranium content in water and water discharge (the average annual amount of water from a unit area of the watershed).

A correlation between the content of dissolved uranium and total mineralisation of surface waters is also observed. The estimations carried out by Baturin and Kochenov (1969) showed that the total amount of dissolved and suspended uranium transported by rivers is the ratio 1:1, with variations from 24:1 to 1:9 for an average content of dissolved uranium in a world river runoff of $(0.50–0.55) \cdot 10^{-6}$ g/l. The above given data concerning the world river runoff were obtained by the authors on the basis of generalisations of data obtained for 16 rivers in the former USSR.

The most comprehensive data on uranium content in river discharge were reported by Sackett et al. (1973). According to their data, the average concentrations of uranium for the 10 deepest rivers, giving 40% of the total world runoff into the ocean, are in good agreement with the data obtained by Baturin and Kochenov (1969).

The uranium content in oceanic water is close to $3 \cdot 10^{-6}$ g/l (Vinogradov 1967). On the basis of experimental data Starik and Kolyadin (1957) showed that in ocean waters uranium occurs in an ionic-dispersion state in the form of the firm uranyl-carbonate complex $[UO_2(CO_3)_3]^{4-}$. Only at pH>7.5 and Eh<0.1 V, which are not typical for oceanic waters, does uranium occur in the form of hydrolyses products, which can be absorbed by colloids and larger suspended particles and removed to the floor. The uranium content is somewhat lower in waters of seas fed by rivers flowing in humid regions (for the Baltic Sea $0.8–2.2 \cdot 10^{-6}$ g/l (Baturin 1968)) and uranium content increases in the waters of closed reservoirs (for Aral Sea $30–50 \cdot 10^{-6}$ g/l, for Caspian Sea $10 \cdot 10^{-6}$ g/l (Kochenov and Baturin 1967)). The distribution of uranium over a basin is significantly homogeneous both with depth and area. And only in the near-bottom layers of the Black Sea does the uranium content markedly decrease down to $n \cdot 10^{-7}$ g/l (Baturin et al. 1966). They accounted for the decrease of uranium content by the processes of absorption of uranium from water during contact with sediments in highly reducible media. One of the main factors, in their opinion, is the occurrence of organic substances that can capture uranium during sedimentation from the water thickness contaminated with hydrogen sulphide. The uranium content in suspension in the Indian Ocean (Kuznetsov et al. 1967) ranges from 0.1 to $2 \cdot 10^{-6}$ g/g. But the amount of uranium in suspension contained in 1 l of water is $0.1–1.2 \cdot 10^{-6}$ g/l, i.e., more than 99.9% of uranium in oceanic water is in dissolved form.

The uranium enters the sea and oceans together with suspended and dissolved portions of river runoff. At present the mechanism of precipitation of dissolved uranium on the oceanic floor is not completely understood. To a great extent this is a result of the inadequacy of experimental material. A correlation exists between the content of organic substances and uranium in precipitation. In Baturin's opinion, the major portion of the dissolved uranium precipitates after the suspended material over shelves and continental slopes. The sedimentation of uranium occurs by its extraction with the organic components of sediments. Since the fractions of dissolved and suspended uranium are on average equal in river runoff and the total content of uranium in the suspended and dissolved material corresponds to its content in the rocks of the Earth's crust, then in modern sediments occurs a regeneration of uranium content, which finally approximates to its content in rocks of the Earth's crust. The rate of precipitation of dissolved uranium on the floor can be estimated by its residence time in a basin, which is usually determined as the quotient of the element's total amount in a basin divided by its average annual input as a component of river runoff (Table 11.3).

It follows from Table 11.3 that the uranium residence time in a basin is closely related to the basin's depth. A number of additional sources of uranium have not been accounted for in the estimations (groundwater discharge, atmospheric precipitation, aeolian deposits) due to coastal erosion.

A question of ultimate importance in nuclear geochronology is that of the constancy of uranium content in the ocean in the Tertiary period, since practically all

Table 11.3 Residence time of dissolved uranium in the water of some seas and the oceans as a whole. (Baturin and Kochenov 1969; Nikolaev et al. 1966)

Basin	Average depth (m)	Dissolved uranium content (kg)	Annually recharged dissolved uranium (kg)	Residence time of dissolved uranium in water (year)
Oceans	3800	$4 \cdot 10^{12}$	$(17–20) \cdot 10^{6}$	$(200–300) \cdot 10^{3}$
Black Sea	1200	$(1–1.65) \cdot 10^{9}$	$2.63 \cdot 10^{5}$	4000
Caspian Sea	188	$4 \cdot 10^{8}$	$3 \cdot 10^{5}$	1300
Baltic Sea	54	$3 \cdot 10^{7}$	$2 \cdot 10^{5}$	150
Aral Sea	16	$4 \cdot 10^{7}$	$(3–4) \cdot 10^{5}$	100–130
Azov Sea	6.8	$7 \cdot 10^{5}$	$4.85 \cdot 10^{4}$	14.5

the methods of nuclear geochronology of sea sediments, which use the elements of the uranium-thorium series, are based on it. Indirect data can be obtained on the basis of uranium content in sea sediments, since the age of the latter is considerably greater than the uranium residence time in the world ocean. A number of data are in agreement with the assumption of its constancy in the past (the approximate constancy of the uranium to carbon ratio in samples of modern and Paleozoic shales, the constancy of uranium distribution in vertical columns of the sea and ocean sediments). However, conflicting evidence also exists. Ancient mollusks contain more uranium than their modern counterparts. The ^{230}Th content should decrease with age down the section of a column, described by smooth exponential curves but peaks are found in the ionium curve. In a number of sea ooze samples an excess of ^{230}Th was found, which does not result from the uranium content in the overlying column of sea water. The problem of uranium content in the ocean in the past and present is far from being solved.

The uranium content in lakes, according to Germanov (Vinogradon 1963), is largely determined by climatic zonality and varies from $3 \cdot 10^{-8}$ to $n \cdot 10^{-4}$ g/l with an average value of about $\sim 1 \cdot 10^{-6}$ g/l. The lowest concentrations of uranium were found in high mountain regions and in the lakes of northern latitudes. The highest uranium concentrations are observed in the shallow lakes of arid steppe zones, in the regions where the concentration of uranium is higher. The balance of uranium was estimated for two lakes in mid-Asia. The uranium residence time in Balkhash Lake is 72 years and in Issyk-Kul Lake it is 5400 years (Baturin and Kochenov 1969).

Uranium content ranges the most in groundwaters. This is accounted for by the difference in hydrodynamical regimes and differences in hydrogeochemical situations that condition the transition of radioactive elements in water. Usually most researchers distinguish three hydrodynamical zones over the range of each artesian basin: those of the intensive, reduced and very reduced water exchange. The waters with dissolved oxygen, possessing high oxidation-reduction potential, occur in the zone of intensive water exchange. In this zone, four-valent uranium is oxidised to a six-valent state with transition into a solution. In the zones of reduced and very reduced water exchange, characterised by reducing conditions, the waters have no oxygen but are enriched with hydrogen sulphide and organic substances. The waters of these zones contain uranium in small amounts. Table 11.4 (Tokarev and Shcherbakov 1956) indicates the average, minimum and maximum uranium contents in waters in different water-bearing rocks and in waters of uranium deposits.

The hydrochemical factors have little effect upon the uranium isotope composition in natural waters, which is determined markedly by the form of uranium distribution in rocks of the water-bearing complex. At present there are several hundreds of determinations of the uranium isotope ratio $\gamma = {}^{234}\text{U}/{}^{238}\text{U}$ in natural waters. Despite the fact that obtained data do not throw light upon all the types of natural waters, some general conclusions can already be drawn. Many works have been devoted to the study of uranium isotopic content in atmospheric precipitation. The precipitation, related to the intensive circulation of marine air masses, falls over the Baltic region in the cold period of the year. The uranium isotope ratio during this period ranges from 1.08 to 1.18 (for sea water the average value is 1.15). In the summer months, the isotope ratio is close to that typical of rocks in the Earth's

Table 11.4 Uranium, radium and radon content in some groundwater. (After Tokarev and Shcherbakov 1956)

Water-bearing rocks	Zone of water exchanger	Uranium (g/l)	Radium (g/l)	Radon (10^{-10} Ci/l)
		Max/min/average	Max/min/average	Max/min/average
Sedimentary	Intensive	$2 \cdot 10^{-7}/8 \cdot 10^{-6}/5 \cdot 10^{-6}$	$1 \cdot 10^{-12}/6 \cdot 10^{-12}/2 \cdot 10^{-12}$	1/50/15
	Restrain	$2 \cdot 10^{-8}/6 \cdot 10^{-6}/2 \cdot 10^{-7}$	$1 \cdot 10^{-11}/1 \cdot 10^{-8}/3 \cdot 10^{-10}$	1/20/6
Acid magmatic	Intensive (crust of weathering)	$2 \cdot 10^{-7}/3 \cdot 10^{-5}/7 \cdot 10^{-6}$	$1 \cdot 10^{-12}/7 \cdot 10^{-12}/2 \cdot 10^{-12}$	10/400/100
	Restrain	$2 \cdot 10^{-7}/8 \cdot 10^{-6}/4 \cdot 10^{-6}$	$2 \cdot 10^{-12}/9 \cdot 10^{-12}/4 \cdot 10^{-12}$	8/400/100
Uranium deposit	Intensive (oxidation zone water)	$5 \cdot 10^{-5}/9 \cdot 10^{-2}/6 \cdot 10^{-4}$	$8 \cdot 10^{-12}/2 \cdot 10^{-9}/8 \cdot 10^{-11}$	50/50,000/1000
	Restrain (reduc-tion zone water)	$2 \cdot 10^{-6}/3 \cdot 10^{-5}/8 \cdot 10^{-6}$	$1 \cdot 10^{-11}/8 \cdot 10^{-10}/6 \cdot 10^{-11}$	50/3000/500

crust (0.89–1.07) and indicates the continental origin of the uranium isotopes in atmospheric fallout. According to Cherdyntsev (1969), the uranium isotope ratio in rivers is 1.1–1.4 (45 determinations) and does not change with time. In salty rivers in the United States it is greater (Thurber 1965), being equal to 1.3–2.0 and in the mountainous river Williams-Green it even reaches 6.35.

In river deposits in the USA (Thurber 1965) in some cases $\gamma < 1$, which is accounted for by the greater mobility of uranium-234 atoms during weathering. There are also cases when $\gamma > 1$. Such sediments contain more organic substances. It is possible that, as in the marine sedimentation cycle, sedimentation of dissolved uranium occurs in river waters with organic substances. Due to the low sensitivity of the apparatus used, Cherdyntsev (1955) could not find the uranium isotope shift in ocean waters. Thurber was the first to discover this shift and to find that $\gamma = 1.15 \pm 0.5$ (Thurber 1963). Later on, the value of the uranium isotope shift was defined more precisely by many researchers. Its most probable value for the world ocean is 1.15 ± 0.1 (Cherdyntsev 1969). According to the data of Cherdyntsev, the value of the uranium isotope shift for water of the Red Sea is 1.18 ± 0.1 and 1.2 ± 0.08 for waters of the Azov Sea. In large river mouth regions the uranium isotope ratio can differ considerably from the mean. So, in the Black Sea water near the Bzib River mouth, it is equal to 1.02 ± 01 and in the central parts of the sea 1.17 ± 0.1.

Detailed studies of the average uranium concentrations in the open regions of oceans and sediments were carried out by Ku et al. (1977). By measurements of more than 100 samples they found that uranium contents vary from $(3.27 \pm 0.05) \cdot 10^{-6}$ in Antarctic to $(3.43 \pm 0.04) \cdot 10^{-6}$ in the Arctic. The values of $^{234}U/^{238}U$ in sediments changed within 0.78–1.07 at an average value of 0.93. In Cherdyntsev's opinion (1969), about 25 % of the uranium is leached by sea water from the sediments, leading to a decrease in the uranium isotope ratio in the sea sediments, i.e., the dissolved uranium does not precipitate to the bottom but is extracted from the deep sea deposits. With the average uranium content in the ooze being $(2–3) \cdot 10^{-4}$ g/g, the extraction from deposits of 25 % of its content releases into water ~ 1 mg of uranium per 1 cm of sea floor sediment. This corresponds to uranium content in the sea water column several kilometres deep. On the whole, the problem of uranium transference in the ocean is far from being solved at present. According to the data of Rona et al. (1965), the value of γ in sea ooze is 1.16–1.17, i.e., is approximately the same as in the sea water with a mean of 1.08 and does not fall below the equilibrium value.

The uranium ratio in lake waters ranges on the whole between the same limits as for rivers, being equal to 1.1–1.4. This is because the sources are the same for both rivers and lakes. The total number of uranium ratio determinations does not exceed several dozen for lake waters. For Sevan Lake $\gamma = 1.72$. The total amount of determinations of uranium isotopic composition in lake waters does not exceed several tens.

The uranium isotope ratio in groundwaters varies greatly depending upon the type of water-bearing rocks. The uranium isotope composition in the Kazakhstan groundwaters was very intensively studied by Syromyatnikov (1961). The main features of uranium migration were ascertained for waters of igneous, sedimentary and metamorphic rocks and for waters of uranium deposits. Figure 11.4 shows the histograms of the uranium distribution in the above-mentioned types of rocks.

In the igneous rocks 117 wells were studied, where the uranium isotope ratio varied from 2 to 5.5 (80 % of the values lie in the range 2.5–4.0). In waters of sedimentary and metamorphic rocks from data for 82 wells, 85 % of the values fall in the range 1.5–2.5. The uranium content in waters of both types of rocks varies greatly. The hydrochemical characteristics also varied over wide ranges but the isotope composition did not vary greatly. In waters of hydrotherms and sedimentary uranium deposits, uranium is practically at its equilibrium value. At the same time, in the deposits characterised by dispersed uranium distribution, ~90 % of the values of the isotope ratio range from 2.5 to 7.0. In waters of the three uranium deposits no

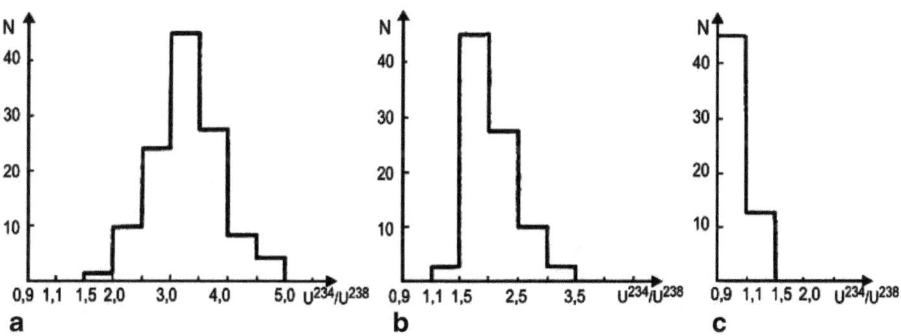

Fig. 11.4 The value of $^{234}U/^{238}U$ ratios in waters from igneous rocks (**a**), sedimentary and metamorphic rocks (**b**) and from uranium deposits (**c**). (Ferronsky and Polyakov 2012)

dependency was found between the uranium isotope ratio and the hydrochemical parameters.

Cherdyntsev (1969) made a summary of the uranium isotope content in some types of waters. In peat soil waters the value of γ varies from 1.12 to 1.78 (12 determinations) and in waters of thermal and carbonic acid springs γ varies from 1.00 to 3.24 (17 determinations). The thermal waters of an active volcanic zone (Kamchatka, Kurile Islands) indicate, on an average of 12 determinations, a uranium isotope ratio of 1.3 (maximum 2.27 and minimum 0.97) (Kuptsov and Cherdyntsev 1969).

11.3.2 Thorium Isotopes in Natural Waters

For convenience of interpretation, the thorium isotope content is sometimes given in the form of the ratios $^{232}Th/^{238}U$, $^{230}Th/^{232}Th$, $^{228}Th/^{232}Th$. There are not many data on these isotope ratios because of technical difficulties involved in thorium extraction and the large amounts of water required for measurements, due to its small content in waters. Thorium is mainly transferred by water in the form of clastic products of destruction and in a colloid-suspended form. Therefore the results of measurements are essentially dependent upon the conditions of its extraction from natural waters.

According to data of Cherdyntsev (1969), the $^{228}Th/^{232}Th$ ratio in rivers of the former USSR ranges on the whole from 0.8 to 1.8. Considerable deviations from the average thorium/uranium ratio are observed for rocks of the Earth's crust (0.8 in activity units). An even greater value of this shift was detected in rivers in the United States (up to 15). The $^{228}Th/^{232}Th$ ratio in 50 % of the studied rivers in the former USSR does not deviate from equilibrium. In all other cases an excess of ^{228}Th (up to 2) is observed. This ratio is markedly different from the equilibrium value (4.6 and 6.8) for the two studied rivers in the USA. The thorium-uranium ratio varies over sufficiently large ranges (0.01–9.2), although it is small for the majority of rivers (in 50 % cases it is less than 0.1). In two cases out of the 27 studied rivers the ratio is greater than equilibrium (> 1.5 and 9.2).

Results of the thorium content measurements in waters of the Sea of Azov were reported in the work by Nikolaev et al. (1961). The thorium concentrations range widely, between $4 \cdot 10^{-9}$ and $219 \cdot 10^{-9}$ g/l. These variations are related to the existence of a large amount of suspended material in the Sea of Azov. Waters sampled from its central part in calm conditions contain $4 \cdot 10^{-9}$ g/l of thorium, which is a feature of the open seas. The maximum thorium content in water sampled from the middle part of the Taganrog Bay, replenished with waters from the Don River, ranges from 218 to $40 \cdot 10^{-9}$ g/l. The thorium concentration in the Black Sea is in the range $(2.2–0.2) \cdot 10^{-9}$ g/l (and remains constant over the studied region) (Starik et al. 1959). According to data reported by Higashi (1959), the thorium content in the Pacific Ocean waters reaches $(2.6–9.1) \cdot 10^{-9}$ g/l but in filtered water thorium was not detected. Assuming, as Vinogradov (1967) has shown, that the average thorium content in ocean water is $n \cdot 10^{-9}$ g/l and that of uranium is $3 \cdot 10^{-6}$ g/l, then in the ocean water the ratio Th/U $\approx$ 0.003, i.e., lower than in the Earth's crust by three orders of magnitude. On the whole the thorium content in water ranges very widely. More accurate measurements of the thorium-232 content in surface ocean waters, carried out by Kaufman (1969), gave a value of $0.07 \cdot 10^{-9}$ g/l.

The ^{230}Th content in the open waters of the Black Sea is $(2.5–4.2) \cdot 10^{-12}$ g/l. The increases to $40 \cdot 10^{-13}$.g/l is near the shores in the delta regions (Starik et al. 1959). In the Pacific Ocean its content is in the range $(6–40) \cdot 10^{-13}$ g/l in surface waters and is $\sim 2 \cdot 10^{-13}$ g/l in deep waters (Higashi 1959). Assuming the average ^{230}Th content is $5 \cdot 10^{-13}$ g/l, then the ocean contains only 1 % of the amount required for equilibrium with uranium. The mechanism of thorium isotope removal to the bottom was studied in detail by Kuznetsov (1962) on the basis of an investigation of the distribution of radioelements in the ocean water suspensions-bottom sediments system. The suspensions were considered to be ocean sediments at the initial stage of their formation. The major portion of thorium is contributed to suspensions as a component of terrigenous material. In addition, absorption extraction of thorium from sea water is also possible. The main processes of ^{230}Th concentration in suspensions are those of biochemical extraction and sorption to particles constituting the suspension.

The geochemical balance of ^{230}Th and ^{232}Th in the basins of the Black Sea and the Sea of Azov was considered by Nikolaev et al. (1966). The residence time of ionium in the Black Sea is 56.6 year, i.e., 66 times less than that in the world ocean. The residence time of ^{230}Th in the Black Sea is 365 year. The authors accounted for the difference in the residence times of ^{230}Th and ^{232}Th in terms of the different forms of their contribution and occurrence both in the river and sea waters and in the bottom sediments. ^{232}Th is considered to be contained in the terrigenous component of the river runoff and ^{230}Th in colloidal form. In the Sea of Azov the ^{232}Th residence time is 160 days and the ^{230}Th 172 days, which is explained by Nikolaev in terms of the shallowness of the basin.

The ratio of ^{228}Th/^{232}Th in waters of the Sea of Azov, according to the data by Nikolaev et al. (1961), is 1.43. In waters of the Pacific Ocean this ratio varies between 10 and 25. It is assumed that the ^{228}Th excess occurs due to ^{228}Ra release during the decay of ^{232}Th, contained in the solid phase (in the ocean sediments and terrigenous material) (Bernat and Goldberg 1969). The cause of deviations is the content of ^{234}Th from equilibrium with ^{238}U (up to 2.8 in activity units) in the upper mixed ocean layers was considered by Bath et al. (1969). The deviations from the equilibrium are explained by the mixing of the upper ocean layers and biological activity.

A vast summary of data on thorium isotope content in groundwater is reported in the work by Syromyatnikov (1961). Usually, the thorium-uranium ratio for groundwaters is considerably lower than that in rivers and lake waters. Radioelements do not typically occur in colloidal form in groundwaters. Sixty water samples from igneous rocks were studied along with 40 from sedimentary rocks and 35 from uranium sediments. For 58 % of water from the igneous rocks, 77 % from sedimentary rocks and 97 % from the uranium ores the thorium-uranium ratio is less than 0.09. In some cases higher values of the thorium-uranium ratio are also indicated. They are likely to occur in waters in contact with those rocks that have a high thorium-uranium ratio. But it has been shown (Titaeva et al. 1973) that the value of this ratio falls at a distance of several dozens of metres from the point of their emergence to the surface, due to decreases of thorium concentration.

The value of the $^{230}Th/^{232}Th$ ratio for groundwaters in general is markedly greater than that typical of the rocks constituting the Earth's crust (~0.8 in the activity units) due to greater ^{230}Th leaching compared with thorium-232. In 95–100% of the studied samples this ratio ranges from < 1 to 4.5. In some cases of rocks with a low thorium-uranium ratio the value of the $^{230}Th/^{232}Th > 15$ (and sometimes it reaches 100). The waters of peats also exhibit a raised $^{230}Th/^{232}Th$ ratio (Cherdynysev 1969), for which a low thorium-uranium ratio is characteristic. The $^{228}Th/^{232}Th$ ratio for groundwaters is practically always > 1. The majority of waters of igneous rocks exhibit a value of this ratio ranging from 4 to 10 and the waters of sedimentary and metamorphic rocks have a ratio ranging from 1.5 to 4.5 (Syromyatnikov 1961).

The protactinium content in natural waters has not been determined in practice. This is due to the fact that the analytical chemistry of the extraction of protactinium from natural objects has only been developed in recent years for dating sea sediments, therefore only data on protactinium content in sea water exist. The content of protactinium in sea water amounts to 3% of its equilibrium amount with uranium (Sackett 1963); in the surface ocean water it is equal to 5%, at a depth of 500 m it is 7%. And of 4000 m it is 5% of the equilibrium (Kuznetsov et al. 1966a). According to Kuznetsov's data the correlation coefficient between the content of protactinium and ^{230}Th in sea sediments is equal to 0.95. Both elements are extracted from sea water by sorption, carried out by poorly soluble particles having highly absorptive surfaces. Protactinium is formed from uranium in sea water and is partly carried there by river runoff.

There are also few data on the actinium content in natural waters. In the Pacific Ocean it amounts to $\sim < 2 \cdot 10^{-15}$ g/l and in the Atlantic Ocean it is $\sim 7 \cdot 10^{-16}$ g/l (when in equilibrium with uranium its content should be $6.4 \cdot 10^{-12}$ g/l) (Sackett 1963). It is transferred from the ocean water into sediments as readily as ^{230}Th and ^{232}Th. The $^{227}Ac/^{235}U$ ratios in water of three springs in the Caucasus neovolcanic region are markedly less than unity (Kuptsov and Chedyntsev 1969). In the Kamchatka thermal waters the ratios range from 0.003 to 12.9.

There are few data on the radium content in river water (they are mainly related to North American rivers). The average radium content is $(0.03–1) \cdot 10^{-12}$ g/l, although in some cases it reaches considerably greater values, for example $(1–3) \cdot 10^{-12}$ g/l in the Mississippi river. The radium content in the Thames (England) is $(0.01–1) \cdot 10^{-12}$ g/l. In the rivers of Germany the radium content ranges from $(0.07–0.8) \cdot 10^{-12}$ g/l. Khristianov and Korchuganov (1971) also reported data on the content of radium in four rivers within the area of the Russian platform (in g/l): the Volga river $0.72 \cdot 10^{-12}$, the Ruza river $0.16 \cdot 10^{-12}$, the Moscow river $0.04 \cdot 10^{-12}$, the Oka river $0.19 \cdot 10^{-12}$. In Vinogradov's (1967) opinion, the average value of radium content in rivers is $0.1 \cdot 10^{-12}$ g/l.

The radium content in the Black Sea amounts to $n \cdot 10^{-13}$ g/l (Grashchenko et al. 1960), $\sim 3.7 \cdot 10^{-14}$ g/l in the surface ocean waters (Broecker et al. 1967) and $1.35 \cdot 10^{-13}$ g/l in the coastal waters (Blanchard and Oakes 1970).

Detailed studies of the uranium content have been carried out (Broecker et al. 1976; Chung et al. 1974; Chung 1974b; Ku and Lin 1976; Chung 1976; Sarmiento and Feely 1976; Li et al. 1973, 1977). These studies have thrown light upon the main principles of radium distribution in the surface and deep waters of all the oceans. The main factors governing the radium content are water circulation and the release of radium from bottom sediments. Radium is contributed to the surface oceanic waters from coastal sediments and to the near-bottom waters from bottom sediments. On the basis of a two-reservoir model (the warm surface waters and the cold deep ones) Li et al. (1973) evaluated the radium flux from coastal sediments, which amounted to $17 \cdot 10^3$ g/year and the flux from the deep bottom sediments, which was $46 \cdot 10^{-3}$ g/year. These processes result in a natural increase of the radium concentrations with depth, which is most noticeable for the Pacific Ocean waters, where the water circulation is retarded compared with the other oceans. In the Arctic and Antarctic waters, where deep waters increase up to $9.3 \cdot 10^{-14}$ g/l (Ku and Lin 1976), concentrations common for surface waters also occur: $(4.2–7.4) \cdot 10^{-14}$ g/l (Chung 1974a). In the surface waters of the Atlantic Ocean the radium content varies over a small range from $(3.0–3.9) \cdot 10^{-14}$ g/l with an average value of $3.5 \cdot 10^{-14}$ g/l. In the near-bottom waters the radium concentration varies over a greater range from $4.4 \cdot 10^{-14}$ g/l to $11.2 \cdot 10^{-14}$ g/l (Ku and Lin 1976).

There are a number of works in which the behaviour of the short-lived radium isotope ^{228}Ra in the oceans has been considered (Moore 1969; Kaufman et al. 1973; Knauss et al. 1978). As well as its long-lived isotope, ^{228}Ra is released into surface waters by diffusion from the continental shelf sediments; it is distributed over the whole ocean. During the decay of the ^{228}Ra the thorium isotope ^{228}Th is formed. Biochemical processes in the surface waters of the ocean result in the separation of the ^{228}Ra and ^{228}Th isotopes. The following relationship always holds for surface waters: $^{228}Ra > ^{228}Th > ^{232}Th$. The radium content in different types of natural waters and in waters of uranium deposits is indicated in Table 11.4. In natural waters the $^{226}Ra/^{238}U$ ratio (in activity units) is usually less than unity, although waters do occur in which radium is in excess compared with uranium.

Extensive studies of the ^{228}Ra content in different types of natural waters were carried out by Syromyatnikov (1961). The majority of samples of rock-waters have a $^{228}Ra/^{226}Ra$ ratio in the range 2–5. Waters of highly destroyed rocks and terrigenous sediments are usually enriched with ^{228}Ra isotopes. A value of the ratio $^{228}Ra/^{226}Ra < 2$ is usually typical of waters surrounding uranium deposits. Very high values of the ratio $^{228}Ra/^{226}Ra$ (up to 380) were found in thermal waters of an active volcanic zone (Kuptsov and Cherdyntsev 1968). A value of the $^{228}Ra/^{226}Ra$ ratio equal to 240, at a thorium-uranium ratio of 3.7, should be expected by the contribution of radium to water by the nuclear recoil mechanism.

Radioactive emanations are released into the atmosphere from the Earth's crust. Their decay products are absorbed by aerosol particles of natural origin and returned to the Earth together with precipitation. The horizontal redistribution of radon relative to its sources in the terrestrial crust occurs in the ocean and atmosphere (Table 11.5).

Table 11.5 Concentration of radon in various regions of the Earth's surface. (Junge 1963)

Region	Concentration of Ra (Ci/cm^3)
Emanation flux, $Ci/cm^2 \cdot sec$	$(0.1–25) \cdot 10^{-17}$ (average $4 \cdot 10^{-17}$)
Soils	$(0.5–10000) \cdot 10^{-13}$ (average $3 \cdot 10^{-13}$)
Near the earth's surface:	
Continental regions	$(70–330) \cdot 10^{-18}$
Continental regions with disturbed	$(400–600) \cdot 10^{-18}$
Soil structure	
South America	$(20–70) \cdot 10^{-18}$
Antarctica	$(0.2–2.0) \cdot 10^{-18}$
Oceans	$(0.5–3.0) \cdot 10^{-18}$
Tn/Rn ratio near the earth's surface	0.01–0.1

The flux of radon emanation depends on the state of the soil. During rain a decrease in its output of 70 % is observed. The temperature and humidity of the soil do not have any effect upon this process. The concentration of radon in the atmospheric layer nearest the Earth depends upon the flux of emanation and turbulent mixing in the overlaying layers. The concentration of radon over inland regions is approximately twice as much as over the ocean, i.e., radon is formed almost completely over the land. The radon concentration in precipitation is about $2 \cdot 10^{-11}$ Ci/cm^3) on average (the minimum value is 0.1, the maximum value is $10 \cdot 10^{-11}$ Ci/cm^3).

Radon content in river waters has not been practically determined. In water supply sources the content of radon is about 100 times greater than in water at equilibrium with radium. In surface continental waters the amount of radon is 10^4 times as much (Khristianov and Korchuganov 1971). On the whole, the concentration of radon in rivers recharged by groundwaters is one order of magnitude greater than in equilibrium with radium. The saturation of groundwaters with radon is conditioned by its high solubility. The major source of radon in groundwaters is that emerging from rocks during emanation. The other sources of emanation (emanation from the atmosphere together with rain, dissolution from the atmosphere, contribution from bottom sediments, from suspensions contained in water, formation from dissolved radium) are relatively unimportant although in a number of cases they should be accounted for. The Rn/Ra ratio (Khristianov and Korchuganov 1971) for the Volga River (Eltsi village) is 96, for the Ruza River (Ruza town) it is 31, for the Moskva River (Zvenigorod town) it is 470 and for Oka River (Shchurovo settlement) it is 42.

In the surface ocean layer, 150 m in depth, the radon concentration gradually increases up to the value relating to equilibrium with radium (Broecker and Kaufman 1970). Table 11.6 indicates the concentrations of radium and radon in the surface ocean waters. The principles of deviations in radium concentration are conditioned by two main factors: the rate of the gas exchange between the ocean and the atmosphere and the rate of vertical mixing. In near-bottom waters the radium concentration decreases with increasing distance from the ocean floor.

Table 11.6 Radium and radon concentrations in the upper (150 m) ocean layer. (Broecker and Kaufman 1970)

Depth	^{222}Rn (dpm/100 kg)	^{226}Ra (dpm/100 kg)
1	5.3	10.1
1	6.0	–
1	5.3	–
50	7.8	–
60	7.5	–
75	–	9.6
150	–	

Bottom sediments with high radium content in the upper layers are important sources of radon, the concentration of which decreases with increasing distance from the sea-floor due to the process of diffusion and mixing (Chung 1974a). In Table 11.7, are the ^{226}Ra and ^{222}Rn concentrations in near-bottom waters based on Chung's data.

The flux of radon from the oceans into the atmosphere is estimated by its relationship with the radium concentration in the surface waters. This flux is rather small, ranging from 11 ± 1 to 260 ± 32 atom/m^{2}sec (Hoang and Servant 1972). According to these authors, the Rn/Ra activity ratios usually amount to 0.2–0.8 in the surface waters.

Radon is released into groundwaters during the emanation of rocks of the water-bearing complex. Diffusion of radon atoms in a solid body is negligible, therefore it mainly releases when radium is located on the surface of the rock (absorbed on the rock). Emanating collectors are thus formed and the radon content in waters can reach large values. In the case of strongly fractured rocks radon escapes from capillaries located deep inside the rocks. The main factors affecting the emanating power are (Tokarev and Shcherbakov 1956): the degree to which the rock is fractured, the temperature (with increasing temperature the emanation power greatly increases), the humidity of the rocks (with increasing humidity emanation drops) and the air pressure (with an increase in pressure emanation decreases).

Despite large variations of the emanation power for individual rock samples (0.01–100 %) the average emanation power in acid magmatic rock ranges from 15 to

Table 11.7 Variation of excess radon concentration with distance from the sea bottom. (Chung 1974a)

Distance from the bottom (m)	^{222}Rn (dpm/100 kg)	^{226}Ra (dpm/100 kg)
100	35.3 ± 0.4	10.8 ± 0.8
80	36.6 ± 0.4	13.6 ± 0.9
60	36.0 ± 0.7	19.3 ± 1.1
50	35.6 ± 0.5	20.7 ± 0.9
40	35.0 ± 0.7	23.1 ± 1.0
30	34.3 ± 0.7	20.0 ± 0.9
20	35.6 ± 0.6	30.4 ± 1.2

30 %. The average for sedimentary and metamorphic rocks is 10–25 %. The higher values of emanating power are in ore deposit formations (32–91 %). The radon content in groundwaters is indicated in Table 11.4 (Tokarev and Shcherbakov 1956).

In some rocks with a low content of radioelements a lower radon content can occur. Thus, in neovolcanic rocks of Caucasus the average uranium content amounts to 3.5) $\cdot 10^{-7}$ g/g (Cherdyntsev et al. 1968). i.e., is almost an order of magnitude lower than the average value typical of rocks of the Earth's crust. The average content of radon in thermal, carbonic acid and fresh waters is extremely low. In the eight water samples studied it is equal to 0.08) $\cdot 10^{-10}$ Ci/l (with a maximum value of $0.29 \cdot 10^{-10}$ Ci/l and a minimum value of $0.01 \cdot 10^{-10}$ Ci/l). An average value of $\sim 1 \cdot 10^{-10}$ Ci/l for the uranium content is observed for the studied products of the active Kamchatka volcanism and Kurile Islands (Kuptsov and Cherdyntsev 1969). But the existence of hot rocks, having high emanating power and liquid magma hearths, lead to raised radon contents (Kuptsov and Cherdyntsev 1968). The average content of radon for 11 fumarol waters of the Kamchatka and Kurile Islands is $4.45 \cdot 10^{-10}$ Ci/l (with a minimum value of 0.05 and a maximum value of $10.8 \cdot 10^{-10}$ Ci/l) (Kuptsov and Cherdyntsev 1968). Considerably higher radon concentrations occur in regions of active volcanism. Thus in the thermal springs of Japanese volcanic regions radon concentrations vary over the wide range (0.3–1200) $\cdot 10^{-10}$ Ci/l, in New Zealand (8–3200) $\cdot 10^{-10}$ Ci/l, in Iceland (3–2000) $\cdot 10^{-10}$ Ci/l and in Kamchatka up to $1000 \cdot 10^{-10}$ Ci/l (Chirkov 1971). The raised radon content can be accounted for by the acid magma hearths being sources of deep emanations. With decreases in temperature the radon content usually drops. The effect of deep emanations is likely to be manifested less in this case.

In the soil air Th/Rn ratios (in equilibrium units) are practically coincidental with the thorium-uranium ratios for water-bearing rocks, since the emanating powers of radon and thoron by rock minerals are approximately equal and the circulation rate of ground air and gaseous currents is not very great. For the eight gas samples from the active volcanic zones of the Kamchatka and Kurile Islands the Th/Rn ratio is low, in all cases being below the sensitivity threshold (the lowest value is less than 0.05). In some volcanic gases of Japan the Th/Rn ratio is elevated to about 10 and sometimes even 370, which may account for the formation of the intermediate collectors of ^{228}Ra–^{228}Ra with a high ^{224}Ra/Ra ratio (Kuptsov and Cherdyntsev 1968).

Radioactive lead ^{210}Pb is formed from radon. The ^{210}Pb concentration in the atmospheric layer near the Earth in England, during the winter months, is about $9 \cdot 10^{-15}$ Ci/kg of air (Pearson et al. 1966). In rain water its concentration from 1961 to 1964 at different points on the globe ranged over (0.2–9.7) $\cdot 10^{-12}$ Ci/l. In the summer months the ^{210}Pb concentration in the near-Earth layers of India increase to about $40 \cdot 10^{-15}$ Ci/kg of air (Joshi and Mahadevan 1967). In the sea water near Cape Town the ^{210}Pb concentration on average is about $38 \cdot 10^{-15}$ Ci/l and varies from 0 to $135 \cdot 10^{-15}$ Ci/l. No variation of ^{210}Pb concentration up to a depth of 600 m has been found (Shannou et al. 1970). It was found that ^{210}Pb is delivered to the surface layer of the Pacific Ocean with a mean velocity of (1–5) $\cdot 10^{-4}$ cm/s (Tsunogai and Nozaki 1971).

More detailed geochemical studies of the behaviour of ^{210}Pb in the oceans were carried out by Tsunogai and Nozaki (1971). The ^{210}Pb isotope arrives at the surface layer of the ocean from the atmosphere, where it is formed during decay of radon. As Nozaki et al. (1976) showed, the flux of ^{210}Pb from the atmosphere along the transect from Tokyo to San-Diego decreases gradually from 1.9 to 0.7 dpm/cm^2 year. The additional source of lead is likely to predetermine the raised value of the $^{210}Pb/^{226}Ra$ ratio. But usually this ratio is considerably lower than the equilibrium one, which is accounted for by the rather short residence time of ^{210}Pb in the surface ocean waters.

In thermal waters of the active volcanic zone of the Kamchatka and Kurile Islands the $^{210}Pb/^{226}Ra$ ratios are subjected to exceptionally large variations (from 2 to 970 + with a mean of 25). Other data of the ^{210}Pb content are not available. There are practically no data on the ^{210}Po content. Between 1961 and 1965 the value of the $^{210}Po/^{210}Pb$ ratio deviated over England in the range 0.05–0.35 (Pearson et al. 1966) and in the sea water near Cape Town from $(8–41) \cdot 10^{-15}$ Ci/l (with a mean of $20 \cdot 10^{-15}$ Ci/l) (Shannou et al. 1970).

In ocean water the $^{210}Po/^{210}Pb$ ratio is usually less than the equilibrium one. The mean value of this ratio in surface ocean waters is equal to 0.5, which corresponds to the average residence time of ^{210}Po in these waters, which is 0.6 year (Nozaki and Tsunogai 1976). On the map plotted by Nozaki et al. (1976), the distribution of the $^{210}Po/^{210}Pb$ ratio is indicated for the surface of the Pacific Ocean. This ratio varies between 0.3 and 0.9 with an average of about 0.5.

11.4 Dating of Surface and Groundwaters

The notion of the age of water is rather ambiguous. The age of water is usually understood to be its residence time in the studied geological object. It is further assumed that either the isotopic composition or radioactive elements change only through radioactive decay or that the law that governs their contribution or removal from water with a definite isotope composition is known. These simple, theoretical suggestions are hard to apply in practice. Therefore, the main criterion of data verification is the comparison of results obtained by different methods, which usually correlate weakly with each other.

11.4.1 Dating of Closed Reservoirs

Cherdyntsev (1969) made the first estimation of groundwater age using radium and radon. The accumulation of radium in groundwaters can be approximately expressed as:

$$n_{Ra} = n_{Ra}^0 (1 - e^{-\lambda_{Ra} t}).$$

When the emanating power is equal to the radium extraction factors, one can put $n_{Ra}^0 = n_{Rn}^0$ (Table 11.8).

Table 11.8 Radon and radium content in some natural waters and their ages. (From Cherdyntsev 1969)

Sampling location and type of sample	Rn, 10^{-10} (Ci/l)	Ra, 10^{-12} (Ci/l)	n_{Ra}/n_{Rn}	Age
Surface waters				
Caucasus:				
Trachiliparites	386	7.3	$1.9 \cdot 10^{-4}$	170 days
Trachiliparites	45	0.5	$1.1 \cdot 10^{-4}$	90 days
Trachiliparites	5.1	0.65	$1.27 \cdot 10^{-3}$	2.8 year
Paleogenic marl	58.0	2.4	$4.1 \cdot 10^{-4}$	0.9 year
Oligocenic lime	18.6	1.3	$7.0 \cdot 10^{-4}$	1.5 year
Quaternary sediments	9.8	0.35	$3.6 \cdot 10^{-4}$	290 days
Quaternary sediments	54.2	0.5	$9.2 \cdot 10^{-5}$	75 days
Kirgizstan:				
Sienites	7.6	0.25	$3.3 \cdot 10^{-4}$	280 days
Deep fresh groundwaters				
Caucasus	3.5	0.79	$2.3 \cdot 10^{-3}$	5 year
Deep thermal groundwaters				
North Caucasus	11.7	9.5	$8.1 \cdot 10^{-3}$	19 year
Dzhermuk, Armenia	0.46	7.0	0.15	380 year
Old Matsesta	5.84	89	0.15	380 year
Old Matsesta	9.9	21	0.21	550 year
Agura	1.27	34	0.27	740 year
Oilfield waters				
Middle Asia	3.0	300	1.0	∞

Then:

$$t = \frac{1}{\lambda_{Ra}} \ln\left(1 - \frac{n_{Ra}^{0}}{n_{Rn}^{0}}\right)$$

This expression is only true as a first approximation. Radium, as a rule, is absorbed by rocks and radon can be lost to the surface or during the movement of gas currents. But this method is undoubtedly useful in distinguishing young waters. Table 16.1 shows the data on ages of a number of surface waters, according to Cherdyntsev.

Chalov (1968) developed a method applicable for the dating of closed basins by disequilibrium uranium. If uranium containing an excess of uranium-234 arrives at some reservoir, then the decay of the excessive amount of the daughter products occurs by the incident component and by the value of the isotope shift in a reservoir, observed at the present time, one can estimate how long a reservoir has been in existence. The disequilibrium uranium arrives at closed reservoirs together with river runoff. In the case of constant contribution or uranium to a reservoir during the whole period of its existence, one can write simple differential equations, reflecting the balance of uranium isotopes, as:

$$\frac{dN_1}{dt} = v(t) - \lambda_1 N_1,$$

$$\frac{dN_1}{dt} = k(t) - \lambda_2 N_2 + \lambda_1 N_1,$$

where N_1 is the number of uranium-238 atoms at time t; N_2 is the number of uranium-234 at time t; $v(t)$, $k(t)$ are the rates of input of the corresponding isotopes into the system.

While solving these differential equations knowledge of the functions $v(t)$ and $k(t)$ is essential.

During his continuous studies, Chalov showed that the ratio of uranium isotopes for one and the same source is a constant. As the object of his studies he chose the rivers of the Issyk-Kul basin. These rivers have shown large seasonal variations in water level at a reasonably steady average annual water discharge, according to yearly observations. Field studies were carried out in the spring, summer and winter, when the hydrodynamic and hydrochemical characteristics are markedly different. The water discharge changed, on average, by nine times and for uranium content by three times. But the deviations of the uranium ratios for a source were within the accuracy of measurements. The average variations of the product of the water discharge and uranium concentrations (being on average 4.8 times in magnitude) prove the seasonal variations in the uranium content to be related not only to the dilution of some waters involved in persistent circulation during the period of melting of glaciers but also (mainly) with the expansion of the domain of uranium leaching from rocks. Thus, Chalov concluded: the uranium isotopic composition of natural waters is only determined by the composition of rocks, subjected to leaching. Therefore, returning to the above-mentioned differential equations we can assume that $v(t)$ and $k(t)$ are alike within a constant factor. In this case these equations can be rewritten as:

$$\frac{dN_1}{dt} = a\varphi(t) - \lambda_1 N_1,$$

$$\frac{dN_1}{dt} = b\varphi(t) - \lambda_2 N_2 + \lambda_1 N_1.$$

In order to solve these equations one further function, $\varphi(t)$, needs to be known. Chalov studied four possible cases: increasing input of uranium into the object over time; time-independent input of uranium; a single contribution of uranium to the object; decrease in the uranium input to the object. In Chalov's opinion, increases in the rate of uranium contribution to the object are unlikely and should be eliminated. The decrease of the rate of uranium contribution to the object can be described by the function $\varphi = ae^{-nt}$, which can describe the two other cases at the following values of the parameter n: at $n \to 0$, which describes the time-independent arrival of uranium and at $n >> \lambda_2$, which is the single contribution of uranium to the object. The parameter n cannot be determined experimentally, since we do not know the law according to which input has occurred in the past. But for the interval over which the assumed age has been varying, $t_{min} = t_n >> \lambda_2$ and $t_{max} = t_{n \to 0}$. Therefore, the age of the studied object can be assumed to be equal to $t = (t_{max} + t_{min})/2$ assuming that a single and unchangeable arrival of uranium is unlikely. For Tertiary (to 1 million year) the error in the age determinations, i.e., deviations of t from t_{max} to t_{min}, increase from 0 to 50 % with increase in age. In some special cases one can evaluate the law by which the arrival of uranium occurred and use the individual formula.

Usually uranium is contributed to a studied object from several sources. The mean value is then determined as:

$$\gamma_0 = \frac{\sum \omega_i \gamma_i}{\sum \omega_i},$$

where ω_i, γ_i are the mass and value of the uranium isotope ratio for an individual source.

The mass is determined by the relative uranium content contributed to the object. The mean annual mass contributed by individual water sources is determined by averaging the seasonal masses P_i:

$$\omega_\iota = \frac{\sum K_i P_i}{\sum P_i},$$

where K_i is the mass of an individual value of P_i, determined by the lifetime during which it can be considered to be constant.

The main formula for the determination of age can be written in the form:

For a single contribution:

$$\frac{\gamma_i - 1}{\gamma_0 - 1} = e^{-\lambda_2 t},$$

and for a time-independent contribution:

$$\frac{\gamma_i - 1}{\gamma_0 - 1} = \frac{1 - e^{-\lambda_2 t}}{\lambda_2 t},$$

where γ_i is the uranium isotopic composition of the studied object; γ_0 is the isotopic composition of uranium arriving at the object.

By this method Chalov et al. (1964) determined the ages of the lakes Issyk-Kul and Chatyr-Kul, Central Asia. The determined age of Issyk-Kul Lake is 110±40 thousand year and that of the Chatyr-Kul is 320±50 thousand year. These determinations are in good agreement with geological considerations. Table 11.9

Table 11.9 Uranium isotopic ratio for waters of Chatyr-Kul Lake and its recharged rivers. (From Chalov et al. 1964)

Sample location	Relative mass of water (%)	$^{234}U/^{238}U$	Average (in wt.) $^{234}U/^{238}U$, recharged into the lake
Kokaigyr River	60	1.21±0.02	
Turgartsu River	40	1.50±0.02	1.33±0.02
Chatyr-Kul Lake	–	1.0	

shows the values of uranium isotope ratios for waters of Chatyr-Kul Lake and its river system with which the basin age has been evaluated.

On the whole, the model assumed by Chalov reflects the process of uranium accumulation in a basin but does not account for some important factors. The main factor is the accumulation of uranium in sediments. The obtained value of the age is not the age of the basin but the residence time of uranium there. Therefore, attempts aimed at age estimation of Aral Sea and Balkhash Lake have failed.

The effect of atmospheric precipitation and aeolian material upon the uranium isotope ratio in Issik-Kul Lake waters was considered by Alekseev et al. (1973). The contribution of the last two uranium sources to lake waters was not considered in Chalov's earlier works.

11.4.2 Dating of Groundwater

An interesting model applicable for the determination of the ages of groundwaters was suggested by Kigoshi (1973). According to his model rain waters are accumulated in a certain reservoir. It is assumed that the rate of contribution of uranium and thorium elements to the liquid phase is permanent during the determined time and the dissolved elements are removed from the liquid phase at a rate proportional to the concentrations of these elements in the liquid phase. With these assumptions, the change in the elements' concentration in a liquid phase can be written as:

$$\frac{dC_{U_8}}{dt} = E_{U_8} - D_U C_{U_8} - \lambda_{U_8} C_{U_8}, \tag{11.1}$$

$$\frac{dC_{U_4}}{dt} = E_{U_4} + \lambda_{T_4} C_{T_4} - D_U C_{U_4} - \lambda_{U_4} C_{U_4}, \tag{11.2}$$

$$\frac{dC_{T_4}}{dt} = E_{T_4} + Q_{T_4} + \lambda_{U_8} C_{U_8} - D_T C_{T_4} - \lambda_{T_4} C_{T_4}, \tag{11.3}$$

where C is the concentration of nuclei in a liquid phase, in atom/ml; U_4, U_8, T_4, T_8 are uranium and thorium, the figure denoting the last figure in the atomic number;

E is the rate of dissolution, in atom/ml; *D* is the rate of nuclei output from the liquid phase, in atoms/year; *Q* is the rate of injection of atoms during the α-decay from the solid phase into the liquid phase by radioactive recoil, in atom/ml year.

In Eq. (11.2) expression $E_{U_4} + \lambda_{T_4} C_{T_4}$ characterises the dissolution of uranium-234, which is in solid phase during the formation of uranium-234 from thorium-234 in the liquid phase. But assuming that all the thorium decay products, being soluble, contribute to the water, the above given expression can be written as $(\lambda_{U_8}/\lambda_{U_4})E_{U_8}/\lambda_{U_4} + Q_{T_4} + \lambda_{U_8} C_{U_8}$. This assumption is rather robust, since it follows from the model experiments that the absorbed uranium is extremely soluble. Therefore, equation (11.2) can be rewritten in the form:

$$\frac{dC_{U_4}}{dt} = \frac{\lambda_{U_8}}{\lambda_{U_4}} E_{U_8} + Q_{T_4} + \lambda_{U_8} C_{U_8} - D_U C_{U_4} - \lambda_{U_4} C_{U_4}.$$

For rather young waters (their age is considerably less than the half-life of uranium-234) the decay process of uranium-234 and uranium-238 can be neglected; uranium is in a relatively steady state in groundwaters and its removal from the liquid phase can also be neglected. Therefore, Eqs. (11.1) and (11.2) become:

$$C_{U_8} = E_{U_8} t, \tag{11.4}$$

$$C_{U_4} = \left[\frac{\lambda_{U_8}}{\lambda_{U_4}} E_{U_8} + Q_{T_4} + \lambda_{U_8} C_{U_8} \right] t. \tag{11.5}$$

Thus, it follows that:

$$\lambda_{U_4} C_{U_4} - \lambda_{U_8} C_{U_8} = \frac{dC_{U_4}}{dt} = \lambda_{U_4} (Q_{T_4} + \lambda_{U_8} C_{U_8}) t. \tag{11.6}$$

The half-life of thorium-234 is small and therefore its concentration can be considered to be unchangeable so, from Eq. (11.3), the expression for the determination of C_{T_4} is:

$$C_{T_4} = \frac{E_{T_4} + Q_{T_4} + \lambda_{U_8} C_{U_8}}{\lambda_{T_4} + D_T}. \tag{11.7}$$

The D_T value can be determined by measuring the activities of ^{228}Ra, ^{228}Th and ^{232}Th in groundwater. In fact, the balance equations for these isotopes are:

$$E_{T_8} + \lambda_{R_8} C_{R_8} = (\lambda_{T_8} + D_T) C_{T_8},$$

$$E_{T_8} = \frac{\lambda_{T_2}}{\lambda_{T_8}} E_{T_2},$$

$$E_{T_2} = D_T C_{T_2}.$$

From the last equations it follows that:

$$\frac{D_T}{\lambda_{T_2}} = \frac{\lambda_{R_8}C_{R_8} - \lambda_{T_8}C_{T_8}}{\lambda_{T_8}C_{T_8} - \lambda_{T_2}C_{T_2}}. \tag{11.8}$$

Assuming that thorium-234 is released into the water phase mainly due to radioactive recoil, $E_{T_4} << Q_{T_4}$, Eq. (16.7) may be rewritten as:

$$(D_T\lambda_{T_4})C_{T_4} = Q_{T_4} + \lambda_{U_8}C_{U_8}. \tag{11.9}$$

Then the expression for the estimation of the age becomes:

$$t = \frac{\lambda_{U_4}C_{U_4} - \lambda_{U_8}C_{U_8}}{\lambda_{U_4}C_{T_4}(D_T + \lambda_{T_4})}. \tag{11.10}$$

Here D_T can be evaluated by Eq. (11.8).

Kigoshi used this model for the determination of the age of groundwaters near Tokyo and Kyoto. The obtained data are in good agreement with those obtained by radiocarbon dating. Table 11.10 shows the main results of the analysis of groundwaters near Tokyo on the basis of which their age has been determined.

The model suggested by Kigoshi has some faults: some of its assumptions are disputable. Kigoshi considers that the excess of uranium-234 in natural waters is conditioned by the large contributions to natural waters of thorium-234. Cherdyntsev and Chalov do not share this belief. But the approach to the problem of groundwater dating by using the isotopes of heavy radioelements is of interest and the attempts being undertaken in this direction will in the future lead to the construction of the perfect model.

Table 11.10 Activity of heavy radioactive elements in groundwaters near Tokyo and water age. (After Kigoshi 1973)

Isotopes	Activity (dpm/80 l water)	Water age by Eq. (16.10) (year)	Radiocarbon age (year)	Radiocarbon age correction (year)
^{234}Th	192±27			
^{232}Th	5.4±7			
^{230}Th	14.7±1.2			
^{228}Th	8.0±0.8	5000±600	6930±140	1350
^{238}U	4.9±0.7			
^{234}U	8.2±0.2			
$^{234}U/^{238}U$	1.68±0.08			
^{228}Ra	8.0±2.0			
$^{234}U_{excess}$	3.3±0.20			

11.5 Dating of Sediments

The isotopes of heavy radioactive elements are most advantageously used in dating sediments of reservoirs. The characteristic feature of the application of heavy radioactive elements is the possibility of dating over a time range up to 1 million year, which cannot be studied by classical radiocarbon, potassium-argon, rubidium-strontium, or lead methods.

11.5.1 Uranium-Uranium Method

As shown above, uranium occurs in natural waters in a disequilibrium state being conditioned by the preferential leaching of uranium-234 as a daughter's product. Therefore, it occurs in precipitation in a disequilibrium state. If subsequent uranium migration does not occur, the system attains a radioactive equilibrium state governed by the radioactive law:

$$\frac{\gamma_t - 1}{\gamma_0 - 1} = e^{-\lambda_2 t},$$

where γ_0 is the initial ratio of the uranium isotopes; γ_t is the uranium isotopes in time t; λ_2 is the decay constant of uranium-234.

The main difficulty in dating while using this method lies in the estimation of the initial uranium isotopic ratio and in proving the absence of the migration of uranium isotopes during the time period considered. Thurber (1963) was the first who used this method for the determination of the age of fossil corals. In fact, during their metabolism, corals accumulate uranium from sea water, characterised by a constant isotope ratio, i.e., the initial uranium isotope ratio is known with sufficient precision. By elevating the uranium isotope shift in the studied specimen of coral one can determine its age. But the first determinations of the uranium isotope ratio showed natural decreases with depth of sampling (Table 11.11).

Experimental data (Cherdyntsev 1969) indicate that at least during the last 200 thousand year the uranium isotope composition of the ocean waters has remained unchanged. There are no reasons to suppose that it differed greatly in more distant times. After death, corals form thick homogeneous accumulations in which uranium

Table 11.11 $^{234}U/^{238}U$ fossil corals of Enivetok Atoll. (After Thurber 1963)

Sampling depth (feet)	$^{234}U/^{238}U$
24–26	1.17±0.01
47–52	1.13±0.03
64–69	1.09±0.01
90–97	1.09±0.01
180	1.07±0.01
700	1.00±0.01

Table 11.12 $^{234}U/^{238}U$ in carbonate material from Bonneville Lake. (After Thurber 1965)

Sampling location	Elevation about lake level (ft)	^{14}C age (10^3 year)	$^{234}U/^{238}U$	$^{234}U/^{238}U_0$	Uranium content (10^{-6} g/g)
Oolites (Tip of cape)	4200	1.5	2.30±0.001	2.30	5.8
Tufa (Peak of Oakveer)	4550	12.9	1.08±0.04	2.11	–
Tufa (Big Gully)	4780	17.7	1.92±0.01	1.95	10
Gastropod (Little valley)	4818	46	1.79±0.03	1.90	10.5
Gastropod (Little valley)	4858	12.4	1.92±0.04	1.94	1.2
Tufa (Leatington)	5040	13	1.90±0.02	1.93	–
Gastropod (Little valley)	4820	14.8	1.79±0.04	1.82	5.0
Gastropod (Big gully)	4800	18.5	1.75±0.05	1.79	6.5
Tufa (Peak of Oakveer)	5100	15.65	1.71±0.03	1.74	–

migration is very improbable. But on the whole the question of uranium migration is not completely understood. According to the data of some authors corals gradually lose uranium due to the recrystallisation of aragonite (Cherdyntsev 1969).

The method of disequilibrium uranium is applicable to the shells of sea mollusks but the probability of uranium migration is greater than in corals since they occur in Tertiary deposits in the form of individual inclusions, where, according to some authors, uranium migrates intensively. The amount of uranium in the shells of living mollusks is small but after their death it increases over several thousands of years. There is a direct possibility of dating the shells of river and lake mollusks, fossil bone and thermal water deposits. But it is hard to evaluate the initial uranium isotope ratio in these cases, since river, lake and groundwaters have lower concentrations than the waters of the world ocean. Thus Thurber, while determining simultaneously the uranium isotope ratio and the age of carbonates in Bonneville Lake, showed that uranium content in lake waters changed in the past, i.e., its water contained uranium of different isotopic composition (Table 11.12)

As was shown earlier, the uranium isotope content is dependent on the composition of the rocks subjected to leaching. Therefore, assuming the water-bearing complex of rocks during the considered interval of time to be unchanged, the uranium isotope composition should also remain unchanged.

Chalov et al. (1966a, b, 1970) showed that in closed reservoirs such as Balkhash Lake and Aral Sea, uranium is carried down to the bottom sediments and securely fixed by them. Knowing the value of the mean uranium isotope ratio contributed by river discharge and the uranium isotope ratio in the lower layers of the bottom sediments one can estimate their age and, therefore, the age of the reservoir. The absolute age of Aral Sea, according to Chalov, is 139± 12 thousand year and that of Balkhash Lake is 37±7 thousand year. Table 11.13 indicates the uranium isotope ratio in the bottom sediments of Balkhash Lake.

The mean value of the $^{234}U/^{238}U$ ratio for waters of the lake is 1.498±0.003. The additional determination of the value obtained for lower layers of the bottom sediments yields 1.462±0.005. Using these two values the age of the lake was determined.

Table 11.13 $^{234}U/^{238}U$ in bottom sediments of Balkhash Lake at maximum depth. (After Chalov et al. 1970)

Sampling depth (m)	$^{234}U/^{238}U$
0–12	1.515±0.013
12–22	1.513±0.013
22–32	1.501±0.015
32–42	1.501±0.009
42–52	1.510±0.014
52–62	1.511±0.013
62–72	1.517±0.012
72–82	1.500±0.017
82–92	1.491±0.008
92–102	1.490±0.009
102–112	1.462±0.009

The main advantage of the disequilibrium uranium dating method is its wide range, including practically all the Tertiary period. The simplicity of the method should also be noted. The age is determined by the age of one isotope, which markedly simplifies the analysis. The sensitivity of the method depends upon the activity of the studied specimens and is to a large extent restricted by the low accuracy available in the determination of the value of the uranium isotope ratio for oceanic waters, 1.15±0.01.

11.5.2 Uranium-Ionium Method

Among the methods based upon uranium-thorium isotope dating the most wide-spread is the uranium-thorium one. The ionium content in natural waters is low and therefore all the secondary formations have practically no ionium at the moment of deposition. But in the course of time uranium-234 is formed in deposits by radioactive decay:

$$\mathrm{Io}(t)=\mathrm{Io}(0)e^{-\lambda_{\mathrm{Io}}t}+{}^{238}\mathrm{U}\left\{\left(1-e^{-\lambda_{\mathrm{Io}}t}\right)+\frac{\lambda_{\mathrm{Io}}}{\lambda_{\mathrm{Io}}-\lambda_{^{238}\mathrm{U}}}\left(\frac{^{234}\mathrm{U}}{^{238}\mathrm{U}}-1\right)\left(1-e^{\left(-\lambda_{\mathrm{Io}}-\lambda_{^{238}\mathrm{U}}t\right)}\right)\right\} \quad (11.11)$$

where Io (U) is the initial ionium content.

The uranium-ionium method is applicable over a time range exceeding that for the uranium-uranium method but within the same range the accuracy and sensitivity of the former is higher, especially over short time intervals. The upper limit of dating provided by this method is dependent upon the value of the measured activity, in practice less than 400–600 thousand year. While estimating the age by the uranium-uranium method the determination of the uranium isotope ratio is required and therefore the age of the studied object can be estimated by the two methods (uranium-uranium and uranium-ionium) simultaneously. A comprehensive summary of the application of the uranium-ionium method was given by Cherdyntsev (1969) in his monograph. Here we shall consider in brief the main objects for which the application of this method is advantageous.

Table 11.14 Uranium and ionium isotope ratios and ionium and uranium-234 ages for fossil corals from the Pacific and Indian oceans. (After Thurber 1965)

Sampling location	$^{234}U/^{238}U$	$Io/^{234}U$	Age (10^3 year)	
			U/U	Io/U
Hawaii, Is. Oahu	1.10±0.015	0.81	140±30	140±50
Tuamotu Is., Anaa	1.12±0.015	0.72	110±20	80±50
Tuamotu Is., Niau	1.09±0.015	0.76	120±20	180±60
Cook Is., Mangaia	1.11±0.014	0.73	110±20	110±50
Western Australia	1.11±0.014	0.81	140±30	110±40
Mauritius	1.11±0.014	0.87	160±40	110±40
Seychelles Is.	1.10±0.014	0.80	140±30	140±50
Seychelles Is.	1.12±0.015	0.78	140±30	80±40

A great number of determinations by the uranium-ionium method have been carried out for sea corals and shells of sea mollusks. In a number of cases the estimations were carried out for one object by two different methods (radiocarbon and uranium-uranium). For the Holocene shells a good coincidence was found between the radiocarbon and uranium-ionium methods but for the ancient samples radiocarbon exhibits lower age values. Comparison of ages determined by the two methods gives a good coincidence (Table 11.14).

Cherdyntsev et al. (1963, 1965) determined the ages of shells of continental mollusks of the Dnestr River and its terraces. The data obtained are in good agreement with geological considerations. Extensive studies of mollusk shells in deposits of lakes Lahontan and Bonneville were carried out by Kaufman and Broecker (1965). The age was determined simultaneously by the radiocarbon, uranium-ionium and radium-uranium methods (Table 11.15).

Table 11.15 The age of the Bonneville Lakes carbonate materials from the $Io/^{234}U$, $Ra/^{234}U$ and ^{14}C methods. (After Kaufman and Broecker (1965))

Sample	$Io/^{234}U$	^{14}C	$Ra/^{234}U$
Big Gully, Utah			
Gastropod	11.0±1.5	17.6±0.6	11.3±2.2
Tufa	21.4±2.2	17.1±0.3	21.2±4.2
Gastropod	10.2±1.1	17.7±0.6	10.6±1.7
Lamington, Utah			
Gastropod	17.5±1.6	14.1±0.31	19.3±1.8
Gastropod	16.8±2.3	14.9±0.3	17.9±3.0
Gastropod	32.8±4.8	17.4±0.4	38.4±5.8
Little Valley, Utah			
Gastropod	8.7±2.2	11.7±0.3	14.3±3.6
Gastropod	16.7±1.4	15.4±0.3	21.4±3.2
Gastropod	18.1±2.3	15.3±0.4	16.9±2.9

The most suitable material for dating by the uranium-ionium method is travertine, which precipitates from natural waters and deposits from karstic caves (Chedyntsev et al. 1965, 1966, 1967). Uranium from carbonaceous waters is transferred into the solid phase with which it becomes strongly bonded. A large number of determinations were carried out by the authors in travertines of Mashuk (Caucasus), Tate and Verteshtsolosh (Hungary), lake tuffs of Karatan (Kazachstan), tuffs of the Leninakan region, Arzni, Djermuk (Armenia), travertines of Ob-I-Garma (Tadjikistan), tuffs of Pleshcheyevo Lake and travertines of Elatma. The majority of age determinations are in good agreement with geological evidence.

The ages determined for travertines and other carbonaceous formations of the karstic caves are in good accord with the archaeological considerations. The karstic caves were inhabited by primitive man. The dating carried out on the carbonaceous formations of the caves was duplicated by determinations of the ages of bones by the utanium-ionium method and radiocarbon dating of charcoal from primitive man's hearths.

A somewhat difficult problem during dating by the uranium-ionium method is the determination of the initial ionium content. Usually it is evaluated by the Io/Th ratio for modern waters in the studied region, or for modern samples. Sometimes the initial value of ionium is determined from the radiocarbon age of a sample.

Cerrai et al. (1965) proposed a method of sediment dating where initial ionium content can be estimated. In fact, by writing down Eq. (16.11) for two different types of deposits (two different minerals) the initial ionium content can be estimated. According to Cerrai's assumption, the Io/Th ratio should be the same for both types of deposits. But it has been shown (Kuptsov and Cherdyntsev 1969) that in some cases, for example for minerals of fumarol fields, the ratio is different for different types of sediments. Therefore, Cerrai's assumption should be investigated further.

The possibility of determining the absolute age of organic river sediments was considered by Titaeva (1966). In order to estimate the possibility of secondary migration of radioelements, studies were carried out in the frozen-rock zone of Lena River terraces. Samples containing no organic substances do not exhibit radioactive disequilibrium. An excess of uranium is observed in the presence of organic substances. In the course of the accumulation of uranium by organic substances it is observed to be in excess relative to the decay product. In order to carry out the dating of sediments, the value of this excessive concentration should be known. The value of thorium concentration in ashes of the upper melted layer of peat was taken as its initial concentration. The ionium content for the ooze sediments of a flood plain was taken to be that determined in modern oozes of the lake, located near the point of sampling. On the basis of these concentrations the equilibrium concentrations were estimated. This value was subtracted from the values of corresponding contents of uranium and ionium in the studied samples, after which their age was determined. The obtained data are in good agreement with the results of radiocarbon dating.

Among other methods of sediment dating in reservoirs, based on the utilisation of heavy elements isotopes, the following methods should be mentioned: radium-uranium (Chalov et al. 1966a, b), ionium-thorium (Starik 1961), ionium-protactinium (Kuznetsov 1962), the method based on the utilisation of lead-210 (Crozaz 1967; Goldberg 1963) and also the decay product tracks (Shukolyukov 1970).

11.6 Radiogenic Isotopes as Indicators of Hydrologic Processes

The value of the uranium isotope ratio in waters is practically independent of hydrochemical factors. It is mainly determined by the uranium distribution in rocks of the water-bearing complex. Therefore, the uranium isotope ratio serves as some kind of natural indicator of water of a certain water-bearing complex. This fact makes it possible to determine patterns of natural water filtration; to distinguish the rock of the water-bearing complex, to construct models of groundwater circulation, to establish the mixing proportions of waters of different water-bearing complexes and to examine the interrelations between waters of different water-bearing complexes. The results of regional studies of uranium isotope ratios in groundwaters in an artesian basin located near Tashkent, being the most intensively studied basin in geological and hydrological respects, were reported by Sultankhodzhaev et al. (1970). On the basis of discovered principles, three zones were distinguished within a basin, which differ from each other by the value of the $^{234}U/^{238}U$ isotope ratios: (1) the zone of infiltration, including the recharge zone and the adjacent area and being characterised by the adequately high isotope ratio 2.0–2.5 (on average 2.1); the waters of this zone belong to the active water exchange region according to hydrological parameters; (2) the transition zone, related to the marginal part of the basin, characterised by uranium isotope ratio values ranging from 1.3 to 3.5 (on average 1.9) and related water motion; (3) the central part of the basin, being the domain of the majority of slow moving waters and characterised by the lowest isotope ratio 1.3–1.8 (on average 1.6).

The above-mentioned zonality in the distribution of the $^{234}U/^{238}U$ ratios corresponds to such hydrochemical indexes as the deuterium concentration, factors of water migration and groundwater age.

Studies of the uranium isotopic composition of the Florida artesian basin were carried out by Kaufman et al. (1969). In their opinion the value of the uranium isotope ratio can be considered as a regional hydrodynamic parameter characterising the aquifer. On the basis of the uranium isotope studies the regional characteristics of permeability of the rocks were determined, the sources of water recharge to the different parts of the basin were found and the vast domain of the Pleistocene waters incorporated into the Florida aquifers was mapped.

Changes in the uranium isotope ($^{234}U/^{238}U$) content in water can only occur during the leaching of uranium from the surrounding rock system and in the course of mixing with other waters containing uranium. Osmond et al. (1968) proposed to apply the value of the content and the isotope of uranium to the determination of the mixing proportions of waters that have different uranium isotope concentrations.

In fact, if water 1, mixing with water 2, gives a mixing flow, then the following expressions are true:

$$V_1 + V_2 = V; \quad M_1 + M_2 = M; \quad M_1\gamma_1 + M_2\gamma_2 = M\,\gamma; \quad M = CV,$$

where V is the volume of water, in l; M is the uranium content, in g; C is the uranium concentration, in g/l; γ is the uranium isotope ratio.

Assuming $V = 1$, one can finally derive the following expression:

$$X = \frac{C}{C_1}\left[\frac{\gamma_1 - \gamma_2}{\gamma_1 - \gamma_2}\right],$$

where X is the fraction of water 1 in the mixed flow.

By measuring the uranium isotope ratios in all three waters and the uranium content in water 1 and in the mixed flow, one can find the proportions in which water 1 and 2 are mixed. This scheme was applied by Osmond et al. (1968) to the construction of groundwater circulation patterns in the Florida artesian basin.

The uranium isotope studies carried out by the authors in waters of the Mirgalimsay complex of ore deposits (Kazakhstan) allowed, using adequately simple assumptions, on estimates it has been assumed that the following waters contribute to the mine waters: surface discharge waters of the rivers Bayaldyr, Birisekh, Kantagi and the water of the Paleozoic shales surrounding the ore body. Analysis of the water mixing proportions was carried out using the above-mentioned formulas. In total, surface discharge amounted to 80%; in a number of mine water tributaries surface waters were absent.

The uranium isotope composition permits one to solve some problems of paleogeography. Thus, estimating the age of Aral Sea (Chalov et al. 1966b) by the uranium disequilibrium method, Chalov managed to obtain an upper limit of the time when the Amu Darya river became a tributary of the Aral Sea (~22 thousand year). Estimating the age of Issik-Kul Lake (Chalov et al. 1964) it was found that the Chu River, one of the main rivers recharging the lake, has not always been contributory. The analysis of the bottom sediments of Lake Balkhash and Issyk-Kul (Chalov et al. 1973) showed that they do not have the same γ values. This fact resulted in the conclusion that both lakes originated and existed independently of each other, despite some geological notions, although the age of the lakes, within the accuracy of the method, is the same.

According to data obtained by Khristianov and Korchuganov (1971) the excess of the radon concentrations over radium concentrations in river waters has been mainly conditioned by groundwater discharge. The groundwaters contribute intensive, natural tracer-radon to the river and therefore studies of the radon content help to solve a number of problems related to deciphering and estimating groundwater discharge into rivers.

The systematic observations of the content of radon in deep groundwaters of tectonically active regions reflect increases in the elastic stresses and changes in the character of the deformations of rocks at great depth. The epicentre of the earthquake represents the focus of the destruction process of rocks when the elastic limit of the medium is exceeded. Besides the main rupture a large number of fissures of various sizes are formed, which result in vibrations of high frequencies. The motion of these elastic waves (of ultrasonic frequencies) through thicknesses of hard rocks and liquids results in the escape of radon into the environment (Gorbushina et al. 1967, 1968). The reasons behind radon ejection are: (1) increase of the emanation factor of rocks with increasing surface of crushed rocks; (2) extraction of gases

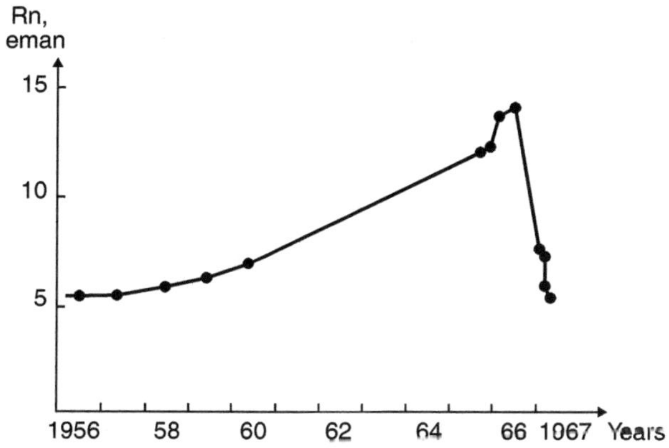

Fig. 11.5 Concentration of radon in thermomineral waters of the Tashkent basin. (Ferronsky and Polyakov 2012)

from rocks effected by ultrasonic and sound vibrations, occurring during and before an earthquake. The amount of escaping gas depends on the frequency and intensity of the sound waves. One of the reasons behind the ejection of gas is its transition from a bounded state to a free state. Under the influence of ultrasonic waves the absorption forces holding the gas to the walls of the pores and fissures become weakened and the gas transfers into a free state. Figure 11.5 indicates the curve of radon content in thermomineral water of the Tashkent basin during 1956–1967. Water samples were taken from a borehole every month. During the 10 years since the beginning of sampling, the radon content increased by a factor of almost 2.5. The radon content dropped sharply after an earthquake on April 26, 1966. Ten days before the 7-ball shock on March 24, 1967, the radon content increased again by 2.5 times and after that dropped again. Besides the variations in the radon content during the Tashkent earthquake sharp changes were also noticed in the value of the uranium isotope ratio in the waters of the Tashkent artesian basin (Gorbushina et al. (1972; Spiridonov and Tyminsky 1971) (Table 11.16).

In those authors' opinion the $^{234}U/^{238}U$ ratios were conditioned by excessive contribution of ^{234}U from rocks to water during the earthquake. The elastic vibrations that occur in rocks during an earthquake favour the enrichment of water in ^{234}U. The consequent decrease of the $^{234}U/^{238}U$ ratio after the earthquake is caused by

Table 11.16 Uranium isotope ratios in groundwaters during and after the Tashkent earthquake. (After Gorbushina et al. 1972)

Sampling well location	$^{234}U/^{238}U$		Ratio II–IV/1967 to VIII/1968
	II–IV/1967	VIII/1968	
Lunacharsky village (Well № 3)	7.0±0.3	1.4±0.1	5.0
Kibray village (Well № 3)	6.2±0.2	1.6±0.1	3.9
Chinbad sanatorium (Well № 7)	3.4±0.1	1.8±0.1	1.9
Tashkent mineral water (Well № 5)	4.2±0.1	1.5±0.1	2.8
Botanik rest home (Well)	3.0±0.1	1.3±0.1	2.3

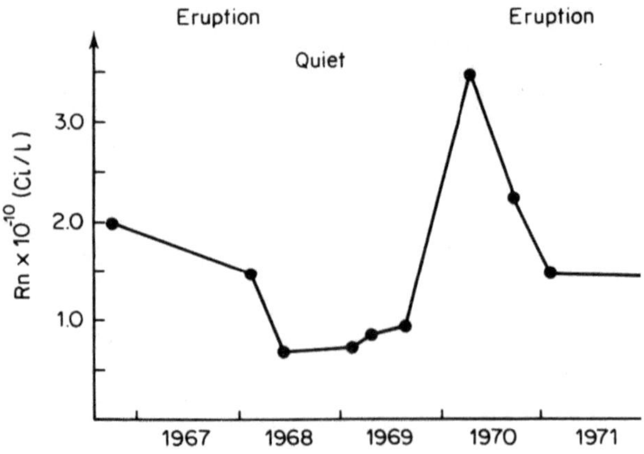

Fig. 11.6 Variation of radon concentration in the thermal spring of the Karymsky volcano. (Ferronsky and Polyakov 2012)

interphase isotope exchange due to which the unsteady equilibrium between uranium in liquid phase and solid phase occurs. Due to this exchange the $^{234}U/^{238}U$ ratio in the waters drops to a value (1.3–1.8) characteristic of waters of a sedimentary complex. The possibility of uranium extraction from rocks into a solution, initiated by ultrasonic vibrations, was discussed by Andreev et al. (1960).

The elastic vibrations of rocks in volcanic interiors that precede an earthquake can result in increases of radon concentrations in the waters of thermal springs of active volcanoes. Chirkov (1971) attempted to study the relationship between volcanic activity and the activity of radon. Systematic observations of radon content were carried out on the spontaneous gas of the thermal spring at the foot of the Karymsky Volcano, Kamchatka. Figure 11.6 shows the change in the radon content in the spring's gases over time.

All the stages of the volcanic activity are included in the period of observations: the final eruption and the transition to the steady stage, the preliminary stage and the beginning of the new eruption. A comparison of this plot with the results of observations on the volcano and data from the Karymskaya seismic station on the incidence of explosive earthquakes have shown that several days after the appearance of the radon anomaly a related anomaly always appears in the volcanic activity. The study of the radon content allows one to forecast both the resumption of the volcanic activity and changes in the type of eruption during activity.

One more problem solved by radiogenic isotopes is paleotermometry based on fission-fragment tracks. The age of minerals estimated by fission tracks is usually lower than in reality. The main cause of this is the burning-off within radioactive damage during geological history. Burning-off is determined by temperature, therefore the content of the fission tracks can provide evidence of the paleothermic conditions in geological objects. The results of the first paleotemperature measurements were reported by Shukolyukov and Komarov (1966). The change in the number of tracks $N\,(t, E)$ in minerals with time can be expressed as:

$$\frac{dN}{dt} = -\lambda_S{}^{238}U + Ce^{-E/RT}N(t,E),$$

where λ_S is the decay constant of uranium-238; E is the burning-off activation energy; R is the gas constant; C is a constant independent of temperature; ^{238}U is the uranium concentration.

Integration of this equation yields:

$$N(t,E) = \frac{\lambda_a{}^{238}U}{\lambda_a - Ce^{-E/RT}} e^{(\lambda_a - Ce^{-E/RT})t} - 1.$$

where λ_a is the uranium α-decay constant; $N(t, E)$ is the measured density of tracks.

There are three unknowns in the expression: age t, temperature T and activation energy E. In order to determine the activity energy, the dependence of the rate of track activation on temperature should be known. Experiments are carried out at two different rates of heating. The burning-off activation energy is estimated from the equation:

$$E = \frac{R(\ln \alpha_1/\alpha_2 + 2\ln T_2/T_1)T_1 T_2}{T_1 - T_2},$$

where T_1 and T_2 are the temperatures at which the track's rate is at a maximum; α_1 and α_2 are the heating rates, in °C/min.

The value C is determined from the expression:

$$\frac{E}{CRd_1 t_1^2} = e^{-E/R\alpha_1 t_1}.$$

After the age of the mineral is determined by means of another method (usually by the potassium-argon method), one can determine the heating temperature T in steady thermal conditions. Using two syngenetic minerals, one can determine both the age and the temperature at steady thermal conditions. In this case one should determine only the value of the activation energy and the frequency factor for each mineral and solve the following two equations (Shukolyukov 1970):

$$t = \frac{2,3}{\lambda_a C_1 e^{-E_1/RT}} \lg\left[\frac{N_1(t,E)}{^{238}U_1}\frac{\lambda_a - C_1 e^{-E_1/RT}}{\lambda_S} + 1\right],$$

$$t = \frac{2,3}{\lambda_a C_2 e^{-E_2/RT}} \lg\left[\frac{N_2(t,E)}{^{238}U_2}\frac{\lambda_a - C_2 e^{-E_2/RT}}{\lambda_S} + 1\right].$$

Against a background of constant low temperatures in nature, short heating periods are evident (magmatic activity, high temperature metamorphism and so on). In this case:

$$N_1 = \left\{\frac{\lambda_S{}^{238}U}{k - \lambda_a}\left[e^{-k\Delta t + \lambda_2 t_2} - e^{\lambda_a t_1}\right] + \frac{\lambda_S}{\lambda_a} U^{238}\left[e^{\lambda_S t} - e^{\lambda_a t_1}\right]\right\} e^{-k\Delta t} + \frac{\lambda_S}{\lambda_a} U^{238}(e^{\lambda_S t_1} - 1),$$

where t_1 and t_2 are the beginning and the end of heating; Δt is the duration of heating; $k = Ce^{-E/RT}$.

Experimental studies of muscovite in the Chai River, Mamsky region, have shown that the age determined by fission tracks ($360 \cdot 10^6$ year) is in good agreement with the age determined by the potassium-argon method. On the basis of the above formulae one can calculate that muscovite has never been heated over 160 °C during a period of $3 \cdot 10^8$ year, or has had temperatures less than 220 °C being heated during 10^7 year or less than 100 °C being heated during the whole period of its existence.

References

Alekseev FA, Bondarev LG, Zverev VL, Spiridonov AI (1973) Influence of precipitation radioactivity on isotopic composition of uranium in the Issyk-Kul Lake in connection with the age determination. Geokhimiya 5:787–780

Andreev PF, Rogozina EM, Rogozin YuM (1960) Extraction of uranium from rocks by ultrasonic action. J Phys Chem 34:2429–2430

Baturin GN (1968) Relationship in forms of uranium migration in some rivers of the USSR territory. Dokl AN SSSR 17:698–701

Baturin GN, Kochenov AV (1969) Uranium migration in rivers and its residence time in waters of oceans, rivers and lakes. Geokhimiya 6:715–723

Baturin GN, Kochenov AV, Kovaleva SA (1966) Some peculiarities of uranium distribution in the Black Seaf water. Dokl AN SSSR 166:698–700

Bath AH, Edmunds WM, Andrews JN (1979) Palaeoclimatic trends deduced from the hydrochemistry of a Triassic sandstone aquifer, United Kingdom. In: Isotope hydrology 1978: proceedings of a symposium, IAEA, Vienna, pp 545–566

Bernat M, Goldberg ED (1969) Thorium isotopes in the marine environment. Earth Planet Sci Lett. 5:308–312

Bhat SG, Krishnaswamy S, Lal D, Moore WS (1969) $^{234}Th/^{238}U$ ratios in the ocean. Earth Planet Sci Lett 5:483–491

Blanchard RL, Oakes D (1970) Relationship between uranium and radium in coastal marine shells and their environment. J Geophys Res 70:2911–2921

Broecker WS, Kaufman A (1970) Near-surface and near-bottom radon results for the 1969 North Pacific Geosecs Station. J Geophys Res 75:7679–7681

Broecker WS, Li YH, Cromwell J (1967) Radium-226 and radon-222 concentration in Atlantic and Pacific oceans. Science 158:1307–1310

Broecker WS, Goddard J, Sarmiento L (1976) The distribution of ^{226}Ra in Atlantic ocean. Earth Planet Sci Lett 32:220–238

Cerrai E, Lonati R, Gazzarini F, Tongeorgi E (1965) Il metodo ionio-uranio per la determinazione della'eta dei minerali vulcanici recenti. Rend Della Soc Mineralog Italia 21:109–115

Chalov PI (1959) Isotope ratio $^{234}U/^{238}U$ in some secondary minerals. Gokhimiya 2:165–170

Chalov PI (1968) Dating by non-equilibrium uranium. Ilim, Frunze

Chalov PI, Tuzova TV, Musin YaA (1964) Isotope ratio of $^{234}U/^{238}U$ in natural waters and its application in nuclear geochronology. Geokhimiya 5:404–413

Chalov PI, Merkulova KI, Tuzova TV (1966a) Absolute age of the Aral Sea determined by non-equilibrium uranium. Dokl AN SSSR 166:89–91

Chalov PI, Merkulova KI, Tuzova TV (1966b) Ratio of $^{234}U/^{238}U$ in water and bottom sediments of the Aral Sea and its absolute age. Geokhimiya 12:1431–1438

Chalov PI, Svetlichnaya NA, Tuzova TV (1970) The results of absolute age determination of Balkhash Lake by nonequilibrium uranium. Dokl AN SSSR 195:190–192

Chalov PI, Svetlichnaya NA, Tuzova TV (1973) Application of nonequilibrium uranium in establishing relationship between continental reservoirs in the past. Geokhimiya 6:897–902

Cherdyntsve VV (1955) Isotopic composition of radioelements in natural objects in connection with their geochronology. In: Annals of the third commission on absolute determination of geologic age. Nauka, Moskva, pp 175–233

Cherdyntsve VV (1969) Uranium-234. Atomizdat, Moscow

Cherdyntsve VV (1973) Nuclear vulcanology. Nauka, Moscow

Cherdyntsev VV, Kazachevsky IV, Kuzmina EA (1963) Isotopic composition of uranium and thorium in zone of hypergeneze. Geokhimiya 3:254–265

Cherdyntsev VV, Kazachevsky IV, Kuzmina EA (1965) The age of Pleistocene carbonate formations by uranium isotopes. Geokhimiya 9:1085–1092

Cherdyntsev VV, Kazachevsky IV, Kislitsyna GI et al (1966) Nonequilibrium uranium in carbonate deposits and their age. Geokhimiya 2:1939–1946

Cherdyntsev VV, Kazachevsky IV, Kuzmina EA et al (1967) Absolute geochronology of Cenozoic deposits. Proc Acad Sci USSR Ser Geol 1:11–20

Cherdyntsev VV, Kuptsov VM, Kuzmina EA, Zverev VL (1968) Radioisotopes and protactinium age of neovulcanic rocks of Caucasus. Geokhimiya 1:77–85

Chirkov AM (1971) ^{222}Rn content in Kamchatka's hydrothems. Dokl AN SSSR 199:202–203

Chung YC (1974a) Transient excess-radon profile in Pacific bottom water. Earth Planet Sci Lett 21:295–300

Chung YC (1974b) Radium-226 and Ra-Ba relationships in Antarctic and Pacific waters. Earth Planet Sci Lett 23:125–135

Chung YC (1976) A deep ^{226}Ra maximum in the northeast Pacific. Earth Planet Sci Lett 32:249–257

Chung YC, Craig H, Ku TL, Goddard J, Broecker WS (1974) Radium226 measurements from three Geosecs intercalibration stations. Earth Planet Sci Lett 23:116–124

Crozaz G (1967) Datation des glaciers par le plomb-210. In: Radioactive dating and methods of low-level count: proc symp. IAEA, Vienna, pp 385–392

Ferronsky VI, Polyakov VA (2012) Isotopes in the Earth's hydrosphere. Springer, Dordrecht

Goldberg E (1963) Geochronology with Lead-210. In: Radioactive dating: pros symp. IAEA, Vienna, pp 121–131

Gorbushina LV, Salmenkova NA, Tyminsky VG (1967) The ages and mixture proportions of mineral waters in the Tashkent artesian basin. Izv Vissh Uch Zaved Ser Geol Razv 2:92–95

Gorbushina LV, Gratsiansky VG, Tyminsky VG (1968) The experience with ultrasound use for recovery of thoron and actinon from solutions. Radiokhimiya 10:495–496

Gorbushina LV, Tyminsky VG, Spiridonov AI (1972) On the mechanism of radiohydrogeological anomalies appearance in seismic regions and their significance in earthquake prediction. Sovetskaya Geologiya 1:153–156

Grashchenko SM, Nikolaev DS, Kolyadin LV et al (1960) Radium content in the Black Sea waters. Dokl AN SSSR 132:1171–1172

Higashi S (1959) Estimation of microgram amount of Th in sea water. J Oceanogr Soc Jpn 15:64

Hoang CT, Servant J (1972) Le flux de radon de la mer. CR Acad Sci 274(24):1321–1349

Joshi LV, Mahadevan TN (1967) Radiochemical determination of lead-210 concentrations in ground level air in India. Nucl Radiat Chem Pros 1:519–523

Junge CE (1963) Air chemistry and radioactivity. Academic, New York

Kaufman A (1969) The ^{232}Th concentration on surface ocean water. Geochim Cosmochim Acta 33:717–724

Kaufman A, Broeker W (1965) Comparison of ^{230}Th and ^{14}C ages for carbonate materials from lakes Lachontan and Bonneville. J Geophys Res 70:4039–4054

Kaufman MI, Rydell HS, Osmond JK (1969) $^{234}U/^{238}U$ disequilibrium as an air to hydrologic study of the Floridian aquifer. J Hydrol 9:374–386

Kaufman A, Trier R, Broecker WS, Feely HW (1973) Distribution of ^{228}Ra in the world ocean. J Geophys Res 78:8827–8848

Khristianov VK, Korchuganov BN (1971) Radon content in the Upper Volga River waters. Geokhimiya 4:492–496

Kigoshi K (1973) Uranium 238/234 disequilibrium and age of underground water. Working paper of the IAEA panel meet, IAEA, Vienna

Knauss KG, Ku TL, Moore WS (1978) Radium and thorium isotopes in the surface waters of the east Pacific and coastal Southern California. Earth Planet Sci Lett 39:235–249
Kochenov AV, Baturin PN (1967) Uranium distribution in the Aral Sea sediments. Oceanology 7:623–627
Ku TL, Lin MC (1976) ^{226}Ra distribution in the Atlantic ocean. Earth Planet Sci Lett 32:236–248
Ku TL, Knauss KG, Mathieu GG (1977) Uranium in open ocean: concentration and isotopic composition. Deep-sea Res 24:1005–1007
Kuptsov VM, Cherdyntsev VV (1968) Radon and thoron in fumarole gases. Dokl AN SSSR 2:436–438
Kuptsov VM, Cherdyntsev VV (1969) Uranium and thorium decay products in the USSR active volcanism. Geokhimiya 6:643–658
Kuznetsov YuV (1962) On the forms of ionium and thorium in the oceans. Geokhimiya 2:177–184
Kuznetsov YuV, Elizarova AN, Frenklikh MS (1966a) Study of sedimentation in oceanic waters by ^{231}P and ^{230}Th isotopes. Radiochemistry 8:459–468
Kuznetsov YuV, Elizarova AN, Frenklikh MS (1966b) Protactinium and thorium content in oceanic waters. Radiochemistry 8:455–458
Kuznetsov Yu, Legin VK, Lisitsin AP et al (1967) Radioactivity of oceanic suspension, 2. Uranium in oceanic suspension. Radiochemistry 9:498–499
Li YH, Ku TL, Mathieu GG, Wolgemuth K (1973) Barium in the Antarctic ocean and implications regarding the marine geochemistry of Ba and ^{226}Ra. Earth Planet Sci Lett 19:352–358
Li YH, Mathieu GG, Biscye P, Simpson HJ (1977) The flux of ^{226}Ra from estuarine and continental shell sediments. Earth Planet Sci Lett 37:237–241
Moore WS (1969) Measurement of ^{228}Ra and ^{228}Th in the sea water. J Geophys Res 74:694–704
Nikolaev SD, Lazarev KF, Grashchenko SM (1961) Thorium isotopes content in the Asov Sea waters. Dokl AN SSSR 138:674–676
Nikolaev SD, Lazarev KF, Korn OP, Drozhin VM (1966) Geochemical balance of radioactive elements in the Black sea and Asov sea basins: 1. Uranium balance. Radiochemiya 11:688–698
Nozaki Y, Tsunogai S (1976) ^{226}Ra, ^{210}Pb, and ^{210}Po distribution in the western North Pacific. Earth Planet Sci Lett 32:313–321
Nozaki Y, Thompson J, Turekian KK (1976) The distribution of ^{210}Pb and ^{210}Po in the surface waters of the Pacific ocean. Earth Planet Sci Lett 32:304–312
Osmond JK, Rydell HS, Kaufman MI (1968) Uranium disequilibrium in groundwater: an isotope delution approach in hydrologic investigations. Science 162:997–999
Pearson DH, Cambray RS, Spiser GS (1966) Lead-210 and polonium-210 in the atmosphere. Tellus 18:427–433
Rona D, Akers LK, Noakes JE, Supernew I (1965) Geochronology in the Gulf of Mexico. Prog Oceanogr 3:289–295
Sackett WM (1963) Geochemistry of ocean water. Trans Am Geophys Union 44:483–485
Sackett WM, Mo R, Sapaldin RF, Exnet ME (1973) A revolution of the marine geochemistry of uranium. In: Radioactive contamination of the marine environment: proc symp, IAEA, Vienna, pp 757–769
Sarmiento JL, Feely HW (1976) The relationship between vertical eddy diffusion and buoyancy gradient in the deep sea. Earth Planet Sci Lett 32:357–370
Shannou LV, Cherry RD, Orren MJ (1970) Polonium-210 and lead-210 in the marine environment cycles. Geochim Cosmochim Acta 34:701–711
Shukolyukov YuA (1970) Uranium nuclear fssion in nature. Atomizdat, Moskva
Shukolyukov YuA, Komarov AN (1966) Possibilities of paleothermometry by uranium fission tracks. Izv AN SSSR Ser Geol 9:137–141
Spiridonov AI, Tyminsky VG (1971) On ^{234}U/^{238}U ratio variation in groundwater. Izv AN SSSR Ser Phys Zem 3:91–93
Starik IE (1961) Nuclear geochronology. Izd AN SSSR. Moscow-Leningrad
Starik IE, Kolyadin LB (1957) On the conditions of uranium existance in oceanic water. Geokhimiya 3:204–213
Starik IE, Melikova OS (1957) Emanation ability of minerals. Trudy Radievogo Inst 5:184–202

Starik IE, Lazarev KF, Nikolaev DS et al (1959) Thorium isotope concentration in the Black Sea waters. Dokl AN SSSR 129:919–921
Styro BI, Shpirkauskayte IK, Kuptsov VM (1970) ^{238}U, ^{232}Th and ^{239}Pu altitude distribution in atmospheric precipitation. At Energ 29:135–136
Sultankhodzhaev AN, Tyminsky GV, Taneev RN (1970) Non-equilibrium uranium in groundwaters of the Tashkent artesian basin. Uzbek Geol J 3:75–77
Syromyatnikov NG (1961) Uranium, thorium and radium isotope migration and interpretation of the radioactive anomalies. Izd AN KazSSR, Alma-Ata
Thurber DL (1963) Natural variations in the ratio $^{234}U/^{238}U$. In: Radioactive dating: proceedings of a symposium, IAEA, Vienna, pp 113–120
Thurber DL (1965) The concentration of some natural radioelements in the waters of the Great basin. Bull Volcanol 28:195–201
Titaeva NA (1966) On the possibilities of orogenic sediments absolute age determination by ionium method. Geokhimiya 10:1183–1192
Titaeva NA, Filonov VA, Ovchenkov VYa et al (1973) Uranium and thorium isotopes behaviour in crystalline rocks-surface water system for cold humid climate conditions. Geokhimiya 10:1522–1528
Tokarev AV, Shcherbakov AV (1956) Radiohydrogeology. Gosgeoltekhizdat, Moskva
Tsunogai Sh, Nozaki Y (1971) Lead-210 and plutonium-210 in the surface water of the Pacific. Geochim J 5:165–173
Vinogradov AP (ed) (1963) The main features of uranium geochemistry. Izd AN SSSR, Moskva
Vinogradov AP (1967) Introduction to ocean geochemistry. Nauka, Moskva
Voytkevich GV (1961) Radiogeology problems. Gosgeoltekhizdat, Moskva

Part IV
Other Applications

Chapter 12
Radioactive Contamination of Natural Waters

Abstract Industrial nuclear technology, which frequently requires the utilisation of very considerable amounts of water, is the main cause of radioactive contamination of natural waters. Surface and underground waters become contaminated with radioactive materials when industrial effluents are discharged into them, or where there is leakage from industrial complexes concerned with the production and enrichment of radioactive materials, treatment of nuclear fuels or the manufacture of fuel rods for industrial and research nuclear reactors. The nature and properties of radioactive effluents, future developments in nuclear technology and disposal of effluents are discussed in this chapter. Migration of radioactive-effluent components through soil and ground, natural minerals and organic sorbents is also analysed.

12.1 Sources of Radioactive Contamination of Water

Industrial technology frequently requires the utilisation of very considerable amounts of water. The associated changes in acidity, salt and bacterial composition, colour, temperature and so on, may be so considerable that the water cannot be used again in the manufacturing cycle and if the regeneration is very expensive or cannot be carried out at all then the water is usually discharged into natural or artificial reservoirs or outlets (seas, lakes, ponds rivers, canals, etc.) or into surface or dipper geologic formations. The surface formations that can be used for this purpose are either natural or artificial basins (enclosed depressions, filtration fields, ditches, shallow drains).

The deeper geological formations that can be used are water-bearing or abandoned oil beds, karst structures, zones of fissuring, crust or weathered crystalline basements and so on. In the latter case, however, special arrangements are necessary (injection wells, deep absorbing drains, and so on). Quite frequently, industrial effluents contain undesirable components in concentrations, even after preliminary purification, considerably exceeding permissible levels. This means that the water must be stored until it has been subjected to the necessary processing. When the volume of this water is very considerable, the only way to store it is to discharge it into geologic formations.

V. I. Ferronsky, *Nuclear Geophysics,* Springer Geophysics,
DOI 10.1007/978-3-319-12451-3_12

In selecting the method for effective and safe storage of industrial effluents, it is first necessary to determine the maximum capacity of the formation for the contamination components and secondly to predict the possible changes in the hydrodynamic, hydrochemical and physiochemical characteristics of the formation due to the presence of the fluid and to estimate the effect of these changes on the behaviour of the contaminants in the medium. Thirdly, it is necessary to develop a storage technology that ensures that the absorbing properties of the formation will be utilised with maximum efficiency. In this chapter we shall largely focus on the determination of the maximum capacity of the formation for the contamination components.

Effluents containing radioactive contaminants, whose toxicity is much higher than that of usual contaminants, occupy a special position among industrial effluents. One of the more important characteristics of water discharged by industry and contaminated with radioactive isotopes is the specific activity. The method of disposal and also the necessity for and the method of preliminary treatment depend on the specific activity and on the isotopic composition of the radioactive contamination.

It is convenient to subdivide effluents into three groups in accordance with their specific activity *a:* (1) low-activity effluents for which *a* is less than 10^{-3} Ci/l; (2) intermediate-activity effluents with *a* between 10^{-3} and 1.0 Ci/l; and (3) high-activity effluents with *a* between 1 and 100 Ci/l.

As a rule, high-activity liquid effluents must be stored in reservoirs specially isolated from the surrounding medium. It is desirable to carry out a preliminary treatment in order to reduce the volume, for example by evaporation. Quite frequently the radioactive contaminants are converted into the solid phase out of concentrated liquid and are then stabilised by cementation, vitrification and so on. They are then removed together with other solid waste materials. Since the volume of intermediate-activity effluents is higher (on average by a factor of five) then the volume of high-activity effluents, the corresponding preliminary treatment designed to reduce the volume and storage become unfeasible in practice. Intermediate-activity effluents that contain short-lived and medium-short-lived components are therefore simply diluted with pure water, or are retained for a sufficient period of time for the specific activity to fall to a level such that the effluent can be discharged into the external medium. If, however, the effluents contain highly toxic long-lived isotopes, the preliminary treatment (precipitation, coprecipitation, absorption, etc.) must be carried out, so that the biologically most dangerous isotopes are partially or completely removed.

Low-activity effluents can be discharged into an external medium, since their volume is higher by a factor of 10–100 than the volume of high-activity effluents. Low-activity effluents are not usually treated, since they contain little or no highly toxic isotopes. If this is not so, then they must be subjected to a preliminary treatment just as the intermediate-activity effluents. Most frequently, the activity of such effluents is due to short-lived isotopes and, therefore, they must be stored for a sufficient period of time for the activity to decay before discharging. When the specific activity of the effluents exceeds by not more than a factor of 100 (the permissible limit for water in open reservoirs in the case of half-lives of up to 60 days) and by a factor of 10 in the case of half-lives in excess of 60 days, the effluents can be discharged into a normal sewage system provided they are diluted tenfold by inactive effluents in the collector of the given enterprise. When the effluents are admitted directly into water, the concentration of the radioactive isotopes should not exceed the permissible limits.

12.1.1 Nature and Properties of Radioactive Effluents

Surface and underground waters become contaminated with radioactive materials when industrial effluents are discharged into them, or when there is leakage from an industrial complex concerned with the production and enrichment of radioactive materials, treatment of nuclear fuels or the manufacture of fuel rods for industrial and research nuclear reactors. Other sources of radioactivity are the reactors themselves, radiochemical centres and “hot” laboratories working on the isolation and concentration of fission and activation products, scientific research centres, individual installations and so on. Nuclear-weapon tests are also known to contribute to the general contamination. As a result of the agreement on the cessation of nuclear tests in the atmosphere, in space and under water, which was signed by the nuclear countries in 1963, the amount of radioactive fallout has decreased considerably but fallout containing long-lived fission products from the pre-1963 tests still continues. Continuing underground nuclear tests may give rise to considerable radioactive contamination of groundwaters under certain hydrogeologic conditions.

The volume and properties of radioactive effluents depend above all on the type of the industry and the technology employed by it. According to Bruce (1960), more than 26 t of uranium were mined and processed per day in the USA in 1959. When uranium is isolated by the acid method (treatment of the ore with H_2SO_4) the amount of fluid effluents per ton of dry ore is about 4 m^3, whereas in the case of the alkaline method (treatment with carbonate or sodium bicarbonate solutions) the effluents amount to above 1 m^3/t of dry ore.

The radioactivity of industrial effluents is determined by the concentration of natural uranium (U_3O_8 up to 0.003–0.01 g/l and occasionally up to 0.1 g/l) and of its daughter products. The most toxic of these are ^{230}Th, ^{234}Th, ^{226}Ra, ^{222}Rn and ^{210}Po. The specific activity of ^{226}Ra in effluents depends on the technology and varies between $8 \cdot 10^{-8}$ and $5.5 \cdot 10^{-6}$ μCi/l. The corresponding ranges for ^{230}Th and ^{234}Th + ^{234}Pa are $9 \cdot 10^{-8}$–$4 \cdot 10^{-5}$ and $1.2 \cdot 10^{-4}$ μCi/l respectively.

From the mining and ore-dressing installations solid concentrates containing about 70% of uranium are transferred to purification plants. During the purification process 1 t of uranium gives rise to about 4.5 m^3 of liquid effluents, which contain natural uranium and its decay products. ^{226}Ra is partly removed from these effluents after treatment with barium sulphate. Liquid effluents from installations manufacturing fuel elements have an analogous isotopic composition but lower specific activity.

Very large volumes of radioactive liquid effluents are produced in the processing of irradiated nuclear fuel elements. For example, in redox-PUREX processes, which are used in the extraction of ^{239}Pu and natural uranium, the volume of high-activity effluents reaches 4.5 m^3/t of uranium, while in the thorax process (extraction of thorium and uranium-235) the volume of the effluents depends on the technology by which the raw material is worked: low salt concentrations are characteristic of effluents from the PUREX process (0.93 mol/l of hydrogen and 0.93 mol/l of nitrates). Higher salt concentrations are characteristic for effluents from other processes. For example, in effluents from the redox process the concentrations of hydrogen, ammonium, sodium and nitrates are 0.3, 1.1 0.23, and 3.05 mol/l respectively. The isotopic composition of the effluents is roughly the same. They include uranium and its fission products, of which the most toxic are ^{90}Sr, ^{137}Cs, ^{95}Zr, ^{95}Ni, ^{144}Ce, ^{144}Pr, ^{106}Ru, ^{147}Pm and ^{147}Sm.

The burn-up of 1 g of nuclear fuel results in about 1 g of fission products, which consist of a mixture of more than 200 radioactive isotopes with mass numbers between 72 and 116. The quantitative yield of a particular isotope depends on its number, the type of fuel and the type of nuclear reaction and may vary between 10^{-7} and 7%. Moreover, the alpha-emitters ^{237}Np, ^{238}Pu, ^{239}Pu, ^{241}Am and ^{242}Cm are produced as a result of neutron irradiation of the nuclear fuel and also of the raw material in breeder reactors. These isotopes can also be present in the effluents, with the exception of ^{238}Pu, which is removed during the precipitation of ^{239}Pu. The effluents' activity varies at various stages of processing from 10 to 15 Ci.l during dissolution and the first extraction to 0.1 Ci/l and less during the salting out and washing of organic solvents.

Some ideas as to the volume of radioactive effluents discharged during the manufacturing and processing of nuclear fuel is given by Table 12.1.

The volume and properties of radioactive effluents from experimental and power reactors depend on the type of reactor. For example, in the case of water-cooled reactors the radioactive effluents consist of water used to cool the fuel elements and to fill the biologic shield, water from basins in which the irradiated fuel elements are stored and water from the various transport channels and circuits not directly connected to the active zone but exposed to a relatively high neutron flux (for example, the second circuit of a reactor with a primary gaseous heat carrier, the steam generator and so on) (Shatts 1964).

Technical or clarified water is usually employed in the technological cycle of reactors because the use of desalinated water is not only expensive but also simply

Table 12.1 Output of effluents in m^3/day of some Western plants. (Gluecauf 1961)

Plant	Processs	High activity		Intermediate activity		Low activity
		Salt free	With high salt content	Salt free	With high salt	
Windscale	PUREX	Up to 10	–	–	10	Up to 20,000
Dounray	Hexane-95	2	2	–	–	10
Hanford	Redox	Up to 10	Up to 10	10	10	10,000

impermissible, owing to the corrosion to which it gives rise. The activity of the effluents originally used for cooling and biological shielding is therefore mainly due to induced activity. In addition to isotopes produced by activation of the stable isotopes, which form the contaminating salts in technical water (^{3}H, ^{16}N, ^{17}N, ^{18}F, ^{19}O, ^{24}Na, ^{27}Mg, ^{32}P, ^{35}S, ^{38}K, ^{38}Cl, ^{45}Ca, ^{47}Ca, ^{49}Ca, ^{51}Cr, ^{52}Mn, ^{56}Mn, ^{59}Fe, ^{64}Cu, etc.), the effluents may also contain activated corrosion products from metals such as ^{51}Cr, ^{52}Mn, ^{56}Mn, ^{59}Fe, ^{58}Co, ^{60}Co, ^{182}Ta, ^{187}W, etc. In high-power reactors the induced specific activity of the coolant may reach up to 10^{-4}–10^{-5} Ci/l and the total gamma activity consists in 76% due to sodium isotopes, 20% due to manganese isotopes and 2% due to cobalt isotopes. After storage for a suitable length time the specific activity falls by a factor of about 10, since the activation products are largely short-lived isotopes (Khonikevich 1964; Bruce 1960).

In the first nuclear power station (built in the USSR) the water in the first circuit has a specific activity of $5 \cdot 10^{-3}$ Ci/l and this is largely due to the short-lived isotopes of oxygen. The overall specific activity is due to intermediate- and short-lived activation products and amounts to 10^{-6} Ci/l (Yakimov 1961). However, the cooling water in the first circuit may be contaminated with nuclear fuel and fission products because the surface of the fuel rod envelopes is contaminated as a result of diffusion through them and similar effects. There is also a certain amount of leakage of fission products and of fuel into the second circuit. The specific activity of the cooling water containing the fuel and the fission products may reach up to 10^{-4}–10^{-3} Ci/l (Gluecauf 1961).

While the reactor is running, the effluents will as a rule contain low-activity contaminants, so that special preliminary treatment before discharge is unnecessary except for cooling over a period of time, which is sufficient for the short-lived isotopes to decay.

In contrast to the radioactive effluents discharged by industry and by reactors, the effluents from scientific-research centres, laboratories and hospitals working with radioactive materials are much more varied and have no definite isotope and salt composition, activity, or volume.

For a salt concentration of up to 1000 mg/l or more the radioactive contaminants in the effluents from scientific-research centres and laboratories may contain the following isotopes: ^{89}Sr (up to 10^{-6} Ci/l), ^{90}Sr ((up to $2 \cdot 10^{-7}$ Ci/l), ^{95}Zr, ^{95}Nb (up to $2 \cdot 10^{-6}$ Ci/l), ^{103}Ru, ^{106}Ru (up to $1.7 \cdot 10^{-6}$ Ci/l), ^{137}Cs (up to 10^{-7} Ci/l) and isotopes of rare earth elements (up to $5 \cdot 10^{-6}$ Ci/l) (Bruce 1960).

Radioactive contaminants in effluents from radiobiological laboratories and hospitals may contain the following isotopes: ^{24}Na, ^{32}P, ^{42}K, ^{51}Cr, ^{82}Br, ^{131}I, ^{198}Au and so on. The specific activities and volumes of the effluents from scientific-research and medical institutions are relatively low.

In all the institutions mentioned above there is one further source of radioactive effluents, namely the water produced during prophylactic deactivation of working rooms, equipment, laundries, sanitation equipment and so on.

The chemical composition of water used for washing is practically constant but the effluents from sanitation devices and especially from laundries contain various complex formers (fluorides, oxalates, sulfonaphthenic preparations, etc.), soaps

Table 12.2 Composition of effluents. (Bol'shakov 1959)

Properties and composition	Laboratory		Laundries
	Radiochemical	Radiobiological	
Specific activity, Ci/l			
a) α-activity	1.10^{-7}	1.10^{-9}	1.10^{-7}
b) β-activity	1.10^{-7}	1.10^{-7}	1.10^{-7}
Acid reaction (pH)	7.8	8.0	5.5
Dry residue, mg/l	639	1050	1755
Including			
Sodium	90	60	360
Potassium	–	30	12
Ammonium	0.2	10	0.2
Calcium	40	80	2
Magnesium	20	20	1
Manganese	–	–	10
Nitrates	100	30	–
Phosphates	–	50	–
Bicarbonates	–	–	10
Oxalates	–	–	70
Sulphates	15	50	90
Sulfonaphthenic acids	40	15	170
Silica	110	100	100
Fatty acids	40	15	160

and other washing agents that considerably complicate preliminary purification. The average characteristics of the effluents from radiochemical and radiobiological laboratories and from special laundries are shown in Table 12.2 (Bol'shakov 1959).

12.1.2 Future Developments in Nuclear Technology and Disposal of Effluents

The development of waste-disposal methods for radioactive industrial effluents is already an acute problem and its importance is likely to increase in the near future.

Some idea about the possible rates of development in nuclear technology can be obtained from data on the current state and the planned developments in nuclear power engineering in the major developed countries.

The first nuclear power plant with a net electrical output of 5 MW started working on June 27, 1954. According to IAEA information, at the beginning of 2013, in 31 countries 437 nuclear power plant units with a capacity of about 372 GW were in operation and 68 plants with a capacity of 65 GW are under construction in 15 countries. The leaders in numbers of operating atomic power plants are USA (104),

France (58), Japan (50), Russian Federation (33), Republic of Korea (23) and India (20). The less developed countries with limited power resources are showing considerable interest in nuclear energy.

The planned increase in the number of nuclear power plants in the future will be accompanied by a corresponding increase in the number of institutions that will mine, process, and enrich radioactive fuel and in the associated problems. This will require a further daily increase of uranium-235 and will produce a yield of a comparable amount of fission products. These are the estimated rates of growth of nuclear power output and the accompanying increase in the yield of radioactive waste in the not too distant future. If we remember the rapid expansion of scientific studies and the applications of radioactive isotopes in medicine, industry, and agriculture, it becomes clear that the volume of radioactive effluents will probably be higher by at least an order of magnitude than the figure just given.

12.2 Migration of Radioactive-Effluent Components Through Soil and Ground

12.2.1 Migration Activity

The migration activity of an element and of its isotopes is determined by the physical and chemical form of the migrant, by the hydrogeologic, hydrochemical and geochemical characteristics of the medium and the hydrophysical, physical and chemical properties of soils, grounds and water-containing rocks.

The most important indicator of the properties of a migration medium is the concentration of the migrating element. Thus, by using the mean concentration by weight of certain elements whose isotopes are present in radioactive effluents, it is possible to estimate their migrational power from the water migration coefficients calculated by Perelman (1961) (Table 12.3).

However, in some localities the migration activity of individual elements or isotopes may be very different from the mean data, depending on the alkali-acid and oxidation-reduction conditions, the type and concentration of typomorphic elements, the concentration, form and properties of organic and inorganic colloids, the form of micro-components under the given conditions and so on. This is confirmed by field and laboratory studies of the mobility of fission products, which give a somewhat different distribution for the migration activity than that determined from

Table 12.3 Migration activity of some elements

Migration activity	Elements
Active migrants	Ca, Fe, Co, Sr, I, Cs, Ba, U
Low-activity migrants	Mo, Sb, Te
Inactive migrants	Y, Zr, Nb, Ru, Ra, Pr, Sn, La

the mean concentration by weight of the corresponding elements (for example, $^{131}I > ^{106}Ru > ^{90}Sr > ^{155}Ce > ^{140}Ba > ^{137}Cs$) (Timofeyeva and Titlyanova 1959; and others).

The ability of natural minerals and organic sorbents to differentiate between various radioactive contaminants is a very important factor governing the rate of migration of the contaminants in the ambient medium. In terms of modern ideas, we can distinguish polar and non-polar sorbents, depending on the nature and specific features of the interaction on absorption.

Absorption by polar sorbents is the stoichiometric replacement of ions capable of exchange and bound by Coulomb forces to the main valences of the sorbent structure by ions in the contacting solution (ion exchange). Ion exchange may be accompanied by certain anomalous processes, which are often very dependent on the magnitude and nature of the absorption and occasionally lead to irreversibility. These processes include chemisorption, non-specific sorption, exchange of the second kind, formation or reduction, enhanced selectivity of the sorbent for a particular ion and so on.

Attempts to estimate quantitatively the role of ion exchange and of the accompanying anomalies during the absorption of radioactive components by natural inorganic and organic sorbents have shown that in most cases the absorption due to ion exchange is considerably greater than absorption due to other processes. This enables us to regard soil-ground media largely as ion exchangers whose absorptive properties are determined by the ion-exchange capacity of their organic and mineral components and can be quantitatively characterised by the exchange capacity per unit weight (Gedroyts 1955). Henceforth we shall use the term 'absorbing power' for soils and grounds to denote the combined set of their ion-exchange and other sorptional properties.

At the present time the main theoretical tools for describing the ion-exchange process are the methods of thermodynamics. However, the application of these methods is restricted by the fact that the various quantities that enter the analysis cannot usually be determined experimentally in an independent fashion. In practice, therefore, it is usual to employ empirical relations borrowed from neighbouring branches of physical chemistry, such as the Langmuir absorption isotherm:

$$\frac{C_{sb}}{S_0} = \frac{aC_{sv}}{1 + aC_{sv}}, \tag{12.1}$$

where C_{sb} and C_{sv} are the equilibrium concentrations of the ion in the sorbent and in the solvent (mole.g); a is a constant that depends on the interaction energy between the ion and the sorbent.

Another useful relationship is the law of mass action. For example, for the exchange of ions having different valences, we have:

$$k = \frac{\bar{C}_A^{Z_B} C_B^{Z_A}}{\bar{C}_B^{Z_A} C_A^{Z_B}}, \tag{12.2}$$

where k is the exchange (equilibrium) constant; $\bar{C}_A$ and $\bar{C}_B$ are concentrations of the exchanging ions in the sorbent (mole/g); C_A and C_B are concentrations of the exchanging ions in the solution (mole/l); and Z_A and Z_B are the ionic valences.

The exchange process is often described with the aid of a distribution coefficient, calculated, for example, from the exchange isotherm. It is numerically equal to the ratio of the ion concentrations in solid and liquid phases:

$$k_d = \frac{\bar{C}_A}{C_A}. \tag{12.3}$$

In the above expression k_d is the distribution coefficient and $\bar{C}_A$ and C_A are the molar equilibrium concentrations of the ion in the sorbent and in the solution respectively.

However, in most cases of exchange on such complicated sorbents as soils there are deviations from the sample ion-exchange regularities. These are due to the specific properties of the sorbent (structural properties, affinity and selectivity, ionic form, tendency to the formation of complexes and so on) as well as the specific properties of the ion (size, charge, degree of hydrogen, form of existence, etc.). Moreover, as already noted, ion exchange is accompanied by parallel and successive reactions: non-specific sorption, chemisorption, sorption of solvent ions, exchanges of the second kind and so on. The effect of these factors is taken into account, for example in the empirical formulas based on the law of mass action, by introducing certain correction coefficients and constants.

For example, the exchange of ions of different valences is described by the following formula:

$$\frac{\bar{C}_{A^{2+}}}{\bar{C}_{B^{2+}}} = k\left[\frac{C_{A^{2+}}}{C_{B^{2+}}}\right]^p, \tag{12.4}$$

where $\bar{C}$ and C are the equilibrium molar concentrations of A^{2+} and B^{+} in the solid and liquid phases respectively and k and p are constants smaller than unity. Another equation, proposed by Gapon (1937), is:

$$\frac{\bar{C}_A}{\bar{C}_B} = k\frac{C^{1/Z_A}}{C^{1/Z_B}}, \tag{12.5}$$

where k is a constant.

It should be noted that most of the existing empirical relations are valid for a sufficiently precise description of the ion-exchange equilibrium only under certain definite conditions and within a restricted range of concentration of the exchanging ions.

A very important factor affecting the absorbing power of soils and grounds is the activity of hydrogen ions in the solution-sorbent system. The dependence of the exchange capacity on pH for a cation exchanger containing one kind of active group is given by Nikol'skiy's equation:

$$S = \alpha + \beta pH, \tag{12.6}$$

where S is the exchange capacity of the ion exchanger and α and β are constants.

However, since soils and grounds form a complicated system of silicate and organic exchangers with different ionogenic groups having different degrees of dissociations, depending on the pH of the medium (for example, alumino-silicates contain mainly the ions of exchange of hydrogen, while humus contains mainly hydroxyl and phenol groups), the linear relation is completely violated and the graph of S as a function of pH may have a stepped form. In general, the cation-exchange properties of soils and grounds are enhanced with increasing pH, since the less acid groups then begin to dissociate and the anion-exchange properties are reduced.

The following assumptions are usually introduced in connection with the absorption of ions present in solution in the form of traces: (a) the absorption of a micro-component depends on the presence and the concentration of a macro-component but it is independent of the presence of other micro-components; (b) the initial linear part of the isotherm corresponds to the distribution of each of the micro-components, so that in this region there is a constant distribution coefficient, which is determined only by the properties of the sorbent and of the ions of the macro- and micro-components; (c) the absorption is directly proportional to the concentration of the micro-component in the solution for concentrations between 10^{-10} and 10^{-3} mol/l. At higher concentrations the capacity of the sorbent may already become saturated (Starik 1959).

However, experiments show that at least two of the above three assumptions are not always valid. Discussing the reasons leading to deviations from the linear Henry isotherms, which is characteristic for the absorption of micro-components, Yegorov (1965) concluded that these include the following: (a) exhaustion of the sorbent's capacity for the given ion, either because the capacity is small or because the concentration of the component can no longer be regarded as a micro-concentration; (b) sorptional inhomogeneities, i.e., centres of different activity in the sorbent; and (c) the presence in the sorbent of micro-component forms exhibiting a different sorption.

The last mentioned factor is particularly important in systems containing micro-concentrations of ions with variable valences, since in these cases a change in pH leads, for example, not only to a change in the absorbing properties of the soil and ground but also to a change in the form of the component in the solution (Grebenshchikov and Prokudina 1965; Nikolaev et al. 1963).

A very important characteristic of the absorption process is the ability of the sorbent to retain the absorbed ions. The strength of binding of the ion to the sorbent increases with increasing ionic charge and, for a given number of charges, with increasing atomic weight. The strength of the binding of ions carrying a given charge is also found to depend on the degree of hydration: it decreases with increasing degree of hydration. The opposite relationship is observed only in the change of trivalent ions. According to Griesbach (1963), the binding strength of cations to natural mineral sorbents increases with increasing R_2O_3/SiO_2 ratio.

The rate of the establishment of the exchange equilibrium is determined by the diffusion rates of the exchanging ions inside grains of the sorbent and in the layers ("film") of liquid surrounding the grains, which is not modified by flowing or mixing of the solution. In any individual case either the first or the second process may predominate (gel or film kinetics respectively).

Gel kinetics are based on the assumption that concentration gradients appear only in the sorbent and, therefore, the diffusion current is proportional to the concentration of the fixed ions and to the coefficient of diffusion within the sorbent and is inversely proportional to the radius of the sorbent grain but is independent of the film thickness, the diffusion coefficient in this film, or the concentration of the component in the solution.

In the case of film kinetics it is assumed that the ion concentrations in the grains are evened out much more rapidly than diffusion in the film and therefore the diffusion current is approximately proportional to the concentration of the solution and to the diffusion coefficient in the film and is inversely proportional to the film thickness but is independent of the concentration of the fixed ions and the diffusion coefficient within the grain. The nature of the kinetic process can be elucidated in practice by considering the coefficient k given by:

$$k = \frac{\bar{C}_B \bar{D}_B \delta}{C_B D_B r_0}\left(5 + 2K_B^A\right), \tag{12.7}$$

where C_B and $\bar{C}_B$ are the equilibrium concentrations of the microcomponent in the liquid and solid phases; D_B and $\bar{D}_B$ are the diffusion coefficients of the microcomponent in the liquid phase; δ is the thickness of the film (10^{-3} and 10^{-2} cm); r_0 is the mean radius of the sorbent grain (cm); and K_B^A is the partition coefficient for exchanging ions:

$$K_B^A = \frac{\bar{C}_A C_B}{\bar{C}_B C_A}. \tag{12.8}$$

12.2.2 Natural Mineral Sorbents

The Wigner classification divides sorbents in accordance with the surface structure into three groups (Griesbach, 1963):

a. Extramicellar-structure sorbents of colloidal or cryotocristalline character (for example, minerals belonging to the following groups: serpentine-kaolinites, halloysite, goethite, hydrargillite, etc.). The particles of these sorbents are usually of ultramicroscopic dimensions and ions capable of exchange lie on the surface. Extramicellar sorbents often exhibit amphoteric properties.
b. Intramicellar-structure sorbents with cation-exchange properties, which include most of the natural mineral sorbents with rigid or swelling skeletons (all primary and most of the secondary minerals).

c. Sorbents with strongly swelling skeletons, which occupy an intermediate position between sorbents in the first and second groups and exhibit amphoteric properties (most organic components of the absorbing complex).

The main mass of the highly dispersed mineral part of soils and grounds, which is responsible for the absorbing power, is due to tiny fragments of primary minerals, above all silicates, which include common minerals belonging to the following groups: feldspars, micas, pyroxenes and amphiboles (about 55 % of the weight of the Earth's crust).

The exchange capacity of minerals belonging to the above groups is low (Gorbunov 1957): (a) feldspars (plagioclases and orthoclases) 1–15 m-eqt./100 g, (b) micas (biotite and muscovite) 3–30 m-eqt./100 g and (c) pyroxenes and amphiboles 10–25 m-eqt./100 g.

Oxides and hydroxides, among which minerals the quartz group is very common, occupy the second place after silicates (about 17 %) as far as abundance is concerned. However, quartz plays a minor role in the absorbing complex, because only a small part of it is comminuted to the colloidal state owing to the considerable strength of its crystalline structure. The exchange capacity of minerals in the quartz group is 1–7 m-eqt./100 g (Remezov 1952).

The secondary minerals, which are most frequently encountered in highly dispersed fractions of soils and grounds, belong to the following groups: montmorillonites, (beidellite, montmorillonite, nontronite), hydromicas (illite, vermiculite, glauconite), halloysites and serpentine-kaolinite (Betekhtin 1956). Owing to their layered structure, high degree of dispersion and weak bounding of exchange cations to the crystal lattice, the minerals in these groups have good exchange properties (Table 12.4).

Montmorillonite and beidellite are characterised by large amounts of highly dispersed fractions. They contain about 80 % of particles with a diameter of less than 0.001 mm including about 60 % of colloidal particles. A relatively high amount of beidellite is characteristic for all types of soil especially serozems (grey soils) on igneous rocks. Montmorillonite is present in small amounts in podzols, while in medium-humus chernozems and brown soils on igneous rocks there are some accumulations of it. Nontronite is found in lower horizons of the crust of weathering (Grim 1959)

Hydromicas are encountered in podzols and chernozems, while in zones of excessive moisture (boggy podzols, tundra, flooded zones etc.) they are found in increased amounts. Illite is encountered more frequently among clays formed during the erosion of mica schists and soils on acidic and intermediate igneous rocks. Glauconite is very common in sandstone in Tertiary deposits, in old and contemporary marine deposits and in friable siliceous and clay-carbonate rocks. Glauconite is also encountered in the upper part of the eluvium of boggy soils.

The main representative of the serpentine-kaolinite group, i.e., kaolinite itself, is largely formed during the weathering of igneous and metamorphic rocks, which are rich in aluminosilicates. A large amount of kaolinite is encountered only in relatively rare boggy-podzol soils and chernozems on granites. The kaolinite then

Table 12.4 Mean composition and exchange properties of some secondary minerals. (Grim 1959; Betekhtin 1956; Gorbunov 1957)

Minerals	Mean chemical capacity									Exchange capacity m.-eqt/100 g
	SiO_2	Al_2O_2	Fe_2O_3	MgO	CaO	H_2O	K_2O	Na_2O	FeO	
Montmorillonite	48–56	11–22	5 or more	4–9	0.8–0.35	12–24	Rare	Rare	–	80–150
Beidellite	45–50	20–28	+	Rare	+	+	Rare	Rare	–	50–70
Nontronite	+	up to 14	Much	up to 8	2	+	+	+	–	40–80
Vermiculite	37–42	10–13	5–7	14–23	–	8–18	up to 5	–	1.3	100–150
Glauconite	48–53	5.5–22	6.1–28	2.4–45	–	5–13.5	4–9.5	0.3	0.8–4.6	50–100
Illite	50–55	25–33	+	+	+	8–9	2–6	–	–	10–40
Halloysite	41	35	+	+	–	24.5	–	–	–	40–50
Kaolinite	46.5	39.5	+	+	+	14	+	+	–	3–15

consists in only about 20–25 % of particles with diameters less than 0.001 mm, with a 5–10 % colloidal fraction (Gorbunov 1957).

Halloyaite, which is the aluminium variety of a large group of mineral forms of complex composition, is encountered largely as colloidal and meta-colloidal formations in the weathering crust of basic and intermediate rocks, in karsting limestones, in zheltozem and in krasnozem.

Among secondary minerals an important role is played by various other compounds with a high absorption capacity associated with a namorphous or cryptocrystalline structure and hence a high degree of dispersion. These include the hydroxides of iron and aluminium (the goethite and hydrargillite groups), opal (quartz group), allophone, etc.

The absorption properties of individual mechanical fractions of soils and grounds are found to vary. Sharply expressed absorption is characteristic for highly disperse mineral compounds having a large specific area. A diameter of 0.001 mm is usually taken as the boundary between fractions that do and do not have well-defined exchange properties (Gorbunov 1957; Remezov 1952). Their investigators took this diameter as 0.0025 mm or even 0.005 mm (Gedroyts 1955). These discrepancies between estimates of the absorbing properties of different fractions of grounds confirm the absence of a direct relationship between the specific area of the particles and their absorption capacity (Fig. 12.1). For example, in addition to its dependence on a specific area, the exchange capacity of an ion exchange is also found to depend on the conditions and nature of the absorption, the properties of the ion and many other factors.

12.2.3 Natural Organic Sorbents

Humus materials are of major importance in the absorbing complex. This was noted by Gedroyts (1955), who was the first to investigate the absorbing complex on the basis of ion exchange.

The amount of humus varies within very broad limits, namely, from 1–2 % in Siberian taiga and up to 10 % or more in the chernozem band on the steppes. In river water the amount of humus compounds is large, 10–15 % mg/l, whereas in lakes it varies from 1 to 150 mg/l. In underground waters (with the exception of boggy areas) the amount of hymus is on average 4–6 mg/l and does not exceed 10 mg/l (Manskaya and Drozdova 1965; Perel'man 1961).

Despite extensive studies, there is still little unanimity about the nature, structure and properties of the organic component of an absorbing complex. The simple average chemical composition of humus material (carbon 58 %, hydrogen + oxygen 30–40 %, nitrogen 3–10 %, ash 2–7 %) cannot be represented in the form of one or several compounds (Table 12.5).

Among the large number (about 40) of well-established compounds in humus, there are three relatively well-defined groups (Rozov 1955): (a) humic acids (humic and ulmic), (b) fulvo acids (crenic and apocrenic) and (c) humans (humin and ulmin).

The structure of all these compounds is basically very similar. They are amorphous polymers consisting in monomers built up from "nucleus-bridge-reaction

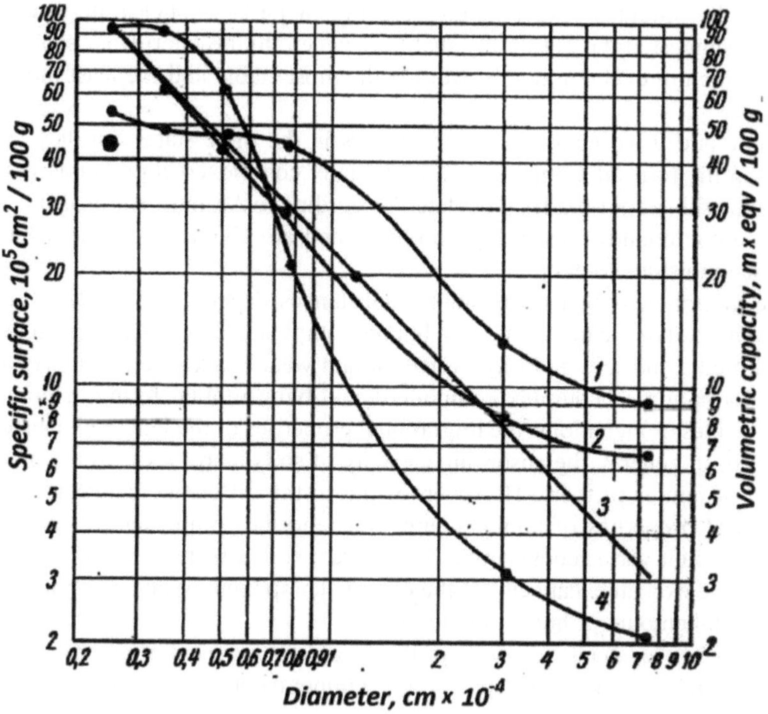

Fig. 12.1 Specific area and exchange capacity as functions of the size of fractions of various types of ground. (*1*) Surface clay, (*2*) losses, (*3*) specific area, (*4*) loam. (Ferronsky et al. 1977)

Table 12.5 Exchange properties of humus materials. (Rozov 1955)

Compounds	Mean chemical composition				Exchange capacity m-eqt./100 g	
	C	H	O		Chernozem	Podzol soil
Humic acids	52–62	52–62	3–3.5	3.5–5	400–500	300–350
Fulvo acids	44–49	3.5–5	44–49	2–4	650–700	650–700

group" units. The structure of the polymers and monomers is not well known but it is known that the humus substances have a loose structure with many internal pores (Kononova 1963).

Depending on the type of the soil-forming process and the mechanical composition of soils, the amount of various humus materials may fluctuate. On average, humic and fulvo acids account for 60–70 % of the total amount of humus material in the soil, whereas humans account for 25–35 %. In soils with a heavy mechanical composition the amount of humans may reach 50 % or more.

Humins are acids that are bound strongly to the mineral (largely highly dispersed) part of the soil ("mull-type" or "argillite-type" humus in some terminologies). The nature of the bonding between humins and mineral parts is not clear but many workers have suggested the possibility that they penetrate the interplanar spaces in the crystal structure of the minerals. Mull is the most stable and the least mobile part of humus.

Humic and fulvo acids are found in soils in the three states but they can also appear in the form of complexed and internally complexed compounds with each other and with sesqui-oxides ("chelates" according to Antipov-Korotaev).

These compounds have a high mobility and are characteristic for zones with increased moisture content.

The migration properties of humus substances depend on the acidity and the redox potential of the medium. Most humus compounds have their highest mobility in acid media and low mobility in neutral and weakly alkaline media. As a result of the formation of complex compounds involving mobile forms of humus and sesqui-oxides or metal ions, the migrational properties of the latter are considerably enhanced. In general, the absorbing capacity of soils increases with the increasing amount of humus in them. However, owing to the difference between the physicochemical properties of humus compounds and the complexes formed by them (whose amount varies considerably in different soils), there is no definite relation between the exchange capacity of the organic component of an absorbing complex and the total amount of humus in the soil.

The specific exchange capacity of the organic component of an absorbing complex is much higher than the capacity of the mineral component [on the average, the ratio is 20:1 (Gorbunov 1957)]. However, even in humus-accumulating horizons of most types of soil, the exchange capacities corresponding to these components are roughly equal because of the relatively low humus content of soils. The mineral component plays the dominant role in the absorbing complex of low-humus or humus-free grounds.

The absorbing complex of soils and grounds is a very complicated system with inhomogeneous mineral and organic components, so that the absorption capacity of the complex is not equal to a simple sum of the component capacities but depends on the component interactions.

12.3 Estimation of Absorbing Properties of Soil and Ground and Migration Activity of Radioactive Micro-Components

12.3.1 Determining the Absorption Capacity

Since natural minerals and organic sorbents can be looked upon as ion exchangers, their absorption properties can be characterised quantitatively by the absorption capacity. This is defined as the number of active groups per unit weight or volume, or by the charge density of the sorbent skeleton and theoretically it is independent of the type and nature of the absorbing ion.

Under natural conditions, the absorption capacity depends on a number of factors: the degree of ionisation of the active groups for a given pH of the medium, the nature, properties and concentration of ions in the solution and so on. It is therefore very important to maintain constant conditions when specimens are prepared for analysis and when measurements are carried out.

In practice, the exchange capacity in physical and chemical studies of natural sorbents is usually referred to as a unit weight of the sorbent (this is the weight exchange capacity) because the exchange capacity per unit volume will not be constant when clay and organic particles swell up (and the degree of swelling depends on the type of sorbent and conditions of the experiment).

The exchange capacity is not independent of the method used to determine it. For example, the exchange capacity determined under static conditions is a measure of the number of ionogenic groups. Dynamical measurements, on the other hand, provide information about the number of groups that participate directly in exchange under given conditions.

Static determinations of the exchange capacity are most frequently carried out in agrochemical studies of soils and grounds, using various modifications of the methods developed by Gedroyts, Bobko, Ashkenazy et al. (Arinushkina 1962). Existing methods can be used to determine the cation-exchange capacity of soils and grounds. No methods are available as yet for the determination of anion-exchange properties. All the methods are based on treating the soil of a ground sample with a solution containing a definite cation (NH_4^+, Ba^{2+}, Ca^{2+}, K^+ and so on) until all the cations are replaced by this definite cation in the absorbing complex. This is followed by quantitative determination of the saturating cations in the solid phase.

Radioactive indicators are now often employed to determine the absorbing properties of soils and grounds. The use of these indicators facilitates considerably the analysis and ensures higher accuracy. There are two main techniques: (1) the ground under investigation is saturated with a given ion, as in the usual agrochemical determinations and is then treated with a solution containing a radioactive isotope of one of the saturating elements. As a result of isotope exchange between the stable and radioactive isotope ions, the activity of the solution is reduced and, after the isotopic exchange equilibrium has been established, the concentration of the radioactive isotope in the solution is measured. This is then used to calculate the concentration of the stable isotope in the solid phase. (2) The radioactive isotope of the saturating ion is present from the beginning in the solution used to treat the sample of the ground under investigation. The absorbing properties of the ground are estimated either directly from the activity of the ground when exchange equilibrium has been reached, or, better, from the residual activity of the solution. Since a single treatment of the soil or ground material with the solution of the saturating ions will not achieve complete replacement of all the ions in the absorbing complex, this operation must be repeated several times. This demands a substantial amount of the indicator and involves a relatively large volume of radioactive waste. The method must therefore be confined to absorbing soils and grounds characterised by a low absorption capacity but it is irreplaceable if the absorption kinetics are to be determined as well as the absorption capacity.

The dynamic absorbing properties of soils and ground are in more precise correspondence with processes occurring in nature. Briefly, the methods used to determine these properties are as follows: a solution containing the displacing cations is passed through a column of the ground under investigation until the absorbing complex is fully saturated. The fact that complete saturation has taken place is established by analysing the filtrate for the concentration of the substituting cation. Pure water is then passed through the column in order to remove the unabsorbed cations in the pore solution. The absorbed cation is then completely extracted from the amount of cation in this extract. Even this brief account will show that such determinations are highly complicated and are therefore very rarely used.

The method is substantially simplified by using a labelled, radioactive cation in the saturating solution. The absorption capacity is then determined from the activity difference between the saturating solution, the filtrate and the washing solution. Extraction is then unnecessary.

12.3.2 Absorption Capacity of Soil and Ground for Components of Contaminants

The absorption capacity is an interesting and important parameter of soils and grounds. However, it is insufficient to predict the behaviour of an individual element or of its radioactive isotope in a solution-ground system, since under real conditions absorption depends on many other factors, including the properties and concentrations of ions present in the solution and in the absorbing complex.

The distribution coefficient is characterised by the absorbing properties of soils and grounds for any given element or isotope under specified conditions, i.e., it can be used as an indicator of its migration activity. There is no direct relationship between exchange capacity and the distribution coefficient for a given cation but, as a rule, large absorption capacities correspond to higher values of the coefficient.

In contrast to the case of exchange-capacity determinations, in the case of measurements of the distribution coefficients for the components of radioactive contaminants that are in contact with the ground under consideration, the solution contains only the isotope under investigation, or a mixture of such isotopes in micro-quantities and there is no stable carrier. The distribution of the micro-component will then depend not only on the properties of the sorbent but also on the properties and concentration of the micro-component. The first problem therefore is the ionic form of the ground and the chemical composition of the contacting solution. Occasionally the soil or ground samples are first completely saturated with some particular cation (NH_4^+, Ba^{2+}, Ca^{2+}, K^+, and so on) and are then mixed with a solution containing the isotope under investigation. The coefficients obtained in this way will of course vary depending on the type of the saturating cation. Preliminary conversion of the ground material into some other ionic form different from the natural form, is inconsistent with the very nature of this determination, since the latter's aim is to estimate the migration activity of the isotope under specific natural conditions. Preliminary saturation should be used in studies of the distribution on pure minerals and in measurements of exchange constants. Comparison of published results on distribution coefficients and their extrapolation is possible only when the experimental methods and conditions of analysis are known in detail, together with the physicochemical properties and the mineralogical and chemical composition of the specimens.

The chemical composition and properties of the experimental solutions must also be chosen from the standpoint of maximum possible approximation to natural conditions.

Particular attention must be paid to the forms of the radioactive isotopes in the industrial effluents and account must be taken of their possible changes during migration. Thus, multiply charged cations may be present in solutions in the form of low-mobility radiocolloids, or they may enter into the composition of highly mobile anionic or neutral complexes. Usually, they are present in solutions in various forms with various migration activities and the quantitative ratios between these forms vary depending on the properties of the medium. For example, ruthenium has a low mobility in cationic form but the effluents of extraction plants contain both the low-mobility nitrate and high-mobility nitrate complexes of ruthenium formed during treatment of the fuel with nitric acid.

The description coefficient (like the absorption capacity of soils and grounds) may be determined both under static and dynamic conditions by methods essentially analogous to those described above. Static methods are commonly used because of their relative simplicity. A spectrum of the ground materials is first dried and the passed through a sieve (aperture diameter 0.5 or 0.25 mm). It is then transferred into a flask and covered with a known amount of solution containing the micro-component under investigation. The flask is shaken until a homogeneous suspension is produced (this is done by special equipment and takes from 20 min to over an hour). The suspension is then allowed to settle (for a day or more) and the solid and liquid phases are separated by filtration or centrifuging (the latter is preferable). The separated solid phase is then again introduced into the flask, the next portion of the original solution is poured in and the whole procedure repeated. The specific activity of the centrifugate is measured at the end of each cycle. The results are calculated from the following formula:

$$k_d = \frac{\left(nN_0 - \sum_{i=1}^{n} N_i\right)}{\sum_{i=1}^{n} N_i} \cdot \frac{nV}{m}, \tag{12.9}$$

where k_d is the distribution coefficient; N_0 and N_i are the specific activities of the original solution and the centrifugate of the i-th cycle respectively (in pulses per minute); *n* is the number of cycles to saturation; *V* is the volume of the solution per cycle (in *ml*); *m* is the sample weight (in *g*).

The number of cycles to saturation depends above all on the absorption capacity of the ground. The higher the absorption capacity the larger is the number of cycles

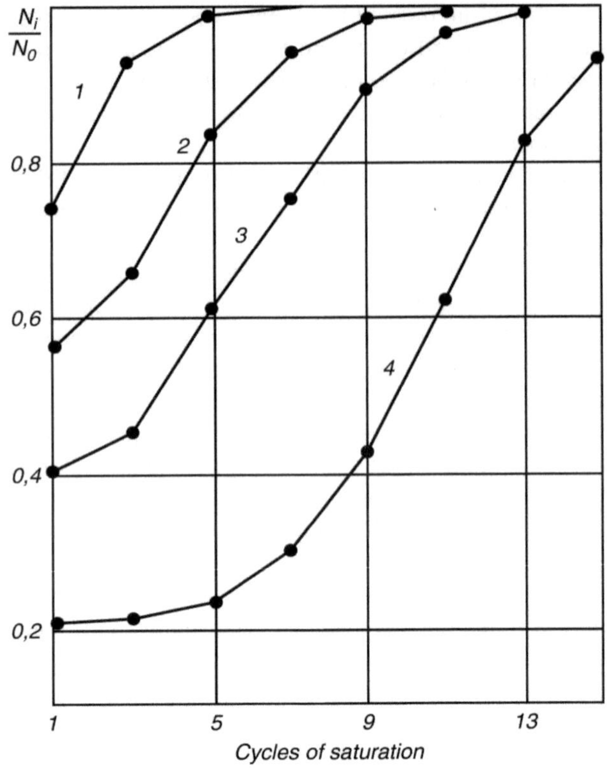

Fig. 12.2 Saturation of different types of ground with ^{90}Sr, deduced from static determinations: *1*—sandy loam, *2*—loam, *3*—covering clay, *4*—montmorillonite clay. (Ferronsky et al. 1977)

to saturation. Experience has shown that the number of cycles may vary from three (light sandy clay, sand) up to 15–20 or more (humus-rich soils, clays). Saturation occurs when the specific activities of the original solution and the centrifugate become equal (Fig. 12.2)

The results of static determinations are very dependent on the size of the ground specimen and the volume of solution per cycle. For example, Prokhorov (1962) showed that when the ratio of the volume of the liquid phase to the weight of the solid sample is less than unity, the distribution coefficient for ^{90}Sr is lower by a factor of about 200 than that obtained by Kokotov (1960) for a ratio of 50. Analogous results have been reported by many workers, including in particular Gorbunov (1963), who investigated absorption processes for ratios approximately equal to the ratio of soil porosity to the bulk density. Other conditions being equal, the distribution coefficient is a non-linear function of the size of the ground sample, the volume of the solution per cycle and the initial concentration of the micro-component in the solution, which is a consequence of the departure of the absorption process in the soils and grounds from the general ion-exchange relation discussed in Sect. 12.2.1.

The distribution coefficients under static conditions are usually determined for values of the above ratio between 5 and 50, since the ground is then well mixed with the solutions and there is no difficulty in separating the phase.

Table 12.6 Absorption of ^{90}Sr by some natural silicates and aluminosilicates. (Spitsyn and Gromov 1959)

Group of minerals	Absorption, %	Group of minerals	Absorption, %
Feldspares	26–55	Montmorillonites	93–99
Micas	75–86	Hydromicas	80–90
Pyroxenes	47–55	Halloysite	88
Quartz	19	Serpentine-kaolinite	33–67

Many static determinations of the absorption properties of natural sorbents have now been published for the following most toxic radioactive isotopes in industrial effluents or radioactive fallout: ^{90}Sr, ^{137}Cs, ^{106}Ru, ^{144}Ce, ^{95}Zr, ^{95}Nb, ^{131}I, ^{239}Pu, etc. However, the absence of a standard method and complete specifications of experimental conditions often means that the various results cannot be compared or used for practical calculations. Most of the published papers were concerned with ^{90}Sr. For example, Spitsyn and Gromov (1959) investigated the absorption of ^{90}Sr by 50 mineral compounds (Table 12.6). They found that the absorption of ^{90}Sr occurs largely as a result of ion exchange, since irreversible absorption was never observed.

These measurements were carried out with the ratio of the volume of the liquid phase to the weight of the solid phase equal to 50, using ^{90}Sr concentrations in the range 10^{-9}–10^{-7} N. The time necessary to reach equilibrium was assumed to be a few days.

The most complete and interesting results were obtained by Kokotov et al. (1960, 1962), who investigated the absorption of ^{90}Sr in different types of soil. The distribution coefficient and the amounts of adsorption were determined for the same ratio, i.e., 50 and the activity of the initial solution was varied between 0.1 and 1.0 μCi/l (^{90}Sr concentration range 10^{-5}–10^{-6} N). The measurements were carried out for different values of pH of the suspension and for different salt backgrounds. It was found that the most complete absorption occurred from neutral and alkaline solutions. Practically no absorption occurred for pH > 3.

Judging by the results published by Spitsyn and Gromov (1959), Kokotov et al. (1960, 1961, 1962) and others, the cations of the macro-component can be arranged in the following series in accordance with the effect on strontium absorption:

$$Al^{3+} > Fel^{3+}Ba^{2+} > Sr^{2+} > Ca^{2+} > Mg^{2+} > H^{+} > NH_4^{+} > K^{+} > Na^{+}.$$

Anions can be arranged in the following series in accordance with their ability to form with strontium sparingly soluble components:

$$PO_4^{3-} > C_2O_4^{2-} >> Cl^{-}; SO_4^{2-} > NO_3^{-}.$$

Extensive systematic observations of the distribution of ^{90}Sr from radioactive fallout in soils and grounds led Schulz (1958) to the conclusion that a proportion of ^{90}Sr (up to 40 % of its total content) is eventually transformed into non-exchange form ("aging" effect). This may be a consequence of the fact that ^{90}Sr becomes part of the insoluble compounds of stable natural strontium through isotopic exchange or a substitution in similar compounds of calcium.

The absorption of the radioactive isotope ^{137}Cs has been investigated fairly extensively. The characteristic features of the absorption of this isotope are: (a) slight dependence of the absorption on the type of soil and on its exchange properties, since in all cases (with the exception of sand) it approaches 100% (static distribution coefficients of the order of 1000 or more; (b) slight dependence of the absorption on the pH of the suspension, since in most cases the absorption is 80 to 100% for pH. 2; (c) irreversibility or poor reversibility of Cs absorption at concentrations up to 10^{-4} g-eqt./g of the ground soil, which is a consequence of the fixation of Cs in the interstices of tetrahedral layers of clay minerals, which can be arranged in the following sequence in accordance with their fixing ability: vermiculite>illite>montmorillonite>kaolonite (Kokotov et al. 1960, 1961; Timofeyeva and Titlyanova 1959; Schulz 1958 etc).

The cations of the macro-components can be arranged in the following sequence in accordance with their effect on the absorption of Cs:

$$Cs^+ > Rb^+ > NH_4^+ > K^+ > H^+ > Ca^{2+} > Na^+.$$

Absorption of radioactive Ce by natural sorbents has not been investigated as extensively as that of strontium and cesium. In general, microquantities of ^{144}Ce are well absorbed by soils and grounds and the absorption increases with increasing concentration of humus and with increasing dispersion of the medium (Nishita 1960). The characteristic features of ^{144}Ce absorption are its clearer dependence on the type of soil or ground and the pH of the suspension as compared with ^{137}Cs. Many workers have noted the high absorption of Ce in the region of weakly acid pH, since radiocolloids of tetravalent Ce may be formed already at pH=2.5 (Nikolaev 1963). In the weakly alkaline range of pH there is a reduction in absorption as a consequence of the formation of poorly absorbed forms due to hydrolysis.

Nikolaev et al. (1963) carried out an electrophoretic study of the state of Ce in a 0.1 M solution of sodium nitrate and detected five types of positively charged and several types of negatively charged forms of Ce.

The isotopes of Ru are similar to other multiply charged ions in that they can enter into the composition of poorly absorbed anionic and neutral complexes, depending on the properties of the solution-ground system. In cationic form, Ru is well absorbed (especially for a pH greater than 5). It is considered that Ru is similar to Ce in that it too is present in neutral solutions in the two forms, one of which is better absorbed than the other.

^{95}Zr and ^{131}I have been investigated in a relatively small number of works but here there is complete unanimity: ^{95}Zr is absorbed by different types of soil and ground to practically 100% in a broad range of pH, which is explained by the fact that ^{95}Z is present in the solution in the form of a radiocolloid (Kokotov et al. 1960). ^{131}I shows practically zero absorption by mineral ionites but is well absorbed by humus. However, the role of the individual humus compounds in the absorption process is not well understood.

Laboratory studies have shown that ^{239}Pu is almost completely absorbed by soils and grounds in a broad range of pH but the presence of the Al ion has an appreciable effect on its absorption and tends to reduce it. The difficulty of removing absorbed Pu with salt solutions indicates that in this case the absorption process is more complicated than ion exchange.

12.3.3 Absorption of Radioactive Components Under Dynamic Conditions

The distribution of radioactive isotopes under dynamic conditions is usually investigated in filtration columns. The method is completely analogous (with the exception of the absorption properties of soils and grounds under dynamic conditions (see 12.3.1)).

The distribution coefficient k_d is calculated from the formula:

$$k_d = \frac{(nN_0 - N_1)V_0}{N_1 m} = \left(\frac{N_0 V_0}{N_f V_f} - 1\right)\frac{V_0}{m}. \qquad (12.10)$$

where N_0, N_1 and N_f are the counting rates corresponding to the specific activities of the initial and equilibrium solutions and of the filtrate respectively (in pulses per minutes), V_0 and V_f are the volumes of the initial solution and of the filtrate respectively (in *ml*) and m is the weight of dry ground in the column (in *g*).

It is however difficult to establish the moment of saturation of the ground by the micro-component from the equality of the specific activities of the filtrate and of the original solution. In spite of the more complicated methods used in dynamic determinations, the results obtained in this way are more representative than the static data. However, the main advantage of dynamic experiments is that they can be used to investigate absorption kinetics and, in particular, to elucidate the dependence of the velocity of the contamination front of the rate of filtration. The equations of ion exchange in columns are used to interpret the data of dynamic experiments.

The filtration of a labelled solution through a column filled with ground material results in the formation of a front that separates it into two zones, namely, one containing the component and another that does not. As a result of further filtration the front is displaced downward in the column with a velocity different from the velocity of the solution. The width of the front depends on the state of the ion-exchange equilibrium and on hydrodynamic characteristics of the sorbent. The ion-exchange equilibrium in its turn is determined by the properties of the sorbent and of the exchanging cations, as well as by exchange kinetics. Thus, reduction in the width of the contamination front is facilitated by selectivity of the sorbent down to a value that does not affect the establishment of equilibrium, improved contact between the solution and the sorbent (e.g., by grinding up the latter) and so on. With very small velocities of the contaminant, the front may however become dispersed as a result of the longitudinal molecular diffusion of the component.

Hydrodynamic factors include, above all, the presence of filtration inhomogeneities in the sorbent, which lead to differences in the velocities of the pore solution and hence to dispersion of the front. If, however, the sorbent does not exhibit selectivity with respect to the component, the contamination front can be predicted either from a preliminary determination of the separation coefficients (see Eq. 12.8) or from the sorption isotherm. If the partition coefficient is greater than unity, or the sorption isotherm is convex, the front will be narrow and will not spread out. If on the other hand, the partition coefficient is less than unity and the isotherm is concave, the front will be broad and will continue to spread out during the motion.

The behaviour of the micro-component front can be predicted only from the magnitude of the partition coefficient, because absorption of the micro-component corresponds to the initial linear part of the isotherm. In general, the rate of displacement of the contamination front decreases with decreasing velocity of the solution concentration and increasing absorption properties of the sorbent.

The velocity of any point of the micro-component front for a partition coefficient greater than unity, or of a point on the front corresponding to the half-value of the initial concentration for a partition coefficient less than unity, can be calculated approximately from the formula (Rachinsky 1964):

$$u = \frac{V}{1 + k_d}, \tag{12.11}$$

where u is the velocity of the front or the half-value concentration point; V is the velocity of the solution and k_d is the distribution coefficient.

Prout (1958) proposed the following alternative formula for the velocity of the concentration front in the ground medium:

$$u = \frac{V_f}{k_d \rho + n}, \tag{12.12}$$

where V_f is the filtration rate in cm/sec; ρ is the bulk density of the ground material in g/cm^3 and n is the porosity (fraction of unity).

Bearing in mind the presence of immobile forms, in the ground moisture, the above formula can be reduced to (Shestakov 1961):

$$u = \frac{V}{1 + k_d \rho / n_0}, \tag{12.13}$$

where n_0 is the active porosity (fraction of unity).

Taking an average bulk density of 1.55 for the different soils and grounds and an active porosity of 0.19–0.35 (large-grain sand and clay respectively), we obtain the following approximate formula from (12.13):

$$\frac{u}{V} = \frac{1}{(9.5 - 9.4) k_d}. \tag{12.14}$$

Finally, the following formulas for the velocities of Sr and Cs fronts in calcium-saturated grounds were proposed:

$$u_{Sr} = \frac{V}{1 + k_{Sr-Ca} - \frac{S}{C_{Ca}} \frac{\rho}{n}}, \tag{12.15}$$

$$u_{Cs} = \frac{V}{1 + \sqrt{k_{Cs-Ca} - \frac{S}{C_{Cs}} \frac{\rho}{n}}}, \tag{12.16}$$

where u_{Sr} and u_{Cs} are the velocities of the Sr and Cs fronts respectively; C is the concentration of strontium or cesium in the liquid phase; and S is the absorption capacity of the ground. These quantities lie in the range $k_{Sr\text{-}Ca} \approx 1.0\text{–}1.6$, and $k_{Cs\text{-}Ca} \gg k_{Sr\text{-}Ca}$. Preliminary laboratory investigations can therefore be used to obtain approximate predictions of the behaviour of radioactive effluents under natural conditions.

More representative results can be obtained from field experiments on the migration of components of natural industrial effluents under conditions approaching those of the proposed waste disposal. The reliability of these results will depend on the suitability of the experimental method, technology and conditions of disposal, representative character of the selected area with respect to the investigated formation and other factors.

For example, it was found that in an artificial sand-gravel deposit the strontium front travelled with a velocity that was lower by a factor of 20 than that of a water front, while under laboratory conditions this ratio should be about 80.

Still greater discrepancies between laboratory and field data were found by Spitsyn and Gromov (1959) who investigated the motion of a neutral solution containing ^{90}Sr, ^{137}Cs, ^{106}Ru and some rare earths. The observations were carried out in fine-grain alluvial sands with intervening layers and lenses of loam. All components were detected at a distance of 1 m from the point at which the solution was introduced, whereas according to laboratory data, cesium and the rare earths should have been absorbed. In this experiment Sr was found to move at the same velocity as water, whereas Ru had a velocity lower by a factor of 2.

Many such experiments could be quoted but there are also examples in which predictions based on laboratory data obtained by close simulation of natural factors were very close to the data obtained under field conditions.

The formulation of experiments under field conditions is, however, quite difficult and the results are frequently interpreted without preliminary laboratory work. Field experiments must therefore be preceded by careful laboratory preparations with maximum possible simulation of the main natural factors that affect the migration activity of contaminants, the properties of the effluents and the method and conditions of waste disposal.

Although extensive experience in the disposal of various industrial effluents and waste, including atomic waste, is already available and despite the large volume of laboratory and field data on the migration of radioactive and other deleterious contaminants in underground capacity, soils and grounds, there are still no adequate criteria that could be used for selecting the location and methods of disposal and for predicting the leakage of radioactivity from the point of burial. The absence of such criteria is due to differences in natural conditions, which have to be taken into account and to the absence of generally accepted methods for laboratory and field experiments. The latter factor means that data obtained by different workers cannot as a rule be directly compared. Moreover, laboratory data cannot be used directly for practical purposes because of the absence of correlative relations between constants characterising the migration process or its individual stages under laboratory and field conditions. This is particularly so in the case of the relation between contaminant migration velocities and directions on the one hand, and moisture dynamics in the aeration zone on the other and also in the case of the absorption of contamination components during the motion of effluents through inhomogeneous grounds, fissured rocks and karst rocks. Further work must therefore be concerned not only with obtaining, collecting and theoretically justifying new experimental data but also with special problems such as standardisation of laboratory and field methods, development of special equipment and technical means and so on. Searches for suitable locations and methods of safe storage of radioactive industrial effluents must be carried out in conjunction with hydrogeologic surveys.

References

Arinushkina YeV (1962) Handbook on chemical analysis of soil. Moscow State University, Moscow

Betekhtin AG (1956) Course of mineralogy. Gosgeoltekhizdat, Moscow

Bol'shakov (1959) Proc. Of the 2nd Int. Conf. on Peaceful uses of atomic energy, Geneva 1958 (Soviet contributions). Vol 4, Atomizdat, Moscow

Bruce FR (1960) The origin and nature of radioactive wastes in the United States Atomic Energy Programme. In: IAEA (ed) Disposal of radioactive wastes. IAEA, Vienna, pp 3–50

Ferronsky VI, Danilin AI, Dubinchuk VT et al (1977) Radioactive investigative methods in engineering geology and Hydrogeology, 2nd edn. Atomizdat, Moscow

Gapon EN (1937) The ions and molecules adsorption by colloidal fraction of soil and structure of the soil colloids. Sel'khozgiz, Moscow

Gedroyts KK (1955) Collected papers, vol. 1. Sel'khozgiz, Moscow

Gluecauf E (ed) (1961) Atomic energy waste, its nature, use, and disposal. US AEC

Gorbunov NI (1957) Absorption capacity of soils and its nature. Izd AN SSSR

Gorbunov NI (1963) High dispersive minerals and methods of their study. Nauka, Moscow

Grebenshchikova VI, Prokudina AF (1965) Studies of the mechanism of the capture of La, Eu, and Sc by ferric hydroxide. In: Coprecipitation and adsorption of radioactive elements. Nauka, Moscow

Griesbach R (1963) Theory and practice of ion exchange. (transl. from Engl). Mir, Moscow

Grim RE (1959) Clay mineralogy (transl. from Engl). Mir, Moscow

Khonikevich AA (1964) Deactivation of effluents from radiochemical laboratories and experimental nuclear reactors. Atomizdat, Moscow

Kokotov YuA (1960) Investigation of the sorption of long-lived fission products by soils and clays. VINITI AN SSSR, N 8 and 9, Moscow
Kokotov YuA (1961) Sorption of long-lived fission products by soils and clay minerals. Radiokhimiya 3:199–200
Kokotov YuA (1962) Selectivity of soils and clays with respect to strontium-60 in different conditions. Radiokhimiya 4:261–273
Kononova MM (1963) Organic matter in soils. Izd. AN SSSR
Manskaya SM, Drozdov TV (1965) Geochemistry of organic matter. Nauka, Moscow
Nikolaev VM (1963) Behavior of radiocolloids of cerium in sorption systems. Radiokhimiya 5:17–24
Nishita H (1960) Soil Sci 89:251–255
Perel'man AI (1961) The geochemistry of landscape. Geografizdat, Moscow
Prokhorov VM (1962) Diffusion of ^{90}Sr in soils and sands. Radiochimiya 4:205–211
Prout WE (1958) Soil Sc I 86:86–89
Rachinsky VV (1964) Introduction to the general theory of sorption dynamics and chromatogtaphy. Nauka, Moscow
Remezov NP (1952) Soils, their properties, and distribution. Uchpedgiz, Moscow
Rozov LP (1955) Meliorative soil science. Sel'khozgiz, Moscow
Schulz RK (1958) On the soil chemistry of radio-strontium. Hilgardia 27:333–339
Shatts A (1964) Activation of water in the secondary and auxiliary circuits of reactors. Atomnaya tekhnika za rubezhom 1:12–18
Shestakov VM (1961) Fundamentals of hydrogeological calculations for filtration from water reservoirs. Vodgeo, Moscow
Spitsyn VI, Gromov VV (1959) Absorption of radioactive strontium by some minerals in soils and grounds. Pochvovedenie 12:45–50
Starik IYe (1959) Fundamentals of radiochemistry. Izd. AN SSSR
Timofeyeva NA, Titlyanova AA (1959) Proceedings of the Ural Branch of the Moscow Society for the study of nature. 2 51–267
Yakimov GV (1961) Radioactive recontamination of water and effluents. Ministry of Communal Economics, Moscow
Yegorov YuV (1965) Departures from Henry's law in radiochemical sorption systems. In: Nauka (ed) Coprecipitation and adsorption of radioactive elements. Nauka, Moscow, pp 69–74

Chapter 13
Induced-Activity Method for Analysis of Rocks and Groundwaters

Abstract Recent developments in nuclear geophysics have led to a fundamentally new method, namely, the method of induced activity or activation analysis. Activation analysis has a very high sensitivity, which for certain elements, exceeds the sensitivity of all older methods. It can often be used to analyse multi-component specimens and to determine simultaneously the contents of several elements in a sample. The principle of the method is based on bombardment of the analysed material with nuclear particles. This process is accompanied by nuclear reactions resulting in formation of radioactive isotopes. Activation analysis is used to determine the amount of a very broad range of elements. Fundamentals of the theory and laboratory analysis of sediments for aluminium and silicon are discussed in this chapter.

13.1 Principles and Range of Application

It is well known that the most rigorous classification of a geologic section can be achieved only by taking into account the chemical and mineralogical composition of the rocks, deduced from data obtained by chemical and mineralogical analysis of cores or samples. In the case of sedimentary rocks it is important to have data on the relative abundance of the main rock-forming elements such as O, Si, Al, Fe, Ca, K, Mg, C, Na and Mn.

The most complete and accurate measure of the content of clay particles in sandy-argillaceous rocks is the ratio of the concentration of Al_2O_3 to that of SiO_2. This ratio is usually referred to as the aluminium-silicate modulus. For example, transition from pure washed sands to clays is accompanied by an increase in the concentration of Al_2O_3 from a few tenths of a percent to 20–40 % and by simultaneous reduction in the amount of SiO_2 from about 100 % to 50–60 % (Rukhin 1957). The corresponding change of the aluminium-silicate modulus is from 0 to 1 and in some cases (for example, in bauxites), the values may be higher still.

The elementary composition of sedimentary rocks has so far been investigated by chemical and spectroscopic methods. As a rule, these analyses are laborious, involve many operations and are extremely difficult and expensive.

Recent developments in nuclear geophysics have led to a fundamentally new method, namely, the method of induced activity or activation analysis. This

V. I. Ferronsky, *Nuclear Geophysics*, Springer Geophysics,
DOI 10.1007/978-3-319-12451-3_13

technique is used increasingly in practical analysis of geologic and geochemical properties, as well as in nuclear geophysics investigations (Belen'kiy et al. 1962).

The principle of the method can be summarised as follows: when a given material is bombarded with nuclear particles, the accompanying wide variety of nuclear reactions results in the formation of radioactive isotopes. The isotopes decay with characteristic half-lives, emitting β and γ radiation of definite energy and this radiation can be used to identify the radioactive reaction products From the amount of induced activity one can then deduce the amount of the parent material participating in the nuclear reactions. Activation analysis has many advantages. In a short time it has conquered a leading position and, in many cases, has superseded other methods.

Activation analysis can be used to determine the amount of a very broad range of elements. For example, most elements in the periodic table have been determined by activation analysis in high slow-neutron fluxes (in reactors). Activation analysis has a very high sensitivity, which for certain elements exceeds the sensitivity of all older methods. It can often be used to analyse multi-component specimens and to determine simultaneously the contents of several elements in a sample.

The analysis can often be carried out without destroying the specimen and can be repeated many times. The analysed sample can be retained for various subsequent studies. It can be analysed with different types of counting apparatus and in different laboratories, so that the reproducibility of the results under different conditions is easily established. Moreover, activation analysis can be used to determine the contents of certain elements in situ under natural conditions of the investigated medium. This last feature is particularly important in geology. Neutron activation analysis is already widely used in nuclear geophysics to determine aluminium and silicon during prospecting for and assessing bauxite deposits (Raipov and Leipunskaya 1962). It is also used for copper determination during prospecting for industrial ores as well as certain other cases.

Neutron activation analysis is the most widely used technique at present. This is because neutrons are uncharged and can readily penetrate the nuclei of different elements. The probability of activation by neutrons, which is represented by the corresponding nuclear reaction cross-section, is therefore higher than the probability for other particles. This is particularly so for thermal neutrons, for which the highest activation-analysis sensitivity has been achieved.

13.2 Activation Reactions in Principle Rock-Forming Elements and Water

Table 13.1 shows the activation parameters for 14 MeV neutrons in the case of the main rock-forming elements in sedimentary rocks. The last column gives the sensitivity calculated by Curtis (1965) for the case of irradiation by neutrons from a generator producing a total output of 5×10^{10} neutrons/s.

Table 13.1 Neutron activation parameters in sedimentary rocks at 14 MeV neutron energy

Sensitivity element	Isotope	Natural abundance %	Activation reaction at 14 MeV	Half-life	Cross section, barn	Macro-cross section (cm^2/g)	β-energy, MeV, yield (%)	γ-energy of main lines, MeV, yield (%)	mg, 5×10^{10} neutrons/s
N	^{14}N	99.835	$^{14}N(n, 2n)^{13}N$	9.96 min	0.55	$2.13 \cdot 10^{-4}$	1.185	–	7.9
O	^{16}O	99.757	$^{16}O(n\ p)^{16}N$	7.35 s	0.09	$3.37 \cdot 10^{-3}$	10.3(20) 4.3(40) 38(40)	7.11(5) 6.14(68.8)	1.5
Na	^{23}Na	100	$^{23}Na(n, p)^{23}Ne$	40.2 s	0.034	$8.8 \cdot 10^{-4}$	3.95(29) 4.40(70)	0.438(29)	2
	^{23}Na	100	$^{23}Na(n, \gamma)^{21}Ne$	14.9 h	0.00033	$8.60 \cdot 10^{-4}$	1.394(100)	2.75(100) 1.368(100)	4.3
Mg	^{24}Mg	78.60	$^{24}Mg(n, p)^{24}Na$	14.9 h	0.19	$3.69 \cdot 10^{-3}$	1.368(100)	1.368(100)	11
	^{25}Mg	10.11	$^{25}Mg(n, p)^{25}Na$	1 min	0.045	$1.12 \cdot 10^{-4}$	3.7(55) 2.7(45)	1.61(6.5) 0.98(14.5) 0.58(14.0) 0.40(14.3)	13
Al	^{27}Al	100	$^{27}Al(n, p)^{27}Mg$	9.54 min	0.07	$1.56 \cdot 10^{-3}$	1.59(42) 1.75(58)	1.015(30) 0.843(70)	1.1
	^{27}Al	100	$^{27}Al(n, \alpha)^{24}Na$		0.11	$2.44 \cdot 10^{-3}$	1.394(100)	1.394(100)	17
	^{27}Al	100	$^{27}Al(n, \gamma)^{28}Al$	2.305 min	0.00053	$1.18 \cdot 10^{-5}$	1.865(100)	1.78(100)	140
Si	^{28}Si	92.16	$^{28}Si(n, p)^{28}Al$	2.305 min	0.22	$4.34 \cdot 10^{-3}$	2.865(100)	1.78(100)	0.4
	^{29}Si	4.71	$^{29}Si(n, p)^{29}Al$	6.7 min	0.10	$1.00 \cdot 10^{-3}$	2.5(70) 1.4(300	2.43(6.2) 1.28(93.8)	17.0
	^{30}Si	3.13	$^{30}Si(n, \alpha)^{27}Mg$	9.54 min	0.080	$5.15 \cdot 10^{-5}$	1.59(42) 1.75(58)	1.015(30) 0.843(70)	30.0
P	^{31}P	100	$^{31}P(n, p)^{31}Si$	2.62 h	0.077	$1.49 \cdot 10^{-5}$	0.47(100)	1.264(0.07)	36
	^{31}P	100	$^{31}P\ (n, \alpha)^{28}Al$	2.305 min	0.150	$2.90 \cdot 10^{-3}$	1.365(100)	1.78(100)	0.6
	^{31}P	100	$^{31}P\ 1(n, 2n)^{30}P$	2.52 min	0.0119		3.24(100)	0.511(200)	

Table 13.1 (continued)

Sensitivity element	Isotope	Natural abundance %	Activation reaction at 14 MeV	Half-life	Cross section, barn	Macro-cross section (cm^2/g)	β-energy, MeV, yield (%)	γ-energy of main lines, MeV, yield (%)	mg, 5×10^{10} neutrons/s
S	^{34}S	4.2	^{34}S l(n, p)^{34}P	12.4 s	0.085	$6.7 \cdot 10^{-3}$	5.1(75) 3.2(25)	2.1(25)	26
	^{34}S	4.2	^{34}S l(n, α)^{31}S	2.64 h	0.0140	$1.10 \cdot 10^{-5}$	0.47(100)	1.264(0.07	65
Cl	^{35}Cl	75.4	$^{35}Cl(n, 2n)^{34m}Cl$	32 min	0.004	$5 \cdot 10^{-5}$	1.3 2.48	3.22(7) 2.10(43) 1.16(18)	44
	^{37}Cl	24.6	^{37}Cl l(n, p)^{37}S	5.04 min	0.028	$1.16 \cdot 10^{-4}$	4.3(10) 1.6(90)	3.09(90)	14
	^{37}Cl	24.6	^{37}Cl l(n, α)^{34}P	12.4 s	0.19	$1.16 \cdot 10^{-4}$	5.1(75)	2.1(25)	2
K	^{39}K	93.08	$^{39}K(n, 2n)^{38}K$	7.7 min	0.01	$1.43 \cdot 10^{-4}$	2.68	2.16(100)	1
	^{41}K	6.91	$^{41}K(n, p)^{41}Ar$	100 min	0.08	$8.5 \cdot 10^{-3}$	1.20(99) 2.48(9)	1.29(99)	62
	^{41}K	6.91	$^{41}K(n, α)^{38}Cl$	37.29 min	0.050	$5.32 \cdot 10^{-3}$	4.81(53) 2.77(16) 1.11(31)	2.15(47) 1.6(31)	43
Mn	^{55}Mn	100	$^{55}Mn(n, α)^{52}V$	3.77 min	0.030	$3.28 \cdot 1^{-4}$	2.47	1.433(100)	5
Fe	^{56}Fe	91.64	$^{56}Fe(n, p)^{56}Mn$	2.578 h	0.110	$1.09 \cdot 10^{-3}$	2.81(50) 1.04(30)	0.845(98.8) 1.81(25.5)	6.5

Table 13.2 Thermal-neutron activation parameters for elements in sedimentary rocks and micro-components of underground water

Element	Isotope	Abundance (%)	Thermal-neutron activation reaction	Half-life of product	Activation cross-section, barn
Na	^{23}Na	100	$^{23}Na(n, \gamma)^{24}N$	15 h	0.53
Mg	^{26}Mg	11.29	$^{26}Mg(n, \gamma)^{27}Mg$	9.45 min	0.025
Al	^{27}Al	100	$^{27}Al(n, \gamma)^{28}Al$	2.27 min	0.215
Si	^{30}Si	3.13	$^{30}Si(n, \gamma)^{31}Si$	2.62 h	0.110
P	^{31}P	100	$^{31}P(n, \gamma)^{32}P$	14.22 days	0.19
S	^{31}S	4.2	$^{34}S(n, \gamma)^{35}S$	87 days	0.26
S	^{35}S	0.17	$^{35}S(n, \gamma)^{36}S$	5.04 min	0.14
Cl	^{35}Cl	75.4	$^{35}Cl(n, \gamma)^{36}Cl$	$4.4 \cdot 10^5$ yrs	40
Cl	^{37}Cl	24.6	$^{37}Cl(n, \gamma)^{38}Cl$	37.3 min	0.56
K	^{39}K	93.08	$^{39}K(n, \gamma)^{40}K$	$1.25 \cdot 10^9$ years	1.87
K	^{41}K	6.91	$^{41}K(n, \gamma)^{42}K$	12.44 h	1.0
Ca	^{40}Ca	95.92	$^{40}Ca(n, \gamma)^{41}Ca$	$1.1 \cdot 10^5$ years	0.22
Mn	^{55}Mn	100	$^{55}Mn(n, \gamma)^{56}Mn$	2.576 h	13.3
Fe	^{54}Fe	5.81	$^{54}Fe(n, \gamma)^{55}Fe$	2.6 years	2.5
Cu	^{63}Cu	69.1	$^{63}Cu(n, \gamma)^{64}Cu$	12.8 h	4.3
Cu	^{65}Cu	30.9	$^{65}Cu(n, \gamma)^{66}Cu$	5.15 min	1.8
Zn	^{64}Zn	48.69	$^{64}Zn(n, \gamma)^{65}Zn$	250 days	0.44
Zn	^{68}Zn	18.56	$^{68}Zn(n, \gamma)^{69}Zn$	52 min	1.00
Zr	^{94}Zr	17.40	$^{94}Zr(n, \gamma)^{95}Zr$	560 days	0.09
Zr	^{96}Zr	1.80	$^{96}Zr(n, \gamma)^{97}Zr$	17 h	0.1
Nb	^{93}Nb	100	$^{93}Nb(n, \gamma)^{994m}Nb$	6.6 min	1.1
Rh	^{103}Rh	100	$^{103}Rh(n, \gamma)^{104m}Rh$	4.4 min	12
La	^{139}La	99.91	$^{139}Lar(n, \gamma)^{140}La$	40.2 h	8.2

Rapid activation analysis under laboratory conditions and in well-logging, using existing and currently available neutron fluxes of 10^8–10^{10} neutrons/s, usually employs only some of the possible fast-neutron activation reactions, namely, $^{16}O(n, p)^{16}N$; $^{28}Si(n, p)^{28}Al$; $^{27}Al(n, p)^{27}Mg$; $^{39}K(n, 2n)^{38}K$; $^{24}Mg(n, p)^{24}Na$; $^{23}Na(n, p)^{23}Ne$; $^{55}Mn(n, \alpha)^{52}V$ and certain others. Owing to the relatively low abundance of elements such as K, Mg, Na and Mn in sedimentary rocks, the reactions can be successfully employed if a fast neutron source with an output of not less than 10^{10} neutrons/s is available.

Table 13.2 shows data on slow-neutron activation of elements in sedimentary rocks and elements occurring as micro-components in groundwater. The following slow-neutron activation processes are suitable, from the point of view of half-lives and activation of a cross-section, for the activation analysis of sedimentary rocks: $^{27}Al(n, \gamma)^{28}Al$; $^{30}Si(n, \gamma)^{31}Si$; $^{35}S(n, \gamma)^{36}S$; $^{55}Mn(n, \gamma)^{56}Mn$.

13.3 Theory of the Method

Consider a target nucleus that becomes radioactive with a decay constant λ when it is bombarded by a constant flux of neutrons. The rate of the formation of radioactive nuclei is equal to the rate of capture, q, of the neutrons by the target nuclei (this rate will be assumed to be constant) minus the rate of the simultaneous radioactive decay, λN_a:

$$\frac{dN_a}{dt} = q - \lambda N_a, \tag{13.1}$$

and hence

$$\frac{dN_a}{N_a - q/\lambda.} = \lambda dt, \tag{13.2}$$

where N_a is the number of radioactive nuclei in the activated sample at a tin\me t and λ is the decay constant of the resulting isotope.

Integrating Eq. (13.2) with respect to t, we obtain:

$$\ln\left[N_a - \frac{q}{\lambda}\right] = -\lambda t + C, \quad \text{or} \quad N_a - \frac{q}{\lambda} = Ce^{-\lambda t}. \tag{13.3}$$

At the initial time (t = 0) the induced activity is zero, i.e., $N_a = 0$. Using this initial condition, we find that $C = q/\lambda$ in Eq. (13.3) and, therefore:

$$N_a = (1 - e^{-\lambda t}). \tag{13.4}$$

At $t \to \infty$, we have $e^{-\lambda t} \to 0$ and $N_a \to N_n = q/\lambda$, where N_n is the number of radioactive nuclei when the activation process saturates. This is the limiting values of the induced activity with decay constant λ. Thus, at any given time t, the number of radioactive nuclei in the irradiated specimen is given by:

$$N_a = N_n(1 - e^{-\lambda t}). \tag{13.5}$$

The rate of the formation of the radioactive nuclei is given by:

$$q = f\sigma \frac{mA}{M} PP_i, \tag{13.6}$$

where f is the neutron flux density in neutron/cm^2 s, σ is the activation cross-section in cm^2, m is the weight of the sample in g, M is the gram-molecular weight of the element under analysis, A is Avogadro's number, P is the relative concentration by weight of the element under analysis in the sample and P_i is the abundance of the activated isotopes.

To obtain the activity induced in the specimen, i.e., the number of decays of the activation product per unit time, we must multiply the number N of radioactive nuclei by the decay constant λ. Using the formula given by (13.6), we obtain the following expression for the induced activity of the specimen at time t_0:

$$N = \frac{f\sigma mAPP_i}{M}(1 - e^{-\lambda t}). \tag{13.7}$$

It follows that the induced activity is directly proportional to the neutron flux, the activity cross-section, the concentration of the activated element and the abundance of the isotope on which the activation reaction takes place. After activation, the sample is introduced into a counting system and the induced radiation is recorded. A certain time is necessary to transport the sample from the neutron flux into the counting equipment. Moreover, it is occasionally convenient to allow certain interfering short-lived isotopes to decay. Let the time spent on these operations be t_1 and let the time of measurement be t_m. Since the radiation from the activated nucleus has a different composition (β and γ) for different isotopes and, in general, different energy, it is necessary to take into account the detection efficiency ε and the relative yield P_y of any particular type of radiation per decay. For example, in the case of activation of ^{27}Al by fast neutrons, the result is the radioactive isotope ^{27}Mg, which emits two γ-rays in parallel stages. Of these, 1.015 MeV gammas are emitted in 30 out of 100 decays, whereas 0.843 MeV gammas are emitted in 70 cases out of 100. Therefore, in this example, $P_{1y} = 0.3$ and $P_{2y} = 0.07$.

It is readily shown that the following number of counts will be recorded during the measuring interval t_m:

$$\begin{aligned} N &= \sum_1^k \frac{f\sigma mAPP_iP_{k,y}\varepsilon_k}{M}(1 - e^{-+\lambda to})e^{-\lambda tm}(1 - e^{-\lambda tm}) \\ &= \sum_1^k \frac{f\sigma mAPP_iP_{k,y}\varepsilon_k}{M}F_1(t_0)F_2(t_m)F_3(t_m), \end{aligned} \tag{13.8}$$

and hence the amount of the element under analysis in the sample is given by:

$$P = \frac{NM\lambda}{\sum_k f\sigma mAPP_{k,y}F_1F_2F_3}. \tag{13.9}$$

For short-lived isotopes it is occasionally possible to ensure that $t_0 \gg 1$ and $\lambda t_m \gg 1$ (for this to be so it is sufficient to take t_0 and t_m to be $(4-5)T_{1/2}$). In that case we have:

$$\begin{aligned} &F_1(t_0) = F2_1(t_m) \approx 1 \text{ and} \\ &N = \sum_k f\sigma mAPP_{k,y}\frac{C_{k,y}\varepsilon_k}{M}e^{-\lambda tm}. \end{aligned} \tag{13.10}$$

Consequently, any initial loss of time leads to a loss of activity and the time must therefore be reduced to a minimum.

Using (13.9), we can estimate the sensitivity with which any particular element can be determined by the activation method. Let us suppose that it is necessary to ensure that N_{min} counts are recorded. The minimum amount of the element under analysis in the sample that can be determined is obviously:

$$P_{min} = \frac{N_{min} M\lambda}{\sum_k f\sigma m A P_i P_{k,y} \varepsilon_1 F_1 F_2 F_3}. \tag{13.11}$$

The sensitivity also increases with detection efficiency. The limiting sensitivity of the method is given by:

$$P_{min} = \frac{N_{min} M\lambda}{\sum_k f\sigma m A P_i P_{k,y}}. \tag{13.12}$$

The values of P_0 calculated from Eq. (13.11) are used to estimate and compare the sensitivity of the method for different elements. The expression given by (13.8) can also be used to estimate the most convenient time intervals for irradiation and measurement. The formula given by (13.9) is unsuitable for direct calculations, because it is unwieldy and the absolute neutron flux cannot always be measured accurately or controlled. The more common procedure in practical activation analysis is therefore to use the relative method of measurement, whereby the activation of the sample is compared with a standard that has a known amount P_s of the element under analysis. Other conditions (t_1, t_m, ε) being equal, the simultaneous irradiation of the sample and standard yields:

$$P = \frac{NP_s}{N_s}, \tag{13.13}$$

where N_s is the number of counts due to the standard material. Moreover, if the sample and the standard are irradiated in different neutron fluxes but the fluxes are controlled in the same way, so that their values are known to within a constant factor, then relative measurements are again possible. Thus, when both the sample and the standard are exposed to the beam from a neutron generator, it is difficult to ensure that the flux intercepted by each is the same. The neutron flux is usually controlled with the aid of α monitor or neutron counter[1]. In this case, it is clear that:

$$P = \frac{NP_s}{N_s}\frac{f_s}{f} = \frac{NP_s}{N_s}\frac{M_s}{M}, \tag{13.14}$$

[1] The α monitor is an end-window Geiger-Muller counter incorporating a thin mica window, which is placed near the generator target and records α particles produced during the reaction.

where M_s/M is the ratio of the neutron-monitor counts due to the standard and the sample.

13.4 Laboratory Activation Analysis for Aluminium and Silicon

Out of the activation reactions of Si and Al by fast neutrons in Table 13.1, the (n, p) reactions are used for practical purposes. The other reactions either require the not very abundant isotopes ^{29}Si and ^{30}Si or their activation cross-sections are too low [$^{30}Si(n, \gamma)^{31}Si$, $^{30}Si(n, \alpha)^{27}Mg$, $^{27}Al(n\ \alpha)^{24}Na$]. Another reason is that the half-life is too long, so that the irradiation time is inconvenient [$^{27}Al(n\ \alpha)^{24}Na$].

The $^{27}Al(n\ \gamma)^{28}Al$ and $^{30}Si(n, \gamma)^{31}Si$ reactions can be used in slow-neutron activation.

The use of Po+Be sources in the analysis for aluminium have one important disadvantage. The sources themselves emit fast neutrons, which are slowed down in the paraffin and therefore the sample is exposed not only to slow but also to fast neutrons. When ^{28}Si is bombarded by fast neutrons the result is ^{28}Al, which is used to determine the amount of Al. If the superposition (interference) of this activation process is not taken into account, the final result may be incorrect, especially since rocks usually contain much more silicon than aluminium. To eliminate this effect, the sample is surrounded with cadmium, which is a strong slow-neutron absorber. This means that ^{27}Al is activated only by fast neutrons. Moreover, Po+Be sources have a relatively short half-life (138 days) and their output does not exceed $3 \cdot 10^3$ neutrons/s.

Controlled neutron sources, or the so-called neutron generators, have become available in recent years. The output of these sources is higher by several orders of magnitude than that of the static isotope sources. Neutron generators are controlled sources of fast monoenergetic neutrons. They are being increasingly used in nuclear geophysics. Laboratory variants were first developed, which are used as stationary sources for neutron-activation analysis of rock specimens. Subsequently, in the late 1950s and early 1960s, new forms of neutron generators were developed, for use in wells. The main advantages of neutron generators are as follows: high output (up to 10^{11} neutrons/s), absence of gamma background, possibility of remote control and operation both under continuous and pulsed conditions and finally, complete radiation safety when the generator is switched off.

The most common reactions used in neutron generators are the (dT) and (dd) reactions. The neutrons are in this case produced as a result of bombardment by accelerated deuterons of internal targets containing deuterium and tritium.

The most widely used reaction is ${}_1^2d + {}_1^3T \rightarrow {}_2^4He + {}_1^0n$, which produces monoenergetic fast neutrons with an energy of 14 MeV. The maximum neutron yield is observed when the deuterons are accelerated to 0.125 MeV.

The neutron generator (Fig 13.1) incorporates an ion source in which deuterium atoms are ionised by a high-frequency field. The deuterium ions are extracted from

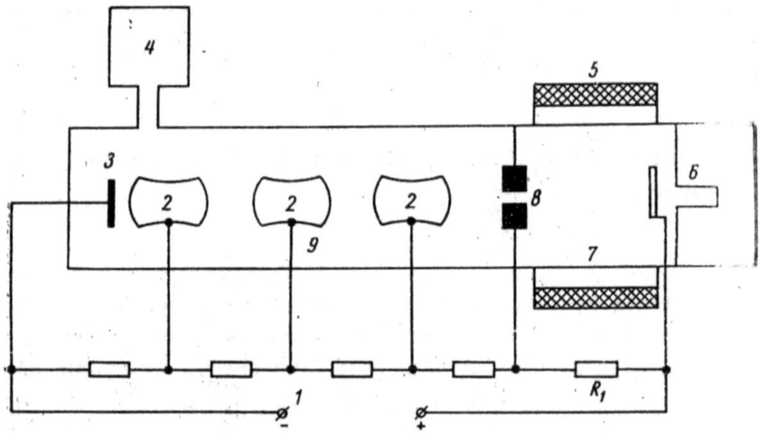

Fig. 13.1 Basic arrangement of a neutron generator: *1* high-voltage source; *2* accelerating electrodes; *3* tritium target; *4* vacuum pump; *5* high-frequency ioniser; *6* deuterium; *7* ion source; *8* exit diaphragm for deuterium ions; *9* accelerating tube. (Ferronsky et al. 1968)

the source by an appropriate set of electrodes and, after focusing, are accelerated by a potential difference of 100–160 keV. The accelerated deuterium ions bombard a titanium or zirconium target saturated with tritium. This results in the emission of fast neutrons with a practically isotropic distribution around the target.

The generator can be operated under conditions that extend its range of application. A vacuum of 10^{-5}–10^{-7} torr is maintained inside the generator for normal operation, while the pressure in the ion source is 10^{-2}–10^{-3} torr.

The sample to be analysed is placed by a pneumatic device in the irradiation position, in line with the generator target and at the end of the activation process is returned to the generator panel for subsequent analysis. The induced radioactivity is measured by a spectrometric scintillation counter incorporating a 70 × 70 mm NaI (Tl) crystal and FEU-58 photomultiplier. Pulses from the photomultiplier are received by a multichannel amplitude spectrum of γ-ray activity induced in the specimen.

In the case of aluminium and silicon, the activated sample is held back for about 40 s to allow the short-lived activity induced in oxygen to decay. The exposure time was varied from a few seconds to several minutes. The maximum neutron yield is $5 \cdot 10^9$ neutrons/s.

The reference samples (standards) were pure semiconducting silicon, pure aluminium (99.999 %) and chemically pure H_2SiO_3.

Figure 13.2 shows the reference spectra of the activity induced in silicon and aluminium. The silicon spectra show the presence of a single photo-peak (total absorption) at 1.78 MeV (see Table 13.1). Figure 13.3 shows the amplitude spectra for the radioactivity induced in some sedimentary rocks. Rocks containing considerable amounts of aluminium (clays) show quite clearly the above peaks. In the case of sand, sandy loam and some loams the photo-peaks corresponding to the activity induced in Al are not well defined owing to the superposition of the Compton distributions associated with gammas from the activated silicon. The following method was therefore employed for the analysis.

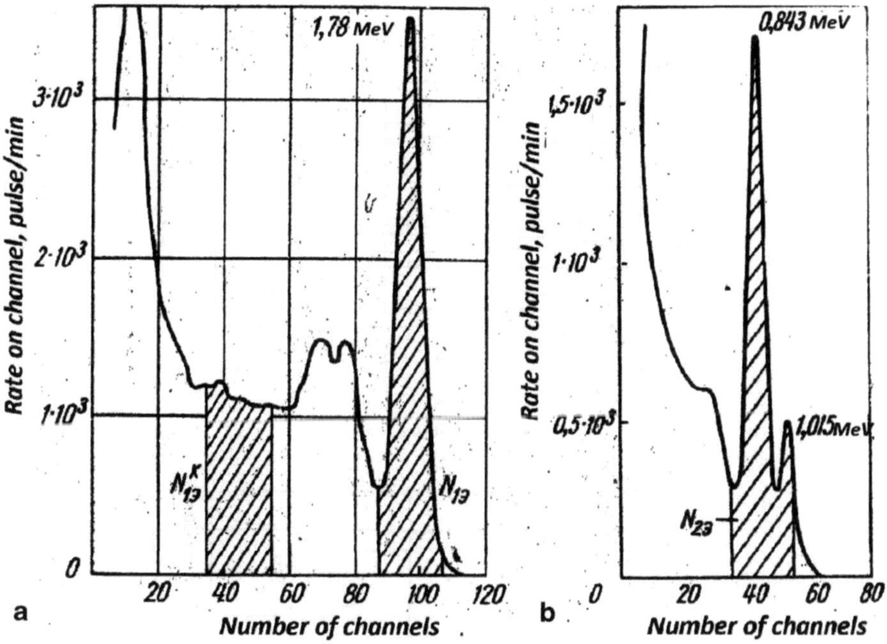

Fig. 13.2 Reference (standard) spectra of the gamma activity induced in Si (**a**) and Al (**b**). (Ferronsky et al. 1968)

The activity corresponding to silicon was determined directly from the amplitude spectrum obtained during the measurement process as the sum of counts in channels lying in the region of the 1.78 MeV photo-peak. The amount of silicon in the sample was then determined from the readings of the neutron-flux monitor using Eq. (13.17).

The contribution of the Compton distribution associated with the 1.78 MeV photo-peak was taken into account in the determination of the activity induced in the aluminium. The true counting rate at 0.843 MeV due to Al activity is obviously equal to the number of counts in the two-component activation spectrum of Al and Si after subtracting the Compton contribution associated with the 1.78 MeV gammas from the activated silicon (Fig 13.3a).

Since the relative Compton contribution for a given counting equipment and given working conditions is a constant quantity, it was sufficient to determine it only once from the reference (standard) spectrum using the formula:

$$\xi_{1k} = \frac{N_{1\varepsilon}^{k}}{N_{1\varepsilon}}.$$

In this expression $N_{1\varepsilon}$ is the number of counts in channels corresponding to the 1.78 MeV photo-peak in the reference spectrum of Si and $N_{1\varepsilon}^{k}$ is the number of

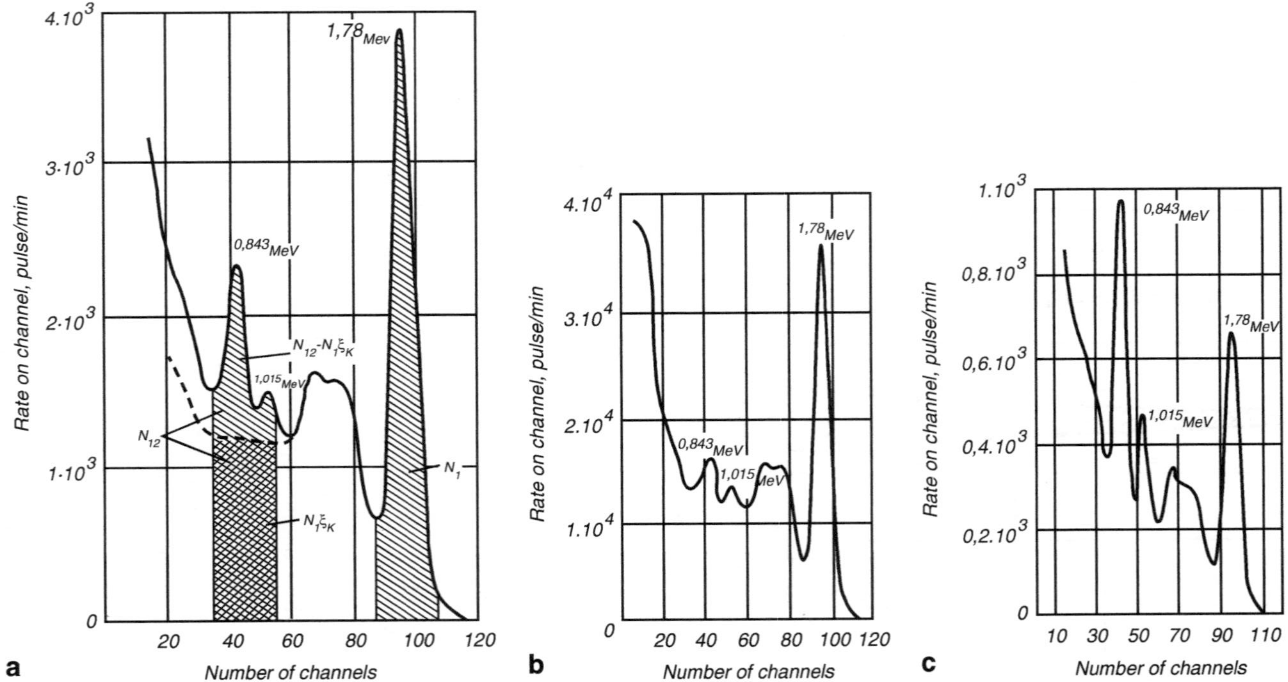

Fig. 13.3 Amplitude spectra of the induced γ-emission; **a** Cambrian clay with $t_0=5$ min, $t_1=40$ s, $t_m=2$ min; **b** monothermite with $t_0=5$ min, $t_1=40$ s, $t_m=2$ min; **c** monothermite with $t_0=5$ min, $t_1=6$ min, $t_m=2$ min. (Ferronsky et al. 1968)

counts in the same spectrum but in channels corresponding to the activity induced in Al (Fig 13.3a). It is then quite easy to determine the number of counts to Al alone:

$$N_2 = N_{12} - N_{1\xi},$$

where N_{12} is the number of counts in the two-component spectrum in the region of the 0.843–1.015 MeV photo-peak and N_1 is the number of counts in the Si channels.

Finally, the amount of Al in the sample was calculated from the formula:

$$P_2 = \frac{P_{2\varepsilon}(N_{12} - N_1\xi_k)M_{2\varepsilon}}{M_2 N_{2\varepsilon}}, \tag{13.15}$$

where $P_{2\varepsilon}$ and P_2 are the amounts of Al in the reference (standard) sample and the working sample, respectively and $M_{2\varepsilon}$ and M_2 are the readings of the monitor when the reference Al sample to be analysed is irradiated.

To estimate the statistical errors of the sample analysis for Al and Si by fast-neutron activation, it was assumed that the amounts of Si and Al in the reference samples were known with high accuracy, i.e., it was assumed that $P_{1\varepsilon}$ and $P_{2\varepsilon}$ were constants. In our case, all this was ensured by using ultrapure silicon and aluminium as the reference samples.

The remaining quantities in Eq. (13.15) are the corresponding counting rates obtained for the neutron flux and the activities of reference and working samples.

The errors in these quantities are described by normal distribution with a variance $D[N]=\sqrt{N}$, root-mean-square deviation $\sigma=\sqrt{D}$ and relative root-mean-square deviation equal to $1/\sqrt{N}$. Using the general rules for the calculation of errors, let as calculate the variance of the silicon content $P_1 = M_{1\varepsilon}N_1/M_1N_{1\varepsilon}$.

We have:

$$D(P_1) = P_{1\varepsilon}^2\left[\frac{M_{1\varepsilon}N_1}{M_1N_{1\varepsilon}}\right]\left[\frac{D(M_{1\varepsilon})}{M_{1\varepsilon}^2} + \frac{D(N_1)}{N_1^2} + \frac{D(M_1)}{M_1^2} + \frac{D(N_{1\varepsilon})}{N_{1\varepsilon}^2}\right]. \tag{13.16}$$

The relative root-mean-square error is:

$$\delta(P_1) = \frac{\sqrt{D\|P_1\|}}{P_1} = \sqrt{\delta^2(M_{1\varepsilon}) + \delta^2(M_1) + \delta^2(N_{1\varepsilon}) + \delta^2(N_1)}, \tag{13.17}$$

$M_{1\varepsilon}$ and M_1 are usually in excess of 10^5 pulses, in the measurement of neutron fluxes by neutron monitors, so that statistical errors in these quantities can be neglected. We therefore finally have:

$$\delta(P_1) = \sqrt{\delta^2(N_{1\varepsilon}) + \delta^2(N_1)} = \sqrt{\frac{1}{N_{1\varepsilon}} + \frac{1}{N_1}}. \tag{13.18}$$

Table 13.3 Results of activation analysis of sandy-argillaceous rocks

Specimen	Sample weight (g)	Content (%)				$\frac{P(Al_2O_3)}{P(SiO_2)}$	$\frac{N_{12}}{N_1}$
		Si	SiO_2	Al	Al_2O_3		
Semiconducting silicon	4.97	100	–	–	–	0	–
Pure quartz sand	9.60	46.7	100	–	–	–	0.268
Sand (Moscow region)	9.58	43.5±0.8	93.2	0.3±0.1	0.6	0.0064	0.297
Moraine loam	9.50	31.3±0.6	67.0	5.26±0.86	10.52	0.152	0.336
Banded clay (Petersburg)	6.3	37.8±0.7	81.0	9.10±1.0	18.2	0.255	0.379
Banded clay	9.45	31.7±0.7	67.0	10.0±1.2	20.0	0.299	0.390
Montmorillonite (Ural)	10.45	29.0±0.5	62.0	13.2±1.26	26.4	0.420	0.383
Montmorillonite (deposit)	6.95	25.4±0.5	54.1	18.0±1.2	36.0	0.668	0.600
Kaolin (Ukraine)	9.20	25.4±1.0	21.0	21.0±2.0	42.0	0.778	0.585

The corresponding formula for P_2 in the case of aluminium can be found in a similar way. The final expression for the relative root-mean-square error, obtained under the same assumptions as for Eq. (13.18), can be shown to be:

$$\delta(P_2)=\sqrt{\frac{N_{12}+\xi_k^2 N_1+\xi_k^2\left[\sqrt{\frac{1}{N_{1\varepsilon}^k}+\frac{1}{N_{1\varepsilon}}}\right]N_1^2}{\left(N_{12}-\xi_k N_1\right)^2}+\frac{1}{N_{2\varepsilon}}} \tag{13.19}$$

In the Compton contribution ξ is known with a high degree of accuracy, which is possible in practice because it is determined from the reference activation spectrum for Si. We have:

$$\delta(P_2)=\sqrt{\frac{N_{12}+\xi_k^2 N_1}{(N_{12}-\xi_k N_1)^2}+\frac{1}{N_{2\varepsilon}}}. \tag{13.20}$$

The expressions given by Eqs. (13.18) and (13.20) can be simplified still further if $N_{1\varepsilon}$ and $N_{2\varepsilon}$ are large, which again can be achieved in practice. We then have:

$$\delta(P_1)=\frac{1}{N_1}; \tag{13.21}$$

$$\delta(P_2)=\frac{\sqrt{N_{12}+\xi_k^2 N_1}}{N_{12}-\xi_k N_1}. \tag{13.22}$$

The statistical error in the determination of Si is thus largely determined by the number of counts recorded during measurements of the activity induced in Si.

The error may be reduced by increasing the irradiation time, the measurement time and the neutron flux. Since the saturation of Si activation occurs after three or four half-lives, i.e., about 10–30 min, there is no point in increasing t_0 and t_m about this figure. Since N_1 is directly proportional to the amount of Si in the specimen, it follows that, other things being equal, a reduction in this quantity will result in $\delta(P_1)$ increasing in inverse proportion to P_1.

The situation is somewhat more complicated in the case of the error in amount of aluminium. This depends not only on the amount of Al but also on that of Si in the specimen, because activity measurements on Al are affected by the Compton distribution due to gamma rays from Si. Moreover the smaller the ratio P_2/P_1 the higher is the error $\delta(P_2)$. To reduce this error we must try to increase the contrast of photo-peaks corresponding to the activity induced in the aluminium against the Compton pulse distribution in the composite spectrum of the specimen. This can be achieved instrumentally by reducing ξ and suitably choosing t_1. The latter method is based on the fact that the activation product of Si with energy of 1.78 MeV decays more rapidly than the activation product of aluminium.

Finally, the cardinal method of reducing errors in analysis is to increase the neutron flux and hence the activity induced in the specimen.

As an example, the results of an activation analysis of some sandy-argillaceous rocks are given below. The specimens were irradiated with neutrons from the NG-160 generator with an output of just under 10^9 neutrons/s. The results are given in Table 13.3. The data were analysed by using the above-quoted formulas. The table also shows the amounts of SiO_2 and Al_2O_3 calculated from the amounts of Si and Al as well as the aluminium-silicate modulus. The last column gives the experimentally determined values of the contrast factor $K = N_{12}/N_1$ i.e., the ratio of the number of counts in the two-component spectrum in channels corresponding to the activation of Al to the number of counts in the Si activation channels.

It is clear that these quantities are linearly related. The intercept of the vertical axis corresponds to $K = \xi_c$ (no aluminium in the specimen). The straight-line relationship can be used to determine the aluminium-silicate modulus without the preliminary determination of Al and Si in the specimen.

The root-mean-square error in the determinations of Si was $\leq 1\,\%$, while the corresponding figure for Al in the case of low Al concentrations was up to 10 %. It is readily shown that by increasing the neutron flux by an order of magnitude it is possible to reduce the error $\delta(P_2)$ by a factor of $\sqrt{10} = 3.2$.

13.5 Conclusions

The different aspects of nuclear radiation and isotopes application in geoengineering and hydrogeology discussed in this book do not cover the full possibilities of their use for scientific study and practical measurements of ground properties and groundwater dynamics.

It follows from the presented material that at present there are three forming and applicable directions of nuclear techniques in engineering geology, hydrology and environmental sciences:

1. Application of nuclear radiation sources for determination of physics and chemical properties of soil and ground.
2. Use of natural stable and radioactive (cosmogenic, radiogenic and technogenic) isotopes as global, regional and local indicators of natural waters and processes;
3. Use of artificial stable and radioactive isotope tracers for study of local natural waters and determination of dynamical properties of soil and pore water;

Substantial economical benefit is obtained in geoengineering exploration by application of nuclear methods by combining with traditional geophysics in penetration logging techniques, which are used by insertion of a penetration logging sound into the ground by means of a hydraulic device. In practice five gauges of gamma, gamma-gamma, neutron-neutron, cone resistance and friction jacket logs are used for measurement of physical and mechanical ground properties. This technique is extremely effective economically and practically for study both in land and sea bottom sediments.

Global and regional studies in hydrology by using natural stable and radioactive isotopes are most effective with international cooperation of specialists under supervision of the International Atomic Energy Agency (IAEA). The isotopic composition of atmospheric moisture and surface continental waters indicates a great variation resulting from considerable latitudinal and altitudinal temperature variations in time. The principles of temporal and spatial distribution of an isotopic composition in atmospheric moisture and surface waters, related to these factors, provide a basis for the interpretation of isotopic data and the elucidation of the condition of global and regional water circulation in nature. At present, comprehensive factual material concerning the isotope composition of atmospheric precipitation has been obtained for the whole world. These data were obtained in the course of systematic measurements of deuterium and oxygen-18 concentrations in precipitation, sampled at the global IAEA/WMO network of stations. They have been published by the IAEA and are available for use by specialists.

Instead of traditional tracers, the radioactive tracer technique for local studies of water filtration properties is well developed and has world-wide use.

References

Belen'kiy BV et al (1962) Spectral activation logging in sedimentary rocks. In: Nuclear geophysics. Gostoptekhizdat, Moscow

Curtis L (1965) Introduction to neutron physics. Atomizdat, Moscow

Ferronsky VI, Danilin AI, Dubinchuk VT et al (1968) Radioactive investigative methods in engineering geology and hydrogeology. Atomizdat, Moscow

Raipov RL, Leipunskaya DI (1962) Application of Po+Be neutron sources for activation logging in aluminum silicon rocks. In: Nuclear Geophysics, Issue 13, Gostopizdat (ed), Moscow

Rukhin LB (1957) Regulations for handling radioactive materials and sources of ionizing radiation. Atomizdat, Moscow

Index

V. I. Ferronsky, *Nuclear Geophysics*, Springer Geophysics,
DOI 10.1007/978-3-319-12451-3

MIX
Papier aus verantwortungsvollen Quellen
Paper from responsible sources
FSC® C105338

FSC
www.fsc.org

If you have any concerns about our products, you can contact us on
ProductSafety@springernature.com

In case Publisher is established outside the EU, the EU authorized representative is:
Springer Nature Customer Service Center GmbH
Europaplatz 3, 69115 Heidelberg, Germany

Printed by Libri Plureos GmbH
in Hamburg, Germany